World Agriculture: Towards 2010

An FAO Study

Edited by

Nikos Alexandratos

Published by the
Food and Agriculture Organization of the United Nations

and

JOHN WILEY & SONS

Chichester · New York · Brisbane · Toronto · Singapore

British Library Cataloguing in Publication Data

A catalogue record for this book is available from the British Library

ISBNs 0-471-95376-8 (hbk)
92-5-103590-3 (pbk)

Typeset in 10/12pt Times by Dobbie Typesetting Limited, Tavistock, Devon
Printed and bound in Great Britain by Biddles Ltd, Guildford and King's Lynn

World Agriculture:
Towards 2010

Contents

Explanatory Notes

SYMBOLS AND UNITS

ha	hectare
kg	kilogram
$	US dollar
tonne	metric ton (1000 kilograms)
billion	thousand million
p.a.	per annum
calories	kilocalories (kcal)

TIME PERIODS

1990	calendar year
1988/90	average for the three years centred on 1989, except if specified otherwise
1970–90	period from 1970 to 1990
1988/90–2010	period from the three year average 1988/90 to 2010

GROWTH RATES

Annual percentage growth rates for historical periods are computed from all the annual data of the period using the Ordinary Least Squares (OLS) method to estimate an exponential curve with time as the explanatory variable. The estimated coefficient of time is the annual growth rate. Annual growth rates for projection periods are compound growth rates calculated from values for the begin- and end-point of the period.

COUNTRIES AND COUNTRY GROUPS

The list of countries and the standard groups used in this book are shown in Appendix 1. In the text, the terms "ex-Centrally Planned Economies" (ex-CPEs), "Eastern Europe" and "Former USSR" are used to denote the countries formerly included in the group "Eastern Europe and USSR". The historical data and projections are presented for these groups, and in the Appendix for the countries under their former names, because at the time this study was being prepared the data available for the new countries were inadequate for a systematic analysis. The term "All Other (or Other) Developed Countries" is used for the countries referred to formerly as "Developed Market Economies", comprising all the developed countries shown in Appendix 1, except the ex-CPEs.

The term "93 developing countries" denotes the 93 developing countries analysed individually in the study. The term "developing countries excluding China" refers to the same 93 countries minus China. The term "all developing countries" refers to the same 93 countries plus all the smaller developing countries not covered individually.

LAND DEFINITIONS

Arable area is the physical land area used for growing crops (both annual and perennial). In any given year, part of the arable area may not be cropped (fallow) or may be cropped more than once (double cropping). The area actually cropped and harvested in any given year is the harvested area. The harvested area expressed as a percentage of the arable area is the cropping intensity. Land with crop production potential consists of all land area which is at present arable or is potentially arable, i.e. is suitable for growing crops when developed (see Chapter 4).

DATA SOURCES

All data are derived from FAO sources unless specified otherwise.

Contributors to the Book

This book is the product of collective work of all the technical units of FAO as well as of persons from outside FAO. It was prepared by a team led by N. Alexandratos under the general direction of H.W. Hjort, Deputy Director-General of FAO. J. Bruinsma was the most senior member of the core team, after the team leader. A Gürkan, D. Brooks and J. Schmidhuber were in the core team, each for about one-half of the period of the preparation of the study and subsequent revisions. M.G. Ottaviani was responsible for data preparation and processing.

The following persons from FAO coordinated the contributions of the different technical units and wrote the relevant chapters or contributed to their writing in various degrees: T. Aldington with P. Turnbull for the contributions of the specialists on the agronomic and natural resources aspects of crop and livestock production for Chapter 4 and, to a smaller extent, Chapters 11 to 13; Ph. Wardle (Chapter 5); B. Dada (Chapter 6); K. Stamoulis (main author of Chapter 7); J. Greenfield with M. de Nigris (coordinated contributions of the commodity and trade specialists for the projections, provided much of the material for, and contributed in drafting Chapter 8); J. Dey (Chapter 9); and T. Contado (Chapter 10). The Food Policy and Nutrition Division contributed to the nutritional evaluation of the food demand projections and, together with the Statistics Division (L. Naiken), provided the estimates of chronic undernutrition, drawing on earlier work for the International Conference on Nutrition of 1992. Many other persons from all technical units of FAO made various contributions to the study.

The Editor wrote Chapters 1 to 3, supervised the preparation of and edited all chapters and wrote or rewrote parts of several chapters. J. Bruinsma contributed to the preparation of nearly all chapters and wholly rewrote Chapter 4 in the revisions. J. Schmidhuber researched selected topics for the revisions, in particular for Chapter 8.

Chapters 11 to 13 were written in their essential by David Norse (ODI). Other persons from outside FAO who wrote or contributed substantially to writing chapters included R. Gaiha and T. Young (Chapter 9) and F. Christy (Chapter 6). S. Johnson, W. Meyers and K. Frohberg cooperated in the projections for the developed countries. G. Fischer (International Institute of Applied Systems Analysis – Vienna) and H. Van Velthuizen, in association with R. Brinkman (FAO), produced the revised evaluation of the crop production potential of land resources of the developing countries and generated the maps for Chapter 4.

Clerical assistance was provided by I. Reyes. M. Brand-Roberti and P. Mascianà were responsible for secretarial work and manuscript preparation.

Foreword by the Director-General of FAO

"WORLD AGRICULTURE:
TOWARDS 2010": WHAT IT IS AND WHAT IT DOES

This report updates, amplifies and extends to the year 2010, the FAO global study "World Agriculture: Toward 2000", last issued in 1987. It assesses the prospects, worldwide, for food and agriculture, including fisheries and forestry, over the years to 2010.

The two most important underlying themes of the study, which are also at the very heart of FAO's activities, concern the prospects for enhanced *food security and nutrition* and for improved *sustainability of agricultural and rural development*. These themes were also the focus of the two major international conferences held in 1992: the United Nations Conference on Environment and Development and the FAO/WHO International Conference on Nutrition. Both conferences provided important signposts for the preparation of the present study, as did the Den Bosch Conference on Sustainable Agriculture and Rural Development held in 1991.

In assessing the prospects for progress towards improved food security and sustainability, it was necessary to analyse in detail many contributory issues. These range from the factors affecting rural poverty and the development of human resources, to those pertaining to the overall economic and international trading conditions, to the status and future of agricultural resources and technology. Among the many issues analysed, it is found that the development of local food production in the low-income countries with high dependence on agriculture for employment and income is the one factor that dominates all others in determining progress or failure in improving their food security. To facilitate the analyses conducted for the study, FAO's experience and expertise in all of the relevant disciplines, from the broad socio-economic to the very specialized and technical ones, have been brought to bear, together with its extensive knowledge of local, national and international situations and policies.

The findings of the study aim to describe the future as it is likely to be, not as it ought to be, and do not aim to wish away problems and challenges. As such they should not be construed to represent goals of an FAO strategy. It is therefore hoped that the findings will contribute to increased awareness of what needs to be done to cope with the problems likely to persist and new ones that may emerge, to guide policies at both national and international levels, and to set priorities for the years ahead.

SOME PROGRESS IN FOOD SECURITY AND NUTRITION LIKELY

The world as a whole has been making progress towards improved *food security* and *nutrition*. This is clear from the substantial increases in per caput food supplies achieved globally and for a large proportion of the population of the developing world. But, as the 1987 study had warned, progress has been slow and uneven. Indeed, many countries and population groups have failed to make significant progress and some of them have even suffered setbacks in their already fragile food security and nutrition situation. Humanity is thus still faced with the stark reality of chronic undernutrition affecting some 800 million people, 20 percent of the population of the developing countries, as many as 37 percent in sub-Saharan Africa and still more in some individual countries. The notion that the world would by now be on a firm path towards eliminating the scourge of hunger and undernutrition by the end of this century has so far proven overly optimistic.

The present study predicts that this uneven path of progress is, unfortunately, likely to prevail also beyond the end of this century. But it also indicates some significant enhancement of food security and nutrition by the year 2010, mainly resulting from increased domestic production but also from some additional growth in food imports. Food exporting countries should face no major problem in supplying the envisaged additional imports, particularly if, as predicted, the former centrally-planned developed countries become much smaller net food importers than was the case until quite recently, or perhaps modest net exporters. However, a number of developing importing countries are likely to continue to face serious foreign exchange constraints and some major logistical problems.

By the year 2010 per caput food supplies will have increased and the incidence of undernutrition will have been further reduced in most developing regions. However, parts of South Asia and Latin America and the Caribbean may still be in a difficult position and much of sub-Saharan Africa would probably not be significantly better off than at present in the absence of concerted action by all concerned. Therefore, the world must brace itself for continuing interventions to cope with the consequences of local food crises and for action to remove permanently their root causes. Nothing short of a significant upgrading of the overall development performance of the lagging countries, with emphasis on poverty reduction, will free the world of the most pressing food insecurity problems. Making progress towards this goal depends on many factors. However, experience amply demonstrates the crucial role of agriculture in the process of overall development, particularly where a large part of the population depends on the sector for employment and income.

PRESSURES ON AGRICULTURAL RESOURCES AND THE ENVIRONMENT TO CONTINUE

On the issue of *sustainability*, the study brings together the most recent evaluation of data on the developing countries' agricultural resources, how they are used now and what may be available for meeting future needs. It does the same for the forestry and the fisheries sectors. The study also provides an assessment of the possible extent and intensity of use of resources over the years to 2010. It concludes that pressure on resources, including those that are associated with degradation, will continue to build up.

The main pressures threatening sustainability are likely to be those emanating from rural poverty, as more and more people attempt to extract a living out of dwindling resources. When these processes occur in an environment of poor and limited resources and when the circumstances for introducing sustainable technologies and practices are not propitious, the risk grows that a vicious circle of poverty and resource degradation will set in. Poverty-related environmental pressures are, however, only part of the story. Agricultural practices, consumption patterns and policies on the part of the rich also contribute to the problem. Responding to environmental pressures from this origin depends on changes in policies to remove incentives for environmentally damaging practices and indeed to introduce disincentives for controlling them, a process already initiated.

However the poverty-related part of environmental degradation is unlikely to be eased before poverty-reducing development has advanced sufficiently to the stage when people and countries become significantly less dependent on the exploitation of agricultural resources. The key concern, therefore, will be how to ensure transition from a world of rapidly growing population and many people chronically undernourished to one of slow or very low demographic growth free from chronic undernutrition, with the least possible adverse effects on resources and the environment. There is considerable scope for improvements in this direction and the study explores a range of technological and other policy options. Provided such improvements are put in place, the prospects are for an easing of pressures on world agricultural resources in the longer term and for minimal further build-up of pressures on the environment caused by agriculture.

I conclude by reiterating the importance of developing local food production in the low-income countries with high dependence on agriculture for employment and income as a key, and often indeed the major, component of any strategy to improve their food security. It is for this reason that this objective is given enhanced priority in the reorientation of FAO's programmes currently under way.

Jacques Diouf

Editor's Preface

This book is a revised and somewhat expanded version of the offical FAO document *Agriculture: Toward 2010* prepared in 1992 and early 1993 for, and considered by, the Twenty-Seventh Session of the FAO Conference in November 1993. It is the latest forward assessment by FAO of possible future developments in world food and agriculture, including the crops, livestock, forestry and fisheries sectors. The assessment is no mere exercise in world food demand and supply projections but rather the product of multidisciplinary work covering the many technical, socioeconomic and policy facts of food and agriculture. The focus is on food security, natural resources and sustainability. The book contains the most comprehensive evaluation of the agricultural potential of the land resources of the developing countries available anywhere.

The time horizon is the period to the year 2010 (in practice and for most variables the period from the three-year average 1988/90 to 2010). The assessment is cast, in the first instance, in terms of a number of key variables, e.g. per caput food supplies and state of nutrition, development of food surpluses and deficits of the different regions and how they may be matched, future gains in crop yields, rate of exploitation of natural resources for agricultural production, etc. Future values for these variables provide the background for exploring the extent to which humankind may be making progress, or otherwise, towards a future free of the most pressing problems related to food and agriculture, e.g. food insecurity for significant parts of world population, threats to the environment and development failures in the low-income countries which depend on agriculture for their development.

In attempting to sketch out the likely evolution of key variables, a (notionally) "positive" approach has been followed, i.e. the objective is to give an idea of the future as it is likely to be rather than as it ought to be from a normative perspective. For example, the conclusion that undernutrition is likely to persist results from this approach to looking into the future. Therefore, the prospective developments presented in, mainly, Chapters 3 to 6 are not goals of an FAO strategy. But they can provide a basis for action to cope with the problems likely to persist and new ones that may emerge. Selected policy issues relevant to such action are discussed in broad terms in the other chapters. The alternative of sketching out a normative scenario in which all problems would be solved in the next 20 years, or in which any predetermined desirable rate of progress would be attained, would not be useful because it would have to be based on assumptions, some of which would be clearly very unrealistic. Some key constraints are likely to stand in the way of achieving desired objectives and cannot be ignored, e.g. the near certainty that population growth rates will be "too high" and the high probability that the rates of overall development will be "too low" in many low-income countries with severe food security problems; or that for large groups of such countries achievement of "required" high agricultural growth rates (e.g. over 4.0 percent p.a.) for long periods is an unlikely prospect.

The field of inquiry is vast, as it covers issues ranging from those in the domain of natural sciences (e.g. biotechnology, soil erosion–crop productivity interactions) to the socioeconomic and political ones useful for understanding why hunger persists in the midst of relative plenty. It is not possible to cover all relevant issues in equal depth and specialists in particular fields may find that their topics are dealt with only in passing or not at all. But the length of the book bears witness to the effort to cover as many aspects as practically feasible. Naturally, the availability or otherwise of systematic comprehensive data was an important criterion in the decision regarding which issues to cover and at what depth. In these circumstances, it is a practical necessity to be selective, treading between the desirable and the feasible. Beyond data availability and the need to keep the overall undertaking within manageable limits, the main criteria used for selecting issues on which to focus in priority were: (a) the relative importance of such issues from a world viewpoint; and (b) the extent to which the 20-year time horizon of the study was appropriate for addressing them.

A 20-YEAR TIME HORIZON: HOW APPROPRIATE IS IT?

The 20-year time horizon is perhaps *too long* for some purposes, in particular for defining with some acceptable degree of confidence the paths of major "exogenous" variables that have a decisive influence on the development of things agricultural, e.g. how foreign debt issues may evolve or what would be the rate of success of the structural adjustment programmes, including their agricultural policy components, now in place in many developing countries. Both condition the prospects for resumption of sustained growth in these countries as well as the capability of governments to invest in agriculture. The situation is complicated further by the uncertainties surrounding the time profile of the reform process in the ex-centrally planned economies, e.g. when would the contraction of their economies bottom out and growth be resumed? It is obvious that developments in this latter group of countries can exert a decisive influence on world markets of agricultural products (will they continue to be major net cereal importers? will their demand for tropical products expand rapidly, stagnate or continue on a slow path?). In relation to the above, it is noted that this study did not, in principle, formulate its own forecasts or assumptions about the overall development outlook for the different countries and regions. It rather relied on the work of other organizations for such forecasts and then attempted to define the probable evolution of food and agriculture, given such exogenously defined overall development environment. Such forecasts are normally not available for periods longer than 10 years, at least not forecasts with the minimum required level of country or country group detail. The exogenous assumptions used for the overall economic outlook are presented in Chapter 3.

Returning to the issue of the 20-year period, it is noted that it is perhaps *too short* for addressing issues which have in recent years leapt to the forefront of the development debate: environment, sustainability, degradation of natural resources, climate change, longer-term capabilities of the earth to cope with increasing demographic pressures, etc. The time horizon of 20 years is *just about right* for addressing issues of the possible development of a number of variables most directly related to technical agriculture, e.g. the rate of diffusion of improved technology or changes in the natural resources environment of the sector, e.g. expansion of irrigation. The same goes for the time required to bring about significant upgrading of human resources in agriculture, e.g. through education and training. However, the rates of uptake by farmers of the opportunities offered by the evolution of agricultural technology to increase productivity and to shift agriculture on to a more sustainable path or the rates at which the opportunities to augment resources through investment in physical assets and in human resources will be exploited are crucially conditioned by overall developments in the economy, society, politics and institutions. It follows that the higher confidence with which one can glean the potentials for technological and resource developments over a 20-year period is watered down when it comes to speculating about their impact on actual food, agriculture and sustainability outcomes. The above are important caveats to be borne in mind when considering the conclusions of this study.

SELECTED ISSUES FOR FOCUSING THE STUDY

There are a multitude of interdependent issues one would want to address in a 20-year assessment of prospects for world food and agriculture. The relative weight of each issue or set of issues in the total problematique varies greatly with the standpoint of the observer. In the *developing countries* the dominant issues relate to reducing undernutrition, enhancing food security, combating rural poverty and achieving rates and patterns of agricultural growth that would contribute to overall economic development. In the *developed countries* more relative weight is attached to managing the transition to a slow-growing agriculture, more responsive to market forces, while economizing on budget costs and safeguarding farm incomes and the livelihoods and life styles of rural communities. The lively debate on the regimes that should govern the conduct of agricultural trade must be essentially seen as a prime external manifestation of these underlying issues in the majority of the industrialized countries plus, of course, as reflecting the more conventional economic concerns (market shares, etc.) of the major exporting countries.

For the *ex-centrally planned* economies (CPEs) the dominant issues in the short to medium term relate mostly to the process of transition to market-oriented economies and are exemplified by such aspects as the reform of the land tenure systems and the shedding of excess labour while minimizing the adverse effects on the social situation associated with the changes affecting the large multi-function socialized units. Of more immediate concern is the need to ensure food supplies to consumers at affordable prices and overhaul the long neglected downstream sector of agriculture (transport, storage, processing, distribution). For the longer term, however, the dominant issues relate to the role visualized for agriculture in the post-reform market-oriented economies. In these countries, the "normal" pain of agricultural adjustment associated with the transition to mature or semi-mature industrial economies is magnified by the process of systemic reform. Policies to cope with this problem may well determine the kind of agricultural sector that will emerge as part of the post-reform market-oriented economies. The empirical evidence as to what the role of agriculture should be in a market-oriented economic system is not very helpful. In the great majority of the market-oriented developed countries, agriculture is far from being a market-oriented sector, though recent policy reforms point in the direction of allowing an enhanced role for market forces in agriculture. Such reforms and their impact on the rules governing trade are important for the ex-CPEs because they define the external environment within which their own transition to market- and, possibly, trade-oriented agriculture will be made.

Cross cutting over all country groups are issues relating to the use made of *agricultural and other environmental resources* in the developing process and how such use relates to the objective of moving towards *long-term sustainability of economies and societies.* Universal as such issues are, they manifest themselves in very diverse forms in the different countries and societies and, what is more important, different people assign them widely differing weights in their hierarchy of objectives when it comes to considering development priorities. The wide ecological and socioeconomic diversity existing in the world ensures that these issues present themselves in the form of a complex mosaic which can easily defy generalizations.

Most of the above issues and many more are addressed in this study in varying degrees, some with the full backing of quantitative analysis, e.g. the evolution of food consumption and nutrition, others in more qualitative terms, e.g. the principles that may, or should, guide policy making for agriculture in the future. However, this being a global study, it endeavours above all to provide the reader with sufficient insights on the few issues in food and agriculture which are of truly global import. There are two such issues that would seem to dominate all others: (a) the persistence of undernutrition and food insecurity for large parts of the developing countries' population; and (b) the process of increasing scarcity and degradation of agricultural and other environmental resources as it relates, directly or indirectly, to the process of meeting the food and income needs of a growing world population. This latter aspect assumes particular significance not only because more food must be produced but also and mainly because a good part of the additional population will continue to depend on the exploitation of agricultural resources, often of poor quality, thus augmenting the pressures on them and heightening the risk of unsustainable exploitation.

ARCHITECTURE OF THE BOOK

Chapter 1 is an extended overview or summary and conclusions. Chapter 2 focuses on the two main themes singled out in the preceding discussion. It starts by taking stock of the broad historical developments and evaluates them from the standpoint of achievements and failures in the quest for improved food security. It moves on then to discuss selected aspects of natural resources and sustainability which are most relevant to analysing food security and agricultural development prospects. Throughout this discussion, as well as in other chapters, an effort is made to highlight the extent to which developments in the main variables at the world level (e.g. per caput production, natural resource scarcities) are relevant to the problem at hand, i.e. the prospects for progress in the countries and population groups with food security problems. Chapter 2 is not a systematic presentation and analysis of historical developments in all variables, or a complete account of the issues relating to the nexus natural resources/sustainability/environment/development/food security. Discussion of these aspects is interspersed in several chapters, as indicated below.

Chapters 3 and 4 present the quantitative projections to year 2010 for the crops and livestock sectors. Chapter 3 presents the possible outcomes for the year 2010 in terms of the main commodities (production, consumption and net trade positions by major region) and draws inferences as to their implications for progress, or failure to make progress, towards the objective of improving food security. The relative emphasis is on the possible future outcomes for the developing countries. The corresponding outcomes for the developed countries are also presented, but in less detail and mostly from the perspective of how they may relate to those for the developing countries in a world which is set to move towards enhanced interdependence in food and agriculture. The less detailed presentation of the developed countries is by and large sufficient to delineate the broad trends which may characterize their agriculture over the time horizon of the study. But the reader will not find here more than passing references to some of the most prominent agricultural policy issues of the developed countries, e.g. budgetary and farm income issues. These issues were the subject of an earlier FAO study on European agriculture (Alexandratos, 1990). Chapter 3 ends with a section on issues relating to the world food–population balance and sustainability in the longer term (beyond 2010). The aim is to provide a framework for thinking about these issues rather than extend the projections to beyond 2010.

Chapter 4 is exclusively on the developing countries and endeavours, as far as possible, to quantify and comment upon the most important variables of technical agriculture underlying the production projections, e.g. land use, irrigation, yields, use of inputs, etc. For reasons explained in the text, this analysis could not be carried out for China. Similarly, it was not carried out for the developed countries mainly because it was considered that natural resource and technology aspects are unlikely to be significant constraints in raising their output (for own consumption and for increasing their net export surplus to the developing countries) at rates that will likely be very low and generally lower than in the past. It is not that the issue is not important. It is very much so, both for the developed countries themselves and for those developing countries with actual or potential high degree of dependence on agricultural trade with the developed countries or the world as a whole. The issue is rather that the resources for this study could not be stretched to cover the natural resources, technology and sustainability issues of the developed countries without cutting down the part devoted to the relevant work for the developing countries. The latter drew on earlier and ongoing work in FAO on the agroecological evaluation of land resources of the developing countries.

Chapters 5 and 6 are on forestry and fisheries. No treatment of food and agriculture futures, particularly one focusing on food security and natural resources and sustainability, would be complete without covering these two sectors. The forest and agriculture are in competition for land use in many countries, particularly those in which the population dependent on agriculture continues to grow. Agricultural expansion has been the major cause of deforestation and pressures for this process to continue will be present in the future. The projections of agricultural land expansion together with rough estimates of the overlap between forest areas and the land with agricultural potential (given in Chapter 4) provide some elements useful for judging the risk of continuation of deforestation.

Food from the sea and inland waters is an important component of total food supplies in many countries, particularly supplies of animal protein in the low-income ones. The fisheries sector is a prime example of a natural-resources-based food production activity whose capacity to increase or even maintain production levels has been stretched to the limit on a global scale. Both forestry and fisheries share characteristics of natural resource property regimes (open access, common property) which are important for analysing issues of sustainability.

Chapter 7 deals with two major themes. The first one encompasses what may be broadly termed economic policies having a bearing on agriculture, with particular reference to the significance for agriculture of the recent thrust of policy reforms aimed at "structural adjustment", i.e. stabilization and resumption of growth in the many developing countries that had experienced growth collapse variously related to unsustainable trends in their macroeconomic equilibria. The second theme concerns the proposition that in the low-income countries with food security problems and high dependence on agriculture, relative shift of emphasis towards agricultural and rural development holds more promise than other strategies of generating benefits both for food security and poverty alleviation as well as for putting the whole economy on to a higher growth path.

Chapter 8 is on issues of agricultural trade. No attempt is made to narrate the policy and political background that had led to what has been termed disarray in world agricultural markets. The reader can find many specialized works on the subject, e.g. Johnson, D. G. (1991) and Tyers and Anderson (1992). It suffices to repeat what was mentioned earlier, i.e. that such disarray and severe distortions were more often than not the result of policies oriented predominantly towards achieving domestic objectives. Such objectives were pursued up to quite recently with vigour and with little regard to their international repercussions by mainly, though not only, the industrialized countries. The focus of Chapter 8 is rather on policy adjustments affecting agricultural trade, in particular those related to the implementation of the Final Act of the recently completed Uruguay Round of Multilateral Trade Negotations. It also discusses more general trade issues of interest to the developing countries and the possible interface of trade policy with that for environmental conservation and sustainability.

Chapter 9 deals with the problem of rural poverty and its possible links to the rate and modalities of agricultural growth. Such links are conditioned by a number of structural characteristics of agriculture and the rural economy as well as by the type and efficiency of policies aimed at promoting agricultural growth and maximizing its poverty-reducing effects. The chapter concludes with a brief survey of the lessons of experience from selected direct anti-poverty interventions.

Chapter 10 concentrates on the important issue of human resources development in agriculture. The focus is on general and technical education and training as well as on extension systems.

Chapters 11 and 12 are in many respects complementary to, and to some extent overlap with, the issues of technical agriculture covered in Chapter 4. But here these issues are examined in a more focused way from the standpoint of impacts on the environment and sustainability and the potential offered by technology to minimize such impacts. Finally, Chapter 13 examines the possible responses to existing and emerging problems of environment and sustainability not only from the standpoint of technologies but also those of institutions and those related to the international aspects of these problems. These last three chapters may be read together with Chapter 4.

HISTORICAL TRENDS AND THE PROJECTIONS

An overview of past developments in world food and agriculture leading up to the present situation is presented in Chapter 2, but only in summary form and in terms of the variables most directly relevant to the task of highlighting the nature and significance of the issues which constitute the main themes of the book. The historical trends are presented in more detail together with the projections in, mainly, Chapters 3 and 4, e.g. production, consumption, trade, yields, land use, etc. This approach reduces duplication and permits direct comparisons to be made between the past and the future for most variables for which projections were made. While such comparisons are useful, they should not lead one to interpret the projections of the many and interdependent agricultural variables as being some form of trend extrapolations, which they definitely are not, as explained in Chapter 3. Use of the term "trends" to describe and interpret the projections can be misleading.

RECORD OF EARLIER PROJECTIONS TO 2000

The predecessor of this study (Alexandratos, 1988) contained projections to 2000 made in 1985–86 from base year 1982/84. The actual outcomes for selected aggregate variables for the period 1982/84–1992 are compared in Chapters 3 and 4 with the trajectories projected for these variables in 1985–86 for the period 1982/84–2000. These comparisons are offered in response to queries about the record of earlier projections. They are not offered as a validation of the projections' methodology.

QUANTITATIVE PROJECTIONS AND DISCUSSION OF POLICIES

The discussion of policy issues in Chapter 7 onwards is not strictly linked to the quantitative projections presented in the earlier chapters. For example in Chapter 7 the general case is made that overvalued real exchange rates discriminate against agriculture but no attempt is made to specify the exchange rates of the different countries that would be compatible with their projections of food and agricultural production, consumption and trade. Likewise, the examination of the role of agriculture in development strategies cannot go beyond the discussion of the general criteria relevant for considering this issue. More specific consideration of such a role can only be attempted in studies for individual countries, not in a study covering the whole world. Even the limited generalizations attempted in the policy chapters should be interpreted with care because they do not apply in equal degree to each and every country situation, and in some cases they may well not apply at all.

In a global study, the validity, or otherwise, of statements on policies and strategies must be judged against the criterion of how well and how objectively they manage to distil and synthesize knowledge and lessons of experience coming from very diverse situations and often from anecdotal sources. The way such evidence is interpreted to draw inferences is often influenced by the existence of particular events at the time of writing. It is not uncommon for current events or those of the recent past to dominate perceptions about problems and required policy responses, e.g. interpreting as a durable change of trends the post mid-1980s declines in world per caput production of cereals or the recent upturn in commodity prices; and, of course, interpretation is subject to influences or biases originating in dominant currents of thinking about development at the time of writing, e.g. the scope for public sector involvement in the economy. No person is free from such influences. It is for the reader to judge whether the objective of interpreting recent events in a proper perspective and of minimizing influences of biases from dogmatic partisan positions has been attained. The conclusions certainly provide no comfort to either the extreme pessimists (impending doom) or to extreme optimists (radical solutions can be had by 2000 or thereabouts).

REVISIONS INTRODUCED IN THE FAO DOCUMENT

The book differs from the original FAO document in that a number of revisions were made after its consideration by the FAO Conference in November 1993. The revisions were made in response to the comments received at the FAO Conference and from several other persons as well as in order to take into account more recent developments, e.g. the conclusion of the Uruguay Round. It did not prove possible to respond in this revision to all requests for including analysis of additional issues, e.g. energy and agriculture or projections of food aid requirements.

Wholly new material includes Section 2.2 in Chapter 2 and Section 3.8 in Chapter 3. The former takes stock of the decline in world per caput production of cereals after the mid-1980s and explores its nature and significance. The latter attempts to provide a framework for thinking about the often raised question of longer-term (beyond the year 2010) developments in the world population–food balance. Chapter 4 has been largely rewritten in response to the many requests for more detailed presentation of the material concerning the natural resources and yield growth paths of the developing countries. The same goes for Chapter 8 in order to take into account the latest information on the Uruguay Round. Significant changes and additions were made in Chapter 7. The methodological note, Appendix 2, is new as is its summarization in Chapter 3. For the rest, revisions were in the main editorial in nature.

REVISED AND MORE RECENT DATA

The basic data of production, trade and utilization available at the start of the work (early 1992) were for up to 1990. Data for 1991 and 1992 as well as revised data for the preceding years became available in the course of the revisions (first half of 1994). For obvious reasons, it was not possible to change the quantitative material of the study to incorporate the new data. But some changes were made when the new data had a significant impact on the argument, e.g. the cereals data and projections of Table 3.17 and those in the Annex to Chapter 3 were adjusted for the shift in reporting the historical production data of the former USSR from bunker to clean weight; and in some tables the data for the 2-year average 1991/92 are shown next to those used for 1988/90, the base year of the study, e.g. Tables 4.7 and 4.15. Otherwise, selected new data, both revised for the base year 1988/90 and for the subsequent two years, are shown for country groups in supplementary tables in the statistical appendix (Appendix 3).

Revisions of the demographic projections need particular mention. Those used in this study come from the Medium variant of the 1990 UN projections. They had a 2010 projected world population of 7.2 billion (a growth rate of 1.55 percent p.a. for 1990–2010) and 915 million for sub-Saharan Africa (3.2 percent p.a.). The latest revision of 1994 reduced the 2010 projected populations to 7.0 billion (1.44 percent p.a.) and 834 million (2.87 percent p.a.), respectively. The drastic downward revision of sub-Saharan Africa's projected population reflects in part the lower rates of improvement of life expectancy due to the impact of the AIDS pandemic (see Chapter 3, note 2). As such, it is far from being the positive evolution that would relax the development constraints represented by rapid population growth in the low-income countries with unfavourable initial conditions. No attempt has been made to revise the agricultural projections following these changes in the population projections, the latest of which became available in late 1994. Obviously, any such revision would be no mean task as it would need to address analytically the complex population–development relationships. Quantification of such relationships for each country (e.g. how the assumed economic growth rates would be affected by the changes in projected population) could well be an impossible task.

N. Alexandratos

CHAPTER 1

Overview

1.1 FOCUS ON NUTRITION/FOOD SECURITY AND AGRICULTURAL RESOURCES/SUSTAINABILITY

This study assesses prospects for world food and agriculture to the year 2010, with special focus on the developing countries. The results present the situation as it is likely to develop and several chapters are devoted to the discussion of ways of tackling both persisting and newly emerging problems. The study covers a wide array of issues in varying degrees of detail as regards their geographic, commodity, resources, technology, and other dimensions as well as related policies. The overall framework for the assessment of prospects is provided by two related major issues in world food and agriculture:

1. Making progress towards elimination of undernutrition and food insecurity of the significant part of the population of the developing countries still subject to these conditions.
2. Safeguarding the productive potential and broader environmental functions of agricultural resources for future generations, the very essence of sustainability, while satisfying food and other needs.

The importance of these issues was underlined by the major international conferences of recent years: the United Nations Conference on Environment and Development (UNCED) and the FAO/WHO International Conference on Nutrition (ICN).

Food and nutrition: progress and failures in the historical period

World per caput supplies of food for direct human consumption are today some 18 percent above what they were 30 years ago. The majority of the developing countries participated in this progress and improved nutrition. However, impressive as this progress has been, it has bypassed a large number of countries and population groups. Many countries continue to have very low per caput food supplies and have hardly made any progress. Indeed, sub-Saharan Africa is today worse off nutritionally than 30 or 20 years ago. In parallel, continuous population growth has meant that the declines in the percentage of the population chronically undernourished did not lead to commensurate declines in the absolute numbers affected. The latter have fallen only modestly and remain stubbornly high at some 800 million persons.

It is now well recognized that failure to alleviate poverty is the main reason why undernutrition persists. This realization, together with the evidence that the world as a whole faced no major constraints in increasing food production by as much as required to meet the growth of effective demand (as shown by the long-term trends for food prices not to rise in real terms, and indeed to decline on balance), contributed to focus attention on ways and means to alleviate poverty and improve the "food entitlements" of the poor, while downplaying the role of increasing per caput food supplies. However, the two aspects cannot be separated in the quest for policy responses to the problem of undernutrition. In the majority of the developing countries, increasing food production is among the principal means for combating poverty. This follows from the fact that the majority of the poor depend on agriculture for employment and incomes. So long as this dependence continues to be high, the growth of food production and of agricultural productivity in the countries with high concentrations of rural poverty will continue to be among the principal means for alleviating poverty and improving nutrition.

The role of world food markets

But global food production capabilities will continue to be important, even if the focus is on the nutrition problem in the developing countries. The fact that success in a number of countries was based on rapidly growing food imports, particularly in the 1970s following the growth of export earnings from the oil boom and easy access to external finance, underlines the role of world food markets in the nutritional developments of the developing countries. In the past, world markets were abundantly supplied by the main cereal exporters, mainly the Western developed countries. This evidence suggests that the world as a whole had sufficient production potential to respond to spurts in import demand without raising prices, apart from occasional shocks. Whether this will be so in the longer term future is another question, which is addressed later on, drawing on the analyses of this study.

In particular the historical evidence needs to be interpreted with care, because the behaviour of world food markets in the past was influenced by the agricultural support policies of major cereals exporting countries. This led to surplus production, stock accumulation, subsidized exports and depressed world market prices. In addition, environmental and resource degradation problems were less of an issue than at present in the decision to support increases in production. Policy reforms under way and in prospect would contribute to all these factors playing a less important role in the future compared with the past in increasing supplies to world markets. Already such policy changes, together with a slowdown in world demand for cereal imports, have led to a decline in recent years in cereals production of the main exporting countries. These declines are behind the fact that world per caput production of cereals is today below its peak of the mid-1980s (see below).

The significance of agricultural resources in the food security *problematique*

In the quest for solutions to the problem of food security and undernutrition, concerns are often expressed about the capability of the world's agricultural resources, technology and human ingenuity to increase food supplies by as much as required to ensure to all people adequate access to food. However, the adequacy of agricultural resources to produce more food is only one part of the resources/environment/sustainability nexus having a bearing on the food problem. Agricultural resources are not only an input into food production but also the major economic asset on which depends a good part of the population in the developing countries for employment and income. Thus, even if the world's resources were adequate to underpin continued growth in food production, the solution of the food problem would still be constrained if the agricultural resources of the poor were insufficient to ensure their livelihood. From this standpoint, the relevant dimension of the perceived growing global imbalance between population and agricultural resources is not so much the need to produce more food globally for more people but rather the fact that the population dependent on agriculture for a living continues to grow.

Some developing countries have made the transition to reduced dependence on agricultural resources for their total employment and income. They include countries which have achieved medium-high levels of per caput food availabilities, even though their agricultural resources per caput (of the total population) have declined to very low levels. Some of them have come to depend increasingly on food imports. For them, the agricultural resource constraints most relevant to their food welfare are those impinging on the global capabilities of the world to produce more food. However, many developing countries are far from this transition. For these latter countries, local agricultural resource constraints will continue to be a major factor in the prospects for solving their food problem. This is because a high proportion of their population, often growing in absolute numbers, depends on these very agricultural resources. Moreover, efforts of growing numbers to make a living out of dwindling resources per caput are sometimes associated with degradation and reduction of the productive potential of these resources. In such cases, there is a high risk that a vicious circle between increasing poverty and resource degradation may be established.

However, it would be incorrect to assume that agricultural resource degradation is exclusively a poverty-related phenomenon. There is sufficient evidence of resource degradation associated with agricultural practices in areas which are certainly not poor, e.g. overuse of agrochemicals in Europe, soil erosion associated with part of grain production in North America and effluents from intensive livestock operations in many countries. Some of these effects are generated or strengthened by policies which provide incentives for unsustainable practices, e.g. support and protection policies which make profitable the excessive use of agrochemicals. Thus, devising policies to safeguard agricultural resources, reduce more general adverse environmental impacts and make progress towards sustainability requires taking account of the factors that determine behaviour *vis-à-vis* the resources of both the poor in the developing countries and the non-poor everywhere.

Notwithstanding the above-mentioned occurrences of increased pressures on agricultural resources generated by the actions of the non-poor, poverty reducing development remains the main hope for easing such pressures in the long term. In the first place, overall population growth slows down with development and agricultural population declines; and secondly, there is less scope for further increases in per caput food consumption when people are well fed. The pressures for increasing food production and for extracting incomes out of agricultural resources in non-sustainable ways become accordingly less intense at higher levels of development. In addition, the objective of resource conservation and environmental protection ranks higher with development in society's hierarchy of preferences, while the means to pursue this objective are also less scarce.

In this context, the question of primary interest for policy is not only how to break the vicious circle between increasing poverty and resource degradation but also how to manage the process of development in ways which minimize the trade-offs between it and the environment. Later sections in this chapter summarize, and other chapters discuss more fully, the environmental pressures likely to emerge in the next 20 years as they can be deduced from the production, resource use and technology projections of this study. They set the stage for examining the options offered by technology and other policies to respond to this challenge.

1.2 PROSPECTIVE DEVELOPMENTS TO YEAR 2010

Continuing, but slower, growth in world population

Over the time horizon of the study the world population may grow to 7.2 billion (or to 7.0 billion according to the latest UN projection; UN, 1994), up from the 5.3 billion of 1990 and the 3.7 billion of only 20 years earlier: 94 percent, or 1.8 billion (1.6 billion in the latest projections) of the total increment in world population will be in the developing countries. Moreover, the regional patterns of population growth are very disparate, e.g. 3.2 percent p.a. in sub-Saharan Africa (reduced to 2.9 percent p.a. in the latest revision of the demographic projections), 1.2 percent p.a. in East Asia. These demographic trends in the developing countries, in combination with their still low levels of per caput food consumption, would require continued strong growth in their food supplies. Not all these additional needs will be expressed as effective market demand. The aggregate increase in the food supplies of the developing countries is likely to be less than required to raise average per caput supplies to levels compatible with food security for all. This is because the general development scene is likely to leave many developing countries and population groups with per caput incomes and potential for access to food not much above present levels.

Better prospects for overall economic growth in the developing countries but with significant exceptions

In the crisis decade of the 1980s, all developing regions experienced declines in per caput incomes, with the important exception of Asia, both East and South. It is likely that these trends will be reversed in the future. The latest World Bank assessment indicates that Asia would continue to perform at fairly high rates of economic growth while the prospects are for modest recovery in both Latin America/Caribbean and the Near East/North Africa. Sub-Saharan Africa would also shift to higher economic growth rates compared with the disastrous 1980s but its per caput income would grow only slightly. These developments in the overall economy already foreshadow the prospect that some regions will continue to make progress towards food security and that others may not make much progress.

The Western developed countries are likely to continue to perform as in the past. The prospects for the ex-centrally planned economies (CPEs) of Europe are shrouded in uncertainty. Their combined GDP is thought to be at present about one-third below that of the pre-reform period. The decline will probably bottom-out in the near future, but it may take a long time before sustained growth re-establishes per caput incomes at the pre-reform levels.

World agricultural growth will continue to slow down

The detailed assessments of this study indicate that the growth rate of world agricultural production at 1.8 percent p.a. will be lower in the period to 2010 compared with that of the past. This slowdown is largely a continuation of long-term historical trends. World production grew at 3.0 percent p.a. in the 1960s, 2.3 percent p.a. in the 1970s and 2.0 percent p.a. in 1980–92. The slowdown is not a negative outcome *per se* to the extent that it reflects some positive developments in the world demographic and development scenes. In the first place, and as noted above, world population growth has been on the decline. Secondly, more and more countries have been raising their per caput food consumption to levels beyond which there is limited scope for further increases. Most developed countries are in this class and they are being gradually joined by some developing countries. To put it in plain language, people who have money to buy more food do not need to do so, though they will probably continue to increase their expenditure on food to pay for the ever increasing margins of marketing, processing, packaging and the services that go with them.

The negative aspect of the slowdown has to do with the fact that it has been happening and will continue to happen while many countries and a significant part of the world population continue to have totally inadequate consumption levels and access to food, with consequent persistence of high levels of undernutrition. In short, the slowdown in world agricultural growth is also due to the fact that people who would consume more do not have sufficient incomes to demand more food and cause it to be produced. World output could expand at higher rates than envisaged in this study if effective demand were to grow faster.

There is in the preceding discussion a notional separation between demand and supply: demand expands independently of supply and causes production to respond. If the additional production is forthcoming at non-increasing prices, one cannot speak of constraints to increasing output. This description fits fairly well the situation in the more advanced countries where incomes and demand originate predominantly in sectors other than agriculture. But it applies much less to many developing countries where incomes of large parts of their population depend, directly and indirectly, on agriculture. In such situations, increasing demand and increasing production are in many respects two faces of the same coin. For if production constraints limit agricultural growth, they act as brakes on both incomes and demand as well as supply. In such situations, one can speak of production constraints limiting progress towards food security, even though such constraints may not apply at the world level.

The policy implication is that in countries with heavy dependence on agriculture, progress towards improved food security depends in major ways on making their own agriculture more productive, at least until such dependence is significantly reduced in the process of development. This self-evident conclusion is not new. It is restated here in order to dispel the notion that agricultural resource constraints do not stand in the way of improving world food security just because there is probably still ample potential for increasing food production in the world as a whole. This notwithstanding, as development proceeds and poor countries reduce their dependence on agriculture for income and employment, and become more integrated into the world economy, the issue of whether there are agricultural resource constraints to making progress towards food security for all will tend to shift from the local level to the global one.

Progress in food and nutrition, but not for all

The implications of the demographic and overall development prospects, together with the assessments of this study for production, consumption and trade are that per caput food supplies for direct human consumption (as measured by the food balance sheets) in the developing countries as a whole would continue to grow, from nearly 2500 calories today to just over 2700 calories by the year 2010. It is likely that by the year 2010 the Near East/North Africa, East Asia (including China) and Latin America/Caribbean regions will be at or above the 3000 calorie mark – significant progress particularly for East Asia. South Asia may also make significant progress but it will still be in 20 years time at a middling position. But the prospects are that per caput food supplies in sub-Saharan Africa will remain at very low levels.

Under the circumstances, the incidence of chronic undernutrition could decline in the three regions with the better prospects. Progress will likely be made also in South Asia, though there could still be 200 million people undernourished in the region by the year 2010. Chronic undernutrition is likely to remain rampant in sub-Saharan Africa, with 32 percent of the population (some 300 million) affected. Thus, the scourge of chronic undernutrition in terms of absolute numbers affected will tend to shift from South Asia to sub-Saharan Africa. These estimates are broad orders of magnitude and relative trends rather than precise predictions of what may happen, subject to the necessary caveats (discussed in Chapter 2). They indicate that it is likely that chronic undernutrition in the developing countries as a whole will persist, perhaps at somewhat lower absolute levels, some 650 million people in the year 2010, against some 800 million people today. Therefore, there will be no respite from the need for interventions to cope with the problem, nor from that of seeking to eradicate poverty, the root cause of undernutrition.

World production of cereals to continue to grow, but not in per caput terms

In the past 20 years, per caput production of cereals for the world as a whole grew from 302 kg in 1969/71 to a peak of 342 kg in 1984/86 but then it declined to 326 kg in 1990/92. It is probable that the average may not grow further and it would still be 326 kg in 2010. This is, however, no cause for general alarm for the reasons discussed earlier in connection with the progressive slowdown in world agricultural growth. In particular, the consumption requirements for all uses in the developed countries (which have 635 kg of per caput total use of cereals and account for 46 percent of world consumption) grow only slowly and may fall in per caput terms. These countries produce collectively as much as needed for their own consumption and to meet the increase in net exports to the developing countries. They could produce more, if more were demanded. These prospects are heavily influenced by possible developments in the ex-CPEs of Europe whose total domestic use of cereals would not only stop increasing rapidly as in the past but may actually fall. This possible development has its origin in the prospect that per caput consumption of livestock products may not recover fully to the pre-reform levels, that there is significant scope for economies in the use of cereals as feed and that post-harvest losses could be reduced significantly.

The recent decline in the world per caput production of cereals has been interpreted by some as indicating a structural change for the worse in the world food trends caused by increasingly binding constraints on the side of production. However, the decline after the mid-1980s was entirely due to falls in the aggregate production of the major net cereal exporting countries. It has not been associated with rises in world market prices and was to a large measure related to policies of some major countries to control the growth of production in that period. Therefore, the decline may not be interpreted as signalling the onset of constraints on the production side which made it difficult to meet the growth of effective demand. The real problem must be seen in the too slow growth of effective demand on the part of those countries and population groups with low levels of food consumption.

The preceding discussion indicates that the world average per caput production has only limited value for measuring trends in world food security. It can also be misleading if it conveys the idea that with the world average constant, any gains in per caput production of one group of countries must be counterbalanced by declines in another group. This need not be the case. It was not so in the 1980s and it will likely not be so in the future. Per caput production is projected to increase in both the developed and the developing countries while the world average may remain at 326 kg (see Table 2.1 in Chapter 2). This paradox is due to the fact that the developing countries start with low per caput production and high population growth rates, and the developed countries are in the opposite situation.

In the event, per caput production of cereals in the developing countries is foreseen to continue growing, from the 216 kg in 1988/90 to 229 kg in 2010. This is a smaller increment than was achieved in the past: 15 kg per decade in the 1970s and the 1980s. But their per caput consumption for all uses may grow faster than production, from 235 to 254 kg, part of it for feed to support the rapidly growing livestock sector. This will require further growth of net imports from the developed countries, which may grow from the 90 million tonnes of 1988/90 to about 160 million tonnes in 2010.[1] The implied rate of growth of the net import requirements is not particularly high judged by the historical record. It is more like that of the 1980s rather than the very rapid one of the 1970s. Financing increased food imports may be considered a normal feature of those developing countries in which both incomes and consumption, particularly of livestock products, grow and other sectors generate foreign exchange earnings. But those developing countries which cannot easily finance increased food imports from scarce foreign exchange earnings will face hardship. It is, therefore, reasonable to foresee a continued role for food aid for a long time to come. If policy reforms towards a more market-oriented international agricultural trade system were to limit the scope for food aid from surpluses, alternative measures will be required to meet the needs. In this respect, the decision included in the Final Act of the recently completed Uruguay Round of Multilateral Trade Negotiations, about measures to attenuate the effects on the food importing developing countries of an eventual rise in world market prices, creating conditions for food security stocks and continuation of food aid flows, assumes particular importance.

Modest growth in the demand for exports of cereals from the major exporting developed regions

Although the prospects for further growth of exports of cereals from the major exporting developed countries to the developing countries offer some scope for further growth of production and exports of the former, the prospects are for their net exports to the rest of the world to grow by much less. This is because the group of the ex-CPEs of Europe would probably cease to be a large net importer in the future and there is a possibility that it could turn into a modest net exporter of cereals by 2010.

There might be significant changes in the market shares in these total net exports of the three major exporting OECD areas, W Europe, N America and Oceania. The policy reforms under way and in prospect, in particular in the context of the provisions of the Final Act of the Uruguay Round, would probably lead to W Europe not increasing further its net exports from the levels of the late 1980s, with all of the additional combined exports of the three groups, and perhaps some more, accruing to North America and Oceania. At least this is what is indicated by the results of most analyses concerning the possible effects of the policy reforms. These findings are, of course, subject to the many caveats attached to the assumptions and models on which these analyses are based.

Continuing strong growth in the livestock sector

The past trends for the livestock sector in developing countries to grow at a relatively high rate are set to continue, though in attenuated form. Part of the growth in their cereal imports will be for increased production and consumption of livestock products. However, the consumption of livestock products in the developing countries will still be well below that of the developed countries in per caput terms in the year 2010. These averages for the developing countries mask wide regional and country diversities, and in both South Asia and sub-Saharan Africa consumption will generally remain at very low levels. The disparities reflect those in incomes as well as production constraints. The latter are a factor in the unfavourable nutritional prospects of some countries in which livestock products, particularly milk, are a major staple food, e.g. in the pastoral societies.

The livestock sector of the developed countries may also grow, but at much slower rates compared with the past, with per caput consumption growing only for poultry meat. This would reflect the prospect that (a) in the ex-CPEs the production and per caput consumption of livestock products may take a long time to recover to near pre-reform levels after the sharp initial declines, and (b) the other developed countries have generally high levels of per caput consumption.

With the continued growth of the livestock sector in the developing countries, their use of cereals as feed will continue to grow fast and it may more than double by the year 2010 to some 340 million tonnes, about 23 percent of their total use. This increasing proportion of total cereals supplies used to feed animals in the developing countries may give rise for concern given the persistence of undernutrition. The concern would be well founded if the use of cereals for feed diverted supplies that would otherwise be available for use by the poor as direct food. This could happen but only in situations where the additional demand for feed would raise prices rather than supplies (whether from domestic production or imports) and price the poor out of the market. There are reasons to believe that this is the exception rather than the rule, as discussed in Chapter 3.

Roots, tubers, plantains: continuing importance in the total food supplies of countries in the humid tropics

Roots, tubers and plantains account for some 40 percent of total food supplies (in terms of calories) for about one-half of the population of sub-Saharan Africa, where overall food supplies are at very low levels. Other countries in both Africa and Latin America/Caribbean also depend significantly on these staples. Production could be increased, and will do, to meet future needs. However, the past trends have been for per caput consumption to decline, at least as far as it can be ascertained from imprecise statistics for this sector. The decline has reflected essentially trends towards urbanization where the high perishability and labour-intensive nature of preparation for consumption make them less preferred foods. With increasing urbanization, it can be expected that there will be further, though modest, declines in average per caput consumption. But dependence of these countries on these products for their total food supplies will continue to be high. The trend towards decline in per caput consumption may be attenuated if imported cereals were to become scarcer, which may well be the case if policy reforms in the developed countries were to raise prices and reduce supplies for concessionary sales and food aid. Likewise, further research into converting starchy roots into less perishable and more convenient food products for the urban population could contribute to attenuate these trends.

The oilcrops sector of the developing countries: continued rapid growth in prospect

In the last 20 years the oilcrops sector of the developing countries grew fast and underwent radical structural change. The oilpalm in East Asia and soybeans in

South America exhibited spectacular growth. The shares of these products and regions in total oilcrop production increased rapidly and those of the other oilcrops of the developing countries (coconuts, groundnuts, cottonseed, sesame) and of the other regions declined accordingly.

The production growth of the sector will continue to be above average compared with the rest of agriculture. Structural change will also continue, but at a much slower pace compared with the past. The expansion of the oilpalm sector will continue to be the most rapid, increasing its share to perhaps 38 percent, up from 32 percent at present and only 16 percent 20 years ago. Soybean production in South America will also continue to grow rapidly, but nothing like the 12-fold increase of the last 20 years, when growth had started from a very low base. The continuation of fairly high growth rates of the oilcrops sector reflects the rapid increase in consumption of the developing countries for both vegetable oils for food and oilseed proteins in support of their rapidly growing livestock sectors. They would also increase further their exports of oils and to a lesser extent those of oilmeals to the rest of the world.

Slower growth in the other main agricultural exports of the developing countries

There are well-known reasons why the generally unfavourable trends in the net exports of the major export commodities of the developing countries to the rest of the world may continue. For *sugar*, the reason is mostly the probable continuation of support and protection policies, market access restrictions and subsidized exports of major developed countries. Then, the ex-CPEs are likely to be much smaller net importers in the future. Therefore, net exports to the developed countries will likely continue to fall. But the developing exporting countries are likely to continue to expand exports because there are growing markets in the net importing developing countries, which increased their net imports nearly four-fold in the last 20 years.

Unlike sugar and some other major export commodities, *coffee and cocoa* are produced only in the developing countries and consumed mostly in the Western developed countries, where per caput consumption levels are already generally high. Therefore, efforts by developing countries to increase supplies in competition with each other translate into small increases in the volume of exports and large declines in prices. For the longer term, there is scope for the situation to improve given the low consumption levels prevailing in the ex-CPEs and the developing countries themselves. But little of this scope may materialize in the form of increased consumption and imports in the next 20 years. Therefore, growth in net exports of about 25 percent, and somewhat higher in production, is a likely outcome. For *tea*, there are somewhat better prospects for production growth, though not for exports, because a good proportion of production is consumed in the developing countries themselves and per caput consumption will continue to increase. Finally, exports of *bananas* have better prospects than those of the tropical beverages since there is still scope for per caput consumption to increase in the developed countries.

In general, for the commodities produced only or mainly in developing countries competing with each other and consumed mostly in developed countries with nearly saturated consumption levels, the prospects for export earnings will continue to be dominated by movements in prices rather than volumes. The very long run remedy to declining prices may be found in the growth of consumption in yet unsaturated markets (ex-CPEs and developing countries themselves) and ultimately in the general development of the producing countries themselves. The latter factor is important because it will create alternative income-earning opportunities and put a floor to how low the returns to labour in these commodity sectors may fall before supply contracts and prices recover.

Finally, the prospects for some agricultural raw materials traditionally exported from the developing countries offer limited scope for growth in net export earnings, though for different, and not always negative, reasons. Thus, net exports of *tobacco* to the developed countries may not grow at all because their consumption is on the decline while it is on a rapid growth path in the developing countries themselves. For *cotton*, the developing countries have recently turned from being net exporters to become net importers and will further increase their net imports in the future. This is, on the whole, a positive development because it reflects their growing and increasingly export-oriented textiles industry. These trends could become even more pronounced if restrictions to textile exports become less stringent or are abolished. Similar considerations apply to the *hides and skins* sector and the associated expansion of exports of leather goods. Finally, *natural rubber* exports to the developed countries would continue to grow, but also here the developing countries will gradually increase their share in world consumption and may, by the year 2010, account for over one-half of the world total, compared with less than one-quarter 20 years ago. Much of the expansion of consumption will be in East Asia.

The developing countries likely to turn from
net agricultural exporters to net importers

The prospective developments presented above for the major commodity sectors indicate that the net imports of the developing countries of the agricultural commodities for which they are or may become net importers will be growing faster than their net exports of their major export commodities. These trends in import and export volumes point firmly in the direction of the developing countries' combined agricultural trade account switching from surplus to deficit. The movement in this direction has been evident for some time in the historical period. The positive net balance of trade on agricultural account shrank rapidly in the 1970s when food imports from the developed countries exploded. Although the trend was somewhat reversed in the 1980s the overall surplus was only $5.0 billion in 1988/90 compared with $17.5 billion in 1969/71 (both at 1988/90 prices).

The prospect that the developing countries may turn into net agricultural importers does not by itself say much about the welfare implications of this turnaround. It is certain that it will have a negative impact on the welfare of those countries which will continue to depend heavily on slowly growing agricultural exports to finance their food and other imports. There are many low-income countries in this situation and they include those which depend heavily on agricultural export commodities with limited growth prospects. However, for other countries these prospects are part and parcel of the development process. These are the countries whose increased imports or reduced exports of agricultural raw materials are more than compensated by growing exports of the related manufactures; and those in which the increased food imports reflect their growing incomes and food consumption and which are financed from export earnings of other sectors.

1.3 FACTORS IN THE GROWTH OF AGRICULTURE IN DEVELOPING COUNTRIES

Further intensification in prospect, with yield growth the mainstay of production increases

The production outcomes presented earlier will depend on further intensification of agriculture in the developing countries: yields will be higher, more land will be brought into cultivation and irrigated, and the existing land will be used more intensively (multiple cropping and reduced fallow periods).

Yield growth has been the mainstay of production increases in the past. It will be more so in the future, particularly in the land-scarce regions of Asia and Near East/North Africa. At present, average yields differ widely among countries. However, comparisons of average yields convey only limited information about the potential for lagging countries to catch up with the ones achieving higher yields. This is because agroecological conditions differ widely among countries and farming environments. For example, the 5.0 tonnes/ha average wheat yield of Egypt reflects the fact that wheat is irrigated. This yield is not achievable by countries in which wheat is, and will continue to be, predominantly rainfed in adverse agroecological conditions.

Therefore, agroecological differences among countries must be taken into account before any judgement can be passed as to the potential for yield growth. It is for this reason that a painstaking assembly and collation of data on yields achieved in the different countries in six agroecological environments (five rainfed and one irrigated – hereafter referred to as "land classes") was undertaken for this study. The resulting data are not perfect and it has not been possible to assemble sufficient information for China. But for the other developing countries these imperfect data can go a long way towards permitting an assessment of yield growth potential which is far superior to that based only on average yields.

With these caveats in mind, the dependence of the production outcomes presented earlier on yield growth, and how credible the yield projections may be, can be illustrated as follows: the average irrigated rice yield in the developing countries is today 3.7 tonnes/ha, but some countries achieve only 1.0 tonne and others 10.0 tonnes. The one-fifth of countries with the highest yields achieve an average of 6.7 tonnes. The country-by-country assessment of prospects for irrigated rice indicate that the average irrigated rice yield of all countries could be 5.2 tonnes/ha in year 2010. This means that in 20 years' time the average irrigated rice yield for all countries may be still below that achieved today by the top fifth of countries with the highest yields. This may appear conservative, but it is a "best guess" outcome of the judgements made for individual countries taking into account both differences among countries in the quality of irrigated lands as well in the socioeconomic environments which condition the pace of adoption of yield-increasing technologies. Similar considerations apply also to the rate at which average yields of other crops in each land class may edge upwards towards those achieved by the best performing countries today. Thus, the average yield of rainfed wheat in sub-humid land may grow from 1.7 to 2.1 tonnes/ha, compared with the 2.3 tonnes/ha achieved today by the top fifth of countries. For sub-humid rainfed maize the corresponding numbers are for the average yield to grow from 1.8 to 2.6 tonnes/ha compared with 2.8 tonnes/ha achieved by the top 20 percent of countries today. And so on for other crops and land classes (for more details see Chapter 4).

Naturally, further growth in yields, even at the lower rates projected here compared with the past, will not come about unless the research effort continues unabated. The effects of research on yield growth may manifest themselves in different ways compared with the past: more impact through lower, evolutionary growth in average yields based on adaptive and maintenance research and less through the achievement of quantum leaps in yield ceilings. As a result the inter-country yield differentials may narrow a little, though they will remain very wide. For example, for wheat and rice, average yields of countries at the bottom of the yield league may still be in 2010 only one-fifth of those achieved by the 10 percent of countries at the top; and those of the largest producers may still be only one-half those achieved by the countries at the top. Moreover, continuing research effort is needed for the crops and unfavourable environments which have been neglected in the past, as well as for preventing declines and maintaining and perhaps increasing yields in those farming conditions where yields achieved are near the ceilings of experiment station yields.

**Land in crop production to expand and to be
used more intensively**

The developing countries (excluding China) have about 2.5 billion ha of land
on which rainfed crops could achieve reasonable yields, depending on the
technology used. Over 80 percent of it is in the two land-abundant regions of
sub-Saharan Africa and Latin America/Caribbean. The differences in land/
person ratios among regions are enormous, with Asia and the Near East/North
Africa region having particularly low land availabilities per caput. Of this total
land, about 720 million ha are currently used in crop production and another
36 million ha of land so used comes from desert land which has been irrigated.
The projections of this study would require increases in the different countries
which sum up to about 90 million ha. Thus, by the year 2010, the total land in
crop production could be some 850 million ha. The expansion would mostly be
in sub-Saharan Africa and Latin America/Caribbean, some in East Asia
(excluding China) and very little in the other two regions.

Of the some 760 million ha in agricultural use at present, only about 600
million ha are cropped and harvested in any given year. This is because land is
being used at very different intensities in the different regions and
agroecological zones. Thus, it is estimated that only about 55 percent of the
land in regular crop production is cropped and harvested in any given year in
sub-Saharan Africa (the rest being fallow), while the average cropping intensity
is about 110 percent in South Asia, reflecting mainly the multiple cropping in
the region's substantial areas under irrigation as well as the region's more
general land scarcity. It is foreseen that the land needs for crop production
growth will come in part from further increases in cropping intensities, and the
average for the developing countries as a whole could rise from 79 percent at
present to 85 percent in the year 2010. Thus, land cropped and harvested in an
average year would increase from 600 million ha at present to about 720
million ha in year 2010, or 120 million ha increase compared with the 90
million ha of new land to be brought into crop production.

Achievement of the increased intensities and higher yields depends crucially
on maintenance of irrigation and its further expansion by 23 million ha or 19
percent in a net sense, i.e. on top of the expansion needed to offset losses of
irrigated land due to salinization, etc. This is a lower rate of expansion than in
the past because of the well-known problems of increasing unit costs of
irrigation investment and scarcity of water resources and suitable sites, as well
as the enhanced attention paid to avoiding adverse environmental impacts.
Given these constraints, but also for reasons of efficiency, the emphasis in the
future will be more on making more efficient use of water and less on
indiscriminate expansion of irrigated areas. The bulk of additional irrigation
would be in South Asia, which now accounts for 52 percent of all irrigation of
the developing countries (excluding China), a share it will maintain in the
future. It is noted that the above-mentioned 23 million ha of additional

irrigation is a net increment. In practice, the gross investment requirements for irrigation will have to cover a much wider area to account for rehabilitation of existing irrigated areas and to substitute for those permanently lost because of degradation.

Would agricultural expansion encroach on the forest?

The FAO Forest Resources Assessment 1990 produced data on the forest land of the tropical countries. Of the developing countries of the study for which the data on land with crop production potential were estimated, the forest area data are available for only 69 countries. The following comments examine the extent to which agricultural expansion may encroach on the forest. They, therefore, refer only to the subset of the 69 countries which account for all but 4 percent of the total tropical forest area in the FAO Forest Resources Assessment. They are also speculative because the extent of overlap between the forest and the land with agricultural potential is not fully known. Only some elements of such overlap can be deduced indirectly.

Subject to the data caveats, the situation in these 69 countries is one whereby 85 million ha are projected to be converted to agriculture in 20 years out of a total 1720 million ha of land with agricultural potential but not in crop production use at present. The extent to which this land overlaps with the forest area is not fully known, but a minimum estimate (derived as explained in Chapter 4) is about 800 million ha and the real overlap is probably much larger.[2] Not much more can be said on this matter, except perhaps that if all the additional land for agriculture were to come from the forest areas, it would imply an annual rate of deforestation of 4.2 million ha, or 0.25 percent p.a. of the total forest area of these 69 countries of 1690 million ha. This compares with the 15 million ha (0.8 percent p.a.) of annual deforestation estimated for the 1980s. This latter figure, however, includes deforestation from all causes, not only from formal expansion of crop production. In particular, deforestation results from expansion of grazing (not included in the estimates of this study) and informal, unrecorded, agriculture using much more land than considered necessary to achieve the crop production increases. It also includes deforestation from logging of areas not yet reforested by natural regrowth and from fuelwood gathering operations. To the extent that expansion of grazing, informal agriculture, overcutting for fuelwood and unsustainable logging continue in the future it must be expected that deforestation will continue at a much greater rate than needed for expansion of formal agriculture.

Other claims on land

Land with agricultural potential is increasingly occupied by human settlements and infrastructure. Rough estimates for the developing countries (excluding

China) indicate that such uses of land may be about 94 million ha, or 0.033 ha per caput (3000 persons/km^2), but with this ratio varying widely among countries, depending on overall population densities. Not all human settlements are on land with agricultural potential, but about 50 million ha probably are in this category. With population growth, more land will be diverted to human settlements and infrastructure, though perhaps not in proportion, because with increasing population densities the land so used per person will tend to decline to perhaps 0.03 ha. This means that land in human settlements may increase to 128 million ha, of which perhaps some 70 million would be land with agricultural potential, an increase of the latter of 20 million ha. This potential use must therefore be added to that for the expansion of crop production proper, discussed above, to obtain an idea on future claims on the land with agricultural potential.

Further growth in fertilizer use and some in pesticide use in the developing countries

The developing countries (excluding China) use some 37 million tonnes of *fertilizer* (in terms of nutrients NPK). Such use increased four-fold in the last 20 years, though the growth rate of the 1980s was much lower than that of the 1970s. At present, the fertilizer use rates have reached 62 kg/ha of harvested area (about one-half the average of the developed countries), but with very wide differences, ranging regionally from 11 kg in sub-Saharan African to 90 kg in Near East/North Africa. The scope for further increases is much less than in the past. This, in combination with the lower rate of growth of agriculture, will tend to make for further declines in the growth rate of fertilizer consumption, to 3.8 percent p.a. in the period to 2010. Thus, projected fertilizer consumption in the developing countries (excluding China) may rise to some 80 million tonnes and the application rate to some 110 kg/ha. The environmental dimensions of this prospective development are discussed in Chapters 11–12. Here it is worth noting that while there are problems from excessive use in some irrigated areas of the developing countries, there are also problems from too little use in other areas, where it is associated with land degradation due to nutrient mining. Sub-Saharan Africa uses only 11 kg/ha. Even a doubling by 2010, as projected here, would still be too little for eliminating nutrient mining in some areas.

Traditional plant protection methods (tillage, burning, crop rotation) remain important in developing countries. However, methods based on the use of chemical *pesticides* have become widely used in recent decades. It is estimated that in the mid-1980s the developing countries accounted for about one-fifth of world consumption of pesticides (active ingredient). They account for about 50 percent of world use of insecticides, but for much smaller proportions of fungicides and herbicides. This reflects both agroecological and economic factors, e.g. higher incidence of insects in the humid tropics and cheaper labour for weed control. With labour costs rising in some countries, it can be expected that chemical herbicides will be used more widely.

The intensification of production and the expansion of agriculture into new areas in the developing countries could translate into further growth of pesticide use. Such growth could be contained at fairly low rates, through a combination of technological change, improved management and incentives and increasing resort to methods of integrated pest management (IPM). These prospects for the developing countries contrast with those for the developed countries where the lower growth of agriculture and the policies for pesticides as well as further spread of IPM could eventually lead to absolute declines in total use.

1.4 FURTHER PRESSURES ON AGRICULTURAL RESOURCES AND THE ENVIRONMENT

The pressures for conversion to agricultural use and human settlements of land with agricultural potential were dealt with in the preceding section. On the whole such claims (110 million ha in all developing countries, excluding China) over the next 20 years would appear small when compared with about 1.8 billion ha of land with agricultural potential not occupied by either of the two uses. However, land scarcities are very acute in some countries and regions, viz. South Asia and Near East/North Africa. Even the small increases foreseen for them are a significant part of their still unused land. For example, the increments for these two uses would claim about 25 percent of the still unused land with agricultural potential in South Asia. There will be little land left for further expansion beyond the year 2010. It is noted that additional land for agriculture in South Asia will be needed even after allowing for further intensification. The latter could raise cropping intensities from 112 percent to 122 percent and double the fertilizer use rate per hectare.

Even though land constraints are severe in some countries and regions, those of *freshwater* supplies for agriculture are even more limiting for many more countries. The increasing claims on agricultural land for non-agricultural uses are minor when compared with those placed on water resources, because the per caput non-agricultural use of water tends to rise very rapidly with urbanization and industrialization. Competition between agriculture and the other sectors for dwindling per caput availabilities of freshwater will become more intense in the future and in most cases it can only be accommodated by increasing the efficiency in water use.

Degradation of soils is estimated to affect some 1.2 billion ha of land worldwide, of which about 450 million ha are in Asia. Among the causes, deforestation and overgrazing probably contribute one-third each, with the bulk of the balance due mostly to mismanagement of arable land. Soil (water and wind) erosion accounts for just over 1 billion ha of total degradation, with the balance due to chemical and physical degradation. Both man-made and

fertilizer

natural processes (e.g. upward movements in the earth's crust) cause soil degradation. Some degradation will continue to occur in the future but the relationship between soil erosion and productivity loss is complex and more work is needed before firm conclusions can be drawn about the impact of soil erosion on yields.

Degradation from nutrient mining is a serious problem, particularly in the semi-areas of sub-Saharan Africa where livestock manure is in short supply and the use of mineral fertilizer is seldom economic. The problem will probably continue to exist over the next 20 years. Degradation from salinization of soils is primarily a problem of irrigated areas, but also occurs in hot dry zones. Available estimates of irrigated land losses from this cause vary widely while 10–15 percent of irrigated land is to some extent degraded through waterlogging and salinization.

Desertification (broadly: land degradation in dryland areas) is estimated to affect some 30 percent of the world's land surface. More recent thinking on desertification points to a growing consensus that the past estimates of area affected were greatly exaggerated. Some of the more extreme estimates were due to weaknesses in the methodology used to produce them. It is now recognized that drylands are much more resilient to drought and to man's abuse than previously thought. However, further expansion of agriculture into fragile soils in the dryland areas would contribute to increasing problems from this source.

Water contamination of agricultural origin (salt concentrations in irrigated areas, contamination from fertilizer and pesticides as well as from effluents of intensive livestock units and fish farms) will likely increase further because of the long length of time required for appropriate corrective action.

As regards *pesticides*, it is assumed that greater emphasis on integrated pest management and concerns about health and ecosystem conservation will tend to reduce the growth rate of pesticide use. But the more intensive use of land (reduced fallows, more multiple cropping) as well as the higher than average growth of the vegetables sector will contribute to further, though modest, increases in pesticide use in the developing countries.

Further expansion and intensification of agriculture will also contribute to intensified *pressures on the environment of a global nature*. Deforestation will affect adversely the dual role of forests as habitats for biodiversity and as major carbon sinks. Biodiversity will also likely suffer from possible further draining of wetlands for conversion to agriculture, even though this conversion may affect only a minor proportion of total wetlands. Additionally, agriculture will continue to contribute to the growth of greenhouse gases in the atmosphere (biomass burning in the process of deforestation, and methane emissions from rice cultivation and from ruminant livestock).

The eventual impacts of climate change are still uncertain, but on present evidence they may affect particularly adversely those regions already vulnerable to present-day climate variation, notably sub-Saharan Africa. The effects of an eventual rise in the sea level would also be severe for some countries and affect a good part of their high quality land resources. For the present and more immediate future, increased CO_2 levels appear to have a positive effect on agriculture in general, because they contribute to higher yields through faster growth of plant biomass and better water utilization in many crops.

1.5 TECHNOLOGICAL AND OTHER POLICIES TO MINIMIZE TRADE-OFFS BETWEEN AGRICULTURAL DEVELOPMENT AND THE ENVIRONMENT

Existing and possible future technologies provide scope for responding, wholly or partially, to the increased pressures of agricultural origin on the environment. Exploring the potential for doing so requires shifting technology from "hardware" solutions requiring large inputs of fixed and variable capital, e.g. machine-made land terraces or pesticides, to solutions based on more sophisticated, knowledge and information-intensive resource management practices which can lower both off-farm costs and environmental pressures. This is not to suggest that a new technological approach is sufficient by itself. Much will depend on policies and institutional measures providing incentives needed for farmers, forest users and fishermen to adopt sustainable technologies and resource management practices. Institutional measures will include the establishment of well-defined property or user rights for public and private resources, as well as enhanced people's participation and decentralized resource management.

It is noted from the outset that the general debate regarding the merits of low or high external input technological development paths for agriculture has run its course, and there is growing acceptance that neither of the two approaches has the whole answer. What is required is a balanced integration of the two systems. For example, the use of mineral fertilizer will continue to grow but it cannot in many situations provide all the inputs necessary to maintain soil fertility and must be associated with organic manures and other biological inputs as part of an integrated plant nutrition system (IPNS).

More generally, the extent to which countries will follow more environment-friendly practices depends on their socioeconomic and natural resource situations. The developed countries are in a better position to do so, and are moving in this direction. In contrast, the developing countries are in much greater need to improve the management of their agricultural resources because their livelihoods depend crucially on them. At the same time, however, they are in greater need than the developed countries to increase production through intensification and have much less access to technologies and resources for more sustainable production. But there is much scope for improvement and for minimizing trade-offs between more production and the environment even under these unfavourable conditions. The important thing is for policies to recognize that the first priority of many farmers is household food security and family welfare. Thus efforts to minimize trade-offs between more production and the environment must be centred on actions that improve household food security and are profitable on time scales which meet the farmers' differing circumstances or risk perceptions.

It is now well recognized that the past heavy dependence of the agricultural development of the developing countries on the transfer of technologies and management practices of the developed countries contributed to raise production and productivity, but it had some undesirable effects, e.g. discouragement of mixed cropping and minimal tillage practices, dominance of mineral fertilizer, emphasis on engineering rather than biological approaches to soil stabilization, neglect of semi-arid areas and crops, etc. Corrective action would require a shift in national and international research priorities, with particular emphasis on technologies which are not too risky and are profitable at early stages in the adoption process. Efforts to build on indigenous technical knowledge hold promise, but there is no guarantee that they will be sufficient in isolation.

In the quest for limiting *land and water degradation*, the wider adoption of known techniques of soil conservation with low external capital requirements could help boost or stabilize yields in the first half of the projection period. Dryland areas in sub-Saharan Africa and Asia could benefit from such techniques, as would slopelands in the humid tropics. Likewise, efforts for dealing with the *salinization problem* could benefit from integration of the standard corrective action (drainage, canal lining) with a more holistic approach to water management, e.g. conjunctive use of surface and groundwater and parallel use of canal and tubewell systems. More generally, the increasing dependence on raising water use efficiency for coping with the growing water shortages will require some radical rethinking of policy approaches to pricing water and to needed institutional changes.

Wider adoption of *integrated plant nutrition,* its further development and improved management of input use provide the main technological way to meeting the challenge of required increases in nutrient supplies in support of more production while minimizing adverse effects on the environment. Likewise, *integrated pest management* is to be the mainstay of efforts in the plant protection area, with priority to the crops accounting for the bulk of pesticide use: cotton, maize, soybeans, fruit and vegetables.

In the *livestock sector*, there is much in the technological pipeline to meet the challenge of moving towards more sustainable production systems. Policies in this direction could have an impact well before 2010. The aims would be to compensate for the lack or poor quality of land through measures to raise pasture and rangeland output and improve management systems; to bring about a closer integration of crop and livestock production; to raise the supply and quality of supplementary feeds; to achieve genetic improvements from conventional breeding and modern biotechnical tools;

and to complement these gains with cheaper and more effective animal health measures.

Biotechnology offers a range of applications for plant and animal production. Some are likely to have an increasing impact well before 2010; others in the longer term. The former include tissue culture of virus-free stocks of cassava and other root crops, and the introduction of microbial plant growth promoters, e.g. mycorrhiza. The latter include cereals with the ability to fix some of their own nitrogen needs, and transgenic tree crops.

Making progress towards the adoption of technologies for sustainable agriculture will depend greatly on increased *agricultural research* efforts with emphasis on: (a) improved management of biological systems, based on a better understanding of their feedback and balancing processes; (b) better information management, implying the need for sound data on natural resources, land use and farming systems, etc., to improve environmental monitoring capability; and (c) better farm–household system management, in order to obtain a better integration of activities in the household and in the field, and on and off the farm. At the operational level, the research effort should be directed at promoting sustainable increases in productivity in the higher potential areas as well as at targeting marginal and fragile environments where current degradation must be reversed and production stabilized or raised. These thrusts must be supplemented by two cross-cutting and highly complementary approaches, that of rehabilitation and restoration of ecology, and that of exploiting the synergism of indigenous technical knowledge and modern science. All four actions must be supported by international efforts to strengthen the national agricultural research systems, both institutionally and financially.

Finally, *international agricultural trade* and policies affecting it can exert influences on the environment and the prospects for sustainable development. Trade may affect the environment if production shifts from places where it is less sustainable to places where it is more sustainable and vice versa. To the extent that trade contributes to shift production to more sustainable locations, more trade would tend to lower global pressures on resources and the environment. Such pressures would be minimized when all trading countries have environmental policies which embody the environmental externalities into the costs of production and the prices of the traded goods. However, environmental externalities need not be valued in the same way in countries with differing resource endowments and levels of development. In particular poor countries should not be denied opportunities for profitable trade because they do not meet the strict, and often inappropriate for them, environmental conditions reflecting values of much wealthier societies.

1.6 FOREST SECTOR PROSPECTS

With the exception of fuelwood, per caput consumption of *forest and forest-industry products* will continue to grow, particularly in the developing countries, with growth being highest for wood-based panels and paper. The developed countries as a whole should face no major problem in increasing production of wood in sustainable ways by as much as required for their own consumption and exports. The developing countries depend currently to a high degree on natural forests for the production of wood, for own consumption and exports. Such dependence and their higher growth of demand will make it more difficult for them to increase production in sustainable ways, unless greatly improved management measures are instituted and forest plantations greatly expanded.

Developing countries, particularly the poorest ones, depend on wood for a major part of their energy supplies. The shortages of *fuelwood* are likely to persist and become more acute as accessible forest and non-forest sources dwindle due to overexploitation and conversion of forest land to other uses. Although much of the growth in energy consumption of the population groups which depend on fuelwood will be met by the continuing trend towards substitution of alternative fuels for wood, some population groups (e.g. the urban poor or rural people in remote locations) are not likely to have ready access to such alternatives. For them, the future outlook is for more work to be put into procuring wood and making do with less energy.

Pressures on the forest for meeting often conflicting demands are bound to increase, mainly in the developing countries, and continue to imperil the forest's essential environmental functions. The highest risk is manifested in terms of *tropical deforestation*. It continued to advance in the 1980s at about 15 million ha p.a., or 0.8 percent of the total tropical forest area. The FAO Forest Resources Assessment 1990 documents that deforestation is observed to radiate out from the populated areas and that the higher the increase in population densities the higher the rate of deforestation, other things being equal. It also notes that much of the deforestation is related to the expansion of agriculture, whether in the form of recorded conversion to arable land or, more often, unrecorded expansion. Fuelwood collection is also a contributing factor. Logging *per se*, on the other hand, need not lead to permanent loss of forest if soundly managed. It may, however, affect other vital environmental functions of the forest, e.g. biodiversity. Moreover, the opening-up of previously inaccessible forest areas by logging operations tends to facilitate settlement and conversion to agriculture.

These findings seem to confirm the common belief that there is a close association between population growth and deforestation. However, for policy purposes the mechanism connecting these two variables has to be understood. This is no simple matter, for the reasons discussed earlier in relation to the build-up of pressures on agricultural resources and the environment. In particular, it is noted that the most relevant aspect of population growth is the extent to which it is associated with increases in the number of people depending on agriculture, and more generally the rural poor. Many developing countries are far from having reached the stage when pressures from this kind of population growth are relaxed. Some of them are not even making progress towards it.

It follows that further deforestation is to be expected in the future. Some speculative comments on the possible deforestation impact of expansion of agriculture and human settlements for the year 2010 were made in Section 1.4. It was noted there that informal and disorderly expansion of agriculture may lead to a higher rate of conversion of land and forest areas than required by the projected growth in crop production. Expansion of grazing, fuelwood production and unsustainable logging may further contribute to deforestation. Under the circumstances, the key issue is how to minimize loss of forest during this rather protracted, though hopefully transitory, phase, until such time as the inherent forces (development, reduction in agricultural population and rural poverty, etc.) making for containment or reversal of deforestation come into play.

The preceding discussion is based on two premisses: (a) that much of the deforestation is caused by expansion of agriculture, and (b) that it is closely related to the growth of population in poverty, and indeed that part which depends on agriculture for a living. True as these premisses are, they tell only part of the story. In particular it may not be concluded that the rate of damage to the forest will slow down at the initial stage of accelerated economic growth and poverty reduction. There is evidence suggesting that the opposite may happen. This is explained in part by the fact that more intensified exploitation of forest resources and expansion of agriculture to exploit profitable opportunities are part and parcel of the very process of accelerated development and poverty reduction. In practice, countries tend to run down their natural capital to increase incomes as conventionally measured, i.e. without netting-out the income gains for the losses of natural capital. The other contributing factor has to do with the limited capabilities of countries to formulate and enforce rules for sustainable exploitation of the forest resources; and in some cases their own sector-specific or economy-wide policies translate into incentives for unsustainable exploitation. Ignoring other causes of deforestation, in particular the complex interactions of activities by both the poor and the non-poor, can lead to wrong policy conclusions as noted earlier.

1.7 INCREASING RESOURCE CONSTRAINTS IN FISHERIES

The historical developments as well as the future prospects of the fisheries sector are conditioned, to a significant extent, by the wild characteristic of the resource and the fact that, for most species, the levels of production are limited by nature. This has three important consequences. First, beyond certain levels, additional investment in fishing effort does not produce additional yields and, in many cases, actually leads to declines in total catch as well as to economic waste. Such an increase in fishing effort is inevitable in those, almost universal, situations where there is ineffective fisheries management. Second, with growing demand and limited supplies, the real prices of fish products inevitably increase. This has important and damaging consequences for low-income consumers, particularly those in the developing countries. The third major, and more positive, result is that limited natural supplies and high prices serve to stimulate increased production through the cultivation of those species that allow it.

World production of fish had been increasing up to 1989 to a peak of 100 million tonnes after which it declined to about 97 million tonnes in the three subsequent years. The share of culture fisheries in total production has been increasing rapidly and it currently accounts for about 12 percent of the total. Marine capture fisheries account for about 80 million tonnes of the total. It is now evident that the yield of this sector is adversely affected at extraction levels beyond about 80 million tonnes.

The natural resource constraints to increasing production in the capture fisheries sector mean that additional fishing effort and investment is unlikely to increase production and may well lead to declines. Better management and other interventions which would favour recovery of fish stocks could make it possible to increase somewhat capture fisheries production (marine and inland) from the present 85 million tonnes to perhaps 90–110 million tonnes. This estimate is hypothetical and subject to many uncertainties. Culture fisheries (marine and inland) have higher growth potential, but even here constraints are present (technology, environment, disease). There is scope for reducing these constraints, particularly for marine environment aquaculture, and it is possible that production could grow at a higher rate than that of capture fisheries, e.g. from 12 million tonnes to 15–20 million tonnes by year 2010.

It follows that total fish production from all sources could be in year 2010 between 10 and 30 percent above present levels. Over the same period, world population is expected to grow by 36 percent. Therefore, per caput fish supplies will likely fall. Consumption by the poor may fall by more and shift in part to species currently used for reduction to fishmeal, as the species currently less-preferred by high-income consumers are diverted to their segment of the market. These prospective developments can have serious nutritional consequences for the poor consumers in countries with high dependence on fish for protein supplies, e.g. many countries in Asia and Africa.

The increasing supply constraints and the associated rise in the real price of fish will tend to stimulate greater investment in fishing effort, thus establishing a vicious circle whereby stock depletion reduces supplies leading to additional price increases. This process has been aided by heavy subsidies granted to fisheries by major countries. With the reforms under way in the ex-centrally planned economies of Europe, a substantial part of their subsidized operations has become openly uneconomic. The consequent reductions in the fleets of these countries are leading to significant structural change in the world fishing industry.

This vicious circle can partly be broken by the establishment of systems of exclusive use rights which provide the fishermen with a stake in the resource and an interest in future returns. However, as many governments have found, this is difficult to achieve. At national levels, fishery administrators generally do not have the mandate to make such decisions. In international areas or areas where stocks are shared by countries (e.g. the northeast Atlantic), negotiators cannot readily agree to controls which limit the rights of their own fishermen. But as the problems become increasingly severe, the issues are raised to higher political levels and, eventually, will force the necessary decisions. Several countries have already taken the basic steps to create exclusive use rights and have achieved significant benefits. Although the systems still contain many imperfections, the improvements that have been produced provide valuable lessons for other countries. There is some hope, therefore, that the management of fisheries will eventually improve. However, although the benefits will be significant in reducing biological and economic waste, they will still not be sufficient to overcome the limits on supply.

Finally, fisheries and more general policies must address the problems increasingly affecting small-scale fisheries: conflict with large-scale operations in the inshore waters and degradation of the coastal environment. This is necessary for social purposes, for shifting production on to a more sustainable base and for minimizing adverse effects on the environment. With regard to the latter, it is noted that the coastal zone receives large amounts of pollutants including: organic wastes from municipalities, chemical wastes from industries, pesticides and herbicides from agriculture and siltation from forest land clearing and road building. In addition, activities within the coastal zone also affect the environment. These include mining of coral reefs and destruction of mangrove swamps. Fishermen themselves contribute to these kinds of damage by converting mangrove swamps to mariculture ponds for shrimp; by excessive use of feed and antibiotics in cage culture; and by using dynamite, poison and other kinds of techniques that destroy coral reefs.

1.8 POLICIES FOR AGRICULTURE AND RURAL DEVELOPMENT IN DEVELOPING COUNTRIES

Policies for agriculture in an economy-wide context

It is now well accepted that policies for agriculture must be viewed as an important component and be an integral part of the wider policy environment. The initial approaches of the post-Second World War period emphasized, at best, benign neglect of agriculture, extraction of a surplus from it and preference for, often import-substitution-based, industrialization. Such approaches have often been proven counterproductive, though practices based on such perceptions persisted for a long time in several developing countries. It is now well recognized that agriculture's role must be upgraded in development strategies, notwithstanding the fact that in the process of development other sectors are bound to grow faster than agriculture. The general policy thrust underpinning this study draws on the lessons of experience and current thinking, and may be summarized as follows:

1. Contrary to the earlier thinking mentioned above, it is now well accepted that in the developing countries with a high weight of agriculture in the total economy and employment, overall development is impeded if agriculture is neglected, starved of resources or discriminated against by the use of policies which affect adversely producer incentives; and that such neglect is not only socially unacceptable, seeing that the majority of the poor, and often of the total population, depend on agriculture, but also economically inefficient.

2. Farmers and agriculture do respond to incentives, and many of the successes and failures in getting agriculture moving can be explained by policies which permitted such incentives to manifest themselves or, on the contrary, affected them adversely, directly or indirectly. Incentives comprise not only better prices for outputs and lower ones for inputs but also the provision to agriculture of public goods such as infrastructure, education, research, etc.

3. Agriculture's performance is affected not only by policies specifically designed for it (e.g. price supports, taxes, subsidies) but also, and often more deeply, by policies affecting the overall macroeconomic environment (e.g. public sector deficits, inflation, interest rate, exchange rate) as well as policies for the other sectors (e.g. the rate of protection accorded to manufacturing if it makes more expensive the manufactured inputs and consumer goods purchased by agriculture). The lesson is that agriculture cannot prosper in an environment of high inflation, overvalued exchange rates and generally in conditions which turn incentives against it. The importance of the macroeconomic factors came in stark evidence in the aftermath of the 1970s, a period of external shocks, easy borrowing and build-up of foreign debt, which was followed by the emergence of strong macroeconomic disequilibria and ushered in the crisis decade of the 1980s. Policy responses to correct such imbalances (going under the generic name of structural adjustment) while restoring incentives to the sector may also have affected the sector negatively due to public spending cuts, less growth of the demand for agricultural produce and fewer opportunities for agricultural labour to move to other sectors. These reforms may not by themselves engineer resumption of growth but they are considered necessary as a step towards setting the economy on an even keel, in the absence of which strategies for long-term growth have a low probability of succeeding.

4. Certain types of public sector involvement in economic life can be counterproductive. The analysis of experiences here draws heavily on

examples from agriculture, as government involvement particularly in marketing of agricultural produce was very diffuse in some countries. The issues related to the proper role of the public sector have still to be settled (and certainly they cannot be settled on dogmatic grounds) as the expected benefits from reforms to correct these perceived structural shortcomings and the often associated macroeconomic imbalances are in many cases slow in coming and of uncertain magnitude and duration. But some degree of consensus can be gleaned. It reaffirms and strengthens the case for an enhanced role of the public sector in agriculture in such areas as provision of infrastructure, education (including technical education for agriculture), research and technology development and transfer, etc.; with the proviso, of course, that success or failure depend greatly on the organizational and managerial capabilities of governments. The case for this sort of public sector role is further strengthened by increasing evidence about the high rates of return to agricultural research and that what matters for development, together with, and perhaps more than, investment in physical assets, is investment in human capital and knowledge. In parallel, the consensus seems to lend support to the proposition that, in a general sense, governments should backstop rather than supplant the private sector in production and marketing by, mainly, creating the institutional framework and enforcing the rules for markets to work efficiently and for prices to play their vital role as incentives and disincentives for guiding such private sector activities.

In conclusion, it can be stated with confidence that the early post-war ideas of squeezing agriculture for the benefit of other sectors are dead and hopefully buried for good. This does not mean that agriculture's role as supplier of resources to the rest of the economy will cease. But it does mean that in many situations priority must be given to increasing agricultural productivity and the incomes of the rural people if markets for the domestic industry are to be expanded and if a surplus is to be created in agriculture and transferred, rather than extracted, to other sectors. Such transfers are seen primarily as spontaneous responses to the normal course of events whereby agriculture grows less rapidly than other sectors. In these conditions, other sectors offer generally higher rates of return and it is natural that resources are directed to them. Here again, the importance of public sector interventions to promote investment benefiting agriculture is emphasized, e.g. in research, education, infrastructure, etc., because the social rate of return on these investments can exceed by far the private rate of return. In the process of development and structural transformation, the initial conditions prevailing in some countries dictate that there is a strong case for priority to agriculture in development strategies to enable the sector to play its vital role in poverty alleviation and in backstopping overall economic growth.

Policies for, or affecting, international agricultural trade

A number of policy changes have been undertaken in recent years or are under consideration, at both the international and the national levels, which can have profound effects on international agricultural trade. All point to the direction of allowing an enhanced role for market forces to determine trade flows. The reforms in the ex-centrally planned economies of Europe belong in this category. Their potential trade effects were noted earlier. Here belongs also the reform of the European Community's Common Agricultural Policy (CAP), with potential trade effects also noted above. These effects of the CAP reform would be for the major temperate commodities in the direction of those that are expected to be forthcoming from the application of the provisions of the agricultural part of the Final Act of the recently concluded Uruguay Round of Multilateral Trade Negotiations.

The Agreement on Agriculture of the Uruguay Round, in combination with other recent policy reforms, will contribute to change the structure of protection towards measures which allow an enhanced role for market forces to determine production, consumption and trade outcomes. But, on balance, the Agreement represents only limited progress towards free trade in agriculture. Its value is to be seen more in terms of the discipline and transparency it implies for the policies which affect trade.

In parallel, the general thrust of policy reforms of the developing countries described in the preceding section is towards more open economies and structural adjustments which would create more favourable conditions for trade. However, some key problems faced by many developing countries in their agricultural trade relations are not being addressed with the same urgency, if at all. These include issues of, among others, the falling and volatile prices of major tropical export commodities or market access restrictions and subsidized export competition for some of their commodities on the part of developed countries. Finally, the concerns for the environment and the related policies have helped bring into the international trade policy debate the issues concerning the interactions between trade and the environment, as discussed earlier.

Issues of rural poverty and rural development

Over 1 billion people in the developing countries are poor, with a substantial majority of them living in rural areas. The development of agriculture may therefore play a direct role in rural poverty alleviation, since the majority of rural poor depend on agricultural activity for providing the main source of their income and employment. The projected growth rates of agricultural production presented earlier are generally above those of the population dependent on agriculture in all developing countries. The implicit growth rates of the average per caput incomes of the agricultural population are, however, modest, though they can be significant in those countries where the agricultural population is on the decline. Reductions in the incidence of rural poverty from agricultural growth depend not only on its rate per caput but also on its impact on distribution; and also on increasing opportunities for non-agricultural employment in rural areas in synergy with agricultural growth.

The impacts of agricultural growth on different socioeconomic categories of rural producers and labourers, as well as the mechanisms through which these impacts are mediated, depend on the nature of the growth processes and the structural factors underlying the social organization in rural areas. The evidence seems to suggest that while, on balance, agricultural growth can be expected to bring about reductions in rural poverty, some parts of the rural population may become worse-off economically. The structural characteristics of the rural economy at the inception of agricultural growth play a predominant role in the distribution of benefits from higher production.

Access to land is a major factor determining the poverty alleviation effects of agricultural growth as well as conditioning the growth process itself. The most recent attempt to take stock of progress in redistributive *land reform* was undertaken in 1991 for the quadrennial FAO report on progress under the Programme of Action of the World Conference on Agrarian Reform and Rural Development. The report concludes that progress has been limited. Yet the case for such reforms remains strong on both efficiency and equity grounds. It is further strengthened when linkages with the non-agricultural rural sector are considered, because a more equal distribution stimulates also rural non-farm employment. In general, the experience seems to suggest that political commitment and strong follow-up support from the public sector to beneficiaries of land reform are essential ingredients of successful land reform policies. Land reform will continue to be a relevant issue in the quest for rural poverty alleviation. This will be particularly so in countries with increasing agricultural population and poor non-agricultural growth prospects.

Beyond reforms affecting the distribution of land ownership, those of *tenancy arrangements* remain important. The lessons here are that past policies restricting sharecropping contracts were sometimes counterproductive. The tenancy reforms pursued in the reforming centrally planned economies of Asia are proving increasingly successful as they shift from socialized farming to household-based arrangements with adequate security of tenure. It is also increasingly accepted that most traditional land tenure systems in Africa can adapt well to changing circumstances and the policy emphasis should be on providing an appropriate legal and institutional environment.

Limited access to *rural finance* by the poor in agriculture has been a major limiting factor in agricultural development and poverty alleviation. The policy orientation favouring provision of formal finance through specialized credit institutions has often been unsuccessful and there is increasing recognition of the need for less formal arrangements to enhance access to credit of the poor, e.g. Rotating Savings and Credit Associations (ROSCAs).

Concerning *marketing*, the attempts to provide marketing services to agriculture, including to the poor, often together with other services, through parastatals have proven generally, though not always, inefficient. Such inefficiencies are among the reasons why reform of the role of the public sector in agricultural marketing figures prominently in structural adjustment programmes. There is a well-recognized role for government in marketing by providing infrastructure, the legal framework and enforcement of rules and generally supporting the functioning of markets. However, the policy thrust is for direct involvement of the state in marketing to be curtailed and for the private sector to be allowed and encouraged to be the main vehicle for this function. The key issue is how to move smoothly from one organizational form to another, because the poor will suffer most if major disruptions in services occur.

In the longer term, the growth of agriculture and the overall economy would tend to alleviate the rural poverty problem, particularly if agricultural and rural development is directed towards more egalitarian patterns by policies like the ones described above. However, in the immediate future, and for some countries for a long time, rural poverty will continue to be a major problem. Therefore *direct interventions* will continue to be needed. *Rural public works* have long been used for this purpose, particularly in emergency situations. They form the core of government anti-poverty strategies in South Asia and other countries. The experience is generally favourable, and anti-poverty impacts are highest under community participation and careful selection and targeting of beneficiaries.

Interventions in the food and nutrition area will continue to have a place in the total arsenal of anti-poverty policies. The lessons here are that attempting to reach the poor through *general food subsidies* is a very costly policy and, in general, tends to benefit the non-poor more than the poor. More targeted schemes are generally superior in achieving their objectives, though often more difficult to administer. They include *ration schemes, food stamps and supplementary feeding programmes.*

1.9 EMPHASIS ON HUMAN RESOURCES DEVELOPMENT IN DEVELOPING COUNTRIES

As noted many times in the preceding discussion, intensification of agriculture will continue to be, and more so than in the past, the mainstay of production growth in the future. It is now well recognized that what matters for a successful transition to more intensive agriculture, more than physical capital, in the capability of farmers to be energetic agents open to and eager to adopt profitable innovations in both technology and management practices. Moreover, the need to shift agriculture to more sustainable technologies and practices will attach an even higher premium to those capabilities. Therefore, a major thrust in policies for agricultural development must be directed to human resources development (HRD), involving all aspects from basic education to technical one, including formal and informal approaches to creating and transferring skills. HRD includes also the upgrading of health and nutrition. These, as well as education, are objectives of development in their own right and not only means for making people more productive economically.

The required HRD effort in agriculture in the developing countries is considerable because the population economically active in agriculture will continue to grow, albeit slowly. Moreover, there is a huge backlog to absorb, given the prevalence of high illiteracy rates in the rural areas, as well as scarcity of trained extension personnel. It is estimated that there is one extension agent per 2500 people economically active in agriculture in the developing countries. The corresponding ratio is one to about 400 in the developed countries. In the latter, the private sector is also more active in providing extension services. Additionally, the proportion of females in the total extension activity of the developing countries is very low and out of all proportion to the relative importance of women in agriculture. There have been some encouraging trends in the developing countries, though not in all regions, as regards both the growth of the number of people involved in extension, and their quality, as more highly trained persons gradually replace those with fewer skills.

1.10 CONCLUDING REMARKS

There emerges a mixed picture about the future of the world food and agriculture from the assessments of this study. Overall, the world appears set on a path of declining growth rates of agriculture as more and more countries reach medium-high or high levels of per caput food supplies and population growth slows down. The trend could be halted or even reversed for some time if the significant part of world population with still unsatisfied food consumption needs were to be in a position to demand and/or produce food at higher rates than estimated here for up to 2010.

There appear to be no unsurmountable resource and technology constraints at the global level that would stand in the way of increasing world food supplies by as much as required by the growth of effective demand. And, on balance, there is scope for such growth in production to be achieved while taking measures to shift agriculture on to a more sustainable production path. However, the need to accept trade-offs between agricultural growth and the environment will persist in many local situations which combine adverse agroecological and socioeconomic characteristics. The above global statements apply much less, or not at all, to marine capture fisheries. This latter sector provides perhaps the major example of global natural resource constraints which cannot apparently be relaxed through substitution by man-made resources and technology, at least not as far as one can tell on the basis of present knowledge. But substitution can and does take place at the consumption level, as more investment and technology produce substitutes of fish in consumption, albeit imperfect ones, e.g. poultry meat.

The findings of the study imply that many countries and population groups will not be able to benefit in per caput terms more than marginally from the further growth in world food production, nor from the potential for this growth to be even higher than projected here. Only a combination of faster, poverty reducing, development and public policy, both national and international, will ultimately improve access to food by the poor and eliminate chronic undernutrition. In the countries with high concentrations of poverty and high dependence on agriculture, success in this area will often require priority to be placed on agriculture for increasing incomes and food supplies locally. If local agricultural resource endowments are unfavourable, the task of bringing about development can prove very arduous indeed. It is in such contexts that one can speak of resource constraints being real obstacles to solving food and nutrition problems, even though resource constraints to increasing global food production may not be serious.

Finally, looking forward to the longer-term future beyond the year 2010, the concluding section of Chapter 3 attempts some speculative estimates of the agricultural growth requirements for up to 2025. They are meant to provide a framework for thinking about longer-term issues of world food–population balance. They are not projections of likely outcomes. It can be expected that the annual rate of growth of world food production required to sustain the growing population will tend to continue to decline. This is because the growth rate of world population will continue to fall, while the proportion of world population with relatively high levels of per caput food consumption will tend to increase, allowing little scope for further increases.

Eventually, world population growth could fall to zero and total population could stabilize. If by that time all people had satisfactory levels of food consumption, there would be little further pressure for increasing agricultural production. The key issue is whether the world can tread a sustainable path to such an eventual situation of a quasi steady-state agriculture.

One aspect of this issue has to do with the capability of the world's agricultural resources to underpin the growth of production to the notional steady-state level and maintain it at that level thereafter. It is not possible to give a straightforward answer to this question but the following considerations are relevant: (a) in a world without frontiers and with free movement of people and/or having conditions for greatly expanded food trade the binding character of natural resource constraints, if they exist, will be greatly diminished; and (b) there are many countries in which both food supplies and an overwhelming part of their economy depend on local agriculture. As already noted, if their agricultural resources are poor, it is entirely appropriate to speak of agricultural resource constraints standing in the way of achieving food security for all even if one knew for certain that the world as a whole had sufficient resources to grow in sustainable ways as much food as required to meet the needs of a much larger world population.

Resource constraints impinging on agriculture and food production are only part of the issue whether the world can tread a sustainable path to a situation free of food insecurity. For such a world must be virtually free of poverty. The issue then becomes one of sustainable paths to poverty elimination. This would require economic growth all round. If poverty is to be eliminated in the not-too-distant future, economic growth must be rapid in the regions with high poverty concentrations. Pressures on the wider ecosystem would increase, e.g. generation of waste from greatly increased use of energy. If the wider ecosystem had only limited capacity to absorb the impact, it is possible that environmental constraints from this origin, rather than from the agricultural resources, would condition the pace at which the world can tread a sustainable path to food security for all.

NOTES

1. These are *net* imports of all the developing countries, after projected exports from the net exporting developing countries of some 30 million tonnes (up from 17 million tonnes in 1988/90 and 14 million tonnes in 1969/71) have been deducted from the projected imports of the net importing countries of some 190 million tonnes (106 million tonnes in 1988/90 and 34 million tonnes in 1969/71).
2. It is noted that data on protected areas for 63 out of these 69 countries indicate that some 380 million ha are in this class, of which some 200 million ha are on land with agricultural potential but not in agricultural use.

CHAPTER 2

Major Themes in World Food and Agriculture at the Beginning of the 1990s

2.1 THE LONGER TERM HISTORICAL EVOLUTION OF THE GLOBAL POPULATION–FOOD SUPPLY BALANCE AND FOOD AND NUTRITION IN THE DEVELOPING COUNTRIES

As noted in the Foreword, the two issues which dominate all others in a global assessment of food and agriculture prospects are: (a) the persistence of undernutrition and food insecurity for large parts of the population of the developing countries; and (b) the increasing scarcity and degradation of agricultural and other environmental resources as they relate, directly or indirectly, to the process of meeting the food and income needs of a growing world population. This chapter is devoted to presenting the present understanding of the nature and significance of these two issues and, as far as possible, to an analysis of the historical developments that have given rise to the present situation. The intention is to give the reader an understanding of what are the real dimensions of these issues, analyse progress and failures in the historical period and identify the factors that may determine progress or failure in the future.

The evolution of the global population–food supply balance provides an appropriate background for reviewing the evolution of the food and nutrition situation in the developing countries. It must, however, be noted from the outset that examining developments at the global level, e.g. by juxtaposing the evolution of world food production and that of world population, offers few analytical insights for understanding the evolution of the food and nutrition situation in the countries and population groups subject to food insecurity. This is not to deny the fact that the global food demand–supply balance influences the incidence of undernutrition. It does so mainly through its impact on the price of food products bought and sold by the poor as well as through its influence on food aid flows. But the fact that hunger persists in the developing countries at a time when global food production has evolved to a stage when sufficient food is produced to meet the needs of every person on the planet bears witness to the multifaceted dimension of the problem in which the evolution of world per caput food production is but one of the many determining variables. Still, the issue of global balance is very alive in the minds of the public, particularly when issues of the capabilities of the earth to support the ever growing global population are considered. Such issues have recently assumed increasing importance in the context of perceived constraints to the global food production capabilities related to natural resource degradation and other environmental problems.

Concerning the evolution of the global population–food supply balance, it is noted that the last few decades witnessed unprecedented increases in world population. Only 30 years ago, world population was 3.0 billion; it was 5.5 billion in 1992. There have been ever increasing annual absolute increments in world population during this period. During 1960–65, 63 million persons were added to world population every year. The annual increment rose to 72 million in the early 1970s, to 82 million in the early 1980s and is estimated to be some 93 million at present. It may not peak until the year 2000. Thereafter, the annual increment (not total population) is projected to start declining very slowly, e.g. it will be some 85 million by 2025, by which time world population will have reached 8.5 billion. (Data are from the 1992 UN population assessment, medium projection variant (UN, 1993b). The population projections used in this study are from the 1990 UN assessment. They are shown in Chapter 3 and, in more detail, in the statistical appendix (Appendix 3).)

How has agriculture responded to these increases in world population? The global picture is shown in Figure 2.1. Production grew faster than population. Per caput production is today about 18 percent above that of 30 years ago. Food availabilities for the world as a whole are today equivalent to some 2700 kilocalories per person per day (referred to as simply calories in the rest of the text), up from 2300 calories 30 years ago.[1] And this is counting only food consumed directly by human beings. In addition, some 640 million tonnes of cereals are fed to animals for producing the livestock products which people consume. Diversion of even one-third of such animal feed of cereals to direct human consumption could raise per caput food availabilities to some 3000 calories in a net sense, i.e. after adjusting for calorie losses due to consequent reductions in the production and consumption of livestock products. This is not to suggest that such potential diversion is a practical or even necessary proposition. But the example serves to illustrate the fact that the existing per caput food availabilities are considered sufficient for everyone on the planet to have adequate nutrition, provided they are distributed equally.

Yet such food availabilities are not distributed equally. At the one extreme, Western Europe's per caput food availabilities stand at some 3500 calories; those of North America at some 3600 calories. At the other extreme they are only 2100 calories in sub-Saharan Africa and 2200 calories in India and Bangladesh together (see Appendix 3). Thus, for a large part of the developing world, food availabilities are far from being adequate for all people to have

Indices (1961 = 100)

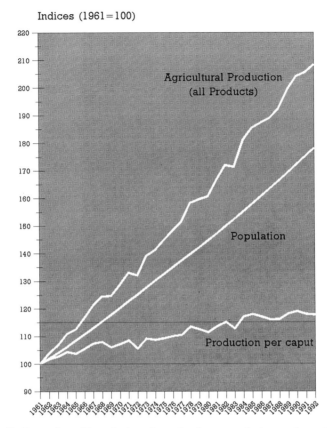

Figure 2.1   Indices of world agricultural production, population and production per
caput

access to sufficient food at all times, in short for their food security. So long as
this situation persists, a world food problem will continue to exist,
notwithstanding adequate production at the global level. The notion that the
problem is not one of production but of distribution is attractive. But while this
notion is correct in a numerical and static sense, it is trivial and can be
misleading. For one thing it entails the notion of drastic redistribution of static
world food supplies as a possible solution. For another, it relegates the need to
increase production to a subsidiary role.

It is increasingly recognized that people with inadequate food consumption
levels are in that condition because they do not earn sufficient incomes to
demand as much food as required to satisfy their needs. One should then be
speaking not of food scarcity but rather scarcity of incomes or purchasing
power, in short, poverty or lack of entitlements to food (Sen, 1987). This way
of emphasizing entitlements to food rather than food supplies has come to
dominate thinking in efforts to understand and explain the prevalence of

undernutrition and prescribe policies to overcome it. The entitlements approach correctly de-emphasizes the role of average per caput food supplies as an indicator for a complete understanding of issues of inadequate access to food by the poor. It should not, however, detract from the fact that ever increasing food supplies will be needed for solving the food problem in the future. Moreover, the level of per caput food production in the countries with high dependence on agriculture for employment and incomes is itself a major determinant of the food entitlements of the poor (Lipton and Ravallion, 1993, p. 50).

On the issue of the role of a potential redistribution of the globally adequate food supplies, it is noted that if the poor countries' incomes were to increase to levels that would put them in a position to raise their solvable food demand significantly, massive redistribution of existing supplies through the market to meet the increased demand would not be necessary, at least not at the scale suggested by present imbalances in inter-country food supplies. This is so because: (a) the increment in demand would not occur in a big spurt overnight but rather gradually over a number of years allowing time for production to respond; and (b) it is probable that much of this additional demand would be met by increases in the poor countries' own production. This latter proposition follows from the fact that, with few exceptions, a more productive agriculture in the poor countries would be an integral part of the process of increasing their incomes. The majority of the world's poor earn their living by producing food and in most poor countries employment and income earning opportunities in all sectors, not just agriculture, are closely linked to how productive agriculture is. Therefore, in most poor countries, increases in incomes that would raise demand and the increases in food supplies generated locally from a more productive agriculture go in tandem.

In conclusion, the present relative abundance of food at the global level and the apparent potential for redistribution of static world food supplies are of more theoretical than practical significance when it comes to thinking of ways and means of improving the food welfare of the poor countries. This being so, policy responses to the food problem will have to address, among other things and in priority, the issue of growth and geographical distribution of food supplies in the future. That is, if consumption in the poor countries is to be raised to "acceptable" levels, additional food must be produced in the right places. In parallel, the scope and the need for transfers of food through trade and food aid will continue to grow.

2.2 DEVELOPMENTS IN THE MORE RECENT PAST

World agricultural growth has been slowing down. The growth rate fell from 3.0 percent p.a. in the 1960s, to 2.3 percent p.a. in the 1970s and to 2.0 percent p.a. in the latest period 1980–92. These developments have given rise to expressions of concern that production constraints are becoming more

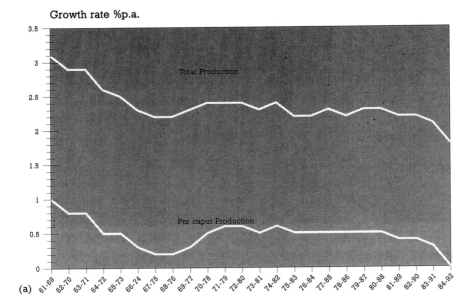

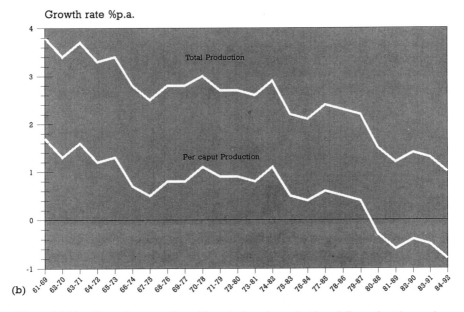

Figure 2.2(a)   Growth rates of world agricultural production (all products), moving eight-year periods, 1961–92. (b) Growth rates of world cereals production (with rice in milled form), moving eight-year periods, 1961–92

stringent and may ultimately threaten world food security. Brown (1994) signals the year 1984 as marking a turning point in world agriculture because world production of cereals in per caput terms is today lower than in the mid-1980s.[2] Attempting an interpretation of these developments requires, in the first place, examination of the facts. Does the post-1984 period (i.e. the eight-year period from 1984 to 1992, the latest year of available data) represent a new experience in the historical evolution of world agricultural production in relation to population?

To answer this question, the growth rate of this latest eight-year period may be compared with the growth rates for all moving eight-year periods from 1961 to 1992. Figure 2.2(a) plots the growth rates of world agricultural production (all products) in these moving eight-year periods, in both aggregate and per caput terms; and Figure 2.2(b) does the same for cereals. It can be seen that 1984–92 is the first eight-year period in the last three decades when the growth rate of world per caput agricultural production (all products) fell to zero. In parallel, these latest eight years signal a new record in the negative growth rate of world per caput production of cereals.

It may be concluded that those who signal the mid-1980s as a turning point in world agriculture have a point. At the same time, one may or may not agree with the interpretation that these developments: (a) represent a turn for the worse; (b) reflect increasingly binding production constraints; and (c) are indicative of things to come. Points (a) and (b) are addressed below. Point (c) is the subject of Chapter 3.

A turning point for the worse?

In principle, a slowdown in world agricultural growth need not translate into a worsening of the trends towards improved food security and nutrition for the majority of world population that characterized the historical period, for two reasons:

1. The proportion of world population which is well fed is today greater than in the past. Therefore, the part of world agriculture which supplies the "well fed" segment of the world population need not continue to grow at the growth rates of the past in order to maintain their per caput food supplies at the "satisfactory" levels already achieved. It happens that this part of the population, which comprises almost all developed countries and parts of the population of the developing countries, consumes well over 50 percent of world food output.

2. In parallel, the growth of world population tends also to slow down, even that part of world population with still very low levels of nutrition.

These two factors in combination indicate that maintenance or increase of the past growth rate of world agriculture is not a necessary condition, much less a sufficient condition, for progress towards improved food security to continue to be made. Attention should really focus on developments affecting the availability of, and access to, food of the countries and population groups subject to food insecurity. The growth rate of world production is just one of the many factors bearing on this problem; but the growth rate of their own production is of primary importance. This issue is examined below.

At issue is what has been happening in the countries with food security problems and whose food security is influenced in a major way by developments in their own agriculture. Most developing countries are in this category. A slowdown in their agricultural growth rate may often be interpreted as a threat to their food security because of their (a) high economic dependence on agriculture, (b) low levels of per caput food supplies and (c) limited capacity to increase supplies through imports. In short, failure of agriculture to grow at a "sufficiently" high rate affects the prospects for food security adversely on both the demand and the supply sides (see Alexandratos, 1992, for estimates of the role of the growth rate of agricultural production in explaining inter-country differences in per caput food supplies). However, the developing countries are a very diversified lot from the standpoint of these criteria. For some of them, the growth rate of their per caput agricultural production may not be an important determinant of food security. It is necessary, therefore, to review historical developments in per caput agricultural production for different groups of developing countries.

No single criterion is entirely appropriate for classifying the different countries according to their food security vulnerability to variations in their agricultural growth rate. However, even a crude distinction is better than treating the whole of the developing countries as a homogeneous group. The share of agricultural labour force in total labour force is one such criterion (see Appendix 3). This criterion emphasizes the demand (or income) side of the food security *problematique* because it measures the share of the population which depends mainly on agriculture for employment and income. The data for this indicator are not very good but they have the merit that guesstimates are available for all 93 developing countries of the study.

Of the 93 countries, 31 may be classified as having relatively "low" dependence on agriculture for employment and income generation because they have less than one-third of their economically active population (EAP) in agriculture.[3] A few countries with reportedly more than one-third of their EAP in agriculture are also included in this "low" dependence group on other criteria, mainly their relatively high per caput food supplies and low share of GDP originating in agriculture. For these 31 countries in the "low" dependence group, the hypothesis is made that their food security is comparatively less vulnerable than in other countries to variations in the growth rate of per caput agricultural production. The group has a population of 730 million or 19 percent of the population of the developing countries, average per caput food supplies of 2910 calories/day and all but five of them are in the medium-high range of this variable (2600 to 3300 calories/day).

42

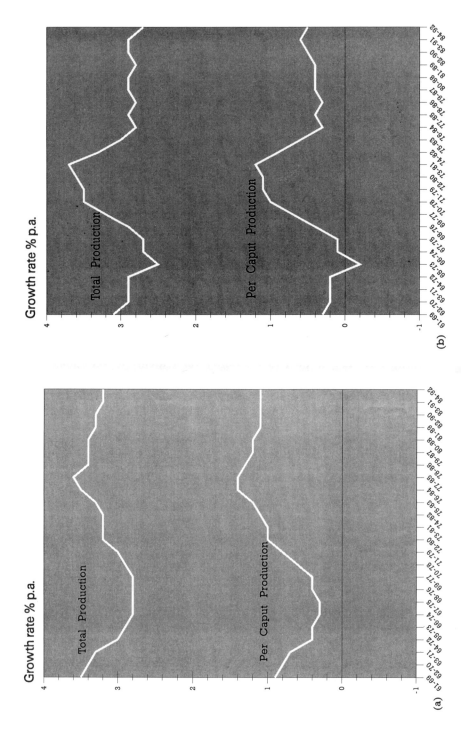

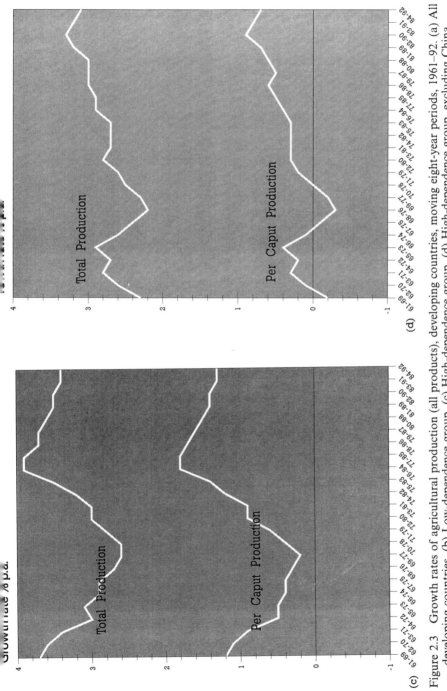

Figure 2.3 Growth rates of agricultural production (all products), developing countries, moving eight-year periods, 1961–92. (a) All developing countries. (b) Low-dependence group. (c) High-dependence group. (d) High-dependence group, excluding China

Of the remaining 62 countries with over one-third of their EAP in agriculture, all except four have per caput food supplies under 2600 calories and the majority of them are nearer the 2000 calorie mark. The group average is 2370 calories or 2230 calories if China is excluded. Moreover, the majority of the countries in this group have comparatively low net food imports per caput and some depend heavily on agricultural exports for their balance of payments, hence for their capacity to import food. It follows that their food security depends heavily on the performance of their own agriculture from both the demand side (incomes of the bulk of their population) and the supply side (food produced or imported from agricultural export earnings).

This group of countries with "high" dependence on agriculture is still too heterogeneous as it includes countries ranging from those with nearly total dependence on agriculture (80–90 percent of economically active population and 55–60 percent of GDP, e.g. Tanzania, Burundi, Nepal, etc.) to the more typical ones of South Asia (60–70 percent of EAP, 30–35 percent of GDP), down to those atypical countries with mineral resources but still the bulk of EAP in agriculture (e.g. Congo, Gabon, perhaps Botswana, all with 60–70 percent of EAP but only 5–10 percent of GDP in agriculture). However, notwithstanding their wide heterogeneity, all these countries share to a considerable degree the characteristic that access to food of significant parts of their populations depends on the performance of their own agriculture for the reasons described above. China belongs to this group, but its large population weight, its past success in diversifying the rural economy and its higher than average per caput food supplies justify showing the data for this group with and without China. Historical developments for these groups of the developing countries are shown in Figures 2.3(a) to 2.3(d).

It is noted that: (a) for the developing countries as a whole, the growth rates of per caput agricultural production (all products) have not been generally lower in recent eight-year moving periods compared with earlier ones (Fig. 2.3(a)); (b) the same observation applies also to the two groups of countries distinguished as described above (Figures 2.3(b), 2.3(c)); and (c) for the most vulnerable group of countries (those with high dependence on agriculture, excluding China) the per caput growth rates of recent years have actually been higher than in earlier periods (Fig. 2.3(d)).

In the light of this evidence, it is difficult to accept the position that developments in recent years have marked a turning point for the worse. It is more appropriate to consider that trends in recent years, just like those of the longer-term past, have been characterized by the persistence of low and totally inadequate growth rates of per caput production in the countries whose food security would profit significantly from higher production. Putting the issue in these terms goes to the heart of any analysis of the nature and significance of agricultural growth trends which must always be evaluated in relation to needs rather than in abstract. This is a theme that permeates much of the discussion in this study. It also helps explain why this study focuses predominantly on the agricultural growth prospects and food security in the developing countries.

In conclusion, the slowdown in world production growth after the mid-1980s did not generally reflect developments in the countries whose food security is vulnerable to falls in the growth rate of their own production. As discussed below, the slowdown reflected mainly production adjustments in the main cereal exporting countries in response to stagnant export demand and the need to control the growth of stocks. The flexibility of the main exporting countries to respond to changes in world market conditions, by increasing as well as reducing supplies, will remain an important factor in world food security. In the past they have generally demonstrated such flexibility and tended to absorb shocks in world food markets through, among other things, the responsiveness of feed use of cereals to world price changes in several major exporting countries and by having policies which incidentally led them to hold stocks which were higher than they would otherwise have been. Such flexibility may be affected in the future by the policy reforms under way or planned. For example, a reduction in production-related agricultural support could lead to permanent reductions in stocks or diversion of cereals land to other quasi-permanent uses, e.g. forest crops. By contrast, changes in policies to make domestic prices more sensitive to changes in world prices could increase the responsiveness of domestic markets in some of the main exporting countries. Environmental restrictions may also play a role, e.g. quasi-permanent withdrawal of land for conservation purposes or limits on the use of agrochemicals. Finally, the policy reforms which reduce support to agriculture may also create an unfavourable environment for support to publicly funded agricultural research to increase yields in the developed exporting countries (see O'Brien, 1994). This may also limit the positive

Table 2.1   Per caput production of cereals (with rice in milled form), three-year averages and projections

	Total production* (million tonnes)			Population[†] (million)			Prod./caput (kg)		
	1979/81	1990/92	2010	1980	1991	2010	1979/81	1990/92	2010
World	1444	1756	2334	4447	5387	7150	325	326	326
Developed countries	793	873	1016	1170	1262	1406	678	692	722
Developing countries	651	883	1318	3277	4125	5744	200	214	229

*Production data and projections revised to account for the changes in the data of the former USSR following the shift in its data from bunker to clean weight.
[†]Population data and projections from the 1992 Assessment (UN, 1993b). These revised data are not used in the analyses for this study. For this reason the projected per caput numbers are somewhat different from those shown in Chapter 3.

spill-over effects of agricultural innovation on productivity in the developing countries.

A final point may be made before concluding this discussion. This has to do with the need to avoid reading too much into the world averages when they are in per caput terms. An illustration is given in Table 2.1. World per caput production of cereals did not increase between the three-year averages 1979/81 and 1990/92. Yet it increased in both the developed and the developing countries. This paradox is due to the widely differing per caput production and relative population shares in the two groups in the initial year and their different growth rates of population. Apparently, from this standpoint, the world should not be viewed as a zero-sum game situation where, if the world average remains flat, gains of one group must imply losses for the other. This is a rather important aspect of the issues considered here because the projections in Chapter 3 (anticipated here in Table 2.1) show that world per caput production of cereals is likely to remain flat, yet it would increase in both the developed and the developing countries.

Has the world production environment become more difficult for agricultural growth?

No straightforward answer can be given to this question from the mere observation of decelerating world agricultural growth. No doubt, there is fragmentary evidence that land and water resources are subject to degradation, in addition to their declining in per caput terms following population growth, and that the momentum of yield growth is coming increasingly under strain. However, the evolution of production potential has to be evaluated in relation to needs for more agricultural output. To the extent that the needs are expressed as effective demand, a tightening of production constraints in relation to the evolution of demand would have signalled its presence by rising prices. There is no evidence that this phenomenon has occurred, at least not in terms of world market prices which have tended to decline in the 1980s rather than rise.[4] Therefore, a prima facie case can be made that the slowdown must be interpreted in terms of a host of factors rather than solely, or even predominantly, as resulting from a change in the fundamentals on the production side, the latter to be viewed always in relation to the growth of effective demand.

In support of this position, it is noted that the slowdown in the production growth rate of cereals originated predominantly in the major net cereals exporting countries (USA, Canada, the EC, Australia, Argentina, Thailand) which accounted for 36 percent of world production in the mid-1980s. Their production was 602 million tonnes in 1990/92 (824 kg per caput), down from 621 million tonnes in 1984/86 (888 kg per caput). In contrast, production in the rest of the world rose over the same period from 1040 million tonnes to 1153 million tonnes and per caput production remained largely unchanged. Some of the major exporting countries had to resort on several occasions during this period to supply management measures and reduction of incentives to producers in order to rein in the growth of production. It is therefore difficult to interpret the decline in their production as indicative of the onset of production "fatigue". It is more likely that the sluggish growth of export demand for their output dominated the picture.

If the above interpretation is correct, it provides an additional reason for focusing attention less on changes in world production, particularly if they reflect the ups and downs in the main exporting countries, and more on the reasons why effective demand, including import demand, is not growing faster in the countries and population groups with severe food security problems. Examination of the food security *problematique* from this standpoint will help bring into focus the role of production constraints at the local level, which indeed play a fundamental role in the perpetuation of food insecurity in many low-income and agriculture-dependent countries. For, as noted earlier, it is in these countries that production growth determines not only the growth of food supplies but also, and to a significant degree, that of incomes and demand.

2.3 THE DEVELOPING COUNTRIES: MAGNITUDE OF THE FOOD PROBLEM AND HISTORICAL DEVELOPMENTS

The overall picture

How big is the food availability problem in the developing countries? How has it developed over time? How many people suffer from undernutrition? These are some of the questions addressed in what follows. First, the magnitude of the problem and its evolution over time. Per caput food supplies in the developing countries as a whole have been increasing, from 1950 calories in the early 1960s to 2470 calories in 1988/90 (revised and more recent data to 1992 are given in the Appendix 3). This happened while their population grew from 2.1 billion to nearly 4.0 billion. Therefore, significant progress has been made. This progress, or lack of it in some cases, can be seen visually in Figure 2.4 where the evolution is shown for the individual regions.

The result of these developments has been that in 1988/90 only some 330 million people, or some 8.5 percent of the developing countries' population, lived in countries where per caput food supplies are extremely low – under 2100 calories (Table 2.2). Thirty years ago these numbers were 1.7 billion or 80 percent of the total. The progress achieved can also be seen by looking at the picture from the other end. Some 650 million people, or 17 percent of the total population of the developing countries, lived in 1988/90 in countries with per caput food supplies over 2700 calories. Again, 30 years ago these numbers were only 35 million or under 2 percent of the total.

It is obvious, however, that, impressive as this progress has been, it has not been fast enough nor of a pattern that would have raised per caput food supplies in all countries to levels usually associated with significant reductions in the population suffering from serious problems of food insecurity and undernutrition. What is the size of the population in this category depends, of course, not only on national average per caput food supplies but also on how equally such supplies are distributed within each country. The empirical evidence here is scant. However, from the few countries with "good" quality distributional data it is found that on average (i.e. taking the simple average of high-inequality and low-inequality countries) the proportion of the population undernourished is around 10 percent when per caput food supplies are around 2700 calories. It is typically in the range 15–35 percent when national average calories are in the range 2200–2500. The data in Table 2.2 indicate that some 3.3 billion people live in countries with under 2700 calories and some 2 billion in countries with under 2500 calories.[5]

The above numbers give an idea of the magnitude of the problem, if progress were to be measured in terms of national average per caput food supplies. However, not all people in countries with low national averages, even very low

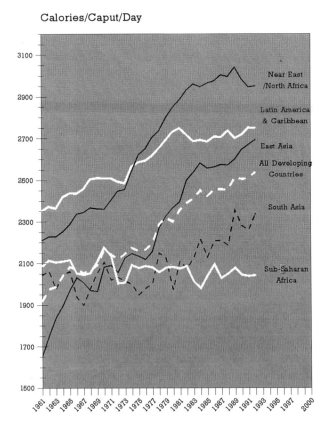

Calories/Caput/Day

Figure 2.4   Developing countries, per caput food supplies (calories per day), 1961–92

Table 2.2 Population living in developing countries* with given per caput food supplies, 1961/63 to 1988/90

Per caput food supplies (cal/day)							
	Three-year averages			Population (million)			
Cal/day	1961/63	1969/71	1979/81	1988/90	1962	1970	1980	1989
Developing countries								
Under 2000	1810	1960	1900	1784	1581†	1046‡	197	123
2000–2100	2045	2030	2100	2040	114	746§	771¶	211
2100–2300	2210	2200	2180	2225	176	338	483	1425**
2300–2500	2400	2395	2345	2405	196	230	1234‖	212
2500–2700	2655	2560	2655	2635	38	176	103	1327††
2700–3000	2785	2870	2780	2795	14	55	293	286
Over 3000	3075	3265	3070	3120	21	24	190	365
Total	*1950*	*2120*	*2330*	*2470*	*2141*	*2616*	*3271*	*3950*
Developed countries	*3030*	*3200*	*3290*	*3400*	*989*	*1074*	*1168*	*1242*
World	*2290*	*2430*	*2580*	*2700*	*3130*	*3690*	*4439*	*5192*

*All countries with food balance sheet (FBS) data. The data in this table and Table 2.3 are before the most recent (1994) revision of the FAO FBS data and the 1992 revision of the population data.
†Includes China (pop. 663 m.) and India (pop. 462 m.).
‡China (pop. 816 m.).
§India (pop. 555 m.).
¶India (pop. 689 m.).
‖China (pop. 978 m.).
**India (pop. 836 m.).
††China (pop. 1102 m.).

ones, are subject to undernutrition. And there are undernourished people even in countries with relatively high national averages. Therefore, a more appropriate estimate of the incidence of undernutrition can be obtained from a combination of the national average food supplies with a distributional parameter and some notion of a nutritional threshold level, i.e. a level of food intake below which a person chronically subjected to it can be classified as undernourished. This method and data (supplemented in many cases by educated guesses for the value of the distributional parameter and the shape of the statistical distribution curve) have been used by FAO to derive rough estimates of the numbers of persons in the developing countries which can be classified as undernourished (FAO, 1992a).

The estimates of the incidence of chronic undernutrition thus obtained were most recently published on the occasion of the International Conference on Nutrition (December 1992). They are shown in Table 2.3. Progress has been made, but the magnitude of the problem remains significant. The largest, though declining, numbers are to be found in Asia but those in sub-Saharan

Table 2.3   Estimates of chronic undernutrition in the 93 developing countries of the study

	Year	Per caput food supplies (cal/day)	Total population (million)	Undernourished*	
				% of total population	Million
Africa (sub-Sahara)	1969/71	2140	268	35	94
	1979/81	2120	358	36	129
	1988/90	2100	473	37	175
Near East/	1969/71	2380	178	24	42
North Africa	1979/81	2830	233	10	23
	1988/90	3010	297	8	24
East Asia	1969/71	2020	1147	44	506
	1979/81	2340	1392	26	366
	1988/90	2600	1598	16	258
South Asia	1969/71	2040	711	34	245
	1979/81	2090	892	31	278
	1988/90	2220	1103	24	265
Latin America	1969/71	2500	281	19	54
and Caribbean	1979/81	2690	357	13	47
	1988/90	2690	433	13	59
Total	1969/71	2120	2585	36	941
	1979/81	2320	3232	26	843
	1988/90	2470	3905	20	781

*Persons who, on average during the course of a year, are estimated to have food consumption levels below those required to maintain body weight and support light activity. This threshold level (ranging from an average of 1760 cal/person/day for Asia to 1985 for Latin America) is set equal to 1.54 times the basal metabolic rate. For more explanations see FAO (1992a).

Africa have been increasing rapidly, both in absolute terms and as a proportion of the region's total population.

The above two alternative approaches to gauging the magnitude of the food problem in the developing countries (population living in countries with given per caput food supplies, Table 2.2, or numbers of undernourished, Table 2.3) provide useful starting points for looking into prospects for the future which are addressed in Chapter 3. That is, one would need to speculate about how average per caput food supplies may develop in the different countries and what may happen to within-country distributions. Unfortunately, the historical data and ability to speculate on prospective changes in socioeconomic and political structures having a bearing on within-country inequalities allow little scope for saying much on the distributional aspect.[6] It is a somewhat less arduous task to address the question how the per caput food availabilities may develop in the future in the different countries. For this purpose, the empirical evidence is analysed in order to, as far as possible, understand what have been the factors responsible for the fact that some countries have made progress in raising per caput food supplies and others have not, or have experienced outright deterioration. One way of addressing this issue is to examine the historical developments in those countries which performed well in the past; and to do the same for those countries which failed to make any significant gains or experienced outright declines (the following discussion draws on Alexandratos, 1992).

The correlates of success

For the purposes of this discussion, the group of countries which performed well in the historical period includes those which, starting from low or very low levels of per caput food supplies 30 years ago (between 1650 and 2300 calories in 1961/63) had reached by the late 1980s high or medium-high levels (between 2600 and 3300 calories in 1988/90).[7] The main characteristics of their historical evolution which probably explain much of their progress in raising per caput food supplies may be summarized as follows:

1. All of them had above average economic growth rates, as evidenced by the growth rates in their per caput incomes. This seems to be the most prevalent common characteristic of these countries.
2. In most countries, there was a spurt in the growth of food imports, particularly in the period of rapid gains in per caput food supplies, as evidenced by the increases in per caput net imports of cereals. This meant rapid declines in their cereals self-sufficiency. But there were exceptions. In particular, China and Indonesia did not follow this pattern as their own agricultures grew to provide the additional food supplies and, most probably, was a key factor in raising per caput incomes.
3. A contributing factor to the nutritional improvement in this group of countries has been the fact that global agriculture provided readily and without much strain the food imports that underpinned the growth of their consumption, mostly in the 1970s. It is noted, however, that the historical experience of responsiveness of world agricultural production to increases in demand must be interpreted with care, in particular for drawing conclusions about the potential of world agriculture to respond to spurts in demand in the future. This is because the data showing that increasing quantities were forthcoming in the world markets at the same time as prices declined are vitiated by the agricultural support and protection policies of major countries (e.g. USA, EC, Japan). In practice, part of the costs of delivering the increased output were covered by the heavy subsidies granted to agriculture in these countries (see Note 4). It is not known how world agricultural production would have reacted to the increasing demand in the absence of such distortions. Some indications can be obtained from the trade liberalization studies which simulate world food markets with removal of such distortions. They generally indicate

that world food prices would have been somewhat, but not much, higher (Goldin and Knudsen, 1990).

4. There emerges a mixed picture concerning the role of domestic agricultural growth in the process of increasing per caput food supplies. All sorts of situations are encountered, with per caput production declining in some countries while increasing in others at moderate or very high growth rates. Drawing the conclusion that agricultural growth does not matter would not be warranted, however. More likely, agricultural growth plays a largely subsidiary role in countries where agriculture is a small sector in the economy with a relatively small share of the population depending on it for a living (see Section 2.2), and much of the gains in economic growth and import capacity derive from the non-agricultural sector, particularly from non-agricultural commodity sectors. This seems to have been the case of many of the oil-exporting countries. But in countries where such conditions are not prevalent, agricultural growth seems to be an essential ingredient in the process of increasing per caput food supplies, through its role in the provision of supplies, income and employment, and in support of economic growth and the balance of payments. China's experience in the post-reform period after 1978 seems to conform to this pattern.

5. In all countries, much of the quantum improvement in per caput food supplies was achieved in a relatively short period of time, in most cases around 10 years. Judging the durability of such gains is more difficult. There are examples of countries where improvement and retrogression of per caput food supplies follow the commodity boom and bust cycles. It is, therefore, possible that the food and nutrition gains will tend to prove more durable in countries in which the circumstances that brought them about are part and parcel of wider economic and social transformations, e.g. China and Korea (Republic). The same probably holds for countries in which the windfalls from commodity booms are put to good use to bring about such transformations, as indeed happened in some of the countries in the group analysed here.

6. Finally, the relationship between the growth rate of population and that of per caput food supplies does not produce any consistent pattern. Fast progress in the latter variable occurred in countries with very high rates of population growth, e.g. Libya and Saudi Arabia. Again, no hasty conclusions may be drawn. In particular the evidence from these countries may not be taken to mean that fast demographic growth is not an obstacle to improving welfare. All countries in this group with high population growth rates experienced special circumstances (the windfall income gains from the oil sector) and indeed in some cases the high demographic growth was partly the result of such special circumstances, e.g. due to labour immigration. The proposition that poor countries with high population growth face a more arduous task in improving welfare than

those with lower population growth is in no way falsified by the experience of the above countries.

The correlates of failure and retrogression

At the other extreme, the study of the experiences of the many countries which, starting from low initial conditions 30 years ago, failed to make progress or suffered outright declines, should provide some insights as to the reasons for failure. The study of the relevant data for a sample of these countries[8] leads to the following conclusions.

For the great majority of these countries one could have predicted that the food situation would be really bad even before looking at the data. Many of them are in sub-Saharan Africa, a fact which by itself tells a lot, given the overall economic and agricultural stagnation that has been plaguing the region for some time now. Add to this the fact that many of these countries, both in Africa and elsewhere, have suffered or are still going through severe disruptions caused by war and political disturbances and one has in a nutshell the explanation for failure and retrogression on the food and nutrition front.

The data do no more than confirm this impressionistic prediction. Indeed, the most common characteristics of these countries are declines in per caput incomes and per caput agricultural production. The two are not, of course, independent of each other. Their per caput food imports did increase, often by means of food aid. However, in contrast to the experiences of the countries in the preceding category, their per caput imports of cereals remained at generally modest levels, while the declines in cereals self-sufficiency were accordingly contained, at the cost, of course, of stagnant or declining per caput food supplies.

Some generalizations and how they can aid the assessment of the future

The preceding discussion, based as it is on an impressionistic examination of the data for a restricted number of countries, is meant to provide some clues as to what have been the main correlates of success or failure in raising per caput food supplies in the different countries. However, it is far from a complete analysis. At best it indicates that success is commonly associated with sustained growth in per caput incomes and with varying combinations of growth in domestic agricultural production and import capacity.

A more complete and formal analysis of the data for all developing countries (similar to that reported in Alexandratos, 1992) confirms that these findings hold in a general way, meaning that these three variables alone (per caput incomes, growth of agriculture, food imports) explain only partly the differentials among countries of per caput food supplies. For example, for the 65 net cereal-importing developing countries thus analysed only 25 had per caput food supplies within a range ± 5 percent of those justified (or predicted) by their levels of the above three variables. Another 24 countries were within a range of between ± 5 and 10 percent. The deviations for the remaining 16 countries were outside the ± 10 percent range. These findings suggest that some countries achieve relatively "satisfactory" levels of per caput food supplies at comparatively low levels of per caput income. And others have per caput food supplies lower than their per caput incomes would lead one to expect. Understanding (or hypothesizing about) the reasons for these discrepancies is probably the most important insight for guiding the quest for policy responses to the problems of food and undernutrition. That is, if some countries have managed to make progress on the food front while remaining essentially poor on the overall income side, would not other countries with a food problem learn something from their experience?

Obviously, other factors beside per caput incomes, growth of production and food imports, such as food prices and the distribution of income or the incidence of poverty, are important determinants of the inter-country differences in per caput food availabilities. Moreover, some of the differences are explained by the fact that the conventional income statistics, expressed in a common currency, usually the dollar, can distort the relative positions of the different countries on the income scale because the internal purchasing power of this somewhat fictitious dollar can vary widely among countries. For example, Egypt, Honduras, Bolivia and Zimbabwe with reported per caput incomes in the range of $580–650 have per caput food supplies of calories 3310, 2210, 2010 and 2260, respectively. These differences are partly explained by the wide disparities in the purchasing power of the dollar in the different countries. When incomes are corrected for this factor and converted to dollars reflecting purchasing power parities, they become $3600 (Egypt), $1820 (Honduras), $2170 (Bolivia) and $2160 (Zimbabwe) (income data from World Bank, 1993a, Tables 1, 30). In addition, public policy which influences access to food, directly (e.g. through public food distribution schemes) or indirectly (e.g. through policies to alleviate poverty), can be an important factor in explaining the observed differences in per caput food supplies among countries.[9]

Important as these other factors are, they are difficult to account for systematically in the assessment of the possible evolution of per caput food supplies in the future. The reason is that not enough is known, nor can it be deduced in any valid and systematic way, about how such things as the income distribution, the incidence of poverty and public policies affecting access to food may evolve in the individual countries, let alone the prospects that peaceful conditions will prevail and when in the countries plagued by war and other disturbances. This being so, the main guide for the projections is the effects of growth in per caput incomes (mostly obtained from other organizations), the evaluation of the agricultural production prospects undertaken for this study and a notion of possible levels of food imports. Yet, the preceding discussion indicated that these three variables alone are not very good predictors of inter-country differences in per caput food supplies and that other country-specific factors must be taken into account. This apparent contradiction is bypassed by starting each country's projections of per caput food supplies from its own base year figure. The latter already embodies the effects on per caput food supplies of these "other" country-specific factors as well as the effects of per caput incomes, the growth rate of agriculture and the level of net food imports. Notionally, ignoring these "other" factors means that they are assumed to continue to play in the future the same role as they played in determining the base year food supply levels. And, naturally, some of these factors, in particular the prevalence, or otherwise, of peace, are supposedly already incorporated in the income growth projections of other organizations which, as noted, are taken as exogenously given in the projections of per caput food supplies, as well as for providing the general overall economic background within which the production and food import prospects are evaluated (see World Bank, 1994a, p. 24). For example, it is difficult to assume that countries projected to be on a low overall economic growth trajectory can mobilize the significant amounts of resources for investment in agriculture, rural infrastructure and human capital needed to underpin significant accelerations in agricultural production.

2.4 ISSUES OF AGRICULTURAL RESOURCES, ENVIRONMENT AND SUSTAINABILITY

General considerations

Concern with the state of the environment and the dwindling quantity (per caput) of land and water resources as well as their degradation requires that the conclusions of the preceding section be amplified to address questions like the following: *To what extent may the resource and environmental constraints impinge on the prospects for increasing food supplies and assuring access to food by all, the very essence of food security? Can such progress be achieved while ensuring that the gains made and the potential for further gains are maintained for future generations, the very essence of sustainability?* The rest of this chapter endeavours to put the overall issue in a proper perspective. More specific discussion is to be found in Chapters 4, 11, 12 and 13. These chapters provide estimates of the pressures that are likely to be put on agricultural resources in the process of increasing production in the period to 2010. They also explore the options for policy responses to minimize the unavoidable trade-offs between increasing production and pressures on the environment.

The preceding section singled out the rate of increase in per caput food supplies of countries and population groups with inadequate access to food as a practical proxy for measuring progress towards solving problems of food security and undernutrition. It also highlighted a number of interdependent factors as being instrumental in increasing per caput food supplies: poverty-reducing economic growth; the multiple role of agricultural growth in the majority of the developing countries (increasing food supplies and providing employment and income-earning opportunities to the poor, both directly and indirectly via the growth linkages of agriculture); enhanced capacity to import food; and public policy.

It follows, therefore, that important as the agricultural resource constraints are in conditioning the prospects for food production and generation of incomes in agriculture, the wider environmental constraints can also affect in important ways the prospects of eliminating undernutrition because of their possible effects in restraining the overall economic growth rate and the potential for reducing poverty. For example, reducing emissions of greenhouse gases and the existence of non-agricultural resource constraints may cause the world economic growth rate to be lower than what it would otherwise be. Those low-income countries which depend, actually or potentially, on a buoyant world economy for their development will find it more difficult to improve their economic growth rates and reduce poverty. Moreover, the adverse environmental impacts (local, but with global implications) which often accompany the accelerated growth in the use of energy in the transition from low- to middle-income levels in the low-income countries will tend to make such transition more difficult.[10] These are examples of how the more general environmental and resource constraints, and not only the agricultural ones, may impinge on the prospects for reducing undernutrition.

Given the above considerations, the extent to which the agricultural resources are adequate or otherwise to produce as much food as required to increase per caput food supplies for a growing population in sustainable ways must be examined in the context of these grander themes concerning the overall resource and environmental constraints. It may well be, for example, that at the global level the binding constraints would not be those impinging directly on the production of food but rather those standing in the way of achieving economic growth rates and patterns adequate to eliminate poverty in the not-too-distant future.

There is another sense in which the agricultural resources may not be the binding constraint to making progress towards elimination of undernutrition, at least not in a global context and in the longer term. It was noted in the preceding section that: (a) progress achieved in the last few decades points to ever increasing per caput food consumption levels; (b) a considerable number of countries have made the transition from low and medium-low levels to medium-high ones; and (c) beyond these levels the growth of per caput consumption of food tends to slow down before stopping altogether when it reaches physiologically maximum levels.

It follows that a combination of continuation of these developments in per caput food consumption and the expected slowdown in population growth will eventually translate into a slowdown in the rate at which pressures are exerted on world agricultural resources for increasing food production. That is, the world may reach at some future date a stage when very little additional growth in global food production would be necessary to maintain adequate food supplies for all. The experience of many developed countries in which there is little scope for further expansion of aggregate agricultural output for domestic use, and in which land has often to be taken out of production, is telling. The gist of the matter is, therefore, whether this stage can be reached while maintaining sufficient agricultural resources for continuing production at a nearly steady-state level in sustainable ways and with enough natural habitat intact for it to continue providing its essential life-support functions. Some further discussion on these matters is given in Chapter 3 in an attempt at speculating on possible developments beyond the year 2010.

The preceding discussion provides the background for discussing in the remainder of this chapter the relative significance of agricultural resource constraints for making progress towards reducing undernutrition. This means essentially the prospects for increasing per caput food supplies for the population groups with inadequate access to food, taking into account both the *supply side* of the problem (e.g. can enough food be produced in sustainable ways?) as well as the *demand side* (e.g. in what ways may agricultural resource constraints play a role in the process of enhancing access to food by the poor?). The link between the two sides is provided by the fact that the great majority of the poor depend for employment and income on the exploitation of those very agricultural resources.

Land and water resources in the quest for sustainable responses to the food problem

In Section 2.3, the food problem was defined in terms of a few measurable variables (per caput food supplies by country, incidence of undernutrition) and analysed in terms of others (agricultural production, per caput incomes, distribution, food imports). In this analysis, the possible role of agricultural resources was not considered explicitly. It can be hypothesized that such role is subsumed in that of some of the variables considered, notably agricultural production and per caput incomes. The widespread concerns with agricultural resources, the environment and sustainability require that an attempt be made to consider such a role more directly. This issue is addressed in terms of: (a) what is known about the land and water resources; and (b) how the constraints relating to these resources may enter the determination of the rate of progress towards solving the food problem.

The state of knowledge on the extent and use of agricultural resources and the historical evolution of such use leaves much to be desired. For example, the data on cropping patterns by agroecological zones used in this study had to be compiled from fragmented sources and supplemented by expert judgements (see Chapter 4). Likewise, the data on the state of degradation of irrigated lands or erosion of rainfed lands are limited and very little is known about how such states have been evolving over time. Furthermore, there is a dearth of systematic information on yet unexploited irrigation potential. The data on water resources (river flows, acquifers), such as they exist, need to be interfaced with those on land (terrain, soil, etc., characteristics) and analysed in the context of a host of socioeconomic factors affecting their use, before comprehensive and credible estimates of the potential for irrigation expansion can be obtained.

Potential of land and water resources for rainfed crop production in the developing countries

Given these data shortcomings, resort may be made to the second best option offered by the rather more systematic data in the soil and climatic inventories used in FAO's Agroecological Zones (AEZ) work for the developing countries. These data (recently re-elaborated for this study) permit the derivation of estimates of *land stocks of varying quality with potential for growing crops under rainfed conditions*. Examples of the kind of data thus obtained are given in Table 2.4 for South Asia and tropical South America (more comprehensive data for the developing countries are shown in Chapter 4).

It is noted that such evaluation indicates the potential for rainfed crop production of land in its natural state, i.e. not taking into account improvements or deteriorations brought about by human activity. The results should, therefore, be read with this caveat in mind, because it is well known that much of the land in agricultural use has been modified in the course of time, for better or for worse, by human intervention. Some account of such alterations is taken in the process of accounting for irrigated land, e.g. in the extreme, desert land with no agricultural potential in its natural state is added to the agricultural land if it has been irrigated.

The above data on land convey also significant information on *water resources for agriculture*. On the latter subject, a distinction is made between water supplies from rainfall which are directly utilizable in rainfed agriculture if falling or ending up in soils with appropriate qualities, in particular the capacity to retain humidity in the root zone for the length of time required by the growth cycles of the different crops; and the part of rainfall which feeds into water bodies like rivers and acquifers and which, together with stocks of fossil water, can be used for agriculture only through human intervention (irrigation). As noted, the data on this latter resource are not adequate for a full evaluation. But the data on supplies from directly utilizable rainfall are part and parcel of the evaluation of the above mentioned AEZ data set on land resources for agriculture. This is because when evaluating any piece of land for its suitability to produce one or more crops at "acceptable" yields under alternative technologies (see notes to Table 2.4), the rainfall regime and the soil's water holding capacity are key elements in the solution.

In the end, declaring that, for example, South Asia has some 50 million ha of land with terrain/soil characteristics "very suitable" or "suitable" for rainfed agriculture which receive rainfall and have waterholding characteristics sufficient for a growing period of 180–269 days, is equivalent to making a statement on water availabilities for rainfed agriculture. This is counting not rainfall in abstract terms (e.g. in millimetres) but more precisely that part which ends up in soils with other desirable characteristics for farming. Barring changes in the rainfall regimes and the quality of soils, it can be assumed that this is a perennial resource. This estimate is perhaps more robust than those of water resources for irrigation. The latter are subject to greater uncertainty concerning their permanence over time because of: (a) possible reduction of the water supplies due to overexploitation; (b) the risk of deterioration of the irrigated lands (waterlogging, salinization) and of infrastructure (siltation, etc.); and (c) possible diversion of water supplies to competing non-agricultural uses.

In practice, therefore, the classification of land with agricultural potential according to the length of growing period (LGP) criterion goes some way towards defining the water constraints for agriculture. It has the added advantage that the estimates thus obtained are less subject to uncertainty compared with those referring to water resources for irrigation, as discussed in Chapter 4. The importance of improving the data and knowledge about these aspects is evident from the fact that at present some 37 percent of the gross value of crop output (and 50 percent of that of cereals) of the developing countries comes from irrigated lands. The estimates thus derived for the land and water availabilities for rainfed agriculture are supplemented with two pieces of additional relevant information, viz.: (a) the extent to which they are irrigated, including an estimation of irrigation of land not suitable for rainfed production in its natural state (rows B, C in Table 2.4); and (b) if they are used currently for crop production (not including land used for fodder, whether cultivated or natural grass for grazing).

This estimate of the current use status makes it possible to obtain as residual the land with crop production potential of varying quality which is not yet in agricultural use. However, its mere existence does not mean that it should be considered as available for expansion of crop production in the future (see below). The stark contrast between the situations in South Asia and tropical South America is evident (Table 2.4). It becomes even starker when expressed in terms of population densities, given that South Asia's population is 1.1 billion and that of tropical South America only 240 million. Moreover, South Asia has 65 percent of its economically active population in agriculture (265 million), while tropical South America has only 25 percent (22 million). There are, therefore, even starker differences in terms of agricultural land actually or potentially available per person in the agricultural labour force. This latter variable is the key one for understanding the forces that may shape the future in terms of the population–resources balance. As already noted, this balance

Table 2.4 Examples of land balance sheet: South Asia and tropical South America

	Moisture regime (LGP, days)*	Land quality†	South Asia††			Tropical South America‡‡		
			Total	In crop prod. use	Balance‡	Total	In crop prod. use	Balance‡
Land with crop prod. potential by class								
1. Dry semi-arid	75–119	VS,S,MS	29.2	22.1	7.1	9.8	3.5	6.3
2. Moist semi-arid	120–179	VS,S	82.4	61.0	21.4	32.2	10.8	21.5
3. Sub-humid	180–269	VS,S	50.6	45.4	5.2	121.5	47.4	74.2
4. Humid	270+	VS,S	6.0	} 25.1	} 3.3	329.5	} 44.0	} 516.2
5. Marginal, moist semi-arid, sub-humid, humid	120+	MS	22.4			230.7		
6. Fluvisols/gleysols	Nat. flooded	VS,S	21.3	} 21.6	} 0.6	65.2	} 8.1	} 99.1
7. Marginal fluvisols/gleysols	Nat. flooded	MS	0.9			42.0		
A. Total 1 to 7			212.8	175.2	37.6	831.0	113.7	717.3
B. —of which irrigated				48.1			4.5	
C. Additional irrig. from hyperarid land			15.3	15.3		0.9	0.9	
D. **Total land with crop prod. potential (A+C)**			228.1	190.5	37.6	831.9	114.6	717.3
E. **Land without crop prod. potential**			204.9			532.7		
E.1 —Hyperarid			45.6			22.7		
E.2 —Other constraints			159.3			510.0		

F. **Total forest area**§	61.1	802.9
F.1 —Could be on land without crop prod. potential¶	55.0	310.2
F.2 —Minimum forest area on land with crop prod. potential‖	6.1	492.7
G. **Total in human settlements and infrastructure****	25.7 (0.023 ha/person)	11.3 (0.046 ha/person)
G.1 —On land without crop prod. potential	8.9	4.7
G.2 —Balance on land with crop prod. potential	16.8	6.6
H. **Protected areas**	15.6	143.6
H.1 —On land without crop prod. potential	10.3	52.1
H.2 —Balance on land with crop prod. potential	5.3	91.5

*LGP = length of growing period is the number of days during the year when temperature and rainfed soil moisture permit plant growth.

†VS = very suitable, in the sense that obtainable yields can be 80% or higher of those obtainable in land without constraints; S = suitable, yields 40–80% of the constraint-free ones; MS = marginally suitable, yields 20–40%.

‡Could be partly in grazing use; additional grazing land is the fallow part of the land used in crop production.

§Data from the FRA90 assessment (see Chapter 5 on forestry).

¶Maximum amount of land on which trees, but not crops, could exist.

‖Balance of forest area (F2 = F − F1) which by necessity must be on land with crop prod. potential. As such, it is the minimum overlap between forest and agricultural land. It could be much larger in reality.

**For method of estimating land in human settlements and infrastructure see Chapter 4.

††Bangladesh, India, Nepal, Pakistan, Sri Lanka.

‡‡Bolivia, Brazil, Colombia, Ecuador, Paraguay, Peru, Venezuela.

has two main dimensions: (a) how much more food must be produced, which is directly linked to the growth of *total population* and the per caput consumption of food; and (b) how many people are, or will be, making a living out of the exploitation of agricultural resources. The relevant variable here is the size of the *population economically active in agriculture*.

As noted, the existence of land with crop production potential does not necessarily mean that such land may be so used. In the first place, part of it is used for *human settlements* (habitation, industry, infrastructures) and more of it will be so used in the future following population growth. There are no reliable data on how much land is occupied by human settlements. Sporadic data for some countries have been used to derive the estimates shown in Table 2.4 (Row G, for methods of estimation see Chapter 4). These estimates are subject to an unknown, though probably very large, margin of error.

Secondly, part of the area with crop production potential overlaps with *forest*. The extent of this overlap is not precisely known. The table provides minimum estimates obtained by first deducting from the total forest area the part which, on agroecological criteria, could exist on the land without agricultural potential (Row F.1 in Table 2.4). The balance of the total forest area must be by definition on the land with agricultural potential (Row F.2). This is a minimum estimate of overlap and the actual one is probably much larger. Considerations of environment and sustainability dictate that a good part of the land with forest should not be considered for prospective agricultural expansion. Indeed some areas, not always under forest, are legally protected (Rows H.1, H.2 in Table 2.4). Moreover, forest lands do contribute to food security and any gains in food production from conversion to agriculture must be netted out for the food security losses incurred. This is because significant numbers of people, much beyond the communities of forest dwellers, depend on sustainably managed forests and trees either as a source of complementary food supplies, or even more importantly, as a source of off-farm income.

It is evident from the preceding discussion that the question "How much more land can be drawn into crop production?" cannot be answered only, or even predominantly, on the basis of the data presented here. For one thing, the extent of, perhaps multiple, overlap between land with crop production potential, forest, human settlements and protected areas is not known with an acceptable margin of error. For another, more of the land with crop production potential (whether presently used or not) will be occupied by human settlements in the future following population growth. Further, a host of other factors (socioeconomic, technological, etc.) will determine the combination of area expansion and growth of yields that will underpin the future growth of production. In the event, and as explained in Chapter 4, the additional land to come under crop production by year 2010 may be about 4 million ha in South Asia and 20 million ha in tropical South America. In addition, continued population growth would probably require additional land for human settlements and infrastructures of some 9 million ha and 3 million ha in the two regions, respectively.

Declining land/person ratios

As noted, the continuous decline of agricultural resources per caput following population growth is one of the major reasons why concern is expressed in relation to the population–food supply balance. The other reason has to do with the deterioration of the quality and food production potential of the resources. The data discussed above may be used to shed some light on the nature and significance of the decline in the resources/person ratio (hereafter referred to as land/person ratio). The values of this latter ratio in the different developing countries span a very wide range, from the very low to the very high. For example, at the very low end are countries like Egypt, Mauritius, Rwanda, etc., with ratios of land-in-use of under 0.1 ha per person in the total population and virtually zero reserves for further expansion. At the other extreme, countries like Argentina or the Central African Republic (CAR) have land-in-use ratios of close to 1 ha per person and considerable reserves.[11]

With population growth, more and more countries will be shifting closer to values of the land/person ratios typical of those encountered currently in the land-scarce countries. Does this matter for their food and nutrition? An approach to obtaining a first partial answer is to examine if the currently land-scarce countries are worse-off nutritionally (in terms of per caput food availabilities) compared with the more land-abundant ones. This is attempted in Table 2.5 with the land/person ratios adjusted as indicated in Note 11. The picture emerging from the table just confirms what is known, i.e. there is no apparent close relationship between the land/person ratios and per caput food supplies. If anything, many land-abundant countries have low per caput food supplies, while most of the nutritionally better-off countries seem to be precisely those with the highest land scarcities. At the same time, most of these latter countries have considerable cereal imports.

Should this evidence be interpreted to mean that the perceived threat of ever declining land/person ratios is misplaced? Not necessarily. In the first place, the national land/person ratio, even if adjusted for land quality differentials, is just one of the many factors that determine per caput food supplies. Its importance cannot be evidenced without an analysis accounting for the role of these other factors (essentially respecting the clause "other things being equal"). Secondly, the high dependence of the land-scarce, good-nutrition countries on imported cereals means that the perceived threat of the declining land/person ratios must be understood in a global context. That is, a decline in an individual country's land/person ratio may not threaten its own food welfare provided there is enough land elsewhere (in the actual or potential exporting countries) to keep

Table 2.5 Distribution of developing countries by per caput land-in-use and food supplies, data for 1988/90

Land per caput* (ha)	Food supplies per caput (cal/day)‡					
	Under 2000	2000–2100	2100–2300	2300–2500	2500–2700	Over 2700
Under 0.10	Rwanda (8)†				Jamaica (140)	T. Tobago (213) Jordan (338) Korea, Rep. (225) Mauritius (190)
0.10–0.19	Burundi (4) Somalia (29) Namibia (49)	Kenya (1) Bangladesh (20) Haiti (36)	Yemen (134) Lesotho (117) Sri Lanka (60) Vietnam (−11) Liberia (47) Guatemala (36) Honduras (33)	Venezuela (126) Dominican Rep. (94) El Salvador (36) Philippines (35) Colombia (27) Laos (14) Gabon (74)	Indonesia (10)	Egypt (163) Lebanon (188) S. Arabia (265)
0.20–0.29	Ethiopia (15)	Peru (65) Malawi (13)	India (1) Panama (53) Nepal (2) Nigeria (5) Ghana (20) Uganda (1) Congo (51)	Myanmar (−4) Ecuador (46)	Malaysia (140)	Costa Rica (120) Algeria (251) Korea, PDR (27) Libya (401)

Land per caput*	Countries (net cereal imports†)
0.30–0.39	Sierra Leone (37), Mozambique (30); Chile (14), Suriname (−60), Mauritania (117); Tanzania (2), Gambia (97), Botswana (148), Pakistan (6), Nicaragua (46), Thailand (−113), Madagascar (9), Cambodia (8), Zaire (12); Swaziland (134); Turkey (14), Cuba (235), Tunisia (219), Mexico (77), Iran (101), Syria (114)
0.40–0.50	Angola (49), Afghanistan (17); Bolivia (19), Sudan (16); Zimbabwe (−40), Togo (21); Uruguay (−158), C.Ivoire (50); Morocco (58), Iraq (223)
Over 0.50	Chad (8), CAR (15); Zambia (15); Niger (26), Cameroon (43), Guinea (40), Mali (10), Burkina Faso (17); Senegal (82), Benin (22), Guyana (10); Paraguay (−70); Brazil (18), Argentina (−289)

*Land per caput adjusted to measure land of roughly comparable production potential (see note 11).
†Numbers in parentheses are net cereal imports in kg per caput. A minus sign denotes net exports.
‡Calorie data for 1988/90 before the latest (1994) revision of the FAO food balance sheets.

the global land/person ratio from falling below (unknown) critical minimum values; and, of course, provided that the people in the land-scarce country do not depend in a major way on the local land and water resources for a living. Countries like Korea (Rep.) and Mauritius are in this class.

It follows from the above that declining land/person ratios can threaten the food welfare of those land-scarce countries which depend on agriculture in a major way for a living. And this irrespective of the fact that their own population growth may not have a significant impact on the global land/person ratios. Most countries in this class are those in the upper left quadrant of Table 2.5. Only a combination of much more productive agriculture (in practice, resort to land-augmenting technologies that would halt or reverse the declines) and vigorous non-agricultural growth will free them from the bondage of the ever declining land/person ratios.[12]

In conclusion, the declining land/person ratios do matter for per caput food supplies in two senses. In the *global context* and for countries with high actual or potential dependence on food imports they matter mainly if the declines threaten to push the global ratio below (unknown) critical values, even after allowing for the reprieve to be had from land-augmenting technologies. Should this happen, the effects would be manifested in terms of rising food prices which would affect mostly the poor. It has not happened so far despite continuous declines in the global land/person ratios. How close the world is to eventual critical values and whether such values are likely to be reached before the world achieves stationary population and acceptable per caput food supplies for all is a matter of conjecture.

In the *local context* declines in the land/person ratios do matter for food supplies, nutrition and incomes, mainly for the countries with limited access to imported food and high dependence on agriculture for the maintenance and improvement of living standards and, consequently, of food welfare. If and when such dependence is reduced, the pressures on the global land/person ratios will assume increasing importance also for them.

The possible role of land-augmenting (in practice, yield-increasing) technologies was referred to above for the reprieve such technologies can afford in relation to the consequences of the inexorable declines of the land/person ratios. However, some of the perceived threats to progress towards solving the food problem have precisely to do with the risks to the productive potential of the agricultural resources stemming from the application of these very technologies, e.g. loss of irrigated land to salinization and waterlogging, loss of yield potential and increased risk of crop failures because of pesticide resistance, etc.[13] In addition, efforts to bring new land into cultivation or to use existing agricultural land more intensively can often be associated with degradation (e.g. from reduced fallows, from exposure of fragile soils to erosion following deforestation) and may not add permanently to total productive potential. In the following subsection an attempt is made to address what are hypothesized to be the more fundamental processes driving human activity towards degradation of the productive potential of agricultural resources.

Agricultural activity and degradation of agricultural resources

As noted, there is sufficient (though not comprehensive, nor detailed) evidence establishing that the productive potential of at least part of the world's land and water resources is being degraded by agricultural activity (e.g. soil erosion, waterlogging and salinization of irrigated lands). In addition, agricultural activity generates other adverse environmental impacts (e.g. threat to biodiversity, pollution of surface and groundwater sources). Chapter 11 presents some evidence of these processes. While recognizing that agricultural activity often contributes to maintaining or restoring the productive capacity of land and water resources, this concluding subsection attempts to provide a framework of thinking through why human activity may end up destroying rather than preserving or enhancing this capacity.

The most commonly held view is that these processes are somehow related to continuing demographic growth, in two senses: (a) more food must be produced and this tends to draw into agricultural use land and water resources not previously so used and/or causes such resources to be used more intensively – both processes may generate adverse impacts on the quality of the resources themselves as well as on the broader environment; and (b) in many developing countries population growth is accompanied by increases in the number of persons living off the exploitation of agricultural resources with the consequence that the amount of resources per person declines.

In the normal course of events the decline in per caput resources would tend to raise their value to the persons concerned (being often their main or only income-earning asset) and would lead to their more efficient use, including maintenance and improvement of their productive potential. The fact that much of the agricultural resource base has been improved for agricultural use by human activity in the historical period is testimony to this process. Yet it is often observed that under certain conditions this caring relationship tends to break down with the result that people destroy rather than conserve and improve the productive potential of the resources (see Harrison, 1992).

Understanding why this happens is the most important insight needed for policy responses to promote sustainable development. When this destructive relationship is observed in conditions of poverty, it is commonly taken for granted that poverty explains the behaviour of people *vis-à-vis* the resources. The hypothesized mechanism works (in economic parlance) via the shortening of the time horizon of the poor. In plain language, this means that in conditions of abject poverty the need for survival today takes high precedence over considerations for survival tomorrow. The poor simply do not have sufficient means to provide for today and also invest in resource conservation and improvement to provide for tomorrow.

However, this proposition is far from being a sufficiently complete explanation of processes at work useful for formulating policy responses. For one thing, there is plenty of empirical evidence that this process is not at work in many situations of poverty. For another, it is often observed that agricultural resource degradation occurs also when such resources are exploited by the non-poor (a matter discussed below). It also occurs, and often more so, in conditions where poverty is declining rather than increasing, e.g. when the opening up of income-earning opportunities outside agriculture leads to the abandonment (because they are no longer worth it) of elaborate resource conservation practices, such as the maintenance of terraces to conserve small poor quality land patches on hillsides, etc. (for examples from the Latin American sierras, see de Janvry and Garcia, 1988). Another example of degradation associated with alleviation rather than aggravation of poverty is given by the opening up of profitable opportunities to grow cassava in some Asian countries for export to the EC where it substituted high-priced cereals in the feed sector. It is thought that part of this expansion of cassava production had adverse effects on the land and water resources which were not "internalized" in the export price (see Chapter 13).

It follows from the above that more complex processes are at work and the simple correlation between poverty and environmental degradation can be an oversimplification. This is well recognized, and research work on understanding the role of other variables which mediate the relationship between poverty and environmental degradation can provide valuable insights. Such work emphasizes, for example, the vital importance of institutions governing access to resources (e.g. to common property or open access resources) and how such institutions come under pressure when population density increases; inequality of access to land and landlessness; policies which distort incentives against the use of technology that would contribute to resource conservation, e.g. by depressing the output/fertilizer price ratio and making fertilizer use uneconomic where increased use is vital for prevention of soil mining; and the knock-on effects of policies which facilitate interactions between the non-poor and the poor in ways which lead to degradation, e.g. when deforestation and expansion of agriculture are facilitated by incentives to logging operations which open access roads into, and make possible agricultural settlement of, previously inaccessible forest areas which may have soils that cannot easily sustain crop production.

Understanding the role of these and other mediating variables and getting away from the simple notion that degradation can be explained by poverty alone is important for formulating and implementing policies for sustainable agriculture and resource conservation. It is important because the policy environment in the future will continue to be characterized by pressures on agricultural resources related, in one way or another, to rural poverty. Indeed, the numbers of the rural poor depending on the exploitation of agricultural resources will probably increase further in some countries and decline in others.

It was noted earlier that both processes can be associated with resource degradation. Therefore, the key policy problem is how to minimize adverse environmental impacts of both processes. Chapter 12 presents what are essentially technological options for policy responses; and Chapter 13 deals with policies that would contribute to minimizing the unavoidable trade-offs between agricultural development and the environment.

Poverty-related degradation of agricultural resources is only part of the story. It is well known that part of the degradation process is related to the actions of people who are not in the poor category. This issue has two aspects: *the first one* has to do with consumption levels and patterns of the non-poor, in both the developed and the developing countries. For example, some 30 percent of world cereals output is used as animal feed and a good part of the production of soybeans and other oilseeds is also related to livestock production. Most of the livestock output produced in concentrate-feeding systems is consumed by the medium- and high-income people. To the extent that production of cereals and oilseeds causes degradation (as indeed it does in some places, though not in others) it can be said that part of the degradation is caused by actions of the rich, not the poor. It would perhaps be more correct to say that it is caused by *interactions between rich and poor,* e.g. when expansion of soybeans production in South America raised the price of land there. This induced small farmers to sell land to large soybean operations and move to other areas to colonize new lands. The example cited earlier concerning the expansion of cassava production for export to Europe is another case. Both cases are to some extent related to policies of other countries which maintained artificially high prices for cereals used in livestock production and increased incentives for production and export of these cereal substitutes (for more discussion, see Alexandratos *et al.*, 1994).

Many other cases could be cited to illustrate the complex interactions between the behaviour of the poor and the non-poor resulting in the building up of pressures on agricultural resources. Without a thorough understanding of such complex processes at work leading to resource degradation, it would be difficult to design and implement appropriate policy responses. Accounting for the factors which determine the actions of the poor and the non-poor alike is required even if, in a poverty-focused strategy, the priority objective were to minimize the degradation of the resources operated by the poor.

The *second aspect* has to do with the fact that resource degradation is also associated with agriculture practised by farmers who are not poor. Soil erosion believed to be associated with some of the grain production in North America is a case in point; excessive fertilizer and other agrochemical use in Europe is another; and effluents from intensive livestock operations are in the same category. These are all examples of actions by the non-poor with adverse environmental effects. It all goes to show that associating resource degradation with poverty addresses only part of the issue.

In the end, the focus of policy has to recognize that resource degradation has different consequences for different countries and population groups. For the poor countries, the consequences can be very serious because their welfare depends heavily on the productive potential of their agricultural resources (Schelling, 1992). Therefore, from a purely developmental and conventional welfare standpoint, it is right that preoccupation with resource degradation problems focuses primarily on the developing countries. At the same time, it must be recognized that resource degradation not only in the developing countries, but anywhere on the planet, particularly in the major food exporting developed countries, can make more difficult the solution of the food security problems of the poor if it reduces the global food production potential. Therefore, controlling resource degradation in the rich countries assumes priority even in strategies focused primarily on the food security of the poor; and this irrespective of the fact that the welfare of the rich countries, as conventionally measured by, for example, per caput incomes, may not be seriously threatened by moderate degradation of their own resources. There are, of course, other compelling reasons why the rich countries give high priority to controlling degradation of their own resources, as an objective in its own right.

NOTES

1. The food available for direct human consumption (Food) is computed from the following food balance sheet equation for each country, food product and year: Total food supply = Production + (imports − exports) + (beginning stocks − ending stocks), and Food = Total food supply − Animal feed − Industrial non-food uses − Seed − Waste (from harvest to retail). The resulting estimate of per caput food availabilities can differ from the amounts of food people actually ingest in a nutritional sense, e.g. because of further waste in the post-retail stage (e.g. at the household level). For discussion of this issue and comparisons see FAO (1983).
2. World production of cereals in per caput terms peaked in the historical period at 342 kg in the three-year period 1984/86 and remained below this level in all moving three-year averages since then. It had fallen to 317 kg in 1987/89 (1988 was the year of the North America drought) and recovered to 325 kg by 1990/92.
3. The developing countries span a very wide range of this indicator, from under 15 percent (Jordan, Argentina, Chile, Trinidad and Tobago) to over 80 percent (many countries in sub-Saharan Africa).
4. It is recognized that world market price movements do not always convey unbiased information for evaluating the issue at hand. This is because they have been influenced, generally downwards, by the agricultural support and protection policies of major developed countries. In addition, the non-incorporation into the product prices of alleged losses of natural capital generated by the production process (reduction of the production potential of land and water resources following their degradation) also tends to bias downwards the market prices. In short, the downward movement of world market prices has in part reflected the rich countries' consumer and taxpayer subsidies (see O'Brien, 1994) and perhaps, but to an unknown extent, also an environmental subsidy of all countries which ignore the costs of the alleged natural capital losses.

5. These indicative numbers of per caput food supplies are distinct from the related concept of national average requirements for dietary energy. The latter are derived as the weighted average of the requirements of the individuals in a population group. Individual requirements vary widely with (mainly) age, sex, body weight and level of physical activity. For example, they are 1900 calories/day for a woman in the 18–30 age bracket, with a body weight of 50 kg and light physical activity. They rise to 3700 calories/day for a man of 70 kg in the same age bracket but with heavy physical activity (FAO/WHO/UNU, 1985, Tables 15, 42, 45). These are still averages, since requirements vary among persons of the same age, weight, etc. National average requirements have been computed for individual countries. For example, they were 2130 calories/day for Haiti for moderate activity level (FAO, 1990c). To make this estimate conceptually comparable to that used here for per caput food supplies it must be raised by a margin to account for food losses at the post-retail level. If the margin were 10 percent, Haiti would have a requirement at the retail level of some 2350 calories/person/day. This would have been adequate for meeting the country's nutritional needs if food were distributed to individuals exactly according to their requirements. Since this is never the case in any country, it results that the national average must be still higher if enough food were to be available for people at the lower end of the distribution to have potentially access to food to meet their energy requirements. In the event, Haiti's per caput food supplies were 2000 calories in 1988/90. The preceding discussion in no way implies that the solution to the food problem is to be pursued by means of policies aimed at raising per caput food supplies to whatever level is suggested by the above considerations. Such policies must be seen as a necessary complement to, not a substitute for, policies which address directly the root cause of the problem, the inadequate food entitlements of the poor. However, in countries where most of the poor are in agriculture, the two policies go in tandem.

6. Physiological considerations (i.e. a person needs a minimum food intake for survival and there is an upper physiological limit to how much food a person may consume) dictate that the scope for distributional inequalities is more limited in the case of food intakes than in other "unbounded" variables, e.g. income. Consequently, it may be hypothesized that very low or relatively high levels of average per caput food supplies will be associated with more equal distribution of food intakes compared with those that could prevail when the overall average is the middle range. This is a useful hypothesis for looking at the distributional issue when we know nothing about prospective developments in the other determining variables of the distribution of food intakes. Thus, it can be expected that the distribution of food intakes will tend to become more equal in those developing countries in which the average per caput food supplies will continue to edge upwards towards the 3000 + calorie level. These considerations are taken into account in Chapter 3 when considering the issue of the possible evolution of the incidence of undernutrition in the future.

7. These countries are (in ascending order of their 1961/63 rank): Libya, China, Algeria, Saudi Arabia, Indonesia, Iraq, Korea (Rep.), Iran, Korea (DPR), Tunisia, Morocco, Egypt. Each of them had increased per caput food supplies by at least 40 percent over the historical period considered. Thus, countries which had in 1988/90 2600 calories are included only if they started with less than 1860 calories in 1961/63. Likewise, those with 3300 in 1988/90 are included only if they started with less than 2350 calories. The data used here are those available before the latest (1994) revision of the FAO Food Balance Sheets. Calorie data for all countries are given in the Appendix 3.

8. All developing countries (among the 93 of this study) with per caput food supplies under 2100 calories in 1988/90 (19 countries, listed in ascending order of calories in 1988/90): Ethiopia, Chad, Afghanistan, Mozambique, Central African Rep., Somalia, Angola, Sierra Leone, Rwanda, Burundi, Namibia, Haiti, Bolivia,

Zambia, Peru, Bangladesh, Sudan, Malawi, Kenya. All of them were in 1988/90 in the range 1700 calories (Ethiopia) to 2065 calories (Kenya). In 1961/63 the range was from 1720 calories (Somalia) to 2300 calories (Chad).

9. The role of the public policy can be appreciated from the following World Bank statement: "Egypt probably has the lowest level of malnutrition among countries with comparable levels of per capita income. Egypt in effect is a vast welfare state which provides the bulk of its people cheap food, cheap oil, and, for those living in rent-controlled housing, cheap though insufficient shelter" (cited in Yitzhaki, 1990).

10. See, for example, Kennedy (1993: 192). A case can be made that some developing countries, but not all of them, would gain from policies to reduce GHG emissions if such policies removed economically wasteful energy market distortions (see OECD, 1992).

11. A country may have plenty of land of poor agricultural quality, but not for this reason should it be classified as land-abundant. For example, it is estimated that Niger has land-in-use of 1.5 ha per person in the total population, but with 95 percent of it in the dry semi-arid category. Other countries have much less land-in-use per person but of better quality, including quality improvements brought about by irrigation. For example, Pakistan has only 0.16 ha/person but 86 percent of it is irrigated. The land/person ratios must, therefore, be adjusted before a meaningful comparison among countries can be attempted. Adjustments are made using the following weights: 1.0 for sub-humid; 0.81 for fluvisols/gleysols and 0.35 for marginally productive fluvisols/gleysols; 0.31 for dry semi-arid; 0.88 for moist semi-arid; 0.85 for humid; 0.35 for the marginal areas in the moist semi-arid, sub-humid and humid zones; and 2.2 for irrigated. These weights roughly reflect potential cereal yields. After such adjustments, Niger's land/person ratio falls to 0.50 ha and that of Pakistan rises to 0.31 ha. This is land of fairly comparable production potential after the adjustments. The comparisons in Table 2.5 are on the basis of the thus adjusted land/person ratios.

12. Reference to land-augmenting technologies and the growth of other sectors just underlines the fact that aggregate resources for producing food and/or incomes cannot be considered as given. In the process of development scarce resources are substituted by less scarce ones and the total is augmented by additions of man-made capital, the most important component of which is human ingenuity. Whether there are ultimate limits to this process is another question (see, for example, Daly, 1992; Daly and Townsend, 1993; and for a less pessimistic view, Pearce and Warford, 1993, whose book goes under the suggestive title *World Without End*).

13. Yield-augmenting technologies, generally those associated with the introduction of modern varieties, have come occasionally under fierce criticism, both because of their adverse effects on resources and the environment as well as because they tend to disrupt traditional farming systems and the associated social structures (e.g. Shiva, 1991). Yet such effects need to be evaluated in relation to those, probably much more severe ones, that would have resulted from the increasing pressures of the growing population on resources, traditional farming systems and social structures if these technologies had not made it possible to achieve quantum jumps in food production. The issue is not one of having or not having yield-augmenting technologies. It is more one of minimizing their adverse effects. Agricultural research is increasingly oriented towards this goal, e.g. breeding resistance to pests into modern varieties so as to reduce dependence on chemical pesticides (see Chapter 4).

CHAPTER 3

World Food and Agriculture: a 20-Year Perspective

3.1 INTRODUCTION

This chapter presents the likely developments in world food and agriculture over the period to 2010, in varying degrees of detail concerning commodities and country groups. As noted in the Foreword, this study attempts to sketch out the prospective developments and *give an idea of the future as it is likely to be rather than as it ought to be from a normative perspective*. For example, the conclusion that undernutrition is likely to persist results from this approach to looking into the future. Therefore, the prospective developments presented here are not goals of an FAO strategy.

Highlights of the projection methodology

All the analyses and projections were carried out for a large number of commodities or commodity groups (52 crop and livestock products for food demand analysis, 32 for demand–supply analysis and 40 for production analysis) and for practically all countries individually (see list of countries and commodities in Appendix 1). This great detail makes it possible to draw on specialized knowledge of the different contributors to the study pertaining to individual countries and commodities. It also generates the results in a form which can be used to address issues at very diverse levels, ranging from statements on total agriculture and main regions to those on single commodities and alternative, and often overlapping, country groups. It is also imposed by the interdisciplinary nature of the study, which requires dialogue and consultation at meaningful levels of detail, as well as by the need to account for nearly the totality of use of land and respect relevant constraints. A good deal of the total work for the study is absorbed by the need to create a consistent data-set for the historical period (1961 to 1990) and to supplement it with often not readily available data for the base year, the three-year average 1988/90. This is particularly true for the data on crop production patterns and existing stocks of land with agricultural potential, both by agroecological zone and for each developing country.

The overall approach is to start with projections of demand using Engel demand functions and exogenous assumptions of population and GDP growth.[1] Subsequently, the entry point for the projections of production is to start with provisional targets for production for each commodity and country derived from simple rules about future self-sufficiency and trade levels. There follow several rounds of iterations and adjustments in consultation with specialists on the different countries and disciplines, with particular reference to what are considered to be "feasible" levels of land use, yields and trade. Accounting consistency controls at the commodity, land resources (developing countries only), country and world levels have to be respected throughout. In addition, but only for the cereals, livestock and oilseeds sectors, a formal flex-price model was used (the FAO World Food Model, FAO, 1993i) to provide starting levels for the iterations and to keep track of the implications for all variables of the changes in any one variable introduced in the successive rounds of inspection and adjustment. The model is composed of single country modules and world market feedbacks leading to national and world market clearing through price adjustments.

It is emphasized that the results of the model projections (whether the single Engel demand functions or the flex-price model) were subjected to many rounds of iterative adjustments by specialists on countries and many disciplines, particularly during the phase of analysing the scope for production growth and trade. Specialist input on some commodities was also obtained from collaborative work with outside (non-FAO) sources, e.g.: the International Sugar Organization, for sugar; the Economic and Social Institute, Amsterdam, for rubber; the International Cotton Advisory Committee, for cotton; the World Bank, for jute, and for the projections of the developed countries, with the Center for Agricultural and Rural Development, Iowa State University, and the Institut für Agrarpolitik, University of Bonn. The end-product may be described as a set of projections which meet conditions of accounting consistency and reflect to a large extent constraints and views expressed by the specialists in the different disciplines and countries. A description and evaluation of the methodology is given in Appendix 2.

It is clear from the preceding discussion that the projections presented here are definitely not "trend extrapolations", whether the term is used to denote the derivation of a future value of any variable by simple application of its historical growth rate to the base year value (exponential trend) or the less crude notion of using time as the single explanatory variable in functional forms other than exponential, e.g. linear, semi-log, sigmoid, etc. For one thing, projecting all interlinked variables on the basis of estimated functions of time is a practical impossibility; for another, projecting any single variable at its historical growth rate (which could be negative, zero or very high) often leads to absurd results. Therefore, the term "trend" or "trend extrapolation" is not appropriate for describing these projections.

3.2 POPULATION GROWTH AND THE OVERALL
ECONOMIC GROWTH OUTLOOK

Population

The world population is projected to grow to 7.2 billion by year 2010, up from
5.3 billion in 1990, an increase of 1.9 billion or 36 percent in 20 years (Table
3.1). This is a higher absolute increment than that of the preceding 20 years (1.6
billion) but a lower one in percentage terms. The growth rate of world
population peaked at about 2.1 percent p.a. in 1965–70 and then declined
progressively to the current level of 1.7 percent p.a. Further declines in the
growth rate are foreseen for the future so that by the final five-year period of
the study's horizon (2005–10) growth will have fallen to 1.3 percent p.a. and on
to 1.0 percent p.a. by 2020–25. But the absolute yearly increments will continue
to be large and growing. They are at present larger than at any time in the past
but are about to peak at around 94 million persons per year (Figures 3.1, 3.2).

What matters most for food and nutrition is the population growth of the
developing countries and indeed of those regions with the highest incidence of
poverty and undernutrition (sub-Saharan Africa and South Asia). The
developing countries are still on a high demographic growth path (2.0
percent p.a.) and will account for over 90 percent of the additional world
population in the projection period. But their growth rate is also on the decline,
though slowly. The region with the highest population growth rate (over 3.0

Billion

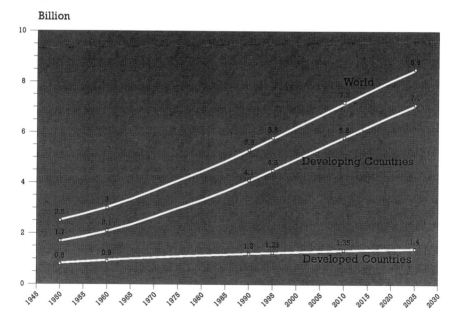

Figure 3.1   World population, 1950–2025

percent p.a.) is sub-Saharan African, though also here the growth rate may peak during the projection period. Still, its population is projected to increase to some 900 million by the year 2010, an average growth rate of 3.2 percent p.a. (revised to 834 million and 2.9 percent p.a. in the latest UN assessment).[2] Making food availabilities grow at an equal rate would still be an achievement and a break from past trends, but even if achieved, it would only serve to prevent further deterioration of the already very inadequate nutrition levels.

Economic growth assumptions

The overall economic growth rates used in this study (Table 3.1) are taken from other sources, mainly the work of other organizations, e.g. the World Bank, and supplemented by judgements only when no projections from other sources were available. The problems associated with the use of exogenous assumptions about the overall economic growth outlook in assessing the food and agriculture prospects were indicated earlier (Note 1).

For the developing countries the outlook is for economic growth to be much better in the 1990s than that of the "crisis" decade of the 1980s. In this latter period, all developing regions except Asia experienced declines in per caput incomes. In the projection period the prospects are for Asia to maintain its high growth rate and all the other regions to shift in varying degrees from negative to positive, though modest, growth rates in per caput incomes. But there are no

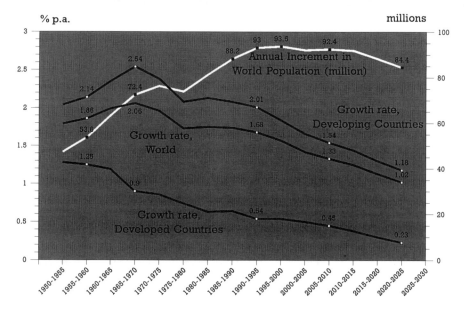

Figure 3.2   Population growth rates and annual increments, 1950–2025. From Medium Variant Projections; 1992 UN Assessment (UN, 1993b)

Table 3.1 Population projections and GDP growth assumptions

	Population (million)		Population (growth rates % p.a.)				GDP growth rates 1989–2010 (% p.a.)		World Bank's latest baseline GDP projections* (% p.a.)			
							Total	Per caput	Total		Per caput	
	1989	2010	1970–80	1980–90	1990–2000	2000–10			1983–93	1994–2003	1983–93	1994–2003
World	5205	7209	1.9	1.8	1.7	1.4			2.9†	3.2		
All developing	3960	5835	2.2	2.1	2.0	1.7			3.8†	5.2		
93 Developing	3905	5758	2.2	2.1	2.0	1.7	5.3	3.4				
Africa (sub-Sahara)	473	915	2.9	3.2	3.3	3.1	3.9	0.7	2.2	3.9	−0.8	0.9
Near East/North Africa	297	493	2.7	2.8	2.6	2.2	4.4	1.9	0.8	3.8	−2.2	0.9
East Asia	1598	2061	1.9	1.6	1.5	0.9	7.0	5.7	7.9	7.6	6.2	6.2
South Asia	1103	1668	2.3	2.4	2.2	1.8	5.1	3.0	5.2	5.3	3.0	3.4
Lat.America + Carib.	433	622	2.4	2.2	1.9	1.6	4.0	2.3	2.2	3.4	0.4	1.7
Developed countries	1244	1373	0.8	0.7	0.5	0.4	2.6	2.1				
ex-CPEs	387	435	0.9	0.8	0.6	0.5	0.5§	0.0	−1.0	2.7		
Western Europe	399	410	0.4	0.3	0.2	0.0	2.7	2.5				
North America	274	311	1.0	0.9	0.6	0.5	2.2	1.6	2.7†‡	2.7‡		
Others	182	214	1.4	1.0	0.9	0.7	4.2	3.4				

Note: Population data and projections are from the 1990 UN Assessment, Medium Variant (UN, 1991). Country level data are shown in the Appendix 3, where regional totals from the 1992 and 1994 UN Assessments (UN, 1993b; 1994) are also shown.
*World Bank (1994a). Regional groups not always identical to those used in this study.
† 1980–93.
‡ High-income OECD countries.
§ The GDP growth rates 1989–2010 are lower than those of the World Bank because they include the deep GDP declines of 1990–94 estimates at some 30%.
The World Bank's growth rates 1989–2010 refer to the region ''Developing Countries in Europe and Central Asia'' which in addition to the ex-CPEs, includes Turkey, Greece and Portugal.

Asia-type growth prospects for these regions, particularly not for sub-Saharan Africa. Although the region's economic growth rate could nearly double compared with the dismal 1980s, this would still leave it with nearly stagnant and very low per caput incomes over the projection period, given its high population growth rate. These developments can hardly lay the foundation for significant improvements in its food and agriculture. Such prospects are indeed reflected in this study's assessments for the region (see below).

The OECD countries are assumed to continue along their growth path of the 1980s, though their low population growth rate would make for rather respectable increases in per caput incomes. The OECD growth rate is an important parameter in the World Bank's assessment of the developing countries' growth prospects (see reference in Table 3.1). The direct link with the latter's food and agriculture is likely to be weak because of the low income and price elasticities of the OECD for the demand of their main traditional agricultural exportables, though the link can be significant for products like fruit and vegetables. However, the indirect links can be important, at least for some regions, if OECD growth stimulates investment flows and their exports of manufactures and this contributes to higher incomes and agricultural demand in the developing countries themselves.

The uncertainties surrounding the prospects for recovery and long-term growth of the ex-centrally planned economies (CPEs) of Europe, particularly of the former USSR, are well known. Their combined GDP had fallen by about one-third between 1989 and 1993. The working hypothesis used in this study is that growth could lead to restoration of their per caput income level to that of the pre-reform period by 2010. The reforms and these growth prospects can be expected to have profound influences on their own food and agriculture and on world food markets. These are discussed later on in this chapter. Here it suffices to mention that world cereals markets may be affected by increased export availabilities from Eastern Europe and greatly reduced import requirements of the former USSR. Both will contribute to continued weakness in these markets as these developments will tend to offset part of the stimulus of continuing growth in the import requirements of the developing countries. In parallel, the import demand of the ex-CPEs for the main traditional agricultural exportables of the developing countries is unlikely to be very buoyant, notwithstanding their still very low levels of per caput consumption of such products (coffee, cocoa, etc.) and the considerable potential for increases.

3.3 AGRICULTURE: PROSPECTIVE DEVELOPMENTS IN AGGREGATE PRODUCTION AND DEMAND

The world as a whole

For the world as a whole, the prospects are that the growth of aggregate (gross) production will continue to slow down. This is in line with the longer term historical developments when the annual growth rate fell from 3.0 percent in the 1960s, to 2.3 percent in the 1970s and to 2.0 percent in 1980–92. The further deceleration of production growth in the post mid-1980s period was noted in Chapter 2. It is foreseen that the growth rate will fall further to 1.8 percent p.a. in the period to 2010 (Table 3.2). A host of factors explain this progressive deceleration. It may not be interpreted to mean that the world as a whole is running out of potential to increase agricultural output, though production constraints at the local level do contribute to this outcome. This happens when local resource constraints are obstacles to development and the growth of incomes with the result that potential demand for food is not expressed as effective demand for imports that would induce production to increase in another part of the world. The decline in cereals production in the main exporting countries after the mid-1980s is telling in this respect (see Chapter 2).

The progressive slowdown in the growth of world production mirrors the similar slowdown in the growth of demand, itself resulting from the lower population growth and the progressive saturation of per caput food consumption levels for parts of the world population. The little scope for further growth in per caput consumption in most developed countries is a case in point. This has a major impact on the scope for further expansion of the world totals because the developed countries account for a high share of gross world consumption of agricultural products, 49 percent in 1988/90, notwithstanding their much lower share in world population (24 percent). At the same time, the considerable scope for further growth in consumption in most developing countries gets expressed as effective demand only gradually, for the well-known reasons of the slow growth in their per caput incomes and, in the countries facing production constraints, import financing capacity.

In conclusion, the further slowdown in world agriculture conveys a composite signal of positive and negative elements for the world's food and agriculture futures. On the negative side is the fact that the potential for producing more to increase food supplies to the people with low consumption levels will be utilized only in part. On the positive side is the slowdown in world population growth and the fact that more and more people attain satisfactory levels of food consumption; and, of course, the less the need for increasing output, the smaller the additional pressures exerted on resources and the environment. But this is not always so. In situations of poverty, a low growth rate of production may be associated with more resource degradation than a higher rate of production. This is because a low rate contributes to perpetuate poverty and can set in motion the vicious circle of poverty–degradation–poverty, as discussed in Chapter 2.

The developing countries

The preceding discussion on the possible outcomes at the world level (slower

Table 3.2   Growth rates of gross agricultural production and domestic demand, all uses (% p.a.)

	Production				Domestic demand (all uses)			
	Total		Per caput		Total		Per caput	
	1970–90	1988/90–2010	1970–90	1988/90–2010	1970–90	1988/90–2010	1970–90	1988/90–2010
World	2.3	1.8	0.5	0.2	2.3	1.8	0.5	0.2
93 Developing countries	3.3	2.6	1.1	0.8	3.6	2.8	1.4	0.9
Africa (sub-Sahara)	1.9	3.0	−1.1	−0.2	2.6	3.3	−0.4	0.1
Near East/North Africa	3.1	2.7	0.3	0.3	4.5	2.8	1.7	0.4
East Asia	4.1	2.7	2.4	1.5	4.1	2.8	2.4	1.6
South Asia	3.1	2.6	0.7	0.6	3.1	2.8	0.8	0.8
Latin America + Carib.	2.9	2.3	0.6	0.6	2.9	2.4	0.6	0.6
Developed countries	1.4	0.7	0.6	0.2	1.2	0.5	0.5	0.0
ex-CPEs	1.2	0.4	0.4	−0.1	1.4	0.2	0.6	−0.4
Other developed countries	1.5	0.8	0.7	0.4	1.2	0.7	0.5	0.2
Memo item Africa (sub-Sahara) Non-food crops (mainly exportables)	0.9	1.9						
Roots/tubers	2.3	2.8						
Other food products	1.9	3.2						

growth than in the past 20 years) applies largely also to the developing countries as a whole, though with less force and important exceptions (see Table 3.2). Also here, the gradual progress towards higher consumption levels in some countries and regions and the overall lower growth rate of population will lead to a slower growth in demand and production compared with the past. East Asia, and to a smaller extent the Near East/North Africa, are the regions closest to this pattern of evolution, though in the latter region the slow growth in income and increasingly stringent foreign exchange constraints will also play a role. Both regions had the highest growth rates of per caput demand (all uses) in the past 20 years, though the sharp slowdown in the growth of per caput demand in the Near East/North Africa region was already evident in the 1980s. East Asia would continue to have the highest, though lower than in the past, growth of per caput demand and production of all regions, given its high projected economic growth. South Asia and Latin America should maintain their middling growth rates of the past 20 years in both per caput demand and production. This leaves sub-Saharan Africa as the only region that could have a higher growth rate compared with the past. This is a significant improvement, but only relatively speaking, since it would only mean that per caput production and demand may just stop falling.

However, these prospective developments for aggregate agriculture cannot be appreciated without a closer look at the prospects for the major commodity sectors. Much of the remainder of this chapter is devoted to such an examination. But an initial idea of how the prospects for the different commodities enter the determination of the agricultural aggregates is given in the lower part of Table 3.2. In it, the projected growth rate of production for sub-Saharan Africa, which is lower than that of projected population, is broken down into three commodity sectors. It is clearly seen that when the sector of the non-food commodities (coffee, cocoa, tobacco, rubber, etc.), which grows slowly because of export market constraints, and that of the roots/tubers (growing less rapidly than population due to long-term declines in per caput consumption) are separated out, the rest of agriculture (essentially the other crop and livestock products, predominantly for domestic use), could grow at 3.2 percent p.a. This growth rate, if achieved, would match that of population and could even exceed it if the latter turned out to be lower as indicated by the latest demographic projections. This is an optimistic assessment of sorts, given the long historical experience of much lower growth rates and falls in per caput production. Even so, it would only prevent further deterioration.

The developed countries

Much of what was said earlier on the world prospects for aggregate production and demand applies to the developed countries. In the OECD countries, there is limited scope for further growth in per caput consumption. The past policies of import substitution in some major countries and the export opportunities offered by the increasing net food deficits of the developing countries and the ex-CPEs had contributed to relax the demand constraints and allow for moderate growth in production. The relief offered by these two factors, particularly that of the state-aided import substitution, had been largely exhausted by the mid-1980s. The scope for further import substitution is very limited for most commodities. Moreover, the policy reforms under way will make it difficult for this group of countries to continue to expand exports with the aid of subsidies. In parallel, the growth of net import requirements of the developing countries for the main agricultural exportables of the OECD area is likely to be slow. Furthermore, part of the associated expansion of the export markets is likely to be counterbalanced by reductions in the net import requirements of the former USSR and some expansion of export availabilities from Eastern Europe. Therefore, the likely outcome for the aggregate agricultural production of the OECD area is that the growth rate will be no higher than the 0.8 percent p.a. achieved in the 1980s, which itself was down from 2.0 percent p.a. in the 1970s.

The uncertainty concerning the prospects for the reforming ex-CPEs was underlined above. Their production fell sharply in the initial years of the reforms, with aggregate 1992 gross output being about 15 percent below that of the three-year average 1988/90. The recovery that will follow is expected to be slow and may lead to the growth rate of aggregate production in the next 20 years (measured from the pre-reform level of 1988/90) being only one half that of the last 20 years (Table 3.2). The sharp reduction in per caput consumption in the initial years of the reforms will probably be reversed, but future levels are unlikely to be above those of the pre-reform period. The commodity composition of consumption may change and the per caput domestic use of all agricultural products may be lower, due to lower waste rates and fewer cereals being used for animal feed per unit of livestock output. Therefore, not all declines in production and total domestic use should be considered as reducing the food welfare of the population (see *The Economist*, 18 June 1994).

3.4 PROSPECTIVE DEVELOPMENTS IN FOOD AND NUTRITION

The implications of the above-described prospects in the demand and production for the evolution of per caput food supplies for direct human consumption by major commodities are summarized in Table 3.3 and those for the incidence of chronic undernutrition in Table 3.4. By and large, it can be expected that the trends towards increasing per caput food supplies in most developing countries will continue. In particular, the average food supplies for the projected 2 billion population of East Asia may reach just over 3000 calories/day. Thus, this region will be edging towards the relatively high levels of Near East/North Africa, with the Latin America/Caribbean region following closely behind. In all three regions, though not in all countries in them, the incidence of undernutrition may fall by year 2010 to the fairly low level of 4–6 percent of the total population. This possible evolution will be helped by a probable fall in inequality of access to food supplies which, as noted in Chapter 2, may be expected to improve as average food supplies edge up towards the 3000 + calorie level. Such an assumption is incorporated in the projections of undernutrition in Table 3.4.

South Asia may also be expected to make significant progress, but the initial conditions are such that by 2010 the per caput food supplies would still be in the low-middle level (2450 calories/day). Accordingly, the incidence of undernutrition is likely to remain high, particularly if measured in terms of the absolute numbers of persons affected. Still, the possible halving of the percentage of the population chronically undernourished to 12 percent by year 2010 denotes significant, though not sufficient, progress.

Concerning sub-Saharan Africa, the earlier discussion of population, economic growth and aggregate agriculture already foreshadowed the conclusions shown in Tables 3.3 and 3.4, i.e. that little progress can be

Table 3.3 Per caput food supplies for direct human consumption (kg/caput)

Commodity	1969/71	1988/90	2010	1969/71	1988/90	2010
	World			**Developing countries**		
Cereals	146.3	164.6	167	145.3	170.5	173
Cereals (all uses)	(305)	(331)	(325)	(190)	(235)	(254)
Roots/tubers/plantains	82.3	65.7	65	80.3	63.1	64
Pulses, dry	7.6	6.3	7	9.3	7.4	8
Sugar, raw equivalent	22.1	22.7	24	14.4	18.0	20
Vegetable oils	6.7	10.1	13	4.7	8.2	11
Meat	26.0	31.9	37	10.5	16.4	25
Milk	74.6	75.3	72	27.4	35.9	42
All food (cal./day)	2430.0	2700.0	2860	2120.0	2470.0	2730
	Africa sub-Sahara			**Near East/North Africa**		
Cereals	115.3	114.3	121	183.3	212.9	210
Cereals (all uses)	(140)	(133)	(140)	(295)	(376)	(386)
Roots/tubers/plantains	209.2	208.0	197	16.1	31.2	30
Pulses, dry	11.9	9.6	10	6.3	7.8	8
Sugar, raw equivalent	7.5	8.2	10	20.5	29.4	31
Vegetable oils	7.5	7.8	8	7.5	13.0	15
Meat	10.5	9.5	10	13.0	18.4	23
Milk	28.1	27.6	26	54.0	59.4	67
All food (cal./day)	2140.0	2100.0	2170	2380.0	3010.0	3120
	East Asia			**South Asia**		
Cereals	151.0	200.5	206	148.0	155.8	163
Cereals (all uses)	(192)	(272)	(319)	(167)	(177)	(181)
Roots/tubers/plantains	94.5	54.9	50	16.7	18.4	17
Pulses, dry	4.8	3.1	3	14.3	11.3	12
Sugar, raw equivalent	5.2	9.3	13	20.3	21.8	24
Vegetable oils	3.2	6.9	11	4.6	7.2	10
Meat	8.6	20.3	40	3.8	4.2	6
Milk	3.5	6.8	9	36.8	53.5	63
All food (cal./day)	2020.0	2600.0	3040	2040.0	2220.0	2450
	Latin America + Caribbean			**Developed countries**		
Cereals	118.8	128.9	139	148.6	146.3	141
Cereals (all uses)	(224)	(260)	(296)	(583)*	(637)*	(633)*
Roots/tubers/plantains	101.2	70.3	67	87.0	73.9	70
Pulses, dry	14.2	10.8	11	3.6	2.9	3
Sugar, raw equivalent	39.3	43.3	43	40.6	37.5	37
Vegetable oils	6.9	12.3	16	11.6	16.2	19
Meat	33.3	39.4	49	63.3	80.5	87
Milk	83.3	92.0	96	188.4	199.1	198
All food (cal./day)	2500.0	2690.0	2950	3200.0	3400.0	3470
	ex-CPEs			**Other developed countries**		
Cereals	195.9	171.5	169	127.5	135.0	128
Cereals (all uses)	(686)*	(828)*	(720)*	(538)	(550)	(590)
Roots/tubers/plantains	121.6	93.2	89	71.4	65.3	61
Pulses, dry	3.8	2.3	2	3.5	3.2	3
Sugar, raw equivalent	40.3	46.2	45	40.7	33.5	34
Vegetable oils	7.3	10.8	12	13.4	18.7	22
Meat	49.9	71.6	70	69.2	84.6	94
Milk	189.0	178.9	176	188.1	208.1	208
All food (cal./day)	3310.0	3380.0	3380	3140.0	3410.0	3510

*Data and projections for the ex-CPEs are before revision of the ex-USSR's cereals production data from bunker to clean weight (see Table 3.17). Revised data for the ex-CPEs: 69/71 652 kg; 88/90 790 kg; 2010 693 kg.

Table 3.4   Per caput food supplies for direct human consumption (calories/day) and possible evolution of the incidence of chronic undernutrition

| | Per caput food supplies (cal/day) | | | Chronic undernutrition | | | | | |
| | | | | Percent of population | | | No. of persons (million) | | |
	1969/ 71	1988/ 90	2010	1969/ 71	1988/ 90	2010	1969/ 71	1988/ 90	2010
World	2430	2700	2860						
93 Developing countries	2120	2470	2730	36	20	11	941	781	637
Africa (sub-Sahara)	2140	2100	2170	35	37	32	94	175	296
Near East/North Africa	2380	3010	3120	24	8	6	42	24	29
East Asia	2020	2600	3040	44	16	4	506	258	77
South Asia	2040	2220	2450	34	24	12	245	265	195
Latin America + Carib.	2500	2690	2950	19	13	6	54	59	40
Developed countries	3200	3400	3470						
ex-CPEs	3310	3380	3380						
Other developed countries	3140	3410	3510						

expected to be made in raising per caput food supplies. Undernutrition will probably remain disturbingly widespread and still affect by year 2010 nearly one-third of the population (300 million people). Indeed, the region will take over from South Asia as the one with the highest number of persons chronically undernourished, no matter that its total population will be by then only about half as large as that of South Asia.

Things may turn out otherwise if sub-Saharan Africa were to lift itself out of quasi-perennial economic stagnation; and if its agriculture were to make even more progress than assessed as probable in this study. In the quest for policy responses to the deteriorating nutritional situation of sub-Saharan Africa, the agricultural assessment of this study indicates that there are certainly potentialities for higher growth than anticipated here in certain commodity sectors. But it would be far fetched to assume that they would materialize in an environment of nearly stagnant per caput incomes which constrain effective demand domestically, while the demand for the main agricultural exportables will also be sluggish. These conditions are hardly propitious for providing incentives to stimulate introduction of feasible technologies and investments in

Table 3.5 Major commodity groups in total gross agricultural production, 93 developing countries

Commodity group	Share % in total value of gross production 1988/90	Growth rates of production*				
		1961–70	1970–80	1980–90	1970–90	1988/ 90–2010
Cereals	30	4.1	3.0	2.8	3.1	2.0
Other basic food crops (roots/tubers, plantains, pulses)	9	2.5	1.5	1.6	1.3	1.7
Other food crops	27	3.1	3.5	3.7	3.7	2.8
Livestock	27	3.9	3.8	4.6	4.3	3.4
Non-food crops (beverages, raw materials)	7	2.7	1.3	3.1	2.6	2.2
Total	100	3.5	3.0	3.4	3.3	2.6

*The growth rates of the historical period are obtained taking all the annual data into account (by statistically fitting exponential curves in the annual data as is the standard practice). It is, therefore, possible for the 20-year growth rate to result as lower or higher than the growth rates of the two component 10-year periods, or their average.

resource improvements that would cause this productive potential to express itself. And stagnant economies will continue to constrain the resources of the public sector for investing in agriculture to boost its production performance (e.g. infrastructure, research, extension, education, etc.). There is certainly scope for more optimistic outcomes in food and agriculture if the widespread policy reforms currently being implemented and the external environment, including resource flows, were to lead to faster overall development than indicated by the assessments of other organizations (see Table 3.1).

3.5 THE DEVELOPING COUNTRIES: PROSPECTS BY MAJOR COMMODITY GROUPS

Differing prospects for the individual commodity sectors

As noted, the prospective developments for the entire agricultural sector reflect the diverging growth prospects for the different commodity sectors. These are summarized in Table 3.5, while the prospects for each major sector are discussed in the subsequent subsections.

Table 3.6 All cereals, 93 developing countries (including rice in milled form)*

A. All 93 countries

Year	Million tonnes				kg/caput				Growth rate of production (% p.a.)	
	Produc-tion	Consump-tion	Net imports†	Self-sufficiency (%)	Consumption				Period	Produc-tion
					Produc-tion	All uses	Direct food	Other uses		
1961/63	350	357	16 (18)	98	165	169	131	38	60-70	4.0
1969/71	480	492	17 (20)	98	186	190	145	45	70-80	3.1
1979/81	650	709	59 (67)	92	201	220	162	58	80-90	2.8
1988/90	845	918	80 (90)	92	216	235	170	65	88/90-2010	2.1
2010	1314	1462	148 (162)	90	228	254	173	81		

B. By region

	Demand				Produc-tion	Net balance†	SSR (%)	Growth rates		
	Food		Feed	Total use				Period	Demand	Produc-tion
	Per caput (kg)	Total								
		........(million tonnes)........								
Sub-Saharan Africa										
1969/71	115	31	1	37	36	−3	97.4	61–90	2.6	1.9
1979/81	113	40	2	48	41	−8	85.5	70–90	2.9	2.1
1988/90	114	54	2	63	54	−8	86.4	80–90	3.2	3.4
2010	121	110	4	128	109	−19	85.5	88/90–2010	3.4	3.4

Near East/North Africa									
1969/71	183	33	10	54	46	−6 (−7)	86.9	61–90	4.0	2.4
1979/81	203	47	19	80	58	−23 (−24)	72.6	70–90	4.3	2.4
1988/90	213	63	32	112	73	−38 (−39)	65.4	80–90	3.9	2.9
2010	210	103	64	190	119	−71 (−72)	62.7	88/90–2010	2.6	2.3
East Asia (incl. China)									
1969/71	151	173	21	220	216	−6 (−9)	98.2	61–90	4.1	4.0
1979/81	181	252	49	334	316	−19 (−25)	94.5	70–90	3.7	3.6
1988/90	201	320	74	435	419	−20 (−27)	96.2	80–90	2.7	3.1
2010	206	424	176	657	635	−22 (−35)	96.7	88/90–2010	2.0	2.0
South Asia									
1969/71	148	106	1	119	116	−5	97.3	61–90	2.8	3.0
1979/81	153	136	2	154	148	−1	96.0	70–90	2.7	3.0
1988/90	156	172	2	196	200	−5	102.0	80–90	2.8	3.0
2010	163	271	4	302	292	−10	96.3	88/90–2010	2.1	1.8
Lat. America + Carib									
1969/71	119	33	22	63	66	+3	104.9	61–90	3.6	2.9
1979/81	128	46	38	94	87	−8	92.9	70–90	3.2	2.4
1988/90	129	56	45	113	99	−10	87.6	80–90	1.8	0.8
2010	139	87	79	184	159	−25 (−26)	86.5	88/90–2010	2.4	2.3

*For an interface of these projections with those for the developed countries and the world total see Table 3.17.

†Numbers in parentheses are the net imports of *all* developing countries, i.e. including those not in the group of 93, some of which are sizeable importers though minor producers. For imports and exports of the developing countries see text.

Cereals in the developing countries

Overall prospects

The trend towards lower growth rates in cereals production is expected to continue (Table 3.6). Notwithstanding this decline, the growth of production should continue to be above that of population in the future. Thus production per caput would continue to grow, but at a slower rate than in the past. For example, per caput production has increased by 30 kg in the last 20 years. It may increase by only 12 kg in the next 20 years. This slowdown is the result of a mix of positive and negative factors. On the positive side, the per caput demand for direct food uses is already at relatively high levels in some countries and demand tends to shift to other food products. This is the case of rice in some East Asian countries. On the negative side, other countries with still low levels of per caput consumption levels may not make much progress because of a combination of low growth in per caput incomes and constraints on increasing production and/or imports. In parallel, the growth of per caput consumption for all uses (+ 19 kg) is likely to exceed that of per caput production (+ 12 kg) with the difference to be met by increasing net imports (Table 3.6). Much of the additional consumption per caput will be for non-food uses, essentially for feed. This would reflect the continued growth in the production and consumption of livestock products (see below).

The result of these possible developments in production and consumption is that the net cereals import requirements of the developing countries would continue to grow, though slowly, more like the path of the 1980s than the explosive one of the 1970s. Thus, net cereals imports may grow from the 90 million tonnes of 1988/90 to some 160 million tonnes[3] in year 2010 and the aggregate cereal self-sufficiency ratio may decline a little to 90 percent by year 2010 (Table 3.6). About one-half of the total increment would be for the Near East/North Africa region and the balance would be mostly for Latin America and sub-Saharan Africa, and only a minor part for South Asia and East Asia, assuming China (Mainland) would continue to be only a small net importer.

The prospect that sub-Saharan Africa's net cereal imports may more than double to nearly 20 million tonnes may be viewed with alarm given the region's difficult balance of payments situation and heavy dependence on food aid. This possible outcome suggests a continued and possibly expanded role for food aid in the future. Still, the region's net import requirements are, and will likely remain, a small part of the total cereals deficit of the developing countries. They are also entirely inadequate for its own needs, as even if its production increased by 3.4 percent p.a. (as assessed by this study) this rate would still be grossly inadequate to raise more than marginally the very low consumption levels. Thus, while it is a matter of concern that the region's net cereal imports may more than double in the next 20 years, the real concern should be how to define a superior outcome for the region, one that would be composed of higher growth of agriculture (though not necessarily of cereals), exports and overall incomes and would lead to higher demand despite, most likely, even higher net food imports.[4]

Production perspectives for the individual cereals

Much of the projected slowdown in total cereals production of the developing countries is due to the prospective developments in wheat and rice, while production growth of coarse grains could be somewhat faster than in the past, though not fast enough to compensate for the slower growth of wheat and rice. This is partly because the bulk of their wheat and rice crops is produced in the two land-scarce regions of Asia and the Near East/North Africa (see Chapter 4); and partly because the demand for wheat and rice is likely to be less buoyant than that for coarse grains for animal feed.

It follows that the scope for increasing production of wheat and rice in the developing countries through area expansion is much more limited than that for coarse grains which have a larger weight in total cereals production in the less land-scarce regions of Latin America and sub-Saharan Africa. This leaves production growth of wheat and rice in the developing countries to depend mainly, and more than in the past, on the growth of yields. In this context, the quantum leaps of yields of these two crops, characteristic of the heyday of the spread of the green revolution, are unlikely to be repeated at the same rate in the future. A slowdown in yield growth in the developing countries (excluding China) is foreseen for both wheat and rice; from an average growth rate of 2.8 percent p.a. in the last 20 years to 1.6 percent p.a. in the next 20 years for wheat; and from 2.3 percent p.a. to 1.5 percent p.a. for rice. Why this would be so is explained in more detail in Chapter 4 where the issue is analysed in terms of the individual agroecological zones, ranging from semi-arid to fully irrigated. A preview of the main parameters of cereals production in the developing countries, excluding China[5], is given in Table 3.7. More complete data and projections for the individual cereals in a world context are given in the Annex.

Food and feed uses of cereals

Total domestic use of cereals in the developing countries in 1988/90 amounted to 930 million tonnes (with rice included in milled terms) of which 90 million tonnes came from net imports from the developed countries. *Direct food* consumption absorbed some 670 million tonnes (72 percent of the total, but with wide regional differences, e.g. from 88 percent in South Asia to 56 percent in Near East/North Africa). *Animal feed* accounted for some 160 million tonnes (17 percent of the total, again with wide regional differences), while the remaining 100 million tonnes were used for seed and industrial non-food uses or represent waste (post-harvest to retail).

As noted (Table 3.6), the *direct food* part of total cereals use is expected to grow at a rate just above that of population and show only marginal increases

Table 3.7   Area, yield and production by major cereal crop, developing countries
(excluding China)

	1969/71	1988/90	2010	Growth rates (% p.a.) 1970–90	1988/90–2010
Production (million tonnes)					
Wheat	67	132	205	3.8	2.1
Rice (paddy)	177	303	459	3.0	2.0
Maize	70	112	196	2.7	2.7
Other coarse grains	68	85	134	1.3	2.2
Total*	322	531	842	2.8	2.2
Yields (kg/ha)					
Wheat	1150	1900	2660	2.8	1.6
irrigated (49%)†		2450	3320		
sub-humid (23%)		1700	2140		
other rainfed (28%)		1260	1810		
Rice (paddy)	1855	2775	3810	2.3	1.5
irrigated (44%)		3690	5165		
naturally flooded (27%)		2415	3125		
other rainfed (29%)		1745	1950		
Maize	1300	1790	2470	1.8	1.5
irrigated (12%)		3690	4550		
sub-humid (48%)		1820	2570		
humid (24%)		1280	1735		
Other coarse grains	730	940	1210	1.3	1.2
dry semi-arid (27%)		480	660		
moist semi-arid (29%)		810	1045		
sub-humid (23%)		1210	1600		
humid (9%)		860	1085		
irrigated (9%)		2220	2750		
Total	1270	1910	2560	2.2	1.4
Area (harvested, million ha)					
Wheat	58	70	77	0.9	0.5
Rice	95	109	120	0.8	0.5
Maize	54	63	80	0.9	1.2
Other coarse grains	93	90	111	0	1.0
Total	300	331	389	0.6	0.8

*In the total production of cereals, rice is included in milled terms.
†Numbers in parentheses denote the area under the crop in each agroecological zone as percentage
of the total harvested area under the crop for 1988/90.

from the present per caput level of some 170 kg for the developing countries as
a whole. The perspectives for the different regions differ from each other and
often from their own historical experiences. *Sub-Saharan Africa* could see some
modest increase in per caput consumption, though this would be far from

adequate for nutritional improvement. Still, it would be an improvement following two decades of no growth, and indeed some decline, in per caput consumption. This "optimistic" outcome for the region depends essentially on the prospect that domestic cereals production (mostly of coarse grains) would grow in the next 20 years at 3.4 percent p.a. This is well above that achieved in the past 20 years (2.1 percent) but equal to that of the 1980s. Likewise, growth in per caput consumption in the *Latin America and Caribbean* region could resume after the no-growth decade of the 1980s. *South Asia* and *East Asia* (excluding China) could continue on their past slow and decelerating growth in the per caput direct consumption of cereals, though, as noted, for different reasons. The saturation/diversification effect will likely dominate the developments in East Asia, while the persistence of low incomes will be more decisive for South Asia.

The end-result is that demand for cereals for direct food consumption may grow at only 1.9 percent p.a. while that for feed would grow nearly twice as fast (3.7 percent p.a.). Why *feed use* should grow at this relatively high rate is explained below in the discussion of the prospective developments of livestock production and consumption. It is noted, however, that this growth rate is below that of earlier periods, when feed use of cereals started from a low base and reached peak growth rates of 7.2 percent p.a. in the 1970s, before slowing down to 3.6 percent p.a. in the 1980s.

Still, feed use may account in year 2010 for some 22 percent of total use, up from 17 percent today, which itself was up from 11 percent 20 years ago. This prospect for an ever increasing share of cereal supplies of the developing countries to be used in the feed sector may give rise to concern in a situation where many countries and population groups are still far from having met their needs for direct food consumption of cereals. The relevance, or otherwise, of this concern depends essentially on the extent to which it can be considered that the use of cereals for feed diverts supplies which would otherwise be accessible to the poor. The answer is less straightforward than would appear at first glance. Increased demand for feed has traditionally originated in the middle- and high-income countries in which both feed and direct food use of cereals have been increasing and the latter is near "satisfactory" levels, though not for all population groups. By contrast, the need to increase per caput direct food use of cereals is to be found predominantly in the low-income countries, where feed use of cereals accounts for only a tiny proportion of total availabilities.

Under these circumstances, the link between the two, often spatially separate, categories of demand (of the middle-income countries for both food and feed, of the poor countries for food) and eventual diversion from the latter to the former could occur, in the first instance, through the operation of the world markets. But the conditions for this to happen are rather stringent: either (a) global supplies are quasi-constant in which case the additional demand for feed would raise prices with the result that part of the demand of the poor for direct consumption would be priced out of the market, or (b) supplies can be increased but only at prices which are above those that would otherwise prevail.

The empirical evidence from the world markets shows that increasing supplies to meet increments in demand (partly originating in the feed sector) have been forthcoming at declining, or non-increasing, real prices, except for occasional short-term shocks. There is, therefore, a prima facie case that increasing demand for feed has not raised permanently world cereal prices. The key question is, of course, whether prices would have been even lower but for the demand for feed. On this, it is more difficult to have firm views. It is noted, however, that one of the key factors behind the declining, or non-increasing, real prices in the presence of increasing demand has been the lowering of production costs following the diffusion of productivity-increasing technology. It is possible that this technological progress was partly linked to the expansion of demand of cereals by the feed sector.

The above considerations seem to suggest that, in a global context, increasing use of cereals for feed was most likely met from additional supplies which were forthcoming at prices not significantly above those that would otherwise have prevailed. There is, however, reason to believe that such food–feed competition can be a significant factor in diverting supplies away from the poor at the level of individual countries or regions within countries. This would be the case of countries which face stringent production constraints and have little scope for increasing supplies through imports; or regions within countries which, because of local production constraints, low incomes or bottlenecks in transport, can be considered to be in a similar position. Such cases have characteristics closely resembling those of the "closed economy" paradigm. In such cases, an increase in the demand for feed would raise prices rather than production and reduce the direct food consumption of the poor.

The key issue is, of course, whether this process actually happens in poor countries in those conditions. It is noted that in the low-income regions of South Asia and sub-Saharan Africa the feed use of cereals remains a minuscule proportion of total use. Moreover, the projections show that this situation would not change much in the future, as nearly all their increases in demand for cereals would be for direct food purposes. Therefore, there is a prima facie case that the food–feed competition may not really be a significant factor in preventing progress towards raising the per caput direct consumption of cereals in the many countries with such consumption still at inadequate levels. If anything, the historical experience of many middle-income developing countries demonstrates that increases in the feed use of cereals occurred in parallel with increases in the per caput consumption for direct food purposes. The general underlying factor has been the growth in per caput incomes in combination with the growth of domestic cereals production and/or improvements in the import capacity. It is the absence of these factors that stands in the way of increasing per caput direct consumption in many low-

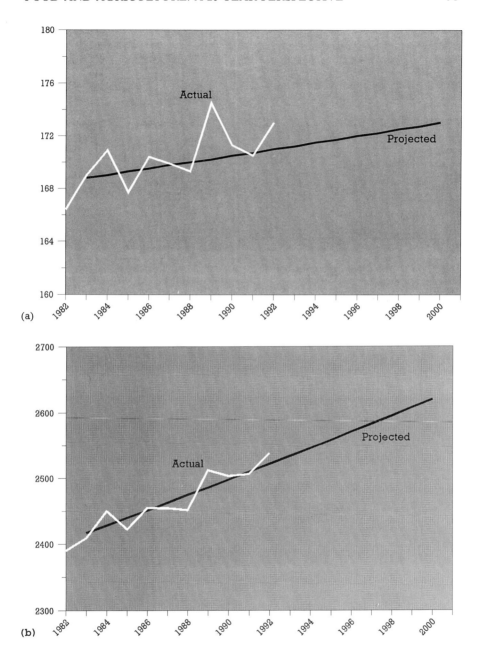

Figure 3.3   Comparison of projections to year 2000 made in 1985–86 with actual outcomes to 1992, developing countries. (a) Direct food consumption, cereals/caput/year (kg). (b) Direct food consumption, all products, calories/caput/day. (c) (*overleaf*) Net imports of cereals (million tonnes). Projections 1982/84–2000 from Alexandratos, 1988: 7, 64, 106.

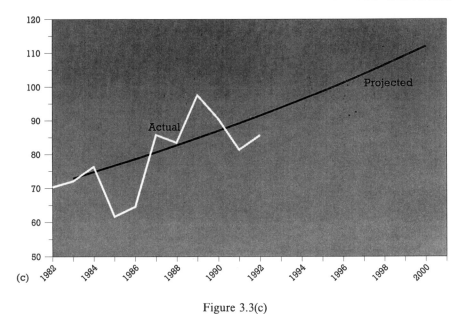

Figure 3.3(c)

income countries rather than the fact that feed use may be increasing in the middle- and high-income countries.

Record of earlier projections

The approach used here to project the likely evolution of the main food and agriculture variables is essentially the same one used in 1985–86 to make similar projections to 2000 (Alexandratos, 1988). The projections of that time for total food (calories) and cereals (direct food, net imports) of the developing countries for the period 1982/84–2000 (the three-year average 1982/84 was the base year) may be compared with the actual outcomes to 1992, i.e. half-way into the 17-year projection period. This comparison is made in Figures 3.3(a)–(c). Chapter 4 presents similar comparisons for areas, yields and production of cereals. By and large, the actual outcomes track fairly closely the projection trajectories of these variables for the developing countries as a whole. These comparisons do not constitute a validation of the projection method. They are offered here in response to the often asked question of how well the earlier projections have performed.

Livestock products in developing countries

Structural change in the food consumption of the developing countries towards more livestock products will continue with significant increases in per caput consumption of meat in all regions except South Asia and sub-Saharan Africa

Table 3.8 Meat production and consumption, 93 developing countries

	1969/71	1979/81	1988/90	2010	Growth rates (% p.a.)	
					1970–90	1988/90–2010
Consumption/demand						
Total (million tonnes)	27.0	42.2	64.0	143	4.8	3.9
Per caput (kg)	10.5	13.0	16.4	25	2.6	2.0
Production (million tonnes)	28.5	42.6	64.8	143	4.6	3.8
By species (million tonnes)						
Cattle	12.1	14.6	18.6	32	2.2	2.7
Sheep/goat	3.0	3.6	4.9	10	2.8	3.1
Pig	9.7	16.8	28.3	64	6.1	4.0
Poultry	3.7	7.6	12.9	37	7.0	5.1

Note: Meat is in carcass weight, excluding offals. Production is that of indigenous meat, with live animal exports counted as domestic meat production and live animals imports as meat imports, in their carcass weight equivalent.

(see Table 3.3). However, their per caput consumption of such products[6] will still be well below those of the high-income countries in 2010. For some developing countries, the consumption of livestock products may not advance even in the longer term future to the stage when it would match the consumption levels of the developed countries, for various reasons including those of ecology and culture. It is noted, however, that in some, though not all, high-income developing countries, meat consumption levels comparable to or a little below those of the developed regions have been attained, e.g. Taiwan (province of China), Singapore, Hong Kong, the United Arab Emirates and Kuwait. Another factor in considering the livestock sector developments is that in some developing countries, including very poor ones, livestock products (mostly milk) are staples, not luxury foods, e.g. in the predominantly pastoral societies of the Sahel.

For *meat* there may be a slowdown in the growth of per caput demand (from 2.6 percent p.a. in 1970–90 to 2.0 percent p.a. in 1988/90–2010), total demand (from 4.8 percent to 3.9 percent) and production (from 4.6 percent to 3.8 percent). The relevant data are shown in Table 3.8. The slowdown would occur in the regions of Near East/North Africa and East Asia. Prospective developments in China decisively influence the total outcome because this country accounts for 40 percent of total meat consumption of the developing countries.[7] For the developing countries without China, the growth rate of per caput demand for meat is likely to be maintained at 1.1 percent p.a., while the growth of their total demand and production is somewhat less than in the past because of the lower growth of population. In addition, there might be a net

Table 3.9   Milk production and consumption, 93 developing countries

	1969/71	1979/81	1988/90	2010	Growth rates (% p.a.) 1970–90	Growth rates (% p.a.) 1988/90–2010
Consumption/demand						
Total (million tonnes)	84.7	122.8	164.0	273	3.7	2.5
Per caput food (kg)	27.4	32.1	35.9	42	1.7	0.7
Production (million tonnes)	78.0	107.3	147.3	248	3.5	2.5
Net trade (million tonnes)	−6.8	−16.2	−16.2	−26	5.8	2.2

Note: All data and projections are for all milk and dairy products in liquid whole milk equivalent. Consumption and trade of butter is not included in the dairy products but in the animal fats. This means that, for example, a country importing or consuming milk powder or cheese is shown as importing or consuming the liquid whole milk equivalent; but if it imports/exports only butter it is not shown as importing/exporting milk or dairy products.

import of about 1 million tonnes, mostly reflecting growing net imports into the Near East/North Africa region, partly offset by increased net exports from Latin America.

Concerning *the milk sector*, there is likely to be a drastic slowdown in the rate of increase in consumption. In contrast to the prospective developments for the meat sector, the slowdown is generalized and all regions may be affected. The reasons for these prospective developments are to be found in a combination of production constraints and reduced availabilities of exports at highly subsidized prices (including food aid) from the main exporting developed countries. It can be seen from the historical data in Table 3.9 that the slowdown has been under way already in the 1980s. In the decade of the 1970s, the growth of net imports provided one-quarter of the increment in consumption. By contrast, there was no growth in net imports in the 1980s, and only slow growth in such net imports is foreseen for the next 20 years.

The trade picture is likely to be dominated by: (a) the reduced scope for subsidized production and exports from the main developed exporting countries, a policy trend likely to hold in the future and be reinforced by developments following the conclusion of the Uruguay Round with consequent upward pressure in world market prices; and (b) the limited import capacity, already apparent in the 1980s, of those countries which fuelled the growth of consumption and imports in the 1970s. For example, in the Near East/North Africa region the growth of imports in the 1970s provided 55 percent of the growth of consumption and the region accounted for nearly 40 percent of the increment in total imports of the developing countries.

An additional factor in the slowdown of consumption and trade growth is the fact that in the fastest growing region, East Asia, the rapid economic growth will be less of a stimulus for increased milk consumption compared with other products because of the dietary habits of the region (presently consuming 6.5 kg per caput compared with 55 kg in the other developing countries). Finally, of particular concern is the prospect that production constraints would permit little growth in per caput consumption in the regions in which milk is a staple food for large parts of the poor population, e.g. in the pastoral societies of sub-Saharan Africa and, to a lesser extent, South Asia.

Implications for the cereals feed sector

As noted earlier, feed use of cereals in the developing countries is becoming an increasingly important component of total cereals use. It grew at 5.6 percent p.a. in the last 20 years (7.2 percent p.a. in the 1970s, 3.6 percent p.a. in the 1980s); this is nearly double the growth rate of the use of cereals in direct food consumption. This growth of the cereals feed use has been one of the main factors behind the rapid increase in the developing countries net cereal imports, particularly during the 1970s. With the exception of China, the middle-income developing countries account for the bulk of cereals feed use. This is, of course, related to their above-average per caput consumption of livestock products. China also shares this characteristic, even though the criterion of per caput income puts the country in the low-income group. Garnaut and Ma (1993) consider that the food consumption statistics of China indicate that the country's per caput income is much above the level given in the conventional national accounts data.[8]

The general mechanism which drives the growth of the cereals feed sector can be simply stated: growth of incomes increases demand for livestock products. The latter causes production to increase, with the bulk of the additional production in most countries coming from the pig and poultry sectors, as discussed in the preceding section. Unlike the case of the ruminant meat sector in the developing countries, substantial increases in pig and poultry meat, and to a lesser extent also dairy, depend heavily on the expansion of intensive and semi-intensive production systems (see Chapter 4). This general pattern would translate into the feed use of cereals growing faster than the volume of livestock output. In parallel, however, gains in feed use productivity and resort to non-grain products in concentrates would tend to attenuate this pattern. Moreover, this pattern is less pronounced or entirely absent in those countries (both developed and developing) with relatively high shares of the cattle/sheep sector in the total growth of livestock production and ample grassland and other non-grain feed resources.

The historical experience of comparative growth rates of cereals feed use and livestock output in the last 20 years (Table 3.10) broadly conforms to the above discussed considerations. In both sub-Saharan Africa and South Asia and to a lesser extent Latin America feed use of cereals grew at or below the growth rate of total livestock output, with the latter measured by aggregating the different

Table 3.10 Use of cereals and oilseed proteins for animal feed, developing countries

| | Quantities (million tonnes) | | | | Growth rates (% p.a.) | | | | | |
| | | | | | Cereal feed or oilseed proteins | | Lifestock output* | | Ratios of growth rates | |
	1969/71	1979/81	1988/90	2110	1970–90	1988/90–2010	1970–90	1988/90–2010	1970–90	1988/90–2010
All developing countries										
All cereals	56	114	160		5.6					
Maize	34	74	104		5.9					
Other coarse grains	16	28	36		4.5					
All cereals										
93 Developing countries	55	110	154	327	5.5	3.7	5.3	3.7	1.0	1.0
Sub-Saharan Africa	1.3	1.6	1.8	4	1.3	4.2	3.2	3.5	0.4	1.2
Near East/North Africa	10.0	18.6	31.6	64	7.2	3.5	4.6	3.7	1.6	0.9
East Asia	20.6	49.8	73.5	176	6.5	4.1	6.9	4.3	0.9	1.0
South Asia	1.2	1.8	2.4	4	3.2	2.7	4.5	3.2	0.7	0.8
Lat. America + Carib.	21.8	37.7	44.6	79	3.8	2.8	3.8	2.8	1.0	1.0
Others (non-study countries)	1.2	4.0	5.7							
Oilseed proteins†										
93 Developing countries	5	9	14	39	6.0	5.0	5.3	3.7	1.2	1.3

* Growth rates computed from a special, feed-specific volume index of livestock output taking into account the fact that in the developing countries much of the beef, mutton and milk production comes from non-grainfed animals, while pigmeat, poultrymeat and eggs come predominantly from grainfed production systems (for relevant data see Tyers and Anderson, 1922: 386). The feed-intensity weighted livestock production index is computed as follows: 0.3 (beef + mutton) + 0.1 (milk) + 1.0 (pig + poultry + eggs).

† Feed use of oilseeds and their meal/cakes converted to 100% crude protein equivalent, e.g. soybeans at 35.6% and soybeanmeal at 45%, cottonseed at 21.1% and 40%, respectively and so on.

livestock products using weights reflecting the differences in their cereals feed intensity (see footnote to Table 3.10). At the other extreme, the experiences of both Near East/North Africa and East Asia (excluding China) seem to conform to a pattern of increasing cereal feed intensity of their livestock sectors, though at varying rates in different periods.

Concerning the future, by and large, and unless there are special country-specific circumstances, the use of cereals feed is likely to grow at roughly the same rate as that of total livestock output, with the latter measured as indicated in the footnote to Table 3.10. This pattern may deviate in some cases from the historical relationships which have often been unstable over time. For example, the data for East Asia show that feed use grew twice as fast as livestock output in the 1970s but half as fast in the 1980s. Moreover, the historical statistics of feed use are often a poor guide to what may happen in the future because they are very unreliable for many developing countries.

The projections of cereals feed use are shown in Table 3.10, together with those of livestock production. A growth rate of 3.7 percent for the next 20 years is well below that of the last 20 years (5.5 percent p.a. in 1970–90) though equal to that of the 1980s. Still, feed use will continue to be the dynamic element in total use of cereals in the developing countries. Its share in total use would increase further, to some 22 percent by year 2010 compared with 17 percent at present and only 11 percent 20 years ago. Finally, it must be noted that cereals feed is only one component, and in many developing countries a minor component, of the (largely unknown) total balance between feed resources and livestock production (see Chapter 4).

Use of oilseed proteins in the livestock sector

The 93 developing countries of this study use oilseed proteins in their livestock production to the tune of some 14 million tonnes. In addition, they use some 1.2 million tonnes of fishmeal protein. Such use grew rapidly in the past two decades at a rate above that of livestock production (Table 3.10), though such growth has tended to slow down in the 1980s. A few developing countries account for some two-thirds of total consumption (China, India, Mexico, Turkey, Pakistan).

The livestock production projections presented here imply that feed consumption of oilseed proteins would continue to grow at rates somewhat above those of total livestock output. The implications of this growth in the livestock sector and the domestic consumption of oilseed proteins is that by the year 2010 the developing countries as a whole would have a net export surplus of oilmeal proteins below present levels. Their present production of oilseeds, after deduction for uses not contributing to actual or potential feed protein supplies (e.g. oilseeds consumed directly as food, seed, etc.), amounts to some 24 million tonnes, i.e. it exceeds their consumption as feed by some 10 million tonnes, in crude protein equivalent. By the year 2010, production (discussed later in this chapter) could reach some 46 million tonnes, leaving some 6 million tonnes for net exports. Latin America would continue to be a large net exporter but at the same time the import requirements of the other regions, particularly East Asia and the Near East/North Africa, would grow rapidly, thus reducing the net export availabilities of the developing countries as a whole.

Other basic food crops (roots, tubers, plantains, pulses)

Roots, tubers, plantains

This category of basic foods comprises a variety of products, the main ones being cassava, sweet potatoes, potatoes, yams, taro and plantains (hereafter referred to as "starchy foods"). For some of them, much of the production and apparent consumption is at the subsistence level. Moreover, they are produced and consumed predominantly in countries with poor statistical services and are subject to above-average rates of waste/post-harvest losses. For these reasons, the data on production and apparent consumption are subject to margins of error which are thought to be much larger than those for other food products.[9]

There are two dimensions of the starchy foods economy which seem to dominate all others when it comes to viewing developments in this sector: ecology and per caput incomes. In other words, these products are the mainstay of diets of poor people in countries characterized by ecological conditions (mainly the forest zones in the humid tropics) making them the predominant subsistence crops. It is precisely in the countries which combine both these characteristics and also have low per caput levels of aggregate food supplies that very high levels of apparent per caput food consumption of roots are encountered. There are 15 countries in sub-Saharan Africa (group "Africa–high" in Table 3.11) each with over 200 kg of apparent per caput consumption (in fresh product weight), but with some having levels of 350–400 kg (Ghana, Gabon, Congo, Uganda) and others above 400 kg (Zaire). Together these countries account for 55 percent of sub-Saharan Africa's population and they depend on roots for some 40 percent of their food supplies. Smaller country groups in both sub-Saharan Africa and the Latin America/Caribbean region have a "medium" level dependence on roots for their total food supplies (100–200 kg per caput) while all the other developing countries have consumption levels well below 100 kg (40 kg on the average), including some with very low levels (typically 5 kg) in sub-Saharan Africa itself, e.g. Mali, Somalia.

It is obvious that the high dependence of diets on these products in some of the poorest countries with high incidence of undernutrition makes this group of crops of prime importance in any assessment of future prospects in nutrition and of policies to improve it. It is also a reminder that the analysis of issues of food security, poverty and undernutrition should not be unduly limited to

Table 3.11   Roots, tubers and plantains (starchy foods) in the developing countries

	Groups with high/medium consumption of starchy foods				
	Africa		Latin America	Other developing countries	93 developing countries
	High*	Medium†	Medium†		
Per caput food supply of starchy foods (kg)					
1969/71	336	205	162	62	80
1988/90	332	174	135	40	63
2010	313	171	125	36	64
Per caput food supply of cereals (kg)					
1969/71	88	121	88	150	145
1988/90	87	119	96	179	170
2010	96	121	112	183	173
Per caput supplies, all food (calories/day)					
1988/90	2160	2190	2280	2500	2470
% from starchy foods	40	21	13	4	6
% from cereals	34	49	36	63	61
2010	2230	2240	2540	2790	2730
% from starchy foods	36	20	11	3	6
% from cereals	35	47	38	58	56
Production growth rates, starchy foods					
1970–90 (% p.a.)	2.4	1.8	1.7	1.0	1.4
1988/90–2010 (% p.a.)	2.8	3.3	1.9	0.9	1.6
Population (million)					
1988/90	262	33	88	3522	3905
2010	509	65	132	5051	5758

*Over 200 kg/caput in 1988/90 (fresh product weight): Benin, Ghana, Côte d'Ivoire, Nigeria, Togo, Angola, Central African Republic, Congo, Gabon, Zaire, Burundi, Mozambique, Rwanda, Tanzania, Uganda.
†100–200 kg/caput, Africa: Guinea, Liberia, Cameroon, Madagascar, Namibia; Latin America/ Caribbean: Dom. Republic, Haiti, Bolivia, Colombia, Ecuador, Paraguay, Peru.

cereals. At the same time, the poor quality of data on this very important source of energy in diets makes difficult the analysis of the food problem and the assessment of the future precisely in some of the countries in which the problem is most acute. The prospects for roots in food consumption are, therefore, presented here for these specific root-dependent countries together with those for cereals and for total food (Table 3.11).

The prospects are that per caput food consumption of starchy foods will continue to decline, partly as a result of continuing urbanization (though its effect will be mainly limited to cassava and yams), but no radical structural change of diets away from the high dependence on these products is to be expected in the African countries. The little growth in per caput incomes and the difficulties in increasing supplies of cereals will permit little improvement in total food supplies and will ensure continued high dependence of the diets of these countries on starchy foods. In the end, countries accounting for over 50 percent of sub-Saharan Africa's population would still depend on starchy foods for some 36 percent of their total food supplies (calories). This assumes that production can grow at rates somewhat above those of the last 20 years, an outcome which is considered feasible. Overall, however, the production of this group of crops in the developing countries as a whole is expected to grow at a rate below that of the population, a pattern well established in the historical period. This reflects essentially the change in diets resulting from urbanization which tends to favour foods (e.g. cereals) which are often cheaper and more convenient for preparation and consumption in urban environments compared with the rural ones. Given the potential for increasing production, policies to stimulate consumption could be important for improving nutrition. Such policies may include research to develop new root-based food products, such as composite flours, noodles, chips, dehydrated products, etc.

No discussion of the starchy roots sector of the developing countries would be complete without a brief examination of the future of *cassava exports*, mainly to Europe for animal feed. Such exports, almost entirely from Thailand and to a lesser extent Indonesia, experienced phenomenal growth in the past, from 5 million tonnes in 1969/71 to 24 million tonnes in 1988/90 in net terms (fresh product equivalent of dried cassava) at which level they peaked in the late 1980s, reflecting, *inter alia*, trade agreements to restrain the growth of such exports. The reasons for these historical developments are well known: imported cassava and oilmeals captured an increasing share of the European feed market substituting for cereals which were uncompetitively priced by the domestic support policies. The situation may change in the future following policy reforms in Europe which would lower cereal prices to the feed sector. It is, therefore, possible that the net exports of the developing countries would decline from the present level of 24 million tonnes, though there is still scope for adjustments in the export price of cassava to retain competitiveness with the lower-priced cereals in the feed markets.

Pulses (dry)

This group of basic foods, often included in the statistics together with cereals in the foodgrains group, comprises a variety of products (beans, peas, chickpeas, lentils, etc.; soybeans and groundnuts are included in the oilseeds, not in pulses) which form an important component of the diet, particularly those of the low-income population groups, in many developing countries (FAO, 1981). Average per caput apparent food consumption of pulses in the developing countries as a whole is some 7.5 kg and it has been falling (it was 12 kg in the early 1960s). There is, however, still a large number of developing countries with relatively high levels (10–20 kg), e.g. Burkina Faso, Uganda, India, Nicaragua, Brazil, etc., and a few with very high ones, e.g. Rwanda, Burundi. With few exceptions (e.g. Brazil, Mexico), these countries have low per caput total food supplies and particularly of livestock products. Pulses are, therefore, an important source of protein, particularly in countries with low livestock/fish consumption levels. Thus, they provide 45 percent of total protein availabilities in Rwanda and Burundi, 25 percent in Uganda, 20 percent in Haiti and 14 percent in India.

The trend towards decline in per caput consumption of the developing countries as a whole was halted in the 1980s, particularly if China is excluded from the total. For the future it is estimated that per caput food consumption may remain at about present levels (9.0–9.5 kg on the average, or 7.5–8.0 kg if China is included). In practice, this is saying that the experience of the 1980s rather than that of the longer historical period may be representative of likely developments in the next 20 years. The dependence of the above-mentioned countries on pulses for their total protein supplies will remain relatively high. The growth rates of total demand (all uses) and production are shown in Table 3.12.

Oilcrops

The world produced in 1988/90 oilcrops which corresponded to some 71 million tonnes of oil equivalent. Actual production of vegetable oil was, however, smaller (60 million tonnes) because some oilseeds are also used for purposes other than oil extraction, e.g. direct food, feed, seed, etc. This latter figure includes some 2 million tonnes of vegetable oil produced from crops (maize) and residues (rice bran) not in the oilcrop category.

Table 3.12   Pulses in the developing countries: growth rates of production and demand (% p.a.)

	1970–80	1980–90	1988/90–2010
Demand (all uses)			
Developing countries	0.4	1.8	2.1
Developing countries (excl. China)	0.4	2.8	2.2
Production			
Developing countries	0.2	1.6	2.0
Developing countries (excl. China)	0.2	2.4	2.2

In the last 20 years the sector experienced above-average growth in production (4.0 percent p.a.) as well as radical structural change as concerns the shares in total production of the individual oilcrops and regions. The rapid expansion of oilpalm products in East Asia, soybeans in South America and sunflower seed and rapeseed mostly in Western Europe and to a lesser extent North America have been the most characteristic signs of this important structural change. Over the same period, some traditional oilcrops (mainly groundnuts, coconuts, sesame and cottonseed) have fallen behind as major sources of world vegetable oils production (their share – in oil equivalent – declined from 38 percent in 1969/71 to 26 percent in 1988/90).

There are a number of characteristics of the oilcrops sector that make its analysis and assessment of future prospects a rather complex undertaking. On the production side, there are both tree crops (oilpalm, coconut, olive trees) characterized by slow supply response to changing market conditions and annual ones with high rates of substitutability with other annual crops (e.g. soybeans for maize) and consequently high supply responses. Then, cottonseed is really a by-product of cotton lint production, whose production response is little affected by oil market conditions. Further, most oilcrops produce joint products of oil and protein meals for livestock.

The demand side is characterized by the fact that for each of these joint products there is a high degree of substitutability in consumption. In addition, the use of vegetable oils in non-food industrial uses (paints, detergents) in which they compete with petroleum products as well as the competition in the food markets with animal fats further complicate the picture; and so does the competition of oilmeals with fishmeal and other protein crops in the animal feeds sector. The possibility of substituting in consumption one product for another has been increasing over time with the evolution of technology. The practical implication is that market forces ensure that prices of the different oils tend to move in unison, and so do the prices of the different oilmeals. At the same time the prices of the joint products of any particular oilseed (e.g. of soybean oil and meal) may follow diverging paths (see World Bank, 1993c).

Oils and fats are a food commodity with high income elasticity of demand in the developing countries. Their per caput consumption of vegetable oils (including both oil proper and the oil equivalent of oilcrop products consumed as food in forms other than oil, e.g. groundnuts as pulses) is around 8 kg, compared with some 16 kg in the developed countries. There is, therefore, considerable scope for further expansion of consumption, though future growth can be expected to be slower than that of the last 20 years when the developing countries had started from very low levels. The relevant projections are shown in Table 3.13.

The demand of the developing countries for oilseed proteins, the joint product of most oilcrops, is expected to grow faster than that for vegetable oils, reflecting the rapid growth of the livestock sector, as discussed earlier. This will

Table 3.13 Vegetable oils: summary data and projections, 93 developing countries (all oilcrops in oil equivalent)

| | Demand | | Total use | Production | Net balance | Growth rates (% p.a.) | | |
| | Food | | | | | | | |
	Per caput (kg)	Total	........................(million tonnes)........................			Period	Demand	Production
93 Developing countries								
1969/71	4.7	12.3	15.8	18.8	2.4			
1988/90	8.2	31.8	39.9	44.8	3.7	1970–90	5.3	4.8
2010	11.3	64.9	80.8	86.9	6.2	1988/90–2010	3.4	3.2

be a continuation and accentuation of the past pattern of the differentials in the
two growth rates, as follows (percent p.a.):

	1970–90	1988/90–2010
Vegetable oils	5.3	3.4
Oilseed proteins	6.0	5.0

As discussed in the section on livestock, these developments are likely to lead
to a situation whereby the developing countries will continue to be large and
growing net exporters of vegetable oils (oils plus the oil equivalent of net
oilseed exports) but their net surplus of oilseed proteins (from oilmeals and
oilmeal equivalent of oilseeds) would decline. Latin America would continue to
be a large net exporter, but the growth in the net deficits of East Asia and Near
East/North Africa will lead to shrinking net exports of oilseed proteins from
the developing countries as a whole. This possible outcome reflects also the
prospect that palm oil (whose production generates little oilmeal in the form of
the palm kernel cake) is likely to continue to expand its share in total oilcrop
production in the developing countries.

The further structural change in the oilseeds sector of the developing
countries that may occur in the future is depicted in Table 3.14. The past trend
for the share of soybeans to increase rapidly could come to a halt, though total
production would still more than double, down from a 12-fold increase during
the last 20 years when it had started from a very low base and expanded rapidly
in South America. The coconut sector will probably see little growth (1.3
percent p.a. and this is probably too optimistic). This reflects, among other
things, the increasing competition from palm kernel oil in the lauric oils
market. Most other oilseeds could maintain their shares at present levels, after
the decline they suffered in the past under the impact of the rapid expansion of
the palm oil and soybeans sectors.

Major agricultural exportables of the developing countries, in brief

Sugar

Sugar consumption in the developing countries as a whole is still well below
that of the developed countries, 18 kg and 37 kg respectively.[10] In the former,
the highest levels are in Latin America/Caribbean and Near East/N Africa
(43 kg and 29 kg, respectively) and lowest in sub-Saharan Africa and East Asia
(8–9 kg). In this latter region, particularly in China, alternative sweeteners
account for a good share of total sweetener consumption. It is foreseen that per
caput sugar consumption in the developing countries would grow by about 1
kg per decade to reach some 20 kg by year 2010, but with the above-indicated
wide regional differentials being maintained, though they would be somewhat
less pronounced than at present. Overall there would be a slowdown in the

Table 3.14 Production of oilcrops: 92 developing countries, excluding China (thousand tonnes, oil equivalent)

By major oilseed		Africa (Sub-Sahara)	Near East/ North Africa	East Asia (excl. China)	South Asia	Latin America + Caribbean	Total	Share by oilseed (%)
				By region				
Palm oil + palmkernel oil	1969/71	1 487		745		270	2 502	16
	1988/90	1 951		8 990	2	806	11 749	32
	2010	3 370		22 020	90	1 640	27 120	38
Soybeans	1969/71	16	3	181	2	372	574	4
	1988/90	46	66	475	314	5 959	6 859	19
	2010	120	200	750	500	12 790	14 360	20
Groundnuts	1969/71	1 562	33	394	1 745	401	4 136	27
	1988/90	1 377	51	617	2 580	228	4 852	13
	2010	2 120	100	1 430	4 830	580	9 060	13
Sunflower	1969/71	12	173		17	425	627	4
	1988/90	58	499	58	212	1 441	2 268	6
	2010	130	1 020	80	370	2 070	3 670	5
Sesame	1969/71	235	47	70	217	155	725	5
	1988/90	207	57	118	326	82	790	2
	2010	320	100	300	640	220	1 580	2
Coconuts	1969/71	138		1 814	912	250	3 114	20
	1988/90	207		3 015	1 262	335	4 820	13
	2010	410		3 390	1 860	480	6 140	9
Cottonseed	1969/71	225	385	13	493	440	1 556	10
	1988/90	273	369	19	1 009	470	2 150	6
	2010	560	700	50	2 310	580	4 200	6
Other	1969/71	192	341	45	992	641	2 211	14
	1988/90	226	549	81	2 104	450	3 409	9
	2010	280	880	120	3 630	960	5 870	8
Total	1969/71	3 867	982	3 263	4 377	2 954	15 445	100
	1988/90	4 345	1 591	13 383	7 808	9 771	36 900	100
	2010	7 310	3 000	28 140	14 230	19 320	72 000	100
Share by regions (%)	1969/71	25	6	21	29	19	100	
	1988/90	12	4	36	21	26	100	
	2010	10	4	39	20	27	100	

growth rate of food consumption of sugar in the developing countries compared with the past (2.5 percent p.a. for 1988/90–2010, compared with 4.1 percent p.a. in the 1970s and 2.8 percent p.a. in the 1980s).

Sugar is a major export commodity of many developing countries. As a whole, the developing countries are large but declining net exporters. Part of the overall decline in their aggregate net exports has been due to the support and protection policies of most major developed countries (see Borrell and Duncan, 1992), and part is due to the emergence of many developing countries as growing net importers. The two developments are not, of course, independent of each other. The lower world market prices resulting from the access restrictions to the import markets of major OECD net importers, including the emergence of the EC as one of the largest net exporters, have probably contributed to the growth of imports into the developing countries. These developments can be appreciated from the data in Table 3.15.

It is expected that there will be continued strong growth of import requirements of the importing developing countries while the OECD area as a whole would continue to be a net exporter, unless policies were to change radically.[11] Developments in the former CPEs of Europe will play a role in limiting the growth of their aggregate consumption and import demand. Their per caput consumption may decline for some time before it recovers. It is likely that in the future more of their consumption will be supplied from domestic production. This means that although the exporting developing countries could expand further their exports, the developing countries as a whole would probably see their net exports to the developed countries decline further from the present level.

The sugar sector provides one example of the increasing role of the developing countries in world commodity markets formerly dominated by the developed countries on the import demand side. If this is considered by some to contain positive elements, it is being brought about for the wrong reasons, e.g. protectionism in the developed countries and to some extent the supply constraints of some developing countries which can ill afford sugar imports, e.g. some countries in sub-Saharan Africa. But it does contain positive elements to the extent that the growth of the import requirements of the developing countries reflects the improvements in incomes, increased consumption of sugar and comparative advantage of the exporting versus the importing countries.

Concerning production of sugarcane and beet in the developing countries, the growth rate (in raw sugar equivalent) is likely to be well below that of the past, perhaps 2.2 percent p.a. to 2010 compared with 3.5 percent p.a. in the last 20 years. This reflects partly the slowdown in the growth of their own consumption of sugar and the likely continued decline in net exports to the developed countries. But it also reflects the fact that the high production growth rates of the past were related to the sizeable production increases of sugarcane for ethanol production in the world's largest sugarcane producer,

Table 3.15 Sugar: net trade positions, five-year averages (million tonnes, raw sugar equivalent)

	1967/71	1977/81	1987/91
Developing countries	*9.3*	*8.5*	*3.7*
Net exporters (in 1987/91)	*12.1*	*17.0*	*15.3*
Cuba	5.5	6.7	6.8
Brazil	1.1	2.4	1.7
Thailand	0.1	1.1	2.5
Mauritius	0.6	0.6	0.6
Dominican Rep.	0.7	0.9	0.4
Others	4.1	5.3	3.3
Net importers (in 1987/91)	−2.8	−8.5	−11.6
China	−0.2	−1.0	−1.4
Algeria	−0.2	−0.5	−0.9
Egypt		−0.4	−0.7
Iran	−0.1	−0.6	−0.6
Iraq	−0.3	−0.5	−0.5
Korea, Rep.	−0.2	−0.4	−0.8
Malaysia	−0.3	−0.4	−0.6
Others	−1.5	−4.7	−6.1
Developed countries	*−9.1*	*−7.6*	*−2.8*
EC-12	−2.3	1.7	3.1
Other W Europe	−0.8	−0.4	−0.3
E Europe + Former USSR	−0.6	−4.7	−4.2
USA	−4.5	−4.0	−1.1
Canada	−0.9	−0.9	−0.8
Japan	−2.0	−2.3	−1.8
Australia	1.7	2.3	2.7
Others	0.3	0.7	−0.4

Brazil.[12] Such use is unlikely to continue to grow at anything like the past rate when it had started from nearly zero base. No further significant growth or some decline is a more probable outcome.

Tropical beverages (coffee, cocoa, tea)

Of these three major export commodities of the developing countries, coffee and cocoa share with some other commodities (e.g. rubber) the characteristic that they are produced exclusively in the developing countries and consumed mainly in the developed countries. The latter account for 71 percent of world consumption of coffee (the same as 20 years ago) and 83 percent of that of cocoa (87 percent in 1969/71). The case of tea is different as the developing countries account for 70 percent of world consumption. Moreover, the developing countries have increased significantly their share in world consumption of tea, but there has been only modest movement in this direction for cocoa and none for coffee.

It follows that for coffee and cocoa the production and export prospects will continue to be determined for a long time by developments in the consumption of the developed countries. The prospects are that there will be only modest further growth in per caput consumption in the main developed country markets for coffee and cocoa and, with the exception of the former CPEs, none for tea (Table 3.16). In parallel, the protracted recession and foreign exchange shortages in Eastern Europe and the former USSR are unlikely to permit growth in the per caput consumption of the region, notwithstanding the generally very low levels and considerable scope for increases. Under the circumstances the growth of production and exports of the developing countries is likely to be very slow. However, things may turn out otherwise if the reforms were to lead to a sizeable part of the population of the former CPEs moving towards consumption patterns more closely resembling those of Western Europe. Therefore, the consumption projections for these countries may be on the pessimistic side, though the recovery from the recent declines may be slow and protracted.

The preceding discussion indicates that the growth of production and exports of the developing countries in these commodities is likely to be slow and below that of the past 20 years. It is noted, however, that producing and exporting more of them, particularly coffee and cocoa, would not lead to commensurate gains in export earnings and welfare if prices were to continue falling. Indeed, the experience of recent years has amply demonstrated that falls in prices led to disastrous declines in the real value of export earnings. There have been periods when exporting more did not bring benefits to the producing exporting countries as a whole, though individual countries did benefit at a cost to others. Competition among exporting countries turned out to be a negative sum game for them (for the analytics of this issue, see Panagariya and Schiff, 1990; Mabbs-Zeno and Krissoff, 1990).

In many ways the origins of the problems facing the producers and exporters of these commodities are not unlike those facing the farmers in the developed countries. Both sell the bulk of their output to markets with low price and income elasticities of demand and slow growth of overall consumption. This means that competition among producers and gains in productivity tend to drive down the prices and benefit consumers rather than raise farm incomes. But farmers in developed countries can benefit from two features of their economies and societies which are not available to the producers in the developing countries. First, in the long-term their incomes cannot be suppressed too much even in the absence of support policies because of the high opportunity cost of their labour, given income-earning opportunities in other sectors (the supply of labour and other non-land factors of production to agriculture become more price-elastic at higher levels of per caput income, Johnson, D. G., 1994b); and second, the society and governments have the means to intervene, and do so, to support farm prices and ease the process of labour transfer to other sectors; and unlike the situation in many developing

Table 3.16   Tropical beverages and bananas

	Per caput consumption (kg)				Net trade ('000 tonnes)			
	Coffee	Cocoa	Tea*	Bananas	Coffee	Cocoa	Tea*	Bananas
Developed countries								
E Europe and								
former USSR								
1969/71	0.3	0.5	0.3	0.2	−120	−180	−33	−64
1988/90	0.6	0.7	0.9	0.4	−233	−260	−230	−165
2010	0.6	0.7	1.1	0.7	−270	−280	−290	−330
Others								
1969/71	4.0	1.4	0.7	7.0	−2940	−1040	−415	−4940
1988/90	4.6	1.9	0.6	8.7	−3910	−1610	−390	−7140
2010	5.3	2.5	0.6	10.8	−4980	−2330	−490	−9600
Developing countries								
1969/71	0.5	0.1	0.3	7.2	3185	1210	460	5300
1988/90	0.4	0.1	0.5	7.3	4240	2100	670	7500
2010	0.5	0.1	0.7	9.0	5250	2610	800	10000

	Growth rate of demand (% p.a.)				Growth rate of production (% p.a.)			
Developing countries								
1970–90	1.8	4.7	4.6	2.3	2.2	2.7	3.7	2.1
1988/90–2010	2.6	2.5	3.3	2.7	1.6	1.4	2.8	2.5
Developed countries								
1970–90	1.6	2.0	1.6	1.5			2.4	1.8
1988/90–2010	1.1	1.7	1.1	1.5			0.8	1.6

*Including mate.

countries, the virtually zero population growth eases the pressure for agriculture to provide more jobs. In addition, agriculture in the developed countries has been aided heavily to seek outlets in the more price elastic export markets for its produce, precisely those of the developing countries. Again, this is not an avenue open to the developing countries, at least not for these products.

Things are different for the farmers producing these commodities in many of the developing countries which depend on the very same markets as the farmers in the developed countries, though for different products. For them, the opportunity cost of their labour is often determined by their productivity in producing food, often for subsistence. So long as this is very low, as indeed it is in many developing countries with adverse resource endowments for food production and continuing rural population growth, the scope for profiting from increased export crop productivity and production remains limited (see Lewis, 1983). And, of course, the scope for government intervention to support incomes is also very limited, no matter that some developing countries have attempted to do so at heavy cost to their macroeconomic equilibria. To some

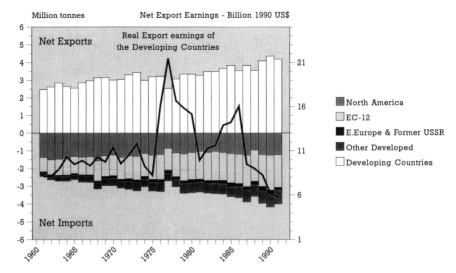

Figure 3.4   Net trade positions and real export earnings of coffee, 1961–91

extent, the above considerations can be seen as establishing an indirect link between agricultural policies in developed countries and the vicissitudes of these commodities. If such policies depress prices in world food markets, they contribute to lower returns to food production in the developing countries and tend to keep the opportunity cost of coffee and cocoa farm labour lower than it would be otherwise. (Part of the text in this subsection and Figures 3.4 and 3.5 are from Alexandratos *et al.*, 1994.)

Developments in export volumes and earnings are depicted in Figures 3.4 and 3.5. The real export earnings of the developing countries from coffee and cocoa were in 1991 at their lowest level of the last 30 years notwithstanding

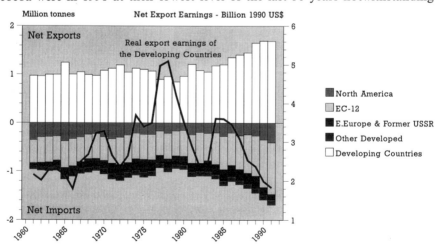

Figure 3.5   Net trade positions and real export earnings of cocoa, 1961–91

significant increases in the volumes exported. (Real net export earnings are the value of net exports of developing countries at current prices deflated by the unit value index of G5 exports of manufactures to the developing countries.) Therefore, the key question is what may happen to the real prices of these commodities and what policies can be put in place to halt and reverse price declines and deal with price volatility. Policy issues are discussed in Chapter 8. Here it is noted that world commodity prices, led by those for coffee, had turned sharply upwards from early 1994 onwards. Is this a signal of a permanent reversal of the price declines? The latest assessment of these trends by the World Bank states that "between 1993 and 2000, [real] coffee prices are projected to rise by nearly 50 percent, cocoa prices by almost 25 percent and tea prices by about 6 percent. Despite these rebounds, the index of beverage prices will likely not reach its 1989 level and will be less than half its peak in 1984" (World Bank, 1994a: 15).

Bananas

The developing countries produced some 45 million tonnes of bananas in 1988/90 and their net exports to the developed countries were 7.5 million tonnes. The bulk of production and exports is in the Latin America/Caribbean region which accounts for 80–85 percent of all exports. There is still scope for growth in consumption in the developed countries, in particular the ex-CPEs where per caput consumption levels are under 0.5 kg compared with nearly 9.0 kg in the other developed countries. Only a small part of this growth potential may materialize and imports into the ex-CPEs may only double. The developing countries' exports will continue to depend almost entirely on the growth of consumption in the other developed countries which account for 90 percent of world imports. The prospects for further growth in their consumption are fair and their net imports may grow from 7.1 million tonnes at present to 9.6 million tonnes by the year 2010, a growth rate of 1.5 percent p.a., the same as that of the last 20 years.

Per caput consumption should also continue to grow in the developing countries themselves, particularly in the fast growing region of East Asia. It is likely that the importing developing countries will increase their imports at a high rate and their share in world imports may more than double from the present level of about 7 percent. This notwithstanding, the high dependence of the developing exporting countries on the import markets of the Western developed countries and on policies affecting such imports will remain overwhelming.

Cotton

Trading patterns in the major non-food agricultural commodities have been undergoing significant changes. These have been due to the diverging trends in consumption of the developed and the developing countries, as well as to changes in the trade of manufactures based on these commodities.

Developments in the cotton sector illustrate this process of structural change. The developed countries used to be by far the major importers of cotton 20 years ago with a share of 75 percent of world gross imports and net imports from the developing countries of 1.6 million tonnes. But their mill consumption remained virtually static (about 6.5 million tonnes) while their production of cotton continued to increase. The result has been that in recent years they turned into small net exporters.

In parallel, mill consumption in the developing countries grew fairly rapidly (3.5 percent p.a. in 1970–90). Their production of raw cotton grew at a slower rate (2.4 percent p.a.) and their sizeable net exports of raw cotton turned into small net imports. The above discussion refers to production, mill consumption and trade of raw cotton. The underlying developments in the final consumption and trade of cotton manufactures have been different and have been instrumental in causing the diverging trends in mill consumption and trade of the raw material.[13] Part of the trends in raw cotton reflects the rapid expansion of exports of cotton manufactures from the developing to the developed countries following the migration of the textiles industry from the latter to the former. This process would have been even more pronounced but for the restrictions in the trade of textiles (see Hamilton, 1990).

The above-described trends would continue, with world production and consumption of cotton growing at about 2.6 percent p.a., to some 31 million tonnes by the year 2010. Although production would continue to rise also in the developed countries and they would still process about 8 million tonnes, mill consumption will be increasingly concentrated in the developing countries, doubling to some 23 million tonnes by the year 2010. The end-result would be that the developing countries would produce some 20 million tonnes but nonetheless would turn into major net importers of raw cotton of about 2.5 million tonnes (and growing exporters of cotton goods). Cotton thus provides an example of transformation of world industrial and trading relationships largely along the lines dictated by comparative advantage (see, for example, Anderson, 1992). To an increasing extent raw materials from the developed (and developing) countries are being combined with low-cost labour from the developing countries to generate industrial development in the latter and cheaper labour-intensive manufactures everywhere.

Natural rubber

Natural rubber, produced exclusively in the developing countries, may be following a similar path as cotton. However, for natural rubber the pace of change is likely to be much slower. The main underpinning factor for rubber would be somewhat different, namely the rapid consumption growth of rubber manufactures in the developing countries.

The share of natural rubber in world consumption of all rubber (natural and synthetic) had fallen to an all-time low of about 30 percent in 1980, but has since recovered to 36 percent, mostly due to the spread in the use of radial tyres and faster consumption growth in developing countries. This trend is expected to continue, and world consumption of natural rubber may grow at 2.6 percent p.a. over the time horizon of the study, nearly the same growth rate as in the last 20 years. By the year 2010 the developing countries (mainly in Asia) may have overtaken the developed countries as the major industrial consumer of natural rubber, with their share rising to 55 percent of world consumption, up from 43 percent in 1988/90 and 24 percent 20 years ago. These developments will essentially reflect the faster growth of the automobile industry and use in the developing countries compared with the developed countries (tyre production accounts for some 50 percent of world natural rubber use.)

Unlike cotton, however, the developing countries will remain growing net exporters, as they are the only producers in the world and the consumption (and net imports) of the developed countries would continue to grow at about 1.3 percent p.a. (against 3.8 percent p.a. in the developing countries). World production of natural rubber may continue to grow at the average rate of the past 20 years (2.5 percent p.a.). Production in East Asia will continue to dominate world production, but its share in the projected world output of 8.6 million tonnes may decline from 84 percent at present to 77 percent in 2010. This would reflect the increasing scarcity of labour in Malaysia and of land in both Malaysia and Thailand. Faster expansion in Indonesia could turn the country into the largest world producer and help keep East Asia's share from declining substantially. These developments may give sub-Saharan Africa one of the few opportunities to expand its share in world exports of a major agricultural commodity, from the present 7 percent to about double that by the year 2010. There is, therefore, scope for significant recovery of rubber production in the main African producing countries (Cameroon, Côte d'Ivoire, Nigeria and Liberia).

Tobacco

Consumption of tobacco in the developed countries has been declining, both total and per caput, with the latter having fallen particularly in the 1980s, from 2.0 kg in 1979/81 to 1.5 kg at present. It may fall further to nearer 1 kg by year 2010. By contrast, the developing countries have been increasing their tobacco consumption relatively rapidly, with per caput consumption having reached 1.1 kg and perhaps growing to 1.4 kg by the year 2010. Growth has been fastest in East Asia. It is ironic that tobacco is the one consumption item in which the developing countries are catching up fast with the developed ones and are likely to surpass them in the future. Apparently, things must get worse (on the consumption and health side) before they can get better. There is perhaps a parallel here with the process of environmental degradation (noted in Chapter 2) which seems to accelerate in some cases at the early stages of development and people endeavour, and can afford, to adopt policies to control it only when development has advanced well beyond these early stages.

The end-result of these prospective developments is that in developed countries both consumption and production of tobacco would probably continue to decline by about 1.0 percent p.a. and their net imports from the developing countries could remain constant at just over 200 000 tonnes (they were 440 000 tonnes 10 years ago). World consumption will be increasingly shifting to the developing countries. They may account for about 80 percent of the world consumption in the year 2010, compared with 70 percent now and only 55 percent 10 years ago. This implies a growth rate in their consumption of 2.3 percent p.a. (lower than in the past) and, given likely stagnant net exports to the developed countries, a growth rate of 1.9 percent p.a. for their production (also lower than in the past).

Conclusions

The preceding review of possible developments, by major commodity sector, has endeavoured to document the extent to which the diverging trends for each of them are likely to shape the future for total food and agriculture. Perhaps the major conclusion is that the developing countries are likely to change from their traditional position of net exporters of agricultural (crop and livestock) products into net importers. This issue is discussed later in this chapter, after a brief review of likely developments in the major commodity sectors of the developed countries.

3.6 THE DEVELOPED COUNTRIES: PROSPECTIVE DEVELOPMENTS IN BRIEF

General considerations

The demographic and overall economic growth prospects of the developed countries, as well as those for aggregate food and agriculture, were presented in Sections 3.2 to 3.4 of this chapter. Possible developments for some commodities were also presented in connection with the discussion of the prospects for the major export commodities of the developing countries. What remains to be reviewed here are the prospects in developed countries for their other major commodity sectors. Cereals and livestock products account for 75 percent of their gross agricultural production, in value terms. Prospective developments in these two sectors are reviewed briefly in the rest of this section.[14]

It is noted that some of the forces that shaped the historical evolution of food and agriculture in most developed countries are likely to persist, e.g. slow growth in domestic consumption and trends towards higher productivity. But there are also major changes under way, or in prospect, which are likely to change the policy environment and affect the individual country groups in very different ways.

These changes have to do with the prospect that in the future there will be much less scope for trade-distorting policies or other interventions to determine the production, consumption and trade outcomes in food and agriculture. Such outcomes will be determined increasingly by market forces or, at least, by less trade-distorting policies. The first impetus in this direction comes from the systemic reforms under way in the ex-CPEs of Europe; and the second from the policy reforms in major Western countries, particularly in Europe. Additional impetus and consolidation of these reforms is provided by the conclusion of the Uruguay Round of Multilateral Trade Negotiations. It is noted, however, that the reforms currently under way, if fully implemented, would tend to generate results for the major temperate zone commodities in the direction and magnitude similar to those that would be generated from the application of the provisions of the agricultural part of the Final Act of the Uruguay Round. These questions are discussed further in Chapter 8, where it is noted that the findings of analytical work attempting to predict the impact of the reforms are subject to many uncertainties.

The cereals sector in the developed countries

The possible developments to year 2010 are summarized in Table 3.17 and interfaced with those for the developing countries. Details by individual cereals are shown in the annex to this chapter. As regards the growth of total use of cereals in the developed countries, the main factors that may play a major role include: (a) the prospect that following the policy reforms in Western Europe cereals will become more competitive with the largely imported cereal substitutes in the feed sector and the region's demand may grow again after the declines experienced in the 1980s; and (b) the possibility that total domestic use of cereals in the ex-CPEs of Europe will be in the future somewhat lower than in the pre-reform period mainly because of developments in the countries of the ex-USSR. For example, the average per caput consumption (all uses) of cereals in this sub-group is projected to be 690 kg compared with 790 kg in 1988/90 (in clean weight). These developments would be the result of slightly lower per caput food consumption of cereals and livestock products, little growth in livestock production, reductions in post-harvest losses, smaller quantities of cereals used as seed and less cereals feed used per unit of livestock output (for discussion, see Brooks, 1993; Johnson, D. G., 1993). In parallel, total use of cereals in North America should continue to grow at about 1 percent p.a. The possibility exists for growth to be somewhat faster if ethanol production in the USA were to exceed the 1.2 billion gallons (5.5 billion litres) expected to be produced mainly from maize in compliance with the requirements of the Clean Air Act of 1990 (House et al., 1993).

Table 3.17 Total cereals, possible developed country outcomes in a world context (million tonnes, with rice milled)

	Main net exporting regions				Other		ex-CPEs				
	Western Europe	North America	Oceania	Sub-total	Western countries	Eastern Europe	Former USSR*	Sub-total	All developed	All developing†	World
Production											
1969/71	143.6	243.3	14.6	401.5	21.2	55.1	168.9 (157.5)	224 (213)	647 (635)	482	1129 (1117)
1979/81	178.5	341.9	21.7	542.1	24.3	68.3	169.6 (158.3)	238 (227)	804 (793)	652	1457 (1445)
1988/90	206.4	313.5	22.9	542.8	22.8	80.5	204.0 (189.0)	285 (270)	850 (834)	847	1697 (1681)
1991/92§	206.8	365.5	22.8	595.1	17.5	72.0	(172.0)	(244)	(856)	889	(1745)
2010				679.0	31.0			(306)	1028 (1016)	1318	2346 (2334)
Total domestic use											
1969/71	166.3	191.6	6.3	364.2	35.5	58.0	169.2 (158)	227 (216)	627 (625)	498	1125 (1113)
1979/81	189.1	202.3	7.3	398.8	46.8	77.6	214.7 (204)	292 (282)	738 (728)	720	1458 (1448)
1988/90	182.6	227.3	8.6	418.5	52.0	81.3	238.8 (224)	320 (305)	791 (775)	931	1721 (1706)
2010	200	279	11	490.0	63.0	85	228 (216)	313 (301)	866 (854)	1480	2346 (2334)
Net trade											
1969/71	−23.8	49.6	9.0	34.8	−14.6	−3.0	5.4	2.4	22.5	−20.4	
1979/81	−11.4	129.4	14.7	132.6	−22.9	−9.2	−31.1	−40.3	69.4	−66.8	
1988/90	24.4	118.7	14.5	157.5	−28.6	−2.2	−34.2	−36.4	92.6	−90.0	
2010				189.0	−32.0			5.0	162	−162.0‡	

*Numbers in parentheses are revised data following the change in reporting cereals production in the former USSR from bunker weight (including foreign material and excess moisture) to clean weight (see Shend, 1993). The projections shown in parentheses have been roughly adjusted to reflect this change.

†Data and projections for *all* developing countries, i.e. including those not in the 93 countries covered individually in this study.

‡Net imports of all developing countries after deducting 30 million tonnes of net exports from the exporting developing from the net imports of the importing developing countries (see text).

§The latest production data for the two-year average 1991/92 are shown here to underline the recovery of North American production after the abnormally low level of 1988 due to drought.

Following these considerations, total demand for cereals in the developed countries is projected at 854 million tonnes, a growth rate of 0.5 percent p.a. for 1988/90–2010 (projection adjusted for the shift of the ex-USSR data from bunker to clean weight – numbers in parentheses in Table 3.17). In parallel, the developed countries must generate a net export surplus of 162 million tonnes for the developing countries, up from the 90 million tonnes in 1988/90. Their required production in the year 2010 is, therefore, 1016 million tonnes. Their latest three-year average annual production for 1990/92 was 873 million tonnes.[15] The implied growth rate for 1990/92–2010 is 0.8 percent p.a. A recent World Bank study comes to similar conclusions. It projects cereals production in the developed countries as a whole to grow at 0.7 percent p.a. between 1990 and 2010 (Mitchell and Ingco, 1993, Tables A3, A10 therein). A growth rate of 0.8 percent p.a. is well below the average growth rate of the preceding 20 years (1.4 percent p.a.), nearly equal to that of the 1980s and well above the negative growth rate registered after the production peak of the mid-1980s.

These possible developments in cereals production in the developed countries may contribute to put in proper perspective the role of environmental constraints in relation to the need for this group of countries to generate collectively growing export surpluses to meet the import requirements of the developing countries. This is asking the question whether such constraints may represent a serious obstacle to achieving a 0.8 percent p.a. growth rate in their cereals production at non-increasing real prices. The analyses conducted for this study, which concentrated predominantly on the developing countries and, in any case, did not attempt to predict future prices, do not permit giving any straightforward answer to this question. But some insights may be gained from the findings of other studies. The above-mentioned World Bank study, which projects to year 2010 world cereals production and consumption levels similar to those of this study, concludes that world market prices would continue to decline as per long-term trend (Mitchell and Ingco, 1993; see Brown and Goldin, 1992, for discussion of food price projections). This finding does not by itself shed light on the issue at hand, i.e. the role of environmental constraints and in particular what prices would be if environmental costs in the developed countries were internalized. But it does indicate that sufficient slack exists in their cereals sector to permit them to increase production at the above indicated growth rate at non-increasing prices.

More to the point, some studies on the USA, by far the world's largest cereals exporter, have explored the impact on future production of what is conceived to be the most important agricultural resource degradation factor, soil erosion. Crosson (1992) cites a number of studies which conclude that "over the next 100 years erosion in the country as a whole would reduce crop yields roughly 3 percent below what they would be in the absence of erosion". He concludes that "by comparison with yield gains expected from advances in technology, the 3 percent erosion-induced loss is trivial" (see also Crosson, 1991).

Returning to the generation of the export surplus for the developing countries, it is noted that it will probably be forthcoming from a combination of more production from the major exporting developed countries and reduction in net imports of the importing ones. That is, not all the increment in the net imports of the developing countries (some 70 million tonnes) will appear as additional demand to be supplied by the main developed exporting regions, i.e. North America, Western Europe and Oceania. This is because more than one-half of this increment will likely be compensated by the disappearance of the ex-CPEs as a major net importer and indeed its emergence as a net exporter of 5 million tonnes, the assumption retained here (see Table 3.17). Other studies consider that this region may emerge by the year 2010 as a much larger net exporter (Mitchell and Ingco, 1993; Johnson, D. G., 1993). Assuming the ex-CPEs region were to shift by year 2010 to a net export surplus of 5 million tonnes (all of it from the countries of East Europe), the net increment of foreign demand faced by the above-mentioned three main exporting developed regions may only be some 32 million tonnes.

This prospect indicates that for these exporting countries, the export markets are unlikely to play the dynamic role of the past, particularly of the 1970s, when export growth fuelled a good part of their production growth. In parallel, policy reforms under way and as foreseen in the commitments for the implementation of the provisions of the Uruguay Round, will make it more difficult for the exporting countries to capture market shares by means of subsidized exports. The net exports of Western Europe are unlikely to be in the future above the average level reached in the last late 1980s. It would then appear that the bulk of the increment in the combined net exports of the three major net exporting developed regions would accrue to North America and Australia.

Livestock sector prospects in the developed countries

There are two major factors that may influence outcomes in this sector. In the ex-CPEs of Europe, the reported per caput consumption of meat had reached levels which were not much below those encountered in other developed countries with much higher incomes. With the price reforms and falls in purchasing power, the per caput consumption is going through a process of decline which may not bottom out for some time. For the longer term, the somewhat optimistic assumption used here is that per caput consumption may just revert to near the reported pre-reform levels of about 70 kg, but with more poultry meat and less beef (Table 3.18). It is an optimistic assumption because even by 2010 the per caput incomes of the region are likely to be well below those of other countries with meat consumption at that level.

For the other developed countries as a whole, the only likely significant change is that further growth in the per caput consumption of meat will come from the poultry sector, while per caput consumption of pork could stabilize after the rapid expansion of the past, which took place mainly in Western Europe. In parallel, per caput consumption of beef would probably remain constant on the whole, with the main shifts comprising expanded consumption in Japan and a reduction in Oceania.

For the milk and dairy sector the most likely outcome is that per caput consumption levels would not change much (all dairy products, in fresh whole milk equivalent). In parallel, there is some scope for further growth of net exports to the developing countries (see Table 3.9) with most of these additional exports supplied (in a net sense) by Eastern Europe and Oceania.

3.7 POSSIBLE DEVELOPMENTS IN THE AGRICULTURAL TRADE BALANCE OF THE DEVELOPING COUNTRIES

Historical developments in the aggregate agricultural trade values of the developing countries are shown in Table 3.19. The quantity indices of total imports and exports indicate that between 1961/63 and 1988/90 imports in real terms increased by 284 percent and exports by only 146 percent. The net export balance in real terms was in 1988/90 less than one-third of its level 20 years earlier. There was a reversal of these trends in the 1980s because of the abrupt slowdown in the growth of the agricultural imports in the crisis decade of the 1980s.

The projected developments in import requirements and export availabilities of the major crop and livestock products indicate that it may not be long before the developing countries as a whole turn from net agricultural exporters to net importers. This would happen if values of exports and imports of the individual commodities were to change *pari passu* with the changes in volumes discussed in the preceding sections. There is nothing in the price trends that would suggest that the average price of the commodities for which they are net exporters would rise faster, or fall less, than that of the commodities for which they are and will remain net importers. As noted earlier, even the projected rises in real prices of beverages are unlikely to bring their price index back to the level of 1989. In parallel, real prices of dairy products and perhaps those of cereals, the main agricultural commodities imported by the developing countries, may not fall and could well rise in the context of policy reforms towards more trade liberalization.

Quantitative and qualitative elements pointing towards the likely turnaround of the developing countries from net agricultural exporters to net importers, are given in Table 3.20. The upper part of the table shows all the commodities with positive net balances in 1988/90, totalling $32 billion. The lower part does so for the commodities with negative net balances, summing up to $27 billion. The likely developments for the future indicate that the negative balances will grow much faster than the positive ones. These likely outcomes point firmly in the direction of the net agricultural trade balance of the

Table 3.18 Summary livestock sector data and projections, developed countries

	E Europe + former USSR			All other developed countries			Total		
	1969/71	1988/90	2010	1969/71	1988/90	2010	1969/71	1988/90	2010
Food/caput (kg)									
Beef	20	27	24	28	28	28	26	27	27
Mutton	3	3	3	3	3	3	3	3	3
Pork	21	29	28	25	31	31	24	31	30
Poultry	5	12	15	12	23	32	10	19	27
Total	50	72	70	69	85	94	63	80	87
Milk	189	179	176	188	208	208	188	199	198
Feed use of cereals *(million tonnes)*	115*	189*	172*	271	294	355	386	482	527

Growth rates (% p.a.)	1970–90	1988/90–2010	1970–90	1988/90–2010	1970–90	1988/90–2010
Feed use of cereals	2.2	−0.4	0.7	0.9	1.2	0.4
Livestock production†	2.1	0.3	1.7	0.8	1.8	0.7

*Before revision for conversion of the ex-USSR cereals production data from bunker to clean weight (see Table 3.17). Revised numbers are as follows: 69/71 111; 88/90 182; 2010 166.
†Growth rates computed as indicated in Table 3.10.

Table 3.19   Values of agricultural trade,* all developing countries, 1961/63 to 1988/90

	1961/63	1969/71	1979/81	1988/90
Current prices ($ billion)				
Exports	13.8	18.7	69.8	87.5
Imports	7.2	10.7	66.0	82.5
Net balance	6.6	8.0	3.8	5.0
Index of volume (1979/81 = 100)				
Exports	55.3	69.7	100	135.8
Imports	34.1	43.5	100	131.0
Implied values at 1988/90 Prices ($ billion)				
Exports	35.6	44.9	64.4	87.5
Imports	21.5	27.4	63.0	82.5
Net balance	14.1	17.5	1.4	5.0

* All crop and livestock products, including both primary and most processed products, but not manufactures based on agricultural raw materials, e.g. textiles or leather goods.

developing countries (crop and livestock products) turning from positive to negative.

As noted in the earlier discussion, some of the increases in net imports of the developing countries, particularly of raw materials, are likely to be more than compensated by growth in the positive net balance of trade in manufactures based on these products. And part of the increased net imports of cereals and livestock products reflect developments in those developing countries which could finance them with export earnings from other sectors. However, these prospective developments are likely to be a heavy burden for those countries which must continue to finance growing food imports from export earnings which are unlikely to be forthcoming at the required rate from other export sectors.

3.8 ISSUES OF FOOD–POPULATION BALANCE BEYOND THE YEAR 2010

General considerations

Concerns are often expressed about the possible adverse developments in the longer-term world food–population balance in the face of: (a) ever increasing population; (b) the need for food supplies to increase faster than population in order to ensure growth in per caput supplies of the countries and population groups with very low and nutritionally inadequate levels; (c) finite agricultural resources, meaning ever declining per caput land and water quantities; (d) degradation of the productive potential of these resources and other adverse environmental effects associated with the process of an expanding and more intensive agriculture; and (e) uncertainties regarding the progress of technology and its potential to generate further growth in yields.

Addressing these issues requires a notion of the magnitudes involved, above all what is the total agricultural production which would be required to ensure that, by a given future year, all countries will have per caput food supplies compatible with an "acceptable" degree of food security. Moreover, estimates of such future magnitudes are needed at a fairly disaggregated level in order to capture the essence of the problem which is the need to increase food supplies in the countries and regions with high population growth rates and low levels of per caput food supplies at present or projected for 2010. This is the subject of this section.

The time horizon of this study is far too short for addressing how the longer term food–population balance may evolve. But some of the findings may help put the examination of the relevant issues in a proper perspective.

Relevant insights from the analyses to 2010

Slowdown in the growth rate of agriculture

The main relevant finding is that world demand for the products of agriculture may grow in the future less rapidly than in the past. The reasons why this may be so are: (a) the progressive slowdown in world population growth; and (b) the fact that a higher proportion of world population is today well fed compared with the past, thus allowing for less scope for further increases in their per caput consumption.

In addition to these two universal forces, the period to 2010 exhibits some particular characteristics which play a role in the projected demand slowdown at the world level. These are: (a) the potential savings from more efficient utilization of total food supplies in the ex-CPEs; (b) the prospect that the most populous region of the world, East Asia, would not need to maintain in the future the very fast growth of consumption it experienced in the last 20 years; and (c) many countries and population groups with low per caput food consumption levels are unlikely to experience sufficient improvements in incomes that would allow them to increase their demand as fast as required for achieving nutritionally adequate levels of food consumption by the year 2010.

Since at the world level demand equals production, the latter's growth rate is also required to be lower in the future compared with the past (1.8 percent p.a. in 1988/90–2010, compared with 3.0 percent in the 1960s, 2.3 percent in the 1970s and 2.0 percent in 1980–92). The projected growth rate in world production is disaggregated as follows: from 1.4 percent p.a. in 1970–90 to 0.7 percent p.a. in 1988/90–2010 for the developed countries as a whole; and from 3.3 to 2.6 percent for the developing countries. It was concluded that for the developed countries, a growth rate half as large as that of the past is unlikely to put much strain on their productive potential; indeed there exists the possibility that production growth would tend to run ahead of that of effective demand

Table 3.20 Developing countries, probable evolution of net agricultural trade balance

	1988/90			Likely changes in net balance, in real terms, 1988/90–2010
	Exports (million $)	Imports (million $)	Net balance (million $)	
Coffee	8 110	566	7 544	+24%
Oilseeds, veg. oils, oilmeals	13 416	9 776	3 640	+50%
Sugar	7 636	4 392	3 244	decline
Rubber	4 382	1 458	2 924	+35%
Cocoa	2 871	660	2 211	+24%
Citrus	2 257	598	1 659	+10–20%
Bananas	2 091	164	1 927	+33%
Other fruit	4 086	2 097	1 989	+100–150%
Vegetables	3 676	1 920	1 756	+50–70%
Tea	2 163	1 108	1 055	+20%
Spices	1 150	580	570	increase, modest
Cassava/other roots	1 424	525	899	−40%
Veg. fibres, excl. cotton	208	117	91	zero or decline
Tobacco	3 696	3 688	8	perhaps zero
Other products (unspecified)	10 495	7 997	2 498	
Sub-total	67 661	35 646	32 015	
Cereals	5 977	21 939	−15 962	+80%
Dairy products	457	5 805	−5 348	+55%
Meat, eggs	6 025	7 162	−1 137	+100% or more
Animal fats	40	729	−689	increase
Hides + skins, excl. leather prod.	745	2 292	−1 547	⎫
Wool, excl. wool textiles	504	1 421	−917	⎬ increase, probably large
Cotton, excl. cotton textiles	3 797	4 062	−265	⎪
Beverages (mostly alcoholic)	1 377	2 329	−952	⎭
Pulses	926	1 120	−194	+100%
Sub-total	19 848	46 859	−27 011	
Grand total	87 509	82 505	5 004	

unless policy reforms allowed a larger role for market forces, rather than support and protection policies, in determining the volumes of production. There is greater uncertainty on these issues for the developing countries. It is for this reason that a detailed evaluation and analysis of their potential for increasing production in terms of land and water resources and of growth in yields, all for the different agroecological zones, was undertaken. The conclusion is that a combination of land expansion not unlike that of the past (bringing into crop production some 90 million ha out of the existing 1.8 billion ha of land with rainfed crop production potential but not in agricultural uses), mainly in the land-abundant regions of sub-Saharan Africa and Latin

America/Caribbean, and significantly lower growth rates than in the past for irrigation, yields and fertilizer use would deliver the 2.6 percent p.a. growth rate of production.

Production growth, environmental resources and sustainability

Lack of comprehensive empirical data on the relationship between the growth of output and the generation of adverse environmental effects precludes the making of any definitive statements in answer to the question whether this growth rate of production and the associated land expansion and more intensification are compatible with sustainability. The conclusion is that more production means more pressures on environmental resources, because almost by definition production of more food for the growing population and for increasing per caput supplies cannot be achieved while leaving the environment intact; for example, converting natural habitat to grow crops or graze livestock or cementing it over to create rural infrastructure subtracts from the total stock of environmental resources. Whether this threatens sustainability, i.e. the potential for increasing food production to the level when all people are well fed and maintaining them at that level, notionally *ad infinitum*, is another question. In principle, it is possible to have reduction of environmental resources without threatening sustainability as defined above. It all depends on relative magnitudes, i.e. on the relative reduction of environmental resources that may come about in the process of transition from the present situation to a steady state of world agricultural production when maintenance of total output rather than continuous growth is required. The issue is whether this transition can be achieved at acceptable environmental costs, that is costs in terms of losses of environmental resources that would not threaten: (a) the sustainability of agricultural production *per se*; and (b) other essential functions of the environment that have an inherent life-support role (e.g. carbon cycle, biodiversity) or have a high ranking in society's hierarchies (e.g. amenity value or mere existence value).

In considering these aspects of long-term development, account must be taken of the fact that, in a fundamental sense, food production is different from other economic sectors because per caput consumption is subject to physiological limits. Once all people have "acceptable" levels of consumption (from the standpoint of nutrition, tastes, social considerations, etc.) world output would need only grow at the rate of population.[16] If the latter is ever declining and tends to zero, so will the growth rate of world agricultural output. The key issue from the standpoint of the world's food production capacity, the environment and sustainability is whether that state of no growth in world output can be reached with: (a) enough environment resources left intact for them to perform their essential life-support functions, notionally *ad infinitum*; and (b) whether even a non-growing world food output much larger than at present (passing through phases requiring it to be some 45 percent above present levels by 2010 and perhaps 75 percent by 2025, see below) is a sustainable proposition. For it will probably be produced with much more intensive production methods using significantly greater quantities of external inputs like fertilizer and pesticides whose supply is finite and/or whose continuous use at high intensities is environmentally damaging.

Those who consider that even present production levels and methods are unsustainable (e.g. Pimentel *et al.*, 1994; Ehrlich *et al.*, 1993) will find no comfort in the proposition that the world could eventually settle in a steady-state position as regards agricultural production. For the future steady-state output will need to be much larger than that of today. In parallel, the notion that the increased production can be achieved in a sustainable manner by carefully managed expansion and intensification, albeit at the cost of some reduction in environmental resources, must be supported by sufficient analysis of the path of getting from here to there. In both cases a notion of future output is necessary. The relevant estimates are attempted in what follows.

Supplies of agricultural products in the year 2025 for given per caput consumption levels

Future output will be the product of future population times per caput consumption. Given population projections, future per caput consumption can be specified by: (a) using some normative yardstick because the issue is the volume of total supplies compatible with progress in food security for all; and (b) accounting for the fact that when per caput consumption in terms of calories grows, the volume of agricultural products used up to generate these calories grows at a faster rate. This happens for two reasons: firstly, higher value products increase their share in the basket of foods consumed directly[17] and, secondly, indirect use of agricultural products in the form of animal feed also increases more than proportionally as diets shift to more livestock products.

It can, therefore, be expected that the achievement of any target that would raise average per caput calories to "acceptable" levels would require more than proportional growth in the total availabilities of agricultural products in the countries concerned and, at the world level, also of production. However, attaching numerical values to these concepts is no simple matter. The following considerations are relevant.

1. The findings of this study project the "most likely" situation for the year 2010. This would still leave the regions of sub-Saharan Africa and South Asia with per caput food supplies of 2170 and 2450 calories respectively, that is levels which are grossly inadequate for food security. As noted earlier, this lack of sufficient progress in raising per caput food supplies in these regions is one of the reasons why world demand and production is projected to continue to slow down. However, if these year 2010 outcomes are at all credible, they must be taken as the basis from which to start

measuring the further growth required to attain per caput food supply levels compatible with some target for the longer term.

2. Population projections by country are available only to the year 2025 (UN, 1993b). Therefore, an exploration of the future in the required country detail can only go to year 2025. Population projections at the country level are required because targets for future per caput food supplies can only be specified in relation to the existing gaps between present (or year 2010) levels and those that would be approximately compatible with significant reductions in chronic undernutrition. The rather arbitrary, and certainly not overgenerous, rules used here to define the targets for after 2010 and up to 2025 are as follows: in 2025 no country should have per caput calories under the current world average of 2700;[18] countries which are projected at or above that level by the year 2010 should have further de-escalating increases subject to a maximum of 3050 calories by the year 2025; and countries which are projected to be above that maximum in 2010, just maintain their 2010 levels up to the year 2025.

3. The resulting per caput calories for 2025 are shown by region in Table 3.21, Column 6. It can be seen that under these assumptions little further growth in per caput food supplies (in terms of calories) is required at the world level (from 2880 in 2010 to 3000 in 2025) and even the required increase for the average of the developing countries is modest (from 2740 to 2900). But a quantum jump is required for sub-Saharan Africa and a smaller one for South Asia.

4. How do these targets for per caput calories translate into growth for the volume of total supplies of agricultural products measured by the conventional volume index presented earlier? In the first place, these targets must be multiplied by the year 2025 projected population which is shown in Table 3.21, Column 3. Then a notion is needed of by how much the volume of agricultural products per calorie may increase for the reasons given earlier. On this point, the empirical evidence shows that the amount of agricultural products used (all domestic uses, all agricultural products) per calorie in direct human consumption varies widely among countries. Putting the present world average at 100, the data for the individual countries span a range from around 200 in most developed countries with high average per caput calories and high shares of livestock products in their diets through about 75–100 in most developing countries with around 3000 calories but much lower livestock products component (e.g. countries in the Near East/ North Africa region) down to about 50–60 in the countries with low calories and insignificant amounts of livestock products in the diet, e.g. most countries in sub-Saharan Africa and South Asia. Taking all these country data into account, the implied average relationship over all countries of the world between per caput calories and products volume per calorie yields an elasticity of 1.6,[19] meaning that a 10 percent rise in per caput calories would be associated with a 16 percent rise in the products used up per calorie. In

principle, this relationship would represent the long-run evolutionary pattern.

However, this relationship may not be entirely appropriate for the period to 2025 examined here, and perhaps not even for well beyond this year, because it implies that if, for example, countries in South Asia were to make the transition from the present 2200 calories to that of the developed countries (3400 calories) they would be adopting the consumption patterns of the latter of over 80 kg of meat and some 630 kg of cereals (for food and feed), compared with present levels of 4 kg and 180 kg, respectively. Those developing countries which have made a good part of this transition in the last three decades, have not adopted the typical developed country basket of agricultural products. For example, a country with a typical cereals-based diet, Egypt, moved from 2290 calories in 1961/63 to 3310 calories in 1988/90 but its per caput consumption of meat grew from 10 kg to 18 kg, and that of cereals (all uses) from 250 kg to 360 kg. The same pattern is exhibited by other countries in the Near East/North Africa region. Even a fast growing country like Korea (Rep.) made the transition from 1960 calories to 2820 calories while increasing per caput consumption of meat from 4 kg to 20 kg and of cereals (all uses) from 190 kg to 350 kg.

This evidence does not, of course, mean that further structural change in favour of livestock products may not be forthcoming in these countries even if their per caput calories were not to increase appreciably. This may happen if economic growth were to raise per caput incomes towards the levels of the lower tier of the developed countries. The experience of Southern European countries, which moved to high levels of meat consumption in the last three decades, is instructive.[20] But this type of transformation may not be in prospect for some time, if ever, for the majority of the developing countries for which the growth path to 2025 must be defined under the assumptions for per caput calories presented earlier. The more relevant experience appears to be that of those countries which achieved quantum jumps in per caput calories in the three decades 1960–90.[21] There are 24 countries in this class, including Greece, Spain and Portugal. Their average per caput calories (simple average) increased from 2130 in 1961/63 to 3020 in 1988/90. The average amount of agricultural products per calorie was 86 in 1961/63 (always measured with world average in 1988/90 = 100) and had risen to 102 by 1988/90. Taking into account all the 30-year annual average estimates for this group of countries, it results that every 10 percent increase in per caput calories was associated with about 5.7 percent increase in the amount of agricultural products per calorie, an elasticity of about 0.6.[22] This relationship is used here to translate the increases in direct consumption in terms of calories to increases in terms of the volume of total domestic use of agricultural products.[23]

Estimates, interpretation, implications

The results are presented in Table 3.21. The following comments apply:

1. For the world as a whole, the trend towards further slowdown in total production (= total use at the world level) would continue under the assumptions used here and the growth rate would fall from the 1.8 percent p.a. projected in this study for 1988/90–2010 to 1.3 percent in 2010–25. This would happen even if: (a) the developing regions with low per caput food supplies in 2010 (sub-Saharan Africa and to a smaller extent South Asia) were to experience the quantum jump required to achieve the minimum target of 2700 calories by year 2025; and (b) the better-off developing regions would continue to register some further gains in per caput food supplies.

2. It follows that if the world as a whole could achieve a growth rate of agricultural production of 1.3 percent in 2010–25, then it would be reasonable to conclude that global capacity to grow more food would not be a major obstacle to achieving the minimum standards of food security implied by the targets assumed here.

3. However, this scenario requires hefty increases in total supplies for some individual countries and regions. If these regions cannot make their domestic production grow at the rates projected for their total domestic use, this scenario would be feasible only if: (a) the production of other regions can grow faster than their respective demand; and (b) trade flows and patterns can be generated to match regional surpluses and deficits.

4. The most obvious case of dependence of this scenario outcome on significant changes in the existing (or projected for 2010) net agricultural trade balances of the different regions is that of sub-Saharan Africa. The total supplies for domestic use in the region are required to grow at 4.9 percent p.a. in 2010–15 (roughly, 2.6 percent for population growth, 1.5 percent to raise per caput calorie supplies to 2700 and 0.8 percent to account for the above-mentioned increase in the amount of agricultural products per calorie). If, for the sake of example, the region could sustain a production growth rate no higher than 3.5 percent p.a. in 2010–25, then the achievement of the target of 2700 would only be compatible with greatly widening agricultural deficits and hefty declines in self-sufficiency. In practice, sub-Saharan Africa would be required to become like Near East/North Africa, with aggregate self-sufficiency (all products) falling from the projected 98 percent in 2010 to 80 percent in 2025 (it was 117 percent in 1969/71 and had fallen to 104 percent by 1988/90). And this is assuming that the rest of the world could generate the required net surplus, a not unreasonable assumption given the relative magnitudes involved (a large deficit by African standards would be a small proportion of the rest of the world output in 2025, about 2 percent).

Table 3.21 Growth of total availabilities of agricultural products to meet specified consumption levels by 2025

	Population (million)*			Calories/caput/day			Volume of agric. products/calorie (index: world 1988/90 = 100)			Growth rates (% p.a.)			
										Total domestic use		Population	
	1989	2010	2025	1988/90	2010†	2025‡	1988/90	2010§	2025¶	1988/90–2010||	2010–25	1989–2010	2010–25
World*	5148	7073	8382	2700	2880	3000	100	101	101	1.8	1.3	1.5	1.1
Developed	1245	1405	1484	3400	3470	3470	170	170	170	0.6	0.4	0.6	0.4
Developing (93 countries)	3903	5668	6898	2470	2740	2900	78	84	86	2.7	1.8	1.8	1.3
Africa, sub-Sahara	471	874	1282	2100	2170	2700	63	62	70	3.1	4.9	3.0	2.6
Near East/North Africa	301	513	674	3010	3130	3180	92	94	95	2.9	2.0	2.6	1.8
South Asia	1092	1617	1941	2215	2450	2700	65	70	74	2.7	2.2	1.9	1.2
East Asia (incl. China)	1612	2070	2307	2600	3045	3060	78	91	91	2.7	0.8	1.2	0.7
Lat. America + Caribbean	427	593	694	2690	2950	3030	116	120	121	2.2	1.3	1.6	1.1

*Data and projections for the 127 countries of this study. The world total population, including the countries not included here is as follows (million): 1989, 5202; 2010, 7150; and 2025, 8472. All the population data and projections in this table are from the 1992 UN Assessment (UN, 1993b) and as such they are somewhat different from those underlying the projections in other parts of this study.
† Per caput calories for 2010 as projected in this study by country. The regional totals are slightly different from those in other tables because for this table they are aggregated using the new population data.
‡Levels specified for 2025 (see text).
§Computed from the projections of total domestic use and calories of this study.
¶Computed using the estimated elasticity (0.566) of volume of agricultural products used per calorie with respect to calories/caput.
||Growth rates for 1988/90–2010 slightly different from those reported elsewhere in this study due to the differences in the population projections.

5. The preceding considerations lead to a number of conclusions. First, if sub-Saharan Africa were not to have the financial resources to emulate the Near East/North African pattern of increasing dependence on imported food, the achievement of the 2700 calories/caput target becomes problematic. Second, this conclusion depends on the assumption that its agricultural growth rate would not exceed 3.5 percent p.a. in 2010–25 (0.9 percent p.a. in per caput terms). This assumption may be challenged, but it is noted that even East-Asian-type performance would not deliver growth of agricultural production of nearly 5.0 percent p.a. for 15 years.[24] Third, the achievement of a growth rate of effective demand for agricultural products of 4.9 percent p.a. for 2010–25 (2.3 percent p.a. in per caput terms) would depend on achievement of growth rates of incomes implying East-Asia-type economic performance for 2010–25,[25] and this after the projected near stagnation in per caput incomes in sub-Saharan Africa up to 2010.

6. The preceding discussion concentrated on sub-Saharan Africa because it is the region in which the period 2010–25 needs to be very much unlike anything that happened in the past or projected for up to 2010. By contrast, the growth rates of demand projected for the other regions for 2010–25 are invariably below, and sometimes well below (e.g. for East Asia), those projected and, by and large considered feasible, for up to the year 2010. The only other region for which demand would need to grow at over 2.0 percent in 2010–25 is South Asia (2.2 percent p.a.). Would South Asia's production be able to grow at about this rate? Its highest historical growth rate on record for 15 years is the one of 1977–92 (3.4 percent), while the production growth rate projected for 1988/90–10 is 2.8 percent p.a. A growth rate of 2.2 percent for 2010–25 would not be too optimistic in the light of historical trends. Still, its achievement should not be taken for granted, given the increasing scarcities of land and irrigation as well as the tendency for yields to grow less fast than in the past. Much will depend on the continuation of the generation and diffusion of technology to maintain and underpin further growth in yields and on the exploitation of the scope for more efficient use of water in irrigation. Finally, it is noted that the overall economic conditions for this configuration of growth rates in demand and production to be feasible are certainly much less stringent for South Asia than the challenging ones for sub-Saharan Africa. For example, South Asia's required GDP growth rates (3.0–4.0 percent p.a. for 2010–25, depending on the size of the demand elasticity) are well within the realm of realism.

7. The Near East/North Africa region starts from relatively high levels of per caput food supplies which are projected to improve further by year 2010. Therefore little further growth in this variable would be required for the period 2010–25 on the assumptions used here. However, the region is

projected to continue to have relatively high population growth rates beyond 2010, 1.8 percent p.a. which is the second highest population growth rate of all regions after sub-Saharan Africa. At the same time, the region has little potential for land expansion and water scarcities are severe, particularly if Turkey, which is part of the region, is excluded from the totals. Maintenance of its low agricultural self-sufficiency, projected at 77 percent for 2010 (79 percent in 1988/90, 98 percent in 1969/71), would require production to grow at 2.0 percent in 2010–25. Land and water scarcities have been present for a long time, yet the region has recorded respectable production growth rates consistently in the past (no 15-year period growth rate has been less than 2.8 percent and the highest one was 3.3 percent in 1977–92) and the projections of this study indicate a growth rate of 2.7 percent p.a. for 1988/90–2010. Therefore, on the evidence of past trends and assessed prospects for up to 2010, the production growth rate of 2.0 percent p.a. does not appear as overly optimistic. The implication is that the region's self-sufficiency should not decline further appreciably beyond the year 2010, provided demand growth were to be contained at 2.0 percent p.a. as assumed here.

8. The other two regions (East Asia, Latin America and Caribbean) are currently net agricultural exporters and are projected to remain so up to the year 2010. Latin America is by far the largest and growing net exporter, East Asia is a much smaller and nearly static one. The further significant declines in their population growth rates beyond 2010 and the nearly 3000 calories/caput projected for 2010 imply fairly low growth rates of domestic demand for after 2010 (see Table 3.21). Their potential for production growth would be sufficient to match these growth rates. Eventually, and especially for Latin America and Caribbean, even higher growth rates of domestic demand and/or net exports could be accommodated by the potential for production growth to be even higher than implied by the demand growth rates.

9. The preceding considerations for the developing regions can help define the possible role of the developed countries' agriculture. The assumption used here is that the 3470 calories per caput projected here for year 2010 would only need to be maintained, rather than increase further, up to year 2025. Their domestic demand would, therefore, need to grow at the projected population growth rate, 0.4 percent for 2010–25. How much faster their production may need to grow to generate a growing net surplus for the developing countries will depend on how large the collective net deficits of the latter may grow to be.[26] If the production of each developing region were to grow at the same rate as its own demand, their collective net import requirements from the developed countries would increase only modestly. Such an outcome would not afford much scope for the production of the developed countries to grow appreciably above the 0.4 percent p.a. growth rate of their own demand. However, if sub-Saharan Africa's production

growth rate were not to exceed 3.5 percent while its demand grew at the rate of 4.9 percent, the resulting deficit could be supplied if the developed countries' combined production grew at 0.8 percent p.a. for 2010–25, that is nearly the growth rate projected for them in 1988/90–2010. If this configuration of growth rates in sub-Saharan Africa were considered unlikely, as well it might be, the demands on the developed countries' agriculture would be more modest; they would be even more so if Latin America's agricultural potential were to assert itself in the form of higher production growth rates than the modest ones assumed here. By and large it seems reasonable to think that the agricultural growth of the developed countries would need to be in the range 0.5–0.8 percent p.a. for beyond 2010.

Overall conclusions

1. The preceding discussion indicates that for the world as a whole the trend for the agricultural growth rate to decline which was established in the last 30 years and is projected for up to 2010 could well continue after this latter year.
2. With the exception of sub-Saharan Africa, the developing countries would probably experience agricultural growth requirements which imply lower growth than in the past and that projected for up to 2010.
3. The demands placed on the agriculture of the developed countries by the possible evolution in the developing countries could be accommodated by a production growth rate in the former which would be lower than in the past.
4. For the most populous region with food problems at present and projected for 2010 (South Asia) progress beyond 2010 towards an improved food security situation could be accommodated within an agriculture growing less rapidly than in the past. If this were not to happen, e.g. if the production growth rate were to be 2.0 percent p.a. rather than the required 2.2 percent, the additional demand for food imports could be accommodated by a marginal rise of the agricultural growth rate of the developed countries.
5. For sub-Saharan Africa, the initial conditions (population growth and present per caput food supplies, assuming the data on the latter do not systematically underestimate food availabilities) and those foreseen for the medium term are very unfavourable and make the achievement of the target for per caput food supplies problematic by 2025. This prospect underlines the urgency for starting an all out effort now with international assistance, so that more progress than projected here can be made by 2010 and lay the foundation for reaching the modest target by 2025.

Environmental issues in a wider perspective

One would like to conclude this discussion with an analytically documented statement on whether future agricultural growth, even at lower rates than in the past, is a sustainable proposition. The insights gained by analyses for up to year 2010 were referred to above. Even these insights were anecdotal and certainly not definitive or sufficiently documented. It is impossible to conduct for up to the year 2025 even the limited analyses attempted for up to 2010 without re-making the entire study. To do so would involve decomposition of the assumed normative levels of per caput food supplies into individual commodities, assumptions about future trade flows that will influence the location of production increases, fairly detailed analyses of the land and water use implications for producing these products, identification of the evolution of the inter-commodity technical coefficients (e.g. the cereals and oilseeds input per unit of livestock output) and speculation on the possible technological breakthroughs that will affect growth of yields, use of agrochemicals, irrigation management and the like. The analyses to 2010 indicate that: (a) pressures on environmental resources will continue to rise leading to further reductions in the stocks of such resources, though the reductions of the natural habitat resulting from agricultural expansion into new lands of the type projected for up to 2010 would not probably be very large in relation to the existing stocks of land with crop production potential; (b) technologies useful for putting agricultural production on to a more sustainable path will probably continue to be developed; and (c) the question of whether the trade-offs between environmental resources and more production will conform to some notion of tolerable environmental cost cannot be answered in a definitive manner, given the present state of knowledge and the uncertainties concerning future developments in technology.

What is a tolerable environmental cost is, of course, an anthropocentric concept which "legitimizes" the use, even "overuse", of environmental resources so long as the benefits generated by such use are perceived (by individuals, by society as a whole) to outweigh those associated with resource conservation. Even if one were to define a tolerable environmental cost as that which: (a) ensures preservation of sufficient potential for agricultural production for future generations; and (b) leaves enough environmental resources intact to perform other essential life-support functions (carbon cycle, biodiversity), or because a mere "existence" value is recognized to them, one would not have moved very far towards defining the concept in an objective manner, e.g. in terms of a minimum amount of natural habitat that must be left intact. For, in the end, it is the relative valuation of present versus future benefits in the utility functions of individuals and society that will determine the outcome. The higher the urgency to satisfy present needs, the more the balance will be tipped in favour of present consumption. Naturally, not all societies will have the same priorities on these matters, even when their living standards are, in some conventional "objective" sense, identical. Different forms of institutional organization of societies for decision making will likely give expression to different priorities.

When it comes to considering the evolution over time of relative preferences juxtaposing food to environmental conservation, the limitations imposed by physiological factors to the growth of per caput consumption of food assume particular importance. The existence of an upper bound to per caput food consumption together with the progressive slowdown in population growth point to the direction of progressively diminishing relative preference of mankind for more food and consequent upgrading of the value attached to the benefits of conservation of land and water resources. We already witness manifestations of the shift of preferences in the policies of high-income countries with high food consumption levels. Population growth, particularly the kind that increases the number of people with low food consumption levels, postpones the timing of the onset of this type of change in relative preferences. This notwithstanding, the operation of the above-mentioned forces may eventually lead to a steady-state world agricultural output with enough per caput production to satisfy the needs of all people, even after accounting for inequalities in distribution and access to food. At present, there is no way of making any definitive statements about the prospect that the path of getting from here to there will be a sustainable one, nor whether the eventual steady-state output can be reproduced year after year in a sustainable manner.

Viewing only the food production aspects of the environmental issue associated with progress towards improved food security can lead to erroneous conclusions as to the possibility of the world treading a sustainable path to a situation of food security for all. This is because the food security problem is often less one of capabilities to increase food production, particularly when considering the world as a whole, and more one of capabilities to increase the incomes of the poor and eventually eliminate poverty. Many countries representing a high proportion of world population are far from having reached a stage of economic development which assures minimum incomes compatible with food security to everyone. The issue of what are sustainable paths to food security must therefore be restated as sustainable paths to economic development and poverty elimination in, mainly, the poor countries. The role of agriculture and environmental constraints related to food production play a role here, but the overall picture may well be dominated by environmental constraints other than those pertaining to agricultural resources. One example is the capacity of the ecosystem to absorb the impact of greatly increased use of energy which accompanies overall economic growth.

The rate of economic growth required for poverty elimination is uncertain. Progress towards poverty reduction which is partly based on a more equal distribution of incomes and appropriate public policies will tend to lower the overall economic growth rate required for poverty elimination. However, the initial conditions and the process of economic growth are such that a high share of the increment of world GDP will accrue to the non-poor. This being so, the issue is not whether the world could generate in an environmentally sustainable way the increment in GDP which, if it accrued wholly to the poor, would be sufficient to eliminate the poverty. It is rather that making the incomes of the poor grow is an integral part of a growth process that cannot generally be sustained unless the incomes of the non-poor also rise, e.g. when the economic growth of the export-oriented developing countries depends on that of the high-income countries. It follows, therefore, that world GDP will have to grow to be much larger than today if poverty is to be eliminated.[27] The environmental consequences and constraints to achieving a much larger world GDP might well be the more binding constraint rather than the more narrow agricultural resource constraints to increasing food production.

Casting the problem in these terms does not absolve one from the need to address the question of whether the agricultural resource constraints are likely to stand in the way of achieving food security for all. It is difficult to provide a straightforward global answer but the following considerations are relevant: (a) in a world without frontiers and free movement of people and/or having conditions for greatly expanded food trade the binding character of such constraints, if they exist, would be greatly diminished; and (b) there are many countries in which both food supplies and an overwhelming part of their economy depend on local agriculture. If their agricultural resources are poor, it is entirely appropriate to speak of agricultural resource constraints standing in the way of solving the food security problem even if one knew for certain that the world as a whole had sufficient resources to grow as much food as required to meet the needs of the growing world population.

Climate change and long-term food prospects

No discussion of longer term aspects of the world food–population balance and food security would be complete without at least a fleeting reference to climate change. It is not the object of these comments to delve into the scientific uncertainties surrounding the extent to which the rising concentrations of greenhouse gases in the atmosphere will eventually raise mean temperatures, nor on how agriculture (in its production aspects, in particular yields) may be affected in the different zones and latitudes. These issues are discussed briefly in Chapter 11 and a full treatment is to be found in the existing specialist literature, including a forthcoming FAO publication (Norse and Sombroek, forthcoming). The object is rather to address the issue of whether valid inferences can be drawn about possible socioeconomic impacts (in particular those related to food security) of the changes in the natural environment of agricultural production which are hypothesized to occur 50 years or more into the future.

As noted in the Editor's Preface, the longer the time horizon of any forward-looking exercise in food and agriculture, the more difficult it becomes for periods beyond 10–20 years to define the exogenous assumptions such as population growth, overall economic development and prevalence of poverty. Yet, any relevant analysis must be based on specification of such variables in the detail required to identify countries and population groups whose food security might be vulnerable to changes in the conditions of agriculture. It is predominantly the future state of these variables that will determine the capacity of different societies to respond to change; and all this even before considering the prospect that technology may be very different in 50 years and indeed how technology may develop in response to climate change. In practice, one would need a credible reference scenario (how the situation would be without climate change) on which to superimpose and explore climate change effects. For example, climate change effects on the productivity of agricultural resources in the middle and low latitudes are commonly predicted to be negative (increased weather variability, higher moisture stress of crops). It is reasonable to predict that if today's low-income countries in these latitudes were to be still low income with widespread poverty and undernutrition and high dependence on their agriculture 50 years from now, the adverse effects on their food security would be severe.

Adverse impacts on them would be generated from two main sources: (a) the deterioration of the production potential of their own agricultural resources, threatening not only their food supplies but also their very livelihoods if such resources were still to be their main economic asset for employment and income (Schelling, 1992); and (b) if the net global effect would be to reduce the food production potential of the world as a whole,[28] world food prices would be higher than they would otherwise be. When this happens it is the poor that are priced out of the market, while the non-poor are affected much less in their economic well-being and even less in their food security. It is for this reason that any investigation of climate change effects on food security must take into account the socioeconomic conditions that will prevail in the future: which countries and population groups will continue to be poor and depend heavily for a living on local agricultural resources.

Some attempts have been made to explore these issues and speculate on the socioeconomic effects of changes in agricultural production and productivity, resulting from a doubling of CO_2 concentrations in the atmosphere and alternative assumptions about changes in mean temperatures and their knock-on effects on agriculture, as expressed mainly in lower or higher cereal yields in the different zones. Rosenzweig and Parry (1994) have elaborated a scenario to the year 2060 in which the net effect of climate change is shown to be negative as concerns world average cereal yields and global production, i.e. both would be lower than they would have been otherwise. Cereal prices would be higher and so would the incidence of undernutrition. However, the essence of these negative effects on food security of the assumed changes in the physical parameters of agriculture[29] depends on the kind of economy and society depicted in their reference scenario (without climate change) and its capability to respond to shocks. In the event, their reference scenario depicts a situation of significantly higher food prices in the future and of greater undernutrition than at present.

The results presented in this chapter, but also those of other studies, paint a different picture of the future to 2010, one with still significant undernutrition, but on the decline, and with food prices not very different from present ones. The further discussion in this section of issues of world food–population balance to the year 2025 indicates that conditions exist for further slow progress to be made in conquering undernutrition. To consider that in the normal course of events things would be getting worse rather than better (albeit painfully slowly in some areas) would be tantamount to ignoring the great strides in development that are being made in much of Asia, the region with over one-half the world population. In conclusion, the world could be in a much better position to respond to climate change shocks than assumed in the reference scenario of Rosenzweig and Parry. Consequently, the climate change effects on food security could be not as dire as predicted in their study.

What is certain is the fact that the effects of climate change will not be confined to the countries and regions in which they occur. Impacts will be diffused throughout the world through trade and the consequent inter-regional adjustments, a factor which will play an expanding role in the future compared with the past and present. This point is made in a study by Tobey et al. (1992) which challenges the hypothesis of serious impacts of climate change on world food production. Unfortunately, their study does not have a sufficiently fine geographical breakdown of the distribution of impacts, e.g. all the developing countries, except Argentina, Brazil, China and Thailand, are lumped together into a "rest of the world" total. Yet it is the impact on particular low-income countries and regions (e.g. South Asia, sub-Saharan Africa) that is of decisive importance for drawing inferences about food security implications of climate change. Still the point is well taken that inter-regional adjustments through trade will act as shock absorbers of climate change impacts affecting the agriculture of the individual regions. It remains to define a reference scenario showing how well the different low-income countries and regions may be integrated into the world economy and trading system in the future so that they can "export" eventual negative effects of climate change on their agriculture.

In conclusion, predicting, or speculating about, socioeconomic or more narrow food security impacts of climate change will continue to tax the ingenuity of modellers. This is because the purely scientific uncertainties about climate change and its effects on agriculture are compounded by the far greater ones surrounding the prospects for socioeconomic and technological evolution of humankind for periods of 50 or more years. This notwithstanding, the statement made above still stands and does not need support from modelling insights. That is, any climate change that would cause the production potential of agricultural resources to deteriorate in the countries with food security problems and high dependence on agriculture can prove catastrophic for their welfare. This would still be the case even if climate change were to bring significant improvements in the productive potential of resources in the higher latitudes and result in the net increase in the global potential. In practice, the underlying theme is the same as that discussed elsewhere in this book: the link between the global production potential and food security is weak for populations trapped in a vicious circle of poverty, poor agricultural resources, high dependence on them and limited or no access to the actual or potential global plenty.

NOTES

1. The use of exogenous assumptions about the overall economy as an input into the analysis of possible developments in food and agriculture is not entirely appropriate for the countries where agriculture has a large weight in the overall economy, in practice many developing countries. Indeed, the development literature tends to place increasing emphasis on the role of agriculture and the rural sector for promoting overall economic growth in these countries (see Chapter 7). There is, however, precious little by way of empirical estimates covering a large enough number of countries as to what the relevant relationships are, concretely how the rates and patterns of agricultural and overall growth influence each other. This being the case, the approach used in this study, just as in many other studies of agriculture, accounts for influences from the overall economy on agriculture (mainly through the income growth–demand growth link) but not the other way round.

2. The UN Population Assessment of 1992 (UN, 1993b, not used in this study) indicates a lower projected population for the sub-Saharan region of this study (874 million rather than the 915 million used here) and a growth rate of 3.0 percent p.a. rather than 3.2 percent. This is not necessarily an optimistic outcome because the projected lower population in some countries reflects the increased mortality rates due to the AIDS pandemic. For the 15 countries in sub-Saharan Africa for which AIDS effects were estimated, the projected life expectancy at birth for the period 2000–05 is 51.2 years compared with 57.7 (without AIDS) and the total projected population for the year 2005 is 297 million (310 million without AIDS). The projections for the other regions are also somewhat different. Revisions of the UN population projections made in 1994 reduced further sub-Saharan Africa's projected 2010 population to 834 million (a growth rate of 2.9 percent p.a. for 1990–2010) and the world total to 7.0 billion (UN, 1994). The population projections of the UN assessments of 1990, 1992 and 1994 are shown in Appendix 3.

3. These are *net* imports of all the developing countries, after deduction of projected net exports of the *exporting developing countries* of some 30 million tonnes (up from 17 million tonnes in 1988/90 and 14 million tonnes in 1969/71). The net imports of the *importing developing countries* are projected at some 190 million tonnes by the year 2010 (up from 106 million tonnes in 1988/90 and 34 million tonnes in 1969/71).

4. Concern is often expressed at the burgeoning food deficits of "Africa". In practice most of the increases in net cereal imports originated in North Africa (Morocco, Algeria, Tunisia, Libya, Egypt) rather than in the sub-Saharan region. Thus, in 1988/90 net imports in North Africa were 19.4 million tonnes (up from 2.7 million tonnes in 1969/71) and in sub-Saharan Africa 8.1 million tonnes (2.7 million tonnes in 1969/71). It is underlined that the estimated import requirements in no way refer to any notion of Africa's "food deficit", a term often used in a normative sense and

variously estimated as the difference between domestic production and total consumption requirements to meet some normative target of per caput consumption.

5. China is not included in this analysis of production prospects by agroecological zone because of lack of the relevant data, i.e. the breakdown of current crop production and of land reserves by zone. In addition, there are some indications that the available data for China understate area and overstate yields in China's agriculture (State Statistical Bureau of the People's Republic of China, 1990: 315). China (Mainland)'s cereals production was 313 million tonnes in 1988/90, double that of 1969/71. The working assumption used in this study is that the growth rate of production would be lower in the future and production in year 2010 would be about 50 percent above that of 1988/90. Cereals include wheat, coarse grains and milled rice. If yields are actually overstated in the official statistics, there is more scope for production growth through increases in yields than commonly thought (see Johnson, D. G., 1994a). This is one of the reasons why, unlike other studies, this study does not project large net cereal imports for China.

6. Only meat, milk and eggs are considered here. Other livestock products are not analysed separately in the study, e.g. wool, hides and skins and animal fats, though the latter are included in the demand and nutrition analysis.

7. There are a number of reasons for the projected slowdown in China. In the first place, rapid growth in per caput meat consumption in the historical period started from very low levels of 30 years ago (4.5 kg in 1961/63) and received new impetus after the reforms of the late 1970s, to reach 23.5 kg in 1988/90. The growth rate of per caput consumption was 5.7 percent in 1970–90. If growth were to continue as per trend, it would reach 75 kg in year 2010. This is nearly the European level and unlikely for a country at the level of development that China may reach in the next 20 years. Moreover, a continuation of trends of meat production at the rate of the last 20 years would probably put an intolerable strain on the cereals and oilseeds sectors, with feed demand translating into large import requirements. Feed now takes 55–60 million tonnes of cereals in China, or 17–18 percent of total use of wheat, milled rice and coarse grains. The per caput meat consumption projected in this study would still be a respectable 49 kg/caput in year 2010. It would translate into a production growth rate of 4.5 percent p.a.

8. Recent valuation of per caput incomes on the basis of the purchasing power of currencies (PPC) by the World Bank raises the "conventional" estimate for China for 1991 from $370 (per caput GNP, World Bank Atlas method) to $2040 (per caput GDP), a factor of 5.5. This factor is considerably lower for the middle-income countries, e.g. 1.4 for Brazil, 2.2 for Turkey and 2.5 for Chile (World Bank, 1993a, Tables 1,30).

9. In particular, the inclusion of plantains in this group creates problems. It is meant to capture the similarities with roots and tubers in food consumption and nutrition in several countries in the humid tropics. However, it is not always easy to distinguish between the role of plantains in food consumption and nutrition from that of bananas proper consumed as fruit. For the purpose of food consumption analysis, plantains may be considered to be bananas that are picked green and cooked before eating (see FAO, 1979, 1990b). Several countries, particularly in Central America are reported as having fairly high levels of per caput consumption of bananas (40–50 kg). The picture is further complicated by the consumption of both bananas and plantains after fermentation in the form of alcoholic beverages.

10. The OECD countries as a whole have gone through a phase of declining per caput consumption. Both health reasons and policies of high sugar prices in many counries have contributed to shift demand to alternative sweeteners, e.g. corn

sweeteners in the USA and non-caloric sweetners in most countries. It is expected that the phase of declining per caput consumption is coming to an end and per caput consumption in the developed countries as a whole would tend to stabilize at about present levels of 37 kg. The most recent projections for the USA indicate a rise in per caput domestic disappearance by about a pound by the year 2000 (USDA, 1993a).

11. The prospects for the sugar sector discussed here are mainly based on work by FAO and the International Sugar Organization and assume no major policy changes towards significant liberalization of sugar trade (FAO, 1992b). If there is significant reduction in support and protection of the sugar sector in major OECD countries, it can be expected that increased sugar production and exports from the low-cost producers (developing counries, but also Australia and South Africa) will tend to substitute for some of the production in the highly protected markets. In the event, the application of the provisions of the Uruguay Round are not likely to have significant impacts on the sugar trade flows of the major OECD countries. If reforms changed preferential market access arrangements (e.g. under the EC's ACP sugar protocol or the USA's Caribbean Basin Initiative), some of the benefiting countries would be adversely affected (see Borrell and Duncan, 1992; Lord and Barry, 1990; Alexandratos et al., 1994).

12. Currently only about one-third of Brazil's sugarcane output is used for sugar production. See discussion in Borell and Duncan (1992), and Buzzanell (1993).

13. For most commodities in this study, the consumption and trade data of processed products were converted into primary product equivalent in order to establish the link between agricultural production and final consumption and trade of each commodity in both primary and processed form. Unfortunately, the complexity of the data referring to consumption and trade in textiles (e.g. the fact that cotton is used in widely differing mixes with other fibres, both natural and synthetic) has precluded a similar conversion of the data. Therefore, the only valid statement for final consumption (= production) of cotton can only refer to world level totals, as follows: 1980–90: 2.5 per annum p.a., projected 1988/90–2010: 2.6 per annum p.a.

14. The quantifications and policy discussion concerning the prospects for the cereals and livestock sectors draw partly on work of outside institutes specializing in policy analysis and long-term agricultural projections for the developed countries. Senior researchers in these institutes cooperated by extending their analyses and medium-term (10-year) projections to year 2010 (Frohberg, 1993; Johnson, S., 1993; Meyers, 1993).

15. Their production of 835 million tonnes for the three-year average 1988/90 (the base year of the study) was abnormally low due to the 1988 North American drought.

16. If either the per caput consumption levels to be maintained or the population growth rates are not equal in all countries, their maintenance could be compatible with a world output growth rate below that of world population (see Chapter 2, Table 2.1).

17. This is picked up by the conventional method of aggregating the diverse products to create a volume index of total consumption, production, etc. In this method, 1 tonne of wheat is given a weight of 1, 1 tonne of milled rice a weight of 2, 1 tonne of coffee 8, 1 tonne of beef 13 and so on (weights derived from the 1979/81 farm prices, see Rao, 1993). The thus obtained index of total volumes is appropriate for some uses but not for others. In particular, it may not be appropriate for analysing the relationship between the growth of agriculture and the building up of pressures on the environment. This is because any hypothesized relationship between these two variables would be based on the incorrect premise that, for example, production of an additional tonne of rice generates pressures which are twice as high as those

caused by an additional tonne of wheat and only one-fourth those of an additional tonne of coffee. It is obvious that for environmental analysis we would need to aggregate the different products using weights more closely representing their environmental impact. This is no simple task given the wide diversity of production conditions and technologies for any one commodity. This notwithstanding, the development of environmental impact weights for aggregating the different commodities, produced with diverse methods and in different farming systems, into a volume index of agricultural output should rank high in the work for creating data appropriate for environmental analyses, in this case for linking agricultural growth to the generation of environmental pressures.

18. Rule for countries with under 2700 calories in year 2010: calories in 2025 = calories in 2010 + 100, subject to a minimum of 2700.

19. Estimated from the following cross-section regression with country data 1988/90 (t values in parentheses):

$$\ln P = \quad -12.75 + 1.62 \ln C, \qquad \overline{R}^2 = 0.64, \qquad N = 126 \text{ countries}$$
$$\qquad\quad (-14.9) \quad (14.9)$$

With P = agricultural products used per 1000 calories (index, with world average = 1);
C = calories per caput.

20. For example, per caput meat consumption increased from 23 kg to 72 kg in Greece and from 24 kg to 88 kg in Spain. The corresponding figures of cereals are from 280 kg to 470 kg in Greece and from 307 kg to 540 kg in Spain.

21. Criteria for singling out these countries: increase in per caput calories between 1961/63 and 1988/90 of at least 20 percent and per caput calories 1988/90 over 2500.

22. The estimated equation is as follows (t values in parentheses):

$$\ln P = \quad -2.19 + 0.566 \ln C, \qquad \overline{R}^2 = 0.97, \qquad N = 30 \text{ years}$$
$$\qquad\quad (-14) \quad (28.4)$$

With P and C as defined earlier, but here measured as the simple averages of the 24 countries in the sample for each year.

23. A conceptually similar approach is used by Bongaarts (1994) to speculate on future output required for a world population of 10 billion by the year 2050. For this purpose he uses as a yardstick the gross calories (i.e. calories embodied in the total domestic use of agricultural products per caput) which range from 4000 for the developing countries to 10 000 for the developed ones and a world average of 6000. This corresponds, roughly, to a method of aggregating agricultural products using calorie content, rather than prices, as weights.

24. The highest growth rate registered in any possible historical 15-year period in East Asia, the developing region which grew the fastest in the past, was 4.3 percent p.a. for 1975–90 or 1976–91. But then this achievement was an integral part of, and mutually reinforcing with, fast overall economic growth. By contrast, the highest 15-year growth rate recorded in sub-Saharan Africa was 2.4 percent p.a., 1977–92.

25. Assuming the income elasticity of demand for food and agricultural products is a generous 0.5, the required growth rate of per caput income would be 4.6 percent and that of total economic growth 7.2 percent.

26. The developed countries have been traditionally net importers of agricultural products, fairly large ones in the past, much smaller ones in more recent years. This process has meant that their production was able to grow faster than their domestic demand. For the reasons discussed elsewhere in this study, this process is projected

to continue and would turn the developed countries into net agricultural exporters. For beyond the year 2010, their net exports to the developing countries would need to continue to grow even if each developing region's production were to grow at the same rate as that of its domestic demand.

27. A recent paper by UNCTAD (1994a) suggests a five-fold increase of world GDP in 50 years if per caput incomes of the developing countries were to grow at 3.0 percent p.a. Naturally, this 3.0 percent p.a. growth may or may not be compatible with poverty elimination. What matters is the extent to which growth takes place in the countries with high concentrations of poverty. The most recent World Bank (1994a) assessment suggests a growth rate of 3.4 percent for per caput incomes of the developing countries as a whole but only 0.9 percent p.a. for those of sub-Saharan Africa for 1994–2003. This type of growth pattern will leave the world with the increased environmental stresses generated by economic growth and still plenty of poverty.

28. There is great uncertainty about this because more favourable production conditions would be created in the northern latitude areas from increases in mean temperatures and everywhere from the CO_2 fertilization and water use efficiency effects.

29. Rosenzweig and Parry's analysis probably overestimates the negative climate change effects on agriculture. For example, they assume a very modest CO_2 fertilization effect of 10 percent (the stimulus to plant growth at increased levels of atmospheric CO_2), while recent research suggests persistent positive effects of 15–25 percent under field conditions. The CO_2 anti-transpiration effect (increased water use efficiency by plants of up to 40 percent) is not accounted for at all, although this may have significant positive effects in semi-arid areas.

ANNEX: CEREALS PRODUCTION AND NET TRADE, DATA AND PROJECTIONS BY COMMODITY
(million tonnes)

| | Production | | | | | | | Net trade | | | | | | |
| | Data used in this study | | | Revised data‡ | | | | Data used in this study | | | Revised data | | | |
	1969/71	1979/81	1988/90	1988/90	1991	1992	2010	1969/71	1979/81	1988/90	1988/90	1991	1992*	2010†
World														
Wheat	327.7	442.7	550.3	544.0	546.9	563.8	712	1.6	1.5	1.2	1.9	5.4		
Rice milled	206.8	264.2	341.4	339.9	345.4	351.0	482	0.0	-0.3	0.7	1.0	1.0		
Maize	283.6	422.3	451.6	449.1	487.9	528.0	698	0.5	-0.1	0.0	0.1	0.9		
Other coarse grains	311.0	327.5	353.9	345.9	325.2	342.0	442	0.1	1.5	0.7	1.0	0.5		
Total cereals	1129.1	1456.6	1697.5	1678.9	1705.3	1784.8	2334	2.1	2.6	2.6	4.0	7.8		
All developing countries														
Wheat	96.5	156.9	224.9	224.7	241.7	244.2	348	-25.4	-48.8	-59.7	-59.0	-57.2	-55.6	-94
Rice milled	191.1	247.3	324.3	323.0	329.2	333.2	461	-1.2	-2.0	0.0	0.1	0.3	0.6	-2
Maize	102.6	150.1	197.0	194.8	209.3	219.3	358	5.9	-11.5	-18.1	-18.1	-11.1	-14.3	-40
Other coarse grains	92.2	98.1	100.9	102.4	95.7	105.2	151	0.3	-4.7	-12.1	-11.8	-11.6	-14.5	-26
Total cereals	482.4	652.3	847.4	844.9	875.9	901.9	1318	-20.4	-66.8	-90.0	-88.8	-79.5	-83.9	-162
Developed countries														
Wheat	231.2	285.8	325.5	319.3	305.2	319.7	364	27.0	50.3	60.9	60.9	62.6		94
Rice milled	15.7	16.9	17.1	16.9	16.1	17.8	21	1.2	1.6	0.8	0.9	0.7		2
Maize	181.0	272.2	254.6	254.3	278.6	308.6	340	-5.4	11.4	18.1	18.3	12.0		40
Other coarse grains	218.8	229.4	253.0	243.5	229.6	236.8	291	-0.3	6.2	12.8	12.8	12.1		26
Total cereals	646.7	804.3	850.2	834.1	829.5	882.9	1016	22.5	69.4	92.6	92.9	87.3		162

(continued)

ANNEX *(continued)*

	Production							Net trade						
	Data used in this study			Revised data‡				Data used in this study			Revised data			
	1969/71	1979/81	1988/90	1988/90	1991	1992	2010	1969/71	1979/81	1988/90	1988/90	1991	1992*	2010†
93 Developing countries														
Wheat	96.3	156.6	224.2	224.0	241.2	243.7	347	−24.0	−47.0	−57.5	−57.0	−55.1	53.5	−92
Rice milled	189.0	245.1	322.8	321.4	327.6	331.6	459	−0.4	−1.0	1.2	1.1	1.2	1.6	−1
Maize	102.5	149.9	196.6	194.4	208.9	218.9	357	6.8	−8.0	−12.8	−13.0	−5.2	−8.4	−31
Other coarse grains	92.1	97.9	100.6	102.1	95.4	105.0	151	0.7	−3.4	−11.3	−11.1	−11.1	−13.9	−24
Total cereals	479.9	649.6	844.5	841.9	873.1	899.2	1314	−16.8	−59.4	−80.4	−80.0	−70.1	−74.2	−148
Sub-Saharan Africa														
Wheat	1.2	1.4	1.9	1.9	2.3	2.2	5	−1.6	−3.7	−4.3	−4.2	−5.3	−5.1	−10
Rice milled	3.1	4.0	5.7	6.1	6.6	6.5	13	−0.6	−2.3	−2.4	−2.4	−2.9	−3.1	−6
Maize	11.8	13.8	20.6	20.2	18.9	15.9	41	0.0	−1.4	−0.4	−0.4	−0.4	−0.4	−2
Other coarse grains	20.3	21.6	26.0	26.8	27.8	27.8	51	−0.4	−0.5	−0.6	−0.4	−0.7	−0.9	−1
Total cereals	36.5	40.8	54.3	55.1	55.6	52.5	109	−2.5	−7.8	−7.7	−7.4	−9.3	−13.1	−19
Near East/North Africa														
Wheat	25.1	34.1	42.9	43.0	51.9	49.4	67	−6.1	−15.8	−23.5	−23.1	−17.4	−14.7	−37
Rice milled	3.0	3.2	3.6	3.5	4.4	4.8	6	0.2	−1.3	−1.8	−1.9	−1.6	−2.7	−5
Maize	4.6	5.6	7.7	7.8	8.8	8.6	15	−0.2	−3.3	−6.0	−6.0	−5.4	−5.8	−13
Other coarse grains	13.1	15.0	19.0	19.1	21.4	19.7	31	−0.2	−2.5	−6.6	−6.7	−6.2	−6.3	−16
Total cereals	45.8	57.9	73.1	73.4	86.4	82.5	119	−6.3	−22.9	−37.9	−37.8	−30.6	−29.4	−71

East Asia														
Wheat	30.0	59.5	92.1	91.8	96.2	101.9	143	-8.0	-17.3	-21.8	-21.7	-24.0	-20.2	-23
Rice milled	117.7	158.2	203.4	202.4	206.0	211.1	276	0.8	1.7	5.5	5.6	5.4	7.1	12
Maize	41.6	75.6	105.3	103.4	117.1	115.2	198	1.3	-2.9	-3.0	-3.2	1.6	1.2	-10
Other coarse grains	26.8	22.9	17.6	18.0	15.2	15.9	18	-0.1	-0.2	-0.6	-0.6	-0.9	-1.0	-1
Total cereals	216.1	316.2	418.7	415.6	434.5	444.0	635	-6.0	-18.8	-19.9	-19.9	-17.9	-12.9	-22
South Asia														
Wheat	28.0	46.6	65.6	65.6	71.5	71.6	104	-4.4	-3.1	-5.0	-4.9	-2.5	-7.1	-10
Rice milled	57.8	69.4	97.8	97.1	99.0	97.2	144	-0.9	1.3	0.4	0.3	1.6	1.6	0
Maize	7.6	8.1	11.4	11.3	10.4	13.0	16	-0.1	0.0	-0.1	-0.2	-0.1	-0.1	0
Other coarse grains	22.4	23.4	24.9	24.9	19.1	27.9	28	-0.1	0.1	0.0	0.0	0.0	0.0	0
Total cereals	115.7	147.5	199.7	199.0	200.1	209.6	292	-5.5	-1.7	-4.8	-4.8	-1.0	-5.5	-10
Latin America and Caribbean														
Wheat	12.0	15.1	21.7	21.7	19.2	18.6	29	-3.9	-7.1	-2.9	-3.0	-5.9	-6.4	-11
Rice milled	7.3	10.3	12.3	12.3	11.5	12.1	20	0.0	-0.4	-0.5	-0.5	-1.4	-1.3	-2
Maize	36.9	46.8	51.5	51.6	53.8	66.2	87	5.8	-0.4	-3.3	-3.2	-0.9	0.2	-6
Other coarse grains	9.7	15.0	13.2	13.2	11.9	13.7	24	1.6	-0.2	-3.4	-3.4	-3.3	-5.8	-6
Total cereals	65.9	87.2	98.6	98.8	96.5	110.6	159	3.5	-8.1	-10.1	-10.0	-11.5	-13.3	-25
Developed countries														
Ex-CPEs														
Wheat	112.1	113.1	130.5	124.5	103.8	112.7	140	2.9	-14.9	-16.0	-16.1	-18.8		4
Rice milled	1.0	1.8	1.9	1.7	1.4	1.3	3	-0.5	-1.1	-0.6	-0.6	-0.6		-1
Maize	25.1	30.2	29.6	29.6	32.1	21.2	40	0.0	-17.5	-15.5	-15.6	-13.3		2
Other coarse grains	85.8	92.9	122.6	114.0	102.3	112.4	123	0.0	-6.8	-4.2	-4.1	-5.8		0
Total cereals	224.0	237.9	284.6	269.7	239.6	247.5	306	2.4	-40.3	-36.4	-36.4	-38.4		5

(continued)

ANNEX *(continued)*

	Production							Net trade						
	Data used in this study			Revised data†				Data used in this study			Revised data			
	1969/71	1979/81	1988/90	1988/90	1991	1992	2010	1969/71	1979/81	1988/90	1988/90	1991	1992*	2010†
Eastern Europe														
Wheat	19.3	23.2	35.0	35.2	31.8	22.8		−2.4	−2.6	−0.3	−0.3	0.5	−0.7	
Rice milled	0.1	0.1	0.1	0.1	0.1	0.0		−0.2	−0.3	−0.2	−0.2	−0.2	−0.2	
Maize	15.1	21.1	15.8	15.8	22.3	14.1		0.3	−4.6	−1.2	−1.3	−0.5	1.9	
Other coase grains	20.6	23.9	29.6	29.6	30.1	22.5		−0.7	−1.7	−0.4	−0.3	−0.3	0.7	
Total cereals	55.1	68.3	80.5	80.7	84.4	59.4		−3.0	−9.2	−2.2	−2.1	−0.5	1.6	
Former USSR														
Wheat	92.8	89.9	95.5	89.3	72.0	89.8		5.3	−12.3	−15.7	−15.7	−19.3		
Rice milled	0.8	1.7	1.8	1.6	1.3	1.3		−0.3	−0.9	−0.4	−0.5	−0.4		
Maize	10.0	9.1	13.7	13.7	9.8	7.1		−0.3	−12.9	−14.3	−14.3	−12.8		
Other coarse grains	65.3	69.0	93.1	84.4	72.2	89.9		0.8	−5.1	−3.8	−3.8	−5.5		
Total cereals	168.9	169.6	204.0	189.0	155.3	188.1		5.4	−31.1	−34.2	−34.2	−38.0		
Western Europe, North America, Oceania														
Wheat	116.8	170.0	191.4	191.2	198.3	204.7	217	29.1	71.3	82.5	82.4	88.4	95	
Rice milled	3.9	6.2	6.7	6.6	6.7	7.7	10	1.3	2.4	1.7	1.9	1.7	4	
Maize	149.1	230.7	215.6	215.5	238.3	284.4	286	−0.7	38.7	48.7	48.7	41.9	55	
Other coarse grains	131.6	135.1	129.1	128.3	126.3	123.8	166	5.1	20.3	24.7	24.7	25.8	35	
Total cereals	401.5	542.1	542.8	541.6	569.6	620.5	679	34.7	132.6	157.6	157.6	157.8	189	

Western Europe												
Wheat	53.5	68.6	92.6	92.6	101.5	92.2	−6.0	5.9	18.9	18.8	21.4	
Rice milled	1.1	1.1	1.4	1.4	1.5	1.5	−0.2	−0.4	−0.6	−0.6	−0.6	
Maize	23.6	32.1	36.0	36.2	40.6	38.3	−14.2	−20.2	−2.7	−2.7	−3.2	
Other coarse grains	65.4	76.7	76.4	76.3	75.4	62.4	−3.4	3.3	8.8	8.8	11.5	
Total cereals	143.6	178.5	206.4	206.5	219.1	194.4	−23.8	−11.4	24.4	24.3	29.2	
EC-12												
Wheat	46.1	61.1	82.1	82.0	90.9	84.9	−5.1	7.1	18.0	17.9	20.1	22.5
Rice milled	1.1	1.1	1.4	1.4	1.5	1.4	−0.1	−0.2	−0.4	−0.4	−0.3	−0.1
Maize	15.4	20.9	26.2	26.4	27.3	29.7	−13.8	−19.4	−2.2	−2.3	−3.1	−0.5
Other coarse grains	55.4	65.1	64.7	64.7	63.6	53.8	−2.6	3.8	8.8	8.8	9.7	13.1
Total cereals	118.0	148.2	174.3	174.5	183.2	169.9	−21.5	−8.7	24.1	24.1	26.5	35.0
Other Western Europe												
Wheat	7.5	7.5	10.5	10.6	10.6	7.2	−0.9	−1.2	0.9	0.9	1.3	
Rice milled	0.0	0.0	0.0	0.0	0.0	0.1	−0.2	−0.2	−0.2	−0.2	−0.2	
Maize	8.1	11.2	9.8	9.8	13.4	8.6	−0.4	−0.8	−0.4	−0.4	−0.1	
Other coarse grains	10.0	11.6	11.7	11.7	11.8	8.6	−0.8	−0.6	0.0	0.0	1.7	
Total cereals	25.6	30.3	32.0	32.0	35.8	24.5	−2.2	−2.7	0.3	0.2	2.7	
North America												
Wheat	53.9	86.7	84.2	84.0	85.9	96.8	27.7	54.6	52.3	52.3	55.1	57.3
Rice milled	2.6	4.6	4.7	4.7	4.8	5.4	1.4	2.5	2.1	2.1	1.9	1.8
Maize	125.3	198.3	179.3	178.9	197.3	245.7	13.4	58.8	51.3	51.3	45.1	42.7
Other coarse grains	61.4	52.3	45.3	44.5	43.3	52.0	7.0	13.5	12.9	12.9	11.0	12.0
Total cereals	243.3	341.9	313.5	312.2	331.2	399.9	49.6	129.4	118.7	118.6	113.1	113.7

(continued)

ANNEX *(continued)*

| | Production | | | | | | | Net trade | | | | | | |
| | Data used in this study | | | Revised data‡ | | | | Data used in this study | | | Revised data | | | |
	1969/71	1979/81	1988/90	1988/90	1991	1992	2010	1969/71	1979/81	1988/90	1988/90	1991	1992*	2010†
Oceania														
Wheat	9.4	14.8	14.6	14.6	10.9	15.7		7.4	10.8	11.3	11.3	11.8	8.0	−5
Rice milled	0.2	0.5	0.5	0.5	0.5	0.8		0.1	0.3	0.2	0.3	0.4	0.5	−1
Maize	0.3	0.3	0.4	0.4	0.4	0.4		0.0	0.1	0.0	0.0	0.0	0.0	−17
Other coarse grains	4.8	6.1	7.4	7.5	7.6	9.4		1.5	3.5	3.0	3.0	3.4	2.9	−9
Total cereals	14.6	21.7	22.9	22.9	19.4	26.2		9.0	14.6	14.5	14.6	15.6	11.4	−32
Other developed countries														
Wheat	2.2	2.7	3.6	3.7	3.1	2.3	7	−5.0	−6.1	−5.6	−5.5	−7.0	−7.0	
Rice milled	10.9	8.9	8.6	8.6	8.0	8.8	8	0.4	0.4	−0.4	−0.3	−0.4	−0.4	
Maize	6.8	11.3	9.4	9.3	8.2	3.1	14	−4.7	−9.8	−15.0	−14.8	−16.7	−19.9	
Other coarse grains	1.4	1.3	1.2	1.3	0.9	0.7	2	−5.3	−7.3	−7.7	−7.7	−7.9	−8.2	
Total cereals	21.2	24.3	22.8	22.7	20.2	14.9	31	−14.6	−22.8	−28.6	−28.4	−32.0	−35.5	

*1992 trade data are not available in the same specification as for the other years (i.e. imports and exports of all cereal products converted into grain equivalent) for the countries of the former USSR and former Yugoslavia because no 1992 supply-utilization accounts could be constructed for them.
†Production and net trade projections not made separately for the country groups with blanks in the 2010 column.
‡The revised data include the revisions for the former USSR indicated in Table 3.17.

CHAPTER 4

Growth of Agricultural Production in Developing Countries

4.1 INTRODUCTION

This chapter discusses the main agronomic factors underlying the projections of production. Section 4.2 presents an evaluation of availabilities of land with agricultural potential, how much of it is used now and how much may be used in 2010, for both rainfed and irrigated crop production. Section 4.3 presents for the major crops the combinations of increases in land use, the cropping intensities and the yields that may be achieved in the future and would be compatible with the projected growth of the crop sector. Section 4.4 discusses the role of modern varieties and related agricultural research for sustaining the growth of yields. Section 4.5 deals with prospects for fertilizers and plant protection. Section 4.6 presents the main livestock sector parameters underlying the projections of production. Several themes related to this chapter, particularly issues of environment and sustainability, are the subject of Chapters 11 to 13. Unfortunately, China could not be dealt with in the same detail as the other developing countries since data on land with crop production potential and on cropping patterns by agroecological land class are missing.[1] China therefore is covered in the same detail as other countries only in Section 4.6 on livestock production.

4.2 AGRICULTURAL LAND AND IRRIGATION

The overall situation

Land currently used in crop production in the developing countries (excluding China) amounts to some 760 million ha, of which 120 million ha are irrigated, including some 35 million ha of arid and hyperarid land made productive through irrigation. These 760 million ha represent only 30 percent of the total land with rainfed crop production potential which is estimated to be 2570 million ha, including the 35 million ha of irrigated hyperarid land. The remaining 1.8 billion ha would therefore seem to provide significant scope for further expansion of agriculture. However, this impression is severely redimensioned if a number of constraints are taken into account, as follows:

1. About 92 percent of the 1.8 billion ha of land with rainfed crop production potential but not yet so used is in sub-Saharan Africa (44 percent) and Latin America/Caribbean (48 percent). At the other extreme, there is little land for agricultural expansion in South Asia and the Near East/North Africa.
2. Two-thirds of the 1.8 billion ha of land not in crop production is concentrated in a small number of countries, e.g. 27 percent is in Brazil, 9 percent in Zaire and another 36 percent in 13 other countries (Indonesia, Sudan, Angola, Central African Republic, Mozambique, Tanzania, Zambia, Argentina, Bolivia, Colombia, Mexico, Peru and Venezuela).
3. A good part of this land "reserve" is under forest (at least 45 percent, but probably much more) or in protected areas and therefore it should not be considered as a reserve readily available for agricultural expansion.
4. A significant part (72 percent, see Table 4.2) of the agricultural land of the two regions (sub-Saharan Africa and Latin America/Caribbean) which have 92 percent of the total "reserve" suffers from soil and terrain constraints. This is a much higher share than encountered in the other regions. Overall, some 50 percent of the 1.8 billion ha land "reserve" is classified in the categories "humid" or "marginally suitable for crop production" (see below). Only 28 percent of land presently in use is in these two categories. It follows that the land in the "reserve" is of generally inferior quality compared with that presently in agricultural use. However, given the widely differing land scarcities among countries, land in agricultural use in any one country can very well be of inferior quality compared to land not-in-use in another country. Therefore, the preceding statement need not apply to any given group of countries considered together. For example, land-in-use in countries X and Y together may be inferior to land in "reserve" if X has good land in "reserve" and Y has poor land-in-use and no "reserve". It is also possible for land-in-use within individual countries to be inferior to that in "reserve" due to health, infrastructural or institutional constraints affecting accessibility of the better quality land in "reserve".
5. Finally, human settlements and infrastructure occupy some of the land with agricultural potential, roughly estimated at some 3 percent of it. This proportion would increase in the future, perhaps to 4 percent by 2010.

It is against this background that the prospects that more land will come into crop production use in the next 20 years must be examined. A process of expansion into new land has characterized the evolution of agriculture in the past and there is no reason to think that it will not be present in the future in the countries in which a combination of potential and need so dictates. The fact that there is little scope for agricultural land expansion in many countries should not lead one to conclude that this applies to the developing countries as a whole.[2] In what follows, an attempt is made to project how much new land may be brought into crop production by the year 2010. *Potential and need* are the main factors that will determine the rate of expansion. The first step is to estimate the potential. This is done using the geo-referenced agroecological zones (AEZ) data base of FAO.

Estimating the extent of land with crop production potential

For each developing country covered in this study (excluding China) an evaluation was made of the suitability of land for growing 21 crops[3] under rainfed conditions and various levels of technology. The method is described briefly below:

1. The raw material for the evaluation consists of two geo-referenced data-sets: (a) the inventory of soil and land form characteristics from the FAO–Unesco Soil Map of the World (SMW); and (b) the inventory of climate regimes in which data on temperature, rainfall, relative humidity, wind speed and global radiation are used together with information on evapotranspiration, to characterize the thermal regimes and length of growing periods (LGP), i.e. the length of the period during a year (in number of days) when moisture availability in the soil permits crop growth. The two digitized inventories were overlaid in FAO's Geographic Information System (GIS) to create a land resources inventory composed of thousands of agroecological cells which are pieces of land of varying size with homogeneous soil, land form and climate attributes.

2. Each agroecological cell with given soil, terrain and LGP characteristics was tested on the computer for its suitability for growing each of the 21 crops under three levels of technology. The latter are: *low*, using no fertilizers, pesticides or improved seeds, equivalent to subsistence farming; *intermediate*, with some use of fertilizers, pesticides, improved seeds and mechanical tools; *high*, with full use of all required inputs and management practices as in advanced commercial farming. The resulting yields for each cell, crop and technology alternative were then compared with those obtainable under the same technology and LGP characteristics on land without soil and terrain constraints (termed the maximum constraint-free yield – MCFY).[4] Any piece of land (agroecological cell) so tested, or part thereof, is classified as suitable for rainfed crop production if at least one of the crops could be grown under any one of the three technology alternatives with a yield of 20 percent or more of the MCFY for that technology. If more than one crop met this criterion, the amount of land classified as suitable was determined on the basis of the crop which utilized the greatest part of the land in the cell. Any piece of land where none of the 21 crops met this criterion was classified as *not suitable* (NS) for rainfed crop production. It is noted, however, that land which is classified as NS on the basis of this

Table 4.1 Land with rainfed crop production potential, developing countries (excluding China)

Class	Name	Moisture regime (LGP in days)	Land quality	Million ha		
				Potential	In-use	Balance
AT1	Dry semi-arid	75–119	VS,S,MS	154	86	68
AT2	Moist semi-arid	120–179	VS,S	350	148	202
AT3	Sub-humid	180–269	VS,S	594	222	372
AT4	Humid	270+	VS,S	598 }	201	915
AT5	Marginally suitable land in the moist semi-arid, sub-humid, humid classes	120+	MS	518 }		
AT6	Fluvisols/gleysols	Naturally flooded	VS,S	258 }	64	259
AT7	Marginally suitable fluvisols/gleysols	Naturally flooded	MS	65 }		
Total with rainfed potential				2537	721	1816
Irrigated not suitable (arid and hyperarid) land				36	36	
Grand total				2573	757	1816

evaluation is sometimes used for rainfed agriculture in some countries, e.g. where steep land has been terraced or where yields less than the MCFY are acceptable under the local economic and social conditions. It is for these reasons that the reported land in agricultural use in some countries exceeds the areas evaluated here as having rainfed crop production potential (see country data in Appendix 3).

3. The land classified as having potential for rainfed crop production is further classified into three suitability classes on the basis of the obtainable yield as a percentage of the MCFY, as follows: *very suitable* (VS) at least 80 percent; *suitable* (S) 40 to 80 percent; *marginally suitable* (MS) 20 to 40 percent. *Not suitable* (NS) land is that for which obtainable yields are below 20 percent of MCFY for all crops and technology levels.[5]

The estimates of the land with rainfed crop production potential were subsequently aggregated for the purposes of this study into seven rainfed land classes (denoted as land classes AT1 to AT7) defined as in Table 4.1. Data by region are shown in Table 4.4 and the geographical location of the land is shown in the maps in the annex to this chapter.[6]

As was noted above, significant parts of the rainfed land with crop production potential are subject to terrain and soil constraints (Table 4.2). This

Table 4.2   Share of land with terrain and soil constraints in total rainfed land with crop production potential (%)

Constraint	Sub-Saharan Africa	Latin America and Caribbean	Near East/ North Africa	East Asia	South Asia	Developing countries (excl. China)
Steep slopes (16–45%)	11	6	24	13	19	10
Shallow soils (<50 cm)	1	10	4	1	1	1
Low natural fertility	42	46	1	28	4	38
Poor soil drainage	15	28	2	26	11	20
Sandy or stony soils	36	15	17	11	11	23
Soil chemical constraints*	1	2	3	1	2	1
Total land in the AT1 to AT7 land classes affected by one or more constraints†	72	72	43	63	42	67

*Salinity, sodicity and excess of gypsum.
†Individual constraints are non-additive, i.e. they may overlap.

land is classified as having crop production potential if the constraints are not prohibitive, i.e. so long as it could produce a yield of 20 percent or more of the MCFY for at least one of the crops under any one of three technology variants. The importance of the various constraints differs among regions. For example, in the Near East/North Africa region, rainfed land with agricultural potential is largely located in the mountainous areas, where the precipitation is sufficient for cultivation, though the "steep slopes" constraint affects a relatively high share (24 percent) of the potential agricultural land. This constraint affects also relatively high shares of the land in East Asia and South Asia. In sub-Saharan Africa and Latin America and the Caribbean, 40–45 percent of the land with rainfed crop production potential has soils with low natural fertility; and in sub-Saharan Africa a high share of the agricultural land is subject to the constraint "sandy or stony soils".

Competing uses of agricultural land:
forest, protected areas, human settlements and infrastructure

As noted earlier, not all the land with crop production potential shown in Table 4.1 is, or should be considered as, available for agricultural expansion. As far as some speculative estimates could be made,[7] perhaps some 94 million ha of land of all types are occupied by *human settlements and infrastructures* in the developing countries (excluding China), i.e. some 33 ha per 1000 persons.

With population growth, more land will be converted to human settlements and infrastructure, though such use will probably increase less rapidly than total population. The rough estimates used here imply conversion of land to such uses in the period to 2010 at the rate of about 21 ha per 1000 persons increase in total population. However, there are wide differences among countries, depending on overall population densities. For example, in Canada, a low population-density country, it is estimated that population growth in the "urban-centred regions" in 1981–86 resulted in land expansion for such purposes at an average rate of 64 ha per 1000 persons increment in the population of these regions, of which 59 percent was land defined as "with prime capability to produce crops".[8] The situation is very different in the land-scarce countries. A recent UN report contains the following statement referring to India: "At the beginning of the 1980s it was estimated that between 1980 and 2000 about 600 000 ha of rural land would have to be converted to urban use" (UN/ESCAP, 1993). India's urban population was projected at the time to increase by 170 million. This implies 3.5 ha of rural land conversion per 1000 persons increase in the urban population. This is a very small number. It probably refers to conversion of agricultural land only.

Of the total of 94 million ha of land of all types currently estimated to be under human settlements and infrastructure, about 50 million ha are probably included in the land evaluated in the preceding section as having potential for rainfed crop production but not so used at present (land balance). This is a comparatively small proportion, around 2.8 percent of the total land balance. It may grow to account for some 4 percent of the balance by 2010. However, the proportions are much higher in the land-scarce regions. For example, in South Asia some 45 percent of the land balance is probably occupied by human settlements and it could grow to 66 percent by 2010. Therefore, population growth in the land-scarce regions is a significant factor in reducing the net land area available for agricultural use. The relevant estimates by region are shown in Table 4.3.

As noted in Chapter 2, data on extent of *forest cover* from the 1990 Forest Resources Assessment – Tropical Countries (FRA90, FAO, 1993f; see also Chapter 5) are available for only 69 of the developing countries of this study (see list of countries in Appendix 1). These 69 countries account for the great bulk of forest of the tropical countries, 1690 million ha out of a total of 1756 million ha. They also account for the bulk (1724 million ha) of the 1.8 billion ha of land with rainfed crop production potential not yet in agricultural use (the land balance) in the developing countries (excl. China). The extent of *overlap between the forest area and the land balance* could not be ascertained because the FRA90 provided the land data only in tabular form by administrative unit. An estimate of the *minimum overlap* was calculated indirectly by first estimating the area of forest that could exist on the land classified as not suitable for crop production (class NS, as discussed above). If that part of the NS land which was good for forest[9] (916 million ha in the 69 countries) was indeed under

	Land with crop production potential not in crop use (land balance)	Human settlement areas (92 countries)			Forest areas (69 countries)†			Protected areas (63 countries)‡		
		Total	Of which on land balance	As per cent* of land balance	Total	Minimum on land balance	As per cent* of land balance	Total	Of which on land balance	As per cent* of land balance
Sub-Saharan Africa										
1988/90	796.5	23.5	14.5	1.8	511.1	199.8	25.1	151.1	77.5	10.0
2010	754.4	35.4	21.7	2.9						
Near East/North Africa										
1988/90	15.9	12.1	1.7	10.7						
2010	13.3	16.6	3.2	24.1						
East Asia (excl. China)										
1988/90	96.7	14.1	7.2	7.4	210.2	24.7	26.6	63.3	20.7	22.3
2010	81.5	17.7	9.1	11.2						
South Asia										
1988/90	37.6	25.7	16.8	44.7	61.1	6.1	16.2	15.6	5.3	14.1
2010	33.7	34.4	22.3	66.2						
Latin America and Caribbean										
1988/90	869.2	18.7	10.4	1.2	907.4	541.6	67.8	155.0	97.2	12.6
2010	842.0	23.5	13.1	1.6						
Developing countries (excl. China)										
1988/90	1815.9	94.1	50.6	2.8	1689.8	773.9	44.9	385.0	200.7	12.0
2010	1724.9	127.6	69.4	4.0						

Note: see Appendix 1 on country classification for a list of countries for which data on forest areas and protected areas are available. The estimation of the overlap between the forest and protected areas with the land with agricultural potential not in agricultural use (land balance) was made only for the countries with data for forest and protected areas.

*NB Areas under human settlements, forests and protected areas can be overlapping.

†The 69 countries account for 1724 million ha (95%) of the total 1988/90 land balance (1816 million ha).

‡The 63 countries account for 1675 million ha (92%) of the total 1988/90 land balance (1816 million ha).

forest, then the difference (774 million ha of tropical forest area) must be by definition on the 1.8 billion ha of land with crop production potential. This is an estimate of the minimum overlap. The real overlap is probably much larger. Still, the estimate of the minimum overlap is a useful figure for putting the analysis of competition between agriculture and the forest in perspective. The relevant estimates are shown in Table 4.3.

Finally, part of the 1.8 billion ha land balance with rainfed crop production potential may not be used for expansion of agriculture because it is in areas legally defined as protected (national parks, conservation forest and wildlife reserves).[10] The relevant data are available for 63 of the developing countries of this study (excl. China). These 63 countries account for 92 percent of the 1.8 million ha land balance. Their protected areas are 385 million ha. The relevant data are geo-referenced[11] and were overlaid on those of the land balance. It results that some 200 million ha of protected areas are located on the land classified as having rainfed crop production potential, covering 12 percent of the land balance of the 63 countries. The relevant data by region are shown in Table 4.3. Agriculture, among other economic activities, is prohibited by law in the protected areas but law enforcement is weak in some countries with the result that some farming activity takes place in them, though the degree of encroachment is not known.

In conclusion, the data and estimates of Table 4.3 give an idea of the main competing, and often overlapping, uses of land with rainfed crop production potential that play a role in limiting the extent to which new land may be brought into cultivation in the future.

Future expansion of land in crop production, rainfed and irrigated

Land in crop production in the developing countries, excl. China, may expand from the 760 million ha in 1988/90 to 850 million ha in 2010, an increase of 90 million ha or about 5 percent of the 1.8 billion ha land balance. The bulk of the increase would be in sub-Saharan Africa (42 million ha or 5 percent of its land balance) and in Latin America/Caribbean (27 million ha or 3 percent of its land balance). The rest of the increase would be mainly in East Asia, and very little of it would be in South Asia and the Near East/North Africa (see Table 4.4).

This rate of expansion of land in crop production has been derived for each of the land classes of Table 4.1 taking into account the following factors in each country: (a) the land use data or estimates in the base year 1988/90 (land in crop production and relative cropping intensities, as well as the harvested areas and yields by crop, all by land class, including irrigated land); (b) the production projections for each crop; (c) the likely increases in yields by crop and land class; (d) the possible increase in irrigation which increases yields and cropping intensities on previously rainfed land and brings otherwise unusable hyperarid and arid land into use; (e) the changes in the cropping intensities through which the land use by crop (defined in terms of harvested area) is translated into physical area (hereafter called arable land); and (f) the land balances derived as described in the preceding section. The values for each of these parameters are projected on the basis of, essentially, expert judgement in several rounds of iterations subject to overall accounting consistency in the land accounts and for the levels of production, consumption and trade for each commodity, country and the world as a whole, as explained in Appendix 2.

The configurations of the main parameters underlying the conclusion that arable land in crop production would need to, and could, increase by 90 million ha are given in Tables 4.4 (land balances) and 4.5 (cropping intensities and irrigation). The other main parameter of the projections (growth of yields) is discussed in a later section of this chapter. The following observations may be made:

1. Although the arable land may expand by 90 million ha, the harvested area could increase by 124 million ha because cropping intensities would rise. This rise is projected to be from 79 percent in 1988/90 to 85 percent in 2010 on the average for all land classes (Table 4.5). The trend for the cropping intensities to rise and for fallow periods to become shorter is a well established phenomenon (though there are no systematic comprehensive historical data on this variable), accompanying the process of agricultural intensification and reflecting among other things increases in population densities and the rising share of irrigation in total land use. Cropping intensities have also been rising in rainfed agriculture and are projected to continue to do so at rates differing among regions and land classes. The rise in cropping intensities has been one of the factors responsible for increasing the risk of land degradation and threatening sustainability when not accompanied by technological change to conserve the land, including adequate and balanced use of fertilizers to compensate for soil nutrient removal by crops. It can be expected that this risk will continue to exist because in many cases the socioeconomic conditions would not be favourable for promoting the required technological change to ensure sustainable intensification of land use.

2. *Irrigated land* in the developing countries may expand by 23 million ha or by 19 percent in net terms, i.e. assuming that losses of existing irrigated land (due to, for example, water shortages or degradation due to salinization) will be compensated, e.g. through rehabilitation or substitution of new areas for the lost ones. It has not been possible to project the rate of irrigated land losses. The few existing historical data on such losses are too uncertain and anecdotal and do not provide a reliable basis for drawing inferences about the future. If it is assumed that 2.5 percent of existing irrigation must be rehabilitated or substituted by new irrigation each year (that is, if the average life of irrigation schemes were 40 years) then the total irrigation investment activity over the period of the study in the developing countries (excl. China) must encompass some 85 million ha, of which over 70 percent

would be for rehabilitation or substitution and the balance for net expansion.

3. The projections of irrigation used here reflect a composite of information on existing irrigation expansion plans in the different countries, potentials for expansion and need to increase crop production. The projections include some expansion in informal (community managed) irrigation, which is important in sub-Saharan Africa. Cropping intensities on irrigated land would continue to grow, particularly in the land-scarce regions. This would result in the harvested irrigated area increasing by 45 million ha, compared with the 23 million ha projected for the arable (physical) area in irrigation. The projected increase in arable irrigated land is well below that of the preceding 20 years when it was 40 million ha (Table 4.5). It is even lower when considered in relative terms, with the projected growth rate being 0.8 percent p.a., compared with 2.2 percent in the 1970s and 1.9 percent p.a. in the 1980s. The projected slowdown reflects the increasing scarcity of water resources, the rising costs of irrigation investment and the projected lower rate of agricultural production growth.

4. Account must be taken of the fact that irrigation expansion does not always subtract from the stocks of land which are suitable for rainfed agriculture, e.g. when irrigation brings arid and hyperarid land into agricultural use. These areas are included in the broader estimate of "land with agricultural potential" if they are presently irrigated (see Tables 4.1, 4.4). In some regions and countries, irrigated arid and hyperarid land forms an important part of total land presently in use (about one-fifth in Near East/North Africa). Overall, it adds another 36 million ha to the estimate of land with agricultural potential in developing countries. It is projected that another 2 million ha of this type of land could be irrigated by 2010 (Table 4.5).

5. The projected expansion of irrigation will probably use up a small part of the total existing (but unknown) potential for such expansion. It was not found practicable for this study to attempt to make an estimate of ultimate potential for irrigation expansion similar to that made for the land with rainfed crop production potential. To estimate such potential the existing data on water resources need to be interfaced with those on land characteristics and also socioeconomic factors need to be taken into account. On each of these counts there are numerous practical and conceptual difficulties. The definition of water resources available for irrigation and their quantification are subject to widely differing interpretations.[12] Such resources can be quantified meaningfully only on the basis of physical hydrological units (watersheds or aquifers) which can cut across international boundaries. The water of any given hydrological unit cannot be automatically allocated to a specific area of land, because it can be carried over considerable distances and there may be international agreements required, or in existence, defining the allocation rules. Furthermore, since surface and groundwater are linked, it is difficult to

assess these two resources independently. Moreover, groundwater resources are not adequately known in all countries, precluding systematic, reliable assessment. The assessment of available water resources is complicated further by inter-basin transfers, inter-annual variability and the use of fossil water (groundwater resources which are not renewable). Perhaps more important still are the socioeconomic aspects of the problem, because the potential for irrigation expansion can only be defined in terms of its economic and social benefits and costs, including environmental costs.

Projected land expansion in relation to past trends

It is clear from the preceding discussion that the projections of land use were not derived as extrapolations of historical trends. Even if one had wanted to extrapolate these trends (and this is not the case here), the historical time-series data in terms of land classes for each country are not available. Even the existing historical data on total land in crop production use (defined as "arable land and land in permanent crops") do not constitute a sufficiently reliable basis for analysing the historical evolution of this variable in a number of countries. This conclusion is based on a comparison between the total harvested area resulting from the summation of the reported harvested areas by crop and the total arable land reported in the land use statistics. The implicit cropping intensities are often not realistic for the type of land in agricultural use. This is particularly the case with the historical data of some sub-Saharan African countries with large shares of their agricultural land in the semi-arid and marginally suitable categories. Country experts and agronomists recommend that in such cases, if the implicit cropping intensities are unrealistically high, they should be adjusted downwards by increasing the arable land estimates to reflect more closely what they believe is the actual situation.

Adjustments to cropping intensities were made for the base year 1988/90 and they yielded the above-reported estimate of 757 million ha of arable land in crop production use. The unadjusted figure was 669 million ha. For 42 of the 91 countries the adjustment exceeded 20 percent, but these were mainly the smaller countries and accounted for 204 million ha of the reported total of 669 million. After adjustment their arable land estimate for 1988/90 was raised to 308 million ha out of the adjusted total of 757 million ha.

It follows that for this "heavily adjusted" group of countries the historical data are not sufficiently reliable to provide a yardstick against which to compare the projections. But for the remaining 49 countries, whose base year data were adjusted by less than 20 percent or not at all, and which account for about 60 percent of the 757 million ha of 1988/90, the historical evidence is more relevant. For this sub-group, the projected rate of arable land expansion is 0.4 percent p.a. This is the same rate as in 1980–91 and lower than those of

Table 4.4 Land with crop production potential, land-in-use and land balances, developing countries (excluding China) (million ha)

	Dry semi-arid (AT1)	Moist semi-arid (AT2)	Sub-humid (AT3)	Humid (AT4)	Marginal moist semi-arid, sub-humid, humid (AT5)	Fluvi-sols/ gleysols (AT6)	Marginal fluvi-sols/gleysols (AT7)	Sum A1–A7	Total land with crop production potential — Total incl. irrig. arid land: Total	% of land area	Ha per caput	Total of comparable quality† : Total	Ha per caput
Sub-Saharan Africa													
Land with crop production potential	89.0	179.0	294.1	171.1	155.2	104.5	15.4	1008	1009	46			
Land-in-use 1988/90	42.9	43.8	59.5		57.6*		7.9*	212	213		0.45	160	0.34
Balance 1988/90	46.1	135.2	234.7		268.7		111.9	797	797		1.69		
Land-in-use 2010	46.8	52.2	70.2		73.5		11.2	254	255		0.28	194	0.21
Balance 2010	42.2	126.7	223.9		252.9		108.7	754	754		0.83		
Near East/North Africa													
Land with crop production potential	18.6	21.5	17.1	0.2	9.6	9.7	0.8	78	92	8			
Land-in-use 1988/90	14.7	17.5	14.9		9.6		4.9	62	77		0.26	83	0.28
Balance 1988/90	4.0	4.0	2.2		0.1		5.6	16	16		0.05		
Land-in-use 2010	15.9	18.6	15.3		9.6		4.9	64	81		0.16	90	0.18
Balance 2010	2.7	2.9	1.8		0.2		5.6	13	13		0.03		
East Asia (excl. China)													
Land with crop production potential	1.2	7.7	48.8	38.0	53.0	31.5	4.1	184	184	39			
Land-in-use 1988/90	0.0	2.5	20.5		45.0		19.6	88	88		0.18	90	0.18
Balance 1988/90	1.1	5.3	28.2		46.0		16.1	97	97		0.19		
Land-in-use 2010	0.0	3.0	25.1		52.8		21.8	103	103		0.15	104	0.15
Balance 2010	1.1	4.7	23.7		38.2		13.8	82	82		0.12		

South Asia

Land with crop production potential	29.2	82.4	50.6	6.0	22.4	21.3	0.9	213	228	54			
Land-in-use 1988/90	22.1	61.0	45.4	25.1			21.6	175	191		0.17	239	0.22
Balance 1988/90	7.1	21.4	5.2	3.3			0.7	38	38		0.03		
Land-in-use 2010	22.1	62.5	44.7	27.9			22.0	179	195		0.12	258	0.16
Balance 2010	7.1	19.9	5.9	0.5			0.2	34	34		0.02		

Latin America and Caribbean

Land with crop production potential	15.9	59.6	182.8	382.7	277.6	91.1	44.0	1054	1059	52			
Land-in-use 1988/90	6.6	22.9	81.1	64.2			9.7	185	190		0.44	175	0.40
Balance 1988/90	9.3	36.7	101.7	596.2			125.3	869	869		2.01		
Land-in-use 2010	7.4	24.7	94.0	68.5			17.0	212	217		0.35	202	0.32
Balance 2010	8.5	34.9	88.8	591.8			118.0	842	842		1.35		

Developing countries

Land with crop production potential	153.9	350.2	593.4	597.9	517.8	258.1	65.2	2537	2573	40			
Land-in-use 1988/90	86.3	147.7	221.5	201.5			63.7	721	757		0.27	747	0.27
Balance 1988/90	67.6	202.5	372.0	914.3			259.6	1816	1816		0.65		
Land-in-use 2010	92.2	161.0	249.4	232.3			76.9	812	850		0.19	848	0.19
Balance 2010	61.7	189.2	344.0	883.5			246.4	1725	1725		0.39		

*Note: "Land-in-use" could not be identified separately for the land classes AT4 and AT5, and AT6 and AT7.

† Land-in-use in each class aggregated with weights reflecting its productive potential (weights range from 2.2 for irrigated through 1.0 for sub-humid to 0.31 for dry semi-arid, see Chapter 2). The weights used for each land class have been adjusted to reflect the fact that part of the land in each class is presently irrigated. In the absence of estimates for the irrigation potential of the land balance, a homogeneous total of comparable quality cannot be defined for the balance.

Table 4.5 Arable land-in-use, cropping intensities and harvested land,* developing countries, excluding China (million ha)

	Total land-in-use			Rainfed use			Irrigated use						
	Arable	Cropping intensity (%)	Harvested	Arable	Cropping intensity (%)	Harvested	On land with rainfed crop production potential (AT1 to AT7)	On arid and hyper-arid land	Total arable	As percent of total arable land-in-use	Cropping intensity (%)	Harvested	As percent of total harvested land-in-use
Sub-Saharan Africa													
(1969/71)	(124)		(98)						(3.6)				
(1988/90)	(140)		(114)						(5.3)				
1988/90	212.5	55	117.7	207.2	55	113.7	4.6	0.7	5.3	2	75	4.0	3
2010	254.7	62	158.1	247.7	61	152.2	6.2	0.8	7.0	3	84	5.9	4
Near East/North Africa													
(1969/71)	(89)		(53)						(16.3)				
(1988/90)	(93)		(62)						(20.1)				
1988/90	76.5	83	63.4	56.4	77	43.7	5.3	14.8	20.1	26	98	19.7	31
2010	80.5	93	74.8	57.9	85	49.0	6.5	16.2	22.7	28	114	25.8	34
East Asia (excl. China)													
(1969/71)	(68)		(64)						(11.0)				
(1988/90)	(82)		(85)						(20.0)				
1988/90	87.5	101	88.8	68.2	96	65.6	19.3	0.0	19.3	22	120	23.2	26
2010	102.8	105	108.4	81.2	100	81.3	21.5	0.0	21.5	21	126	27.1	25
South Asia													
(1969/71)	(197)		(187)						(44.8)				
(1988/90)	(204)		(205)						(65.2)				
1988/90	190.5	112	213.0	127.1	109	138.4	48.1	15.3	63.4	33	118	74.6	35
2010	194.9	122	237.0	118.6	113	133.6	60.5	15.8	76.3	39	136	103.4	44

Latin America and Caribbean													
(1969/71)	(117)		(88)					(10.0)					
(1988/90)	(150)		(113)					(15.4)					
1988/90	189.6	61	115.6	174.6	58	101.5	9.9	5.1	15.0	8	94	14.1	12
2010	216.8	67	145.0	198.4	64	127.0	13.2	5.1	18.3	8	98	18.0	12
Developing countries													
(1969/71)	(595)		(488)					(85.7)					
(1988/90)	(669)		(579)					(126.1)					
1988/90	756.7	79	598.5	633.6	73	462.9	87.1	35.9	123.0	16	110	135.6	23
2010	849.7	85	723.3	703.8	77	543.1	108.0	37.9	145.9	17	124	180.2	25

* Data in parentheses are the historical data before adjustment. Adjustments were made only for 1988/90 (see text).

Table 4.6 Arable land-in-use, comparison of projections with the historical data (before and after adjustment)

	1961/63	1969/71	1979/81	1989/91	1988/90	1988/90 adjusted*	2010	1961–70	1970–80	1980–91	1988/90 adjusted* to 2010
	(million ha)					(million ha)		(annual growth (% p.a.))			
Developing countries of the study, excl. China	559	595	639	671	669	757	850	0.8	0.7	0.5	0.6
Sub-Saharan Africa	116	124	133	140	140	212	255	0.9	0.7	0.5	0.9
Near East/North Africa	86	89	91	93	93	77	80	0.5	0.1	0.3	0.2
East Asia (excl. China)	63	68	75	82	82	88	103	1.0	1.1	1.0	0.8
South Asia	191	197	202	205	204	190	195	0.4	0.3	0.1	0.1
Latin America and Caribbean	104	117	138	151	150	189	217	1.5	1.7	0.9	0.6
Memo item											
Developing countries of the study excl. sub-Saharan Africa and China	443	471	506	531	529	545	595	0.8	0.7	0.5	0.4
Developing countries with small data adjustments†	408	428	450	465	465	449	492	0.6	0.5	0.5	0.4

China	104	102	100	96	95	−0.3	−0.2	−0.5
All developing countries§	668	702	745	773	‡	0.6	0.6	0.4
Developed countries	666	674	672	668		0.1	0.0	−0.1
North America	224	234	236	234		0.4	0.1	−0.1
Western Europe	108	103	100	98		−0.5	−0.2	−0.2
Ex-CPEs	282	275	272	269		−0.4	−0.1	−0.1
Others	52	61	63	67		2.5	0.4	0.5
World	1335	1376	1416	1441		0.4	0.3	0.2

Note: The historical data are the official ones supplemented by unofficial data and FAO estimates in the standard FAO data on arable land and land in permanent crops.

* Adjusted for this study, as explained in the text.

† Countries for which the 1988/90 land area was adjusted by less than 20%.

‡ Some sources indicate that the land in agricultural use in China is 125 million ha.

§ Including the smaller countries not covered in this study.

the earlier decades (Table 4.6, Row 8). The conclusion is that the growth rate may not continue to fall as in the past, though at 0.4 percent p.a. it will be very low, implying a 10 percent increase over 20 years for this group of countries. This result is compatible with the projections of production and with the prospect that yields would grow at a lower rate than in the past, when their rapid growth made possible high growth rates of production while the expansion rates of arable land were falling.

There is not much more one can say except perhaps that for three out of the five developing regions the projected growth rates of land expansion are lower than those derived from the unadjusted data of the historical period. For South Asia the projected growth rate is not lower than in the 1980s, but remains at only 0.1 percent p.a. This leaves sub-Saharan Africa as the only region for which future agricultural land expansion is projected to be more rapid than in the past, though, as noted, the historical data are not a good guide to what has been happening in the region. Sub-Saharan Africa is also the only region for which a higher growth rate of agriculture is projected compared with the past. Given the unfavourable agroecological conditions prevailing in part of the region for rapid yield increases of many crops, it follows that a higher rate of land expansion than in the past will be required to support a significant acceleration in the growth of production.

Table 4.6 presents an overall picture of the world agricultural land use, including China, the residual (non-study) developing countries and the developed ones. The near constancy of agricultural land use in the developed countries is noted, as are the uncertainties regarding the land use data of China. Assuming some decline of agricultural land use in the developed countries (for which no land projections were made) it can be hypothesized that there will be only modest expansion of land in agricultural use for the world as a whole.

4.3 LAND–YIELD COMBINATIONS, MAJOR CROPS

The aggregate crop production is projected to grow at 2.4 percent p.a., down from the growth rate of 2.9 percent p.a. in 1970–90. The reasons why future growth may be lower than in the past were explained in Chapter 3. The overall combinations of harvested area expansion and yield increases underlying the projections for major crops are shown in Table 4.7. The contributions to total growth of increases in area (and, within it, those of expansion of the physical arable-land area and those obtained from higher cropping intensities, i.e. expansion of multiple cropping and shorter fallows) and in yields are shown in Table 4.8. It is noted that these average yields (all crops, aggregated on the basis of the 1979/81 price weights, all agroecological zones, all countries) are not very useful for understanding the agronomic factors underlying the projections.

The data and projections in Tables 4.7 and 4.8 are, however, useful for

Table 4.7 Area and yields for major crops: developing countries (excluding China)

	Production (P) (million tonnes)				Harvested area (A) (million ha)			Yield (Y) (tonnes/ha)			Growth rates (% p.a.)					
											1970–90			1988/90–2010		
	1969/71	1988/90	1991/92*	2010	1969/71	1988/90	2010	1969/71	1988/90	2010	P	A	Y	P	A	Y
Wheat	67	132	144	205	58	70	77	1.2	1.9	2.7	3.8	0.9	2.8	2.1	0.5	1.6
Rice (paddy)	177	303	309	459	95	109	120	1.9	2.8	3.8	3.0	0.8	2.3	2.0	0.5	1.5
Maize	70	112	117	196	54	63	80	1.3	1.8	2.5	2.7	0.9	1.8	2.7	1.2	1.5
Barley	16	22	24	35	15	17	19	1.1	1.3	1.8	1.8	0.8	1.0	2.3	0.6	1.8
Millet	19	22	21	32	35	32	38	0.6	0.7	0.8	0.4	−0.6	1.0	1.8	0.9	1.0
Sorghum	28	37	37	62	38	37	50	0.7	1.0	1.2	1.7	0.3	1.5	2.5	1.4	1.1
Total cereals	381	631	657	995	299	331	389	1.3	1.9	2.6	2.8	0.6	2.2	2.2	0.8	1.4
Cassava	95	153	149	223	11	15	18	8.3	10.1	12.2	2.4	1.3	1.1	1.8	0.9	0.9
Sugarcane	486	882	939	1365	9	15	18	52.0	59.6	75.4	3.4	2.5	0.8	2.1	1.0	1.1
Pulses	24	30	32	48	46	52	61	0.5	0.6	0.8	1.3	0.7	0.6	2.2	0.7	1.5
Soybeans	3	38	37	79	3	22	33	1.0	1.7	2.4	11.8	9.4	2.1	3.6	1.9	1.7
Groundnuts	14	16	15	30	17	17	21	0.8	1.0	1.4	0.4	−0.4	0.9	3.0	1.2	1.7
Coffee	4	6	6	8	9	11	12	0.5	0.5	0.7	2.2	1.5	0.7	1.5	0.1	1.4
Seed cotton	16	21	22	42	22	19	22	0.7	1.1	1.9	1.3	−0.9	2.2	3.2	0.7	2.5

Note: Sometimes the changes in the annual growth rates between the historical and projection period appear to be large. Often this is a continuation of a change already begun in the historical period or an expected change in one country which has a large weight in the total. For example, annual growth in sugarcane production in the developing countries excluding Brazil is projected to remain the same as in the historical period, namely 2.2 %. Likewise, the area allocated to soybeans in Brazil (currently more than half of total soybean area in developing countries) grew at 21.2% annually in the 1970s, but this growth rate fell to 3.5% in the 1980s.

*Revised data for 1991/92 as known in May 1994, but not used in this study.

Table 4.8 Sources of growth in crop production and in harvested area, developing countries, excluding China (%)

	Crop production				Harvested land	
	1970–90 Contribution of increases in:		1988/90–2010 Contribution of increases in:		1988/90–2010 Contribution of increases in:	
	yields	harvested land	yields	harvested land	arable land	cropping intensity
Developing countries	69	31	66	34	62	38
Africa (sub-Saharan)	53	47	53	47	64	36
Near East/North Africa	73	27	71	29	31	69
East Asia	59	41	61	39	82	18
South Asia	82	18	82	18	22	78
Latin America/ Caribbean	52	48	53	47	60	40

forming an overall idea of the extent to which the production projections depend on the further expansion of land and irrigation, their more intensive use (increasing cropping intensities) and the continuation of yield growth. In particular, they shed some light on the question whether the future may or may not be like the past, although, as noted in the preceding section, the historical data do not always provide a sufficient basis for making this comparison. It is emphasized that these land and yield projections are definitely not extrapolations of the historical trends. The reader is invited to contemplate what the projections would have been if this study had just extrapolated the explosive area growth rates of the historical period for soybeans and sugarcane in a major country like Brazil, which were 10.8 percent p.a. and 7.4 percent p.a., respectively, in the period 1970–90.

The overall conclusions are that: (a) for the major crops (e.g. cereal, soybeans) the growth rates of average yields can be expected to be much below those of the last 20 years, e.g. 1.6 percent p.a. for wheat compared with 2.8 percent p.a. in the past, 1.5 percent p.a. versus 2.3 percent p.a. for rice, etc. (Table 4.7); (b) harvested area expansion will continue to play a significant role in total crop production growth, though, like in the past, much less significant than that of yield increases. In parallel, increases in the cropping intensities of, mainly, the irrigated areas will play a predominant role in the land-scarce regions (South Asia, Near East/North Africa); and (c) as noted in the preceding section, the expansion of irrigated land will probably proceed at a much slower pace than in the past.

Table 4.9 Production of major cereals by land class in developing countries (excluding China)*

	All land classes†			Dry semi-arid (AT1)			Moist semi-arid (AT2)			Sub-humid (AT3)			Humid (AT4+AT5)			Fluvisols and gleysols (AT6+AT7)			Irrigated		
	A‡	Y	P	A	Y	P	A	Y	P	A	Y	P	A	Y	P	A	Y	P	A	Y	P
Wheat																					
1988/90	69.7	1.9	132.4	3.1	0.7	2.2	10.0	1.2	12.3	16.0	1.7	27.3	6.2	1.6	10.3	0.5	0.7	0.4	33.8	2.4	80.0
2010	77.1	2.7	205.0	3.2	1.0	3.3	11.0	1.8	20.3	17.4	2.1	37.1	5.2	2.3	12.1	0.5	1.0	0.5	39.7	3.3	131.7
Rice (paddy)																					
1988/90	109.2	2.8	302.7							10.5	2.1	22.4	21.4	1.6	33.3	29.7	2.4	71.7	47.5	3.7	175.3
2010	120.5	3.8	458.7							5.7	2.4	13.4	24.5	1.9	45.4	32.6	3.1	101.9	57.7	5.2	297.9
Maize																					
1988/90	62.6	1.8	112.2	0.8	0.6	0.5	7.6	1.2	9.0	30.1	1.8	54.9	15.3	1.3	19.5	1.6	1.0	1.7	7.2	3.7	26.7
2010	79.6	2.5	196.6	1.0	0.9	0.9	8.8	1.5	13.6	38.1	2.6	97.7	19.3	1.7	33.5	1.6	1.2	1.9	10.8	4.5	49.0
Barley																					
1988/90	17.2	1.3	21.9	4.7	0.7	3.3	3.9	1.2	4.8	2.8	1.7	4.8	2.3	1.4	3.2	0.6	0.7	0.4	2.9	1.8	5.3
2010	19.4	1.8	35.5	5.2	1.1	5.7	4.2	1.9	7.9	3.3	2.4	7.7	2.9	1.9	5.4	0.8	1.0	0.8	3.1	2.6	8.0
Millet																					
1988/90	31.9	0.7	21.7	10.2	0.4	3.9	9.8	0.6	6.4	6.3	0.9	5.8	2.9	0.5	1.5	0.9	0.8	0.8	1.7	1.9	3.3
2010	38.2	0.8	31.7	12.1	0.5	6.1	12.1	0.8	9.9	7.2	1.1	8.1	3.7	0.6	2.4	1.3	1.2	1.6	1.8	2.0	3.5
Sorghum																					
1988/90	37.1	1.0	36.9	8.8	0.5	4.1	11.3	0.8	8.9	9.4	1.3	11.9	2.5	0.7	1.9	2.0	0.8	1.7	3.0	2.8	8.4
2010	49.7	1.2	61.8	11.7	0.6	6.9	14.5	1.0	14.2	12.7	1.7	21.5	4.3	0.9	3.9	2.8	1.1	3.1	3.8	3.3	12.3

* Data on land and yields by land class at the country level do not exist in any systematic form. They have been assembled for this study based on whatever information was available (country/project reports, expert judgement, etc.). They should therefore be interpreted with care.
†For an explanation of the land classes see Section 4.2 of this chapter.
‡A, Y, P: Area in million ha; Yield in tonnes/ha; Production in million tonnes.

Land–yield combinations in the cereals sector

In the developing countries (excl. China), rice (in terms of paddy) is the most important cereal crop, accounting for 48 percent of total cereals production in 1988/90, followed by wheat (21 percent), maize (18 percent), sorghum (6 percent) and barley and millet (each less than 4 percent). The growth in production of wheat and rice is expected to slow down considerably over the projection period as compared with growth in the last two decades. Coarse grains would maintain their past annual growth in the future reflecting in part the strong growth of demand for cereals used for feed.

By far the greater part (82 percent) of the production of wheat in the developing countries (excl. China) is located in South Asia and Near East/North Africa. Rice production is concentrated in South and East Asia (89 percent) and barley is mainly produced in Near East/North Africa. These are land-scarce regions with higher than average dependence on irrigation. Both necessity and potential dictate that much of the production increases of these three cereals will come from higher yields. Maize and sorghum are mainly produced in Latin America and sub-Saharan Africa, while millet production is evenly distributed between sub-Saharan Africa and South Asia. By and large, the predominance in the production of coarse grains (except barley) of the two land-abundant regions with rainfed agriculture indicates that area expansion will play a comparatively larger role, than for wheat and rice, in the growth of production.

The data and projections in Table 4.7 demonstrate this prospect. For example, a 2.0 percent p.a. increase in rice production may be achieved by a 0.5 percent p.a. increase in harvested rice area (and much less in arable area under rice). At the other extreme a 2.5 percent p.a. growth rate in sorghum production will be based on the harvested sorghum area growing at 1.4 percent p.a. The fact that the production of these coarse grains is overwhelmingly rainfed with, for millet and sorghum, a high proportion being in the two semi-arid land classes, explains why growth of yields may be expected to make a less important contribution to production growth than for wheat and rice. The overall picture of possible area–yield combinations by *agroecological land class* underlying the production projections of cereals is shown in Table 4.9.

These agroecological distinctions convey significant information about the sources of growth in average yields, because they make it possible to distinguish between "genuine" yield increases (due to more fertilizer, better varieties, management, etc.) in a production environment with given physical characteristics, from increases in average yields due to the relative shift of production from the lower to higher potential land classes. For example, part of the 1 tonne/ha increase in the average rice yield is due to the possibility that the share of irrigated land in total rice area will rise from 43.5 percent p.a. in 1988/90 to 47.9 percent in 2010. When this happens, it is possible for the average yield to increase even if irrigated and rainfed yields remained the same.

However, by far the most important factor in raising average yields will be the yield growth in each and every land class. The extent to which such yield increases in a given agroecological zone may depend on the generation by the research system of new varieties (e.g. those that would make possible quantum jumps in the rates of yield growth) or of varieties contributing to slower (evolutionary) growth of yields and for the periodic replacement of existing ones subject to erosion of their yield potential, is central to the issue of research requirements and priorities to sustain the growth of production (for a useful discussion see Byerlee, 1994 and Plucknett, 1994).

The agroecological characteristics used to classify agricultural land into categories in this study provide some useful background for addressing this issue. How useful they are depends on whether the resulting land classes may be considered to be fairly homogeneous as to their potential for yield growth. For example, this condition would be fulfilled if land in the rainfed category "sub-humid" (AT3) had the same physical and biotic characteristics (LGP, soil, terrain) in the different countries, say Argentina and Mexico. In this land class Argentina achieves maize yields of 3.5 tonnes/ha but Mexico only 1.4 tonnes/ha. If the physical production environments were the same it would be reasonable to consider that existing varieties provided considerable scope for yield growth in Mexico. In this case, increasing maize production in Mexico would depend less on new research breakthroughs to increase maize yield ceilings for this type of land and more on a combination of adaptive research and the socioeconomic factors that would enable the Mexican farmers to raise yields by adopting the varieties which produce the higher yields in Argentina.

However, the agroecological criteria used in this study are not sufficiently fine to permit the derivation of firm conclusions based on this type of argument. For example, the rainfed land class "sub-humid" comprises land with: (a) moisture regimes ranging from LGP 180 days to 270 days; (b) soil/terrain characteristics in classes "very suitable" and "suitable"; and (c) the yield potential of "very suitable" land goes from 80 percent to 100 percent of the maximum constraint-free yield (MCFY) and that of "suitable" from 40 percent to 80 percent of the MCFY. It follows, therefore, that the land class "sub-humid" spans a wide range of production environments. In the event, land classified as sub-humid can have potential maximum yields from as low as 40 percent of the MCFY to as high as 100 percent. In conclusion, if most of the "sub-humid" land of country X is nearer the lower limit of the range and that of country Y is nearer the upper limit, the potential for raising yields in country X mainly by means of policies favouring the transfer, adaptation and diffusion of existing technology and varieties used in country Y would be severely circumscribed. It is noted that even land in the irrigated class is far less homogeneous than suggested by the apparent flexibility afforded by irrigation to relax the moisture (LGP) constraints. This is because, the quality of the irrigated land can range from "fully equipped and

Table 4.10 Cereal yields in major agroecological land classes and inter-country differences, developing countries (excluding China)

Product/land class	Percent of production coming from the given land class, 1988/90	Yields* (tonnes/ha)				
		Average (weighted)		Country range†		
		1988/90	2010	1969/71	1988/90	2010
Rice (paddy), all land classes	100	2.8	3.8	0.9–4.6	0.9–6.6	1.5–7.2
Irrigated	58	3.7	5.2		1.7–7.2	3.4–8.0
Fluvisols and gleysols	24	2.4	3.1		1.0–3.6	1.4–4.0
Wheat, all land classes	100	1.9	2.7	0.5–2.7	0.8–5.1	1.2–6.4
Irrigated	60	2.4	3.3		1.1–5.4	1.9–6.7
Rainfed, sub-humid	21	1.7	2.1		0.9–2.9	1.2–4.1
Maize, all land classes	100	1.8	2.5	0.6–3.1	0.6–4.9	1.1–6.0
Irrigated	24	3.8	4.6		1.6–7.9	2.2–8.4
Rainfed, sub-humid	49	1.8	2.6		0.6–3.7	1.2–4.1
Rainfed, humid	17	1.3	1.7		0.4–2.8	0.8–3.6
Millet, all land classes	100	0.7	0.8	0.4–1.3	0.3–1.4	0.6–1.7
Rainfed, dry semi-arid	18	0.4	0.5		0.1–0.6	0.3–0.8
Rainfed, sub-humid	27	0.9	1.1		0.6–1.8	0.7–2.2
Sorghum, all land classes	100	1.0	1.2	0.3–2.8	0.4–3.4	0.6–3.7
Rainfed, dry semi-arid	11	0.5	0.6		0.3–1.0	0.4–1.2
Rainfed, sub-humid	32	1.3	1.7		0.6–3.5	0.9–3.9

* Yields of countries with at least 50 000 ha in the land class and crop shown.
†Simple averages of the yields of the bottom 10% and top 10% of the countries ranked by yield level (not always the same countries in the top or bottom deciles in each year).

Table 4.11 Inter-country gaps in average yields for wheat and rice, developing countries, excluding China* (tonnes/ha)

	Wheat 1969/71		Wheat 1988/90		Wheat 2010	Rice (paddy) 1969/71		Rice (paddy) 1988/90		Rice (paddy) 2010
No countries		32		33	34		44		47	50
Top decile	Mexico	2.92	Zimbabwe	5.73		Egypt	5.27	Egypt	6.65	
	Egypt	2.74	Egypt	5.00		Korea (Rep.)	4.63	Korea (Rep.)	6.41	
	Korea (Rep.)	2.31	Saudi Arabia	4.65		Korea (PDR)	4.25	Korea (PDR)	8.11	
						Peru	4.14	Peru	5.16	
	Average	2.65	Average	5.12	6.37	Average	4.57	Average	6.58	7.25
Bottom decile	Algeria	0.61	Algeria	0.68		Ghana	1.00	Liberia	1.14	
	Myanmar	0.55	Bolivia	0.70		Tanzania	1.00	Mozambique	0.87	
	Libya	0.26	Libya	0.90		Guinea	0.89	Guinea	0.83	
						Zaire	0.76	Zaire	0.91	
	Average	0.47	Average	0.76	1.16	Average	0.91	Average	0.94	1.55
Decile of largest producers (by area)	Turkey	1.32	Turkey	2.02		Indonesia	2.35	Indonesia	4.22	
	India	1.23	India	2.12		Thailand	1.93	Thailand	2.00	
	Pakistan	1.11	Pakistan	1.81		Bangladesh	1.68	Bangladesh	2.57	
						India	1.67	India	2.63	
	Average	1.22	Average	1.98	2.84	Average	1.91	Average	2.86	4.20
Yield of top decile = 100		100		100	100		100		100	100
Bottom decile		18		15	18		20		14	21
Largest producers		46		39	45		42		43	58
Simple average, all countries		43		53	57		47		45	53

* Data and projections for countries with over 50 000 ha under wheat or rice in the year shown. Average yields are simple averages (not weighted by area).

not subject to water shortages" to that of "partly equipped and subject to water shortages".

These limitations of the agroecological classifications notwithstanding, their use in the analyses of this study in association with the distinction of individual cereals, rather than cereals or coarse grains as groups, is a useful step in the direction towards enlightening the debate on the extent to which future yield growth depends on new research breakthroughs. It certainly provides a sounder basis for judging this issue than the mere comparison of inter-country differentials in average yields, let alone differentials among large country groups, e.g. between average yields of the developed and the developing countries. For example, average country yields for wheat in the Near East/ North Africa region range from nearly 5 tonnes/ha in Egypt and Saudi Arabia (all irrigated) to 0.8–0.9 tonnes/ha in Algeria and Iraq (mostly rainfed and in semi-arid conditions). Obviously, these differentials in average country yields convey no useful information for the issue at hand. But it is more relevant to know that rainfed wheat yields in the "sub-humid" land class are twice as high in Turkey as in most other countries of the region. This land-specific yield gap comes closer than the average one to identifying the scope for yield gains in the lagging countries through a catching-up process (for a discussion of this issue based on an analysis of the inter-country differences in *average* yields see Plucknett, 1993).

The preceding discussion is relevant for examining the factors underlying the yield projections of this study. The dominant factor is the realization that the scope for raising yield ceilings by quantum jumps through the introduction of new varieties is more limited now than it was in the past (Ruttan, 1994). Therefore, much of the growth in average yields must come less from raising yields of the countries with the highest yields today and more from raising those of the countries, particularly the large ones, at the middle and lower ranges of the yield distribution. This is why the yield projections imply a narrowing of the inter-country yield differences *for each land class*. The relevant data and projections by land class are shown in Table 4.10.

Does this pattern conform to the historical experience? It is not possible to investigate this issue for individual land classes because there are no relevant historical data. Such data exist only for average yields (over all land classes) in each country. They show that the gap between the countries with the highest and the lowest yields (simple averages of the top and bottom deciles of the countries ranked by yield level) had widened between 1969/71 and 1988/90 (Table 4.10). This occurred mainly through a process whereby the yields of the countries in the top decile in 1969/71 rose by more than those of the countries in the bottom decile. The projections for average yields (over all land classes) imply that the future may be unlike the past and the yield gap may become narrower because the scope of yield growth in the top decile countries of 1988/ 90 is more limited than it was 20 years ago.

This pattern is illustrated in Table 4.11 with data of individual countries for wheat and rice (data for similar comparison for other cereals can be found in Appendix 3). For wheat, the countries in the top decile of the distribution had in 1988/90 yields which were nearly twice as high as the countries in the top decile of 1969/71. In contrast, there was much less yield growth in the countries at the bottom decile. These developments are even more pronounced for rice.

The dependence of the future growth of aggregate production of the developing countries on the narrowing of the inter-country yield gap (as measured here, i.e. the difference in the simple average of yields in the top and bottom deciles of countries) should not, however, be exaggerated. This is because the countries at the two ends of the distribution (top and bottom deciles) account for a relatively small part of the total production of the crop examined. This is true even when, as has been done for Tables 4.10 and 4.11, countries with less than 50 000 ha under the crop (and for Table 4.10 also under the given land class) are excluded from the analysis. In practice, the realism, or otherwise, of the projections of total production of the developing countries, depends crucially on that for yield growth in the countries which account for the bulk of the area under each crop.

For this purpose Table 4.11 also shows the relevant historical data and projections of the 10 percent of countries with the largest areas (top decile of countries ranked by area under the crop). It is seen that: (a) these countries have yields which are less than one-half those in the countries with the highest yields; (b) for wheat, their (simple) average yield is projected to grow by 43 percent which is below the 62 percent increase of the last 20 years; (c) for rice the corresponding percentages are 47 percent and 50 percent; and (d) even with these increases these countries, whose performance carries a large weight in the total, would still have in 2010 (simple) average yields around one-half those projected for the countries in the top decile. Thus, although the gap may narrow, particularly for rice, it would be the result of the more limited scope for yield growth of the countries in the top decile, not because the large countries with middle yields are projected to have higher growth than in the past.

The preceding rather lengthy discussion was considered necessary in order to provide the reader with sufficient material for thinking about the issue of the potential for further growth in yields to underpin the growth of production. This issue is discussed further in Section 4.4. No attempt is made here to translate these projected yields into concrete proposals for agricultural research (magnitude, modalities, priorities). No doubt, further growth in yields, even at the more modest rates projected for the future compared with the past, will not come about unless the research effort continues unabated. It is just that the effects of research on production growth may manifest themselves in different ways: more impact through the results of evolutionary, adaptive and maintenance research and less through achievement of quantum jumps in yield ceilings.

Evaluating the area–yield projections

The approach used in this study to project harvested areas and yields for the different crops is essentially the same one used in 1985–86 to make the projections to 2000 of the 1987 edition of this study (Alexandratos, 1988). Being now halfway into the projection period 1982/84–2000 (the three-year average 1982/84 was the base year of the 1987 study), it is interesting to examine how well those projections fared when confronted with the actual outcomes up to 1992, the latest year with firm data. Relevant comparisons for the major cereals of the developing countries (excluding China, for which, as in this study, no area–yield projections were made then) are presented in Figures 4.1 and 4.2. By and large, the projections tend to overpredict harvested area for all three cereals shown and underpredict yields for wheat and rice, while maize yields followed closely the predicted path. The net result of these pluses and minuses is that the aggregate production for all three cereals (with rice in milled form) is right on track, with the actual production of 462 million tonnes of the three-year average 1990/92 being equal to that obtained by interpolating areas and yields for each of the three cereals for 1991 on their projection trajectories 1982/84-2000. Similar comparisons for the other coarse grains (16 percent of the total cereals production of the developing countries, excluding China) cannot be made because subsequent revisions changed radically the base year (1982/84) data which had formed the basis for the projections, i.e. the area data have been revised from 91.9 million ha to 84.6 million ha and those for yields from 860 kg/ha to 933 kg/ha.

4.4 AGRICULTURAL RESEARCH AND MODERN VARIETIES

The development and diffusion of improved cultivars constituted the mainstay of increasing yields and production in the recent past and, subject to the considerations discussed in the preceding section, will probably continue to do so over the next two decades. Research efforts related to improved cultivars can be categorized in three classes (FAO/TAC Secretariat, 1993): research focusing on raising the absolute yield ceilings, research related to closing the gap between average farm yields and the yield ceiling, and productivity maintenance research. As discussed in the preceding section, current average yields in most countries and land classes are well below potential yields and in most cases this also holds for projected year 2010 yields. After initial quantum jumps in yield ceilings in the 1950s and 1960s following research breakthroughs, further research led to small yearly increments in yield ceiling. The accumulated effect of such yearly changes over time resulted in many cases in yield gains that were a multiple of the initial one-time increases (Byerlee, 1994; Duvick, 1994). With the possible exception of rice (Box 4.1), the prospects are that production growth based on quantum jumps in cereal yield ceilings will be rarer than in the past. However, the slow evolutionary progress

---

Box 4.1   Hybrid rice

Hybrid rice has shown a quantum leap in yield in China and hybrid rice seed technology is progressing in other major rice producers such as India, Indonesia and Vietnam. Hybrid rice was introduced to farmers in China in the mid-1970s, with adoption accelerating in the 1980s. In 1992–93 it was grown on 19 million ha in China (65 percent of the country's total harvested area of rice), 100 000 ha in Korea DPR, and 20 000 ha in Vietnam. It is expected that India's hybrid rice area will leap from 10 000 ha in 1993 to over 500 000 ha in 1996.

Hybrids in irrigated conditions generally yield 1 tonne/ha (paddy) more than the semi-dwarf modern high-yielding varieties with the same amount of or sometimes even fewer inputs. Many Indian and Vietnamese hybrids have shown a yield advantage of 1.5 to 2.5 tonnes/ha. Evidence suggests that the yield advantage of heterosis is relatively greater in lower productivity areas. The International Rice Research Institute (IRRI) hybrid variety IR64616H has been recommended for on-farm evaluation in the Philippines after three seasons as the highest-yielding early-maturing variety in national trials. At 10.7 tonnes/ha it has the highest yield ever recorded at the IRRI farm.

Progress on hybrid rice seed production technology has dramatically lowered the price of seed in China. Yields of hybrid seed were only 1–1.5 tonnes/ha in 1970s but increased to 2.3 tonnes/ha in the 1980s and 4.5 tonnes/ha in 1991–92 (with a peak of 6.35 tonnes/ha). Because of the high cost of hybrid rice seed, hybrids were only profitable in transplanted rice and in irrigated and favourable lowland rice ecologies during the 1980s. With seed costs having fallen, it is now estimated that hybrids would be profitable on 70 million ha worldwide (nearly half of the 145 million ha total rice harvested area). If the cost of hybrid rice seed, as is expected, falls even further, hybrid cultivation could become profitable in direct seeding (which requires 10 times more seeds than in transplanting) in irrigated and favourable lowland ecologies.

Hybrid rice programmes are underway in Colombia, Brazil and Guyana. A programme to develop less labour-intensive ways to produce hybrid rice has been formulated in Indonesia and Vietnam based on the discovery of photo- and thermo-sensitive genetic male sterility. Different types of hybrid rice are under development and results have been impressive. The future of hybrid rice seed technology could be based on apomixis–asexual reproduction, requiring the cooperation of upstream research (genetic engineering and protoplast fusion) in China and developed countries. With apomictic hybrids, farmers can retain seeds from their crops for many seasons whereas with conventional hybrids farmers must renew seeds for each planting cycle.

---

will likely continue and can underpin the growth of average yields on the basis of research for adapting existing high-yielding varieties to local conditions by overcoming site-specific constraints, e.g. by breeding for increased tolerance to biotic stresses (diseases, insects, weeds) and abiotic ones (soil toxicity, temperature, water).

Seeds produced by the formal seed sector (public agricultural research and the private sector) are referred to as "modern varieties" (MVs), though they are often also described as "high-yielding" or "improved" seeds. The genetic material produced by the system formed the basis of the green revolution.

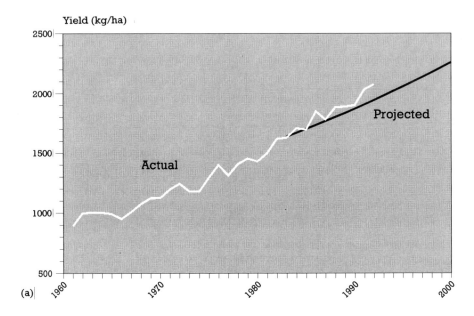

(a)

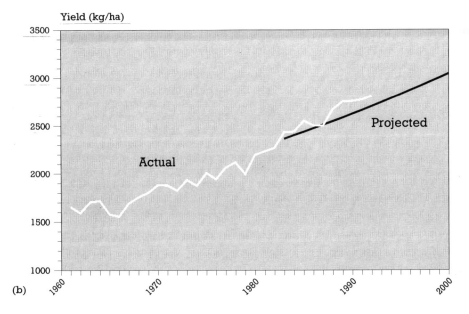

(b)

Figure 4.1 Yield and production: comparison of actual outcomes to 1992 with projections made in 1985–86, developing countries, excluding China. (a) Wheat yield. (b) Rice (paddy) yield. (c) Maize yield. (d) Production of wheat + rice milled + maize. Projections 1982/84–2000 (from Alexandratos, 1988, Table 4.1)

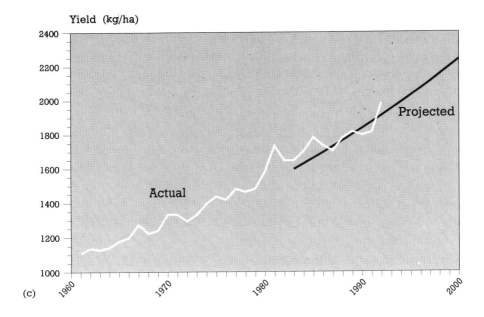

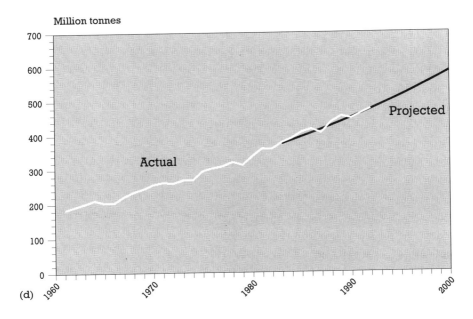

Figure 4.1(c) and (d)

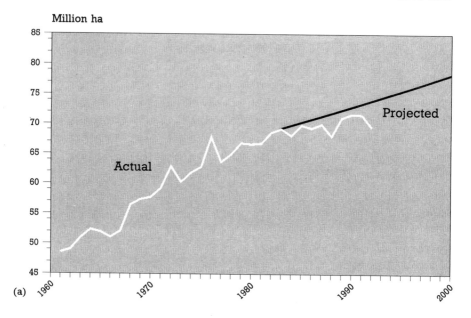

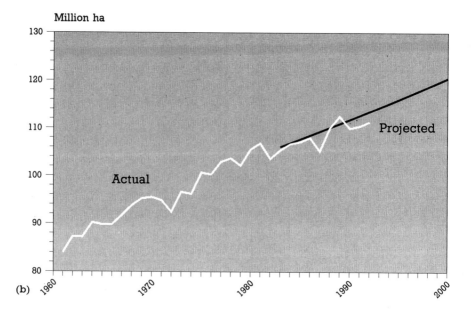

Figure 4.2 Harvested areas: comparison of actual outcomes to 1992 with projections made in 1985–86, developing countries, excluding China. (a) Wheat area. (b) Rice area. (c) Maize area. (d) Wheat + maize + rice area. Projections 1982/84–2000 (from Alexandratos, 1988, Table 4.1)

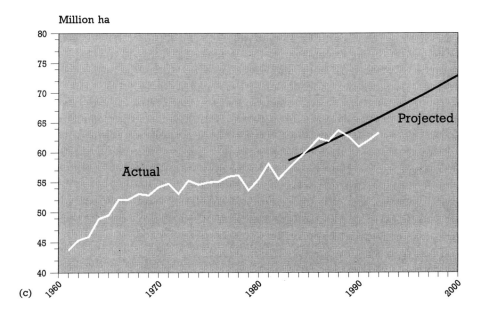

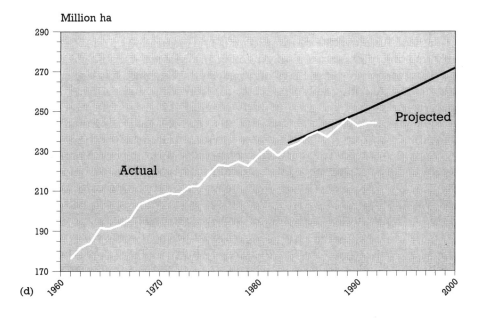

Figure 4.2(c) and (d)

"Traditional", "local" or "farmers'" varieties (TVs) are indigenous types but include those with some foreign material. Between TVs and MVs there exists a wide range of intermediate varieties embodying improvements in various degrees. The share of modern varieties in total seed use has been increasing rapidly (see below). Nevertheless, traditional varieties remain important because of their particular characteristics as well as a source of genetic diversity. They sometimes have superior performance in particular locations, especially in marginal environments. On-farm selection of traditional varieties has led to substantial improvements in their yields. The loss of varieties and individual genes has in certain cases been accelerated by the application of the products of the modern plant breeding methods and a number of analysts fear that this could be a threat to long-term sustainability of agriculture. On the other hand, a number of authors hold that the development and diffusion of modern varieties have had less impact than commonly claimed on diversity and in the post-Green Revolution period they contributed to increase it (Byerlee, 1994). It is argued that in many cases, in particular in the more productive agroecological areas, genetic diversity in the pre-green revolution period was already low since farmers concentrated on a limited number of varieties which were subsequently replaced by a similar number of modern varieties. In the later stages of the green revolution period and afterwards, the number of varieties that has been and is still being developed has increased rapidly and so has genetic diversity as a result of efforts to breed varieties with desirable traits.

Earlier MVs often lacked resistance to many diseases and insects. Since then, research has concentrated on coping with this problem and the more recent MVs incorporate resistance to multiple pests. In this respect, it is held that they are superior to both traditional and the earlier modern varieties and the effort has generally improved yield stability (Byerlee, 1994). As discussed in the section on plant protection, this trend is expected to continue. Of crucial importance in this respect is the success of maintenance research to continually renew pest resistance in the face of evolving new pest biotypes. Naturally, of equal importance are the economic factors and policies that would induce farmers to replace periodically the varieties in favour of the ones with new sources of resistance.

In the last few decades, MVs of cereals have been rapidly replacing the more traditional varieties (Table 4.12). Nevertheless there are a number of situations where the adoption of modern varieties has been slow or non-existent, e.g. the unfavourable agroecological environments. Despite the progress made in adapting MVs to some of these environments, particularly the success in developing MVs resistant to soil stresses, MVs have not been widely adopted in areas with frequent drought stress or, in the case of rice, in areas with poor water control (FAO/TAC Secretariat, 1993). For the future, little progress in raising yield potentials in these environments can be expected from new breeding techniques given the low rates of return to the relevant research effort. Likewise, MVs have not been widely adopted in areas with poor infrastructure

Table 4.12   Share of area (%) planted to modern varieties of rice, wheat and maize in developing countries

	Rice[‡]			Wheat[‡]				Maize
	1970	1983	1991	1970	1977	1983	1990	1990
All developing countries	30	59	74	20*	41*	59*	70	57
Africa (sub-Sahara)	4	15	n.a.	5	22	32	52	43
West Asia/ North Africa	0	11	n.a.	5	18	31	42	53
Asia (excluding China)	12	48	67	42	69	79	88	45
China	77	95	100	n.a.	n.a.	n.a.	70	90
Latin America	4	28	58	11	24	68	82	46

Source: Reproduced by permission from Byerlee (1994).
*Excluding China.
[†]Excludes tall varieties released since 1965. If these varieties are included, the area under MVs increases, especially for rice in Latin America.

and lack of market access. These constraints are particularly pronounced for hybrid maize (in sub-Saharan Africa and the hillside systems of Latin America and Asia), the diffusion of which depends heavily on an efficient system for producing and marketing seed.

Sometimes yield advantages of MVs are more than offset by quality advantages of traditional varieties, e.g. Basmati rice in Asia, white flint maize in Southern Africa. The slow spread of MVs, and sometimes return to TVs, are partly attributed to their quality advantages. However, the breeding of desirable quality traits into the MVs is gradually contributing to their wider diffusion. For example, a high-quality MV of Basmati rice was almost completely adopted within a three-year period in Pakistan, after it became available in the mid-1980s. Another feature limiting the adoption of MVs in marginal areas has been the higher straw yields and fodder values of TVs. In dry areas in India, for example, the high value of fodder of the traditional varieties is often considered to have inhibited the adoption of modern sorghum varieties.

Finally, there are areas where traditional varieties still have a yield advantage over MVs. Such areas are often characterized by insufficient local research capacities to do the necessary adaptation work to the local production conditions as, for example, in the eastern and southern savannas of Africa where rainfall is mainly bimodal and drought is a serious problem. For these drought-prone areas, only limited appropriate material is available (Oram and Hojjati, 1994). On the other hand, there are situations where farmers may adopt a lower-yielding modern variety, if it confers other advantages.

Genetic potentials of many staple crops (other than wheat, rice and maize) have increased little in developing countries. The formal seed sector generally has not been responsive to the needs of small farmers and marginal environments. Catering to their needs is often a difficult and expensive undertaking for national seed systems. Recognizing this, local level seed supply is receiving increased support from governments and development agencies. Likewise, in the last decade, national and international breeders have been paying more attention to genetic improvement of the crops that benefited little from the green revolution, the so-called "orphan crops". This shift of emphasis reflects the increased awareness of the importance of these staples to food security in many developing countries, particularly in marginal areas. Efforts to raise yields of "orphan crops", notably sorghum, millet, pulses, roots and tubers, have been frustrated in part because of the unfavourable environments (socioeconomic as well as physical) in which these crops are typically grown. Recent research has offered improved varieties for some environments, usually those with adequate moisture. However, under conditions of low and uncertain rainfall the risk in using such seeds is high and this often discourages farmers from doing so, especially when other inputs are needed to benefit from the improved varieties. Efforts to increase yield stability are therefore often of higher priority than raising potential yield.

Summary review by major food crops

Rice

As discussed earlier, in many countries rice yields realized by farmers are still relatively low. Table 4.12 shows that the adoption of MVs of rice is not yet complete. MVs developed so far have been mostly suitable for irrigated and favourable lowland areas. Differences in adoption rates of MVs are strongly influenced by the degree of water control. For example, in the Philippines and Indonesia production environments are generally favourable and adoption rates are among the highest in Asia, while in India adoption rates are lower than in the Philippines because a greater proportion of its rainfed areas are drought-prone, flood-prone or both. In some countries, including the major exporters, yield advantages of modern varieties are more than offset by the higher prices of traditional varieties (such as Basmati rice) due to their superior palatability. Recent progress in breeding higher-yielding aromatic types may therefore have a substantial impact on rice production in South Asia. In Pakistan and Thailand, MV adoption is relatively low; even in irrigated areas MVs have not been completely adopted because the quality of available MVs has been poor. Semi-dwarf Basmati varieties have only recently been released in Pakistan and India.

The gap in rice yields between irrigated and non-irrigated areas has widened, and is expected to widen further (Table 4.10). Some progress in raising yield potentials in more marginal environments is expected but the main impact on production will be from adaptation of high-yielding semi-dwarf and hybrid types of rice to local conditions in non-marginal environments. Hybrid rice will make possible further yield increases in East Asia and it is also being introduced in India, Vietnam and the Philippines. It is expected that hybrid rice will largely be in substitution for, rather than incremental to, existing high-yielding varieties. The extension of hybrid rice to other areas can have a profound influence on production before 2010 (Box 4.1).

Wheat

Semi-dwarf varieties of wheat with their high responsiveness to fertilizer were key elements of the green revolution and are now estimated to account for some 70 percent of the wheat area in developing countries (Table 4.12). Virtually all irrigated wheat is now based on semi-dwarfs. The spread of these types has been slower in rainfed areas, though their diffusion was facilitated by breeding varieties with resistance to major fungal diseases in the 1970s. About half the area sown to semi-dwarfs in developing countries is now rainfed but the use of these varieties is generally correlated with moisture: they cover around 60 percent of the area in good rainfall environments but only 20 percent in drier areas.

The possibility to lift the current yield ceiling much further appears limited and no hybrid wheat is foreseen. Following the advantages conferred by the first widely successful varieties, more modest but still significant increases in yield have occurred. Since semi-dwarf wheats were released in the 1960s, plant breeders have maintained a long-term average yield gain estimated at 0.7 percent per year. Much current breeding effort is maintenance research, i.e. maintaining crop resistance to changes in insect pests and diseases.

Maize

Hybrid maize has been established for at least three decades in developing countries and no other breakthrough is foreseen in the short term to shift the yield ceiling of this crop. Open-pollinated varieties are less expensive to develop and received more attention in the 1970s but improved germplasm for tropical hybrids recently shifted the emphasis in national breeding programmes back towards hybrids. MVs have not yet been adopted everywhere, partly because breeders still have to find improved varieties which meet the needs of particular groups of farmers. More location-specific adaptive breeding of improved varieties and hybrids is still needed. Some progress in reducing the vulnerability to drought during the crucial silking period is however probable in the near future, which would be of major importance in much of the semiarid tropics.

The MVs of maize have been adopted mostly in the more favourable environments and among more commercially oriented farmers. The pattern of adoption is typically a progression from local varieties to improved open-pollinated varieties and then to hybrids. The highest rates of adoption are in southern Latin America and the East Asian countries, using genetic material derived mainly from the United States and which is suited to temperate climates. High rates of adoption are also found in East and Southern Africa. The lowest rates of adoption are in Western Africa and the Andean countries of South America, where maize is grown under low fertility conditions and primarily by poor farmers for home consumption. Intermediate levels of adoption are reported in Mexico and in most countries of South Asia.

Although maize benefited from the green revolution, it has shown pronounced geographic variability in yield gains in developing countries which is indicative of the difficulties in raising yields of the so-called "orphan" crops. While wheat and rice are grown under relatively homogeneous agroclimatic conditions where new technologies can be disseminated more easily, maize is grown under a wide range of agroclimatic conditions and improved varieties cannot always be diffused rapidly. Much of the developing world's maize is grown in marginal environments characterized by unreliable rainfall or low soil fertility. Hybrid maize seed generally needs specialized production and distribution facilities which are lacking in many developing countries.

Recent progress in breeding, however, promises to deliver the technical knowledge required to cope better with the difficulties of adapting modern maize varieties to the extraordinary variety of potential growing environments for maize (Byerlee and López-Pereira, 1994). Examples include CIMMYT's work in Mexico's highlands to produce traits suited to highland areas of Asia and Africa (Oram and Hojjati, 1994), and success in adapting cultivars to more intensive farming systems as in the case of rice–maize rotations with transplanted or directly seeded maize as a dry season crop. Progress has also been made in the development of cultivars resistant to multiple species of insects. In general, the technical knowledge to tackle the adaptation problem of maize varieties is already available or forthcoming.

Sorghum and Millet

Sorghum hybrids are widely used in South Asia and Latin America. Hybrids have shown good yield potential also in Africa. Hybrid yields are up to 50 percent higher than yields of control varieties, but seed production is still in the experimental stage. Improved higher-yielding cultivars are available but they do not seem to express themselves in unfavourable environments. It is widely held that improved soil moisture and fertility status are indispensable for higher-yielding varieties. Breeding in these grains is therefore often aimed at stabilizing rather than maximizing yields. Sorghum varieties with resistance to the parasitic weed striga, which is a severe problem in Africa, have recently been released, and lines are now available with tolerance to acidity and aluminium toxicity, which is especially important on some Latin American and African soils. Identification of the mechanisms of drought tolerance are producing improved *millets* and some new varieties are less vulnerable to downy mildew, which is a severe disease of pearl millet.

Roots and tubers

The value of roots and tubers lies in their ability to produce large quantities of dietary energy and in the stability of their production under conditions where other crops may fail. They can be grown under a wide range of adverse conditions such as acid soils, fluctuating soil moisture content, high rainfall and infertile marginal lands. Dramatically higher yields can be achieved even with non-improved cultivars through better soil management and crop husbandry. Yields have been shown to double or triple by introducing a single factor such as fertilizer, though fertilizers are not commonly used on such crops despite demonstrated gains with moderate doses of nutrients. Stagnant or even declining yields in developing countries have resulted not only from the lack of high-yielding cultivars and poor cultural practices but also from rapid degeneration due to diseases, especially viruses. With healthy planting material, large yield increases can be realized.

The potential to produce improved varieties is considered high and so far these crops have remained underexploited in genetic terms. In the past two decades, high-yielding cultivars have been bred for cassava, sweet potato and potatoes with resistance to major diseases and insect pests. The production of planting materials from those varieties with stable resistance is relatively easy. However, certain root and tuber crops (yams, aeroides, plantains and cooking bananas) remain vulnerable to insects and diseases. The problem of disease transmission and multiplication through vegetative propagation is perhaps the most important constraint to production increases of starchy staples. The dissemination of healthy planting materials requires establishment of nurseries with efficient distribution systems, which has proven to be difficult to implement. A more realistic approach would be to develop cultivars with inherent resistance to diseases and insect pests which could be easily multiplied and used by growers with traditional methods. True seed, rather than cuttings, generally does not transmit viruses and diseases and can be used where quality considerations, particularly regularity in size and shape, are of less importance.

4.5 FERTILIZERS AND PLANT PROTECTION AGENTS

Fertilizers

Typical fertilizer response ratios (kilograms of additional crop produced per kilogram of additional plant nutrient applied) in countries with low to middle level yields and areas not subject to severe water constraints, range from 8 to 12 for cereals, 4 to 8 for oil crops and 30 to 50 for roots and tubers (FAO,

1989c). Such high physical response ratios do not always translate into economic incentives for fertilizer use when its costs are high, product prices low and marketing opportunities limited. About one-third of the fertilizer used in developing countries is imported and shortages of foreign exchange affect costs and availability. Likewise, inefficient distribution systems and infrastructure constraints within countries raise the cost to the farmer and limit availability.

Estimates on fertilizer use by crop (FAO/IFA/IFDC, 1992) suggest that for developing countries as a whole (excluding China), about 60 percent of fertilizer is allocated to cereals, with rice alone accounting for one-third of total use and wheat for one-sixth. The allocation of fertilizer to cereals roughly corresponds to the share of cereals in total harvested area (55 percent), though some of the coarse grains, especially sorghum and millet, use little fertilizer. The share of non-food crops and of fruits and vegetables in total fertilizer use is large relative to the share of these crops in harvested area. Sugarcane and cotton are also major consumers accounting for 9 percent and 4 percent respectively of total fertilizer use. Roots and tubers and in particular pulses receive little fertilizer relative to their shares in harvested area.

Table 4.13 shows the present and projected fertilizer use per hectare in the different developing regions. The projections have been derived by applying crop-specific fertilizer use coefficients to the projected harvested areas presented earlier. The coefficients vary with land class and yield level for each crop (see Appendix 2). The fertilizer use per hectare is highest in the Near East/North Africa region and it is projected to remain so. It is likely that by 2010 the region's use per hectare, and to a smaller extent also Asia's, will exceed today's average use of the developed countries, though it would still be well below current use rates in the EC. At the other extreme, sub-Saharan Africa's use will continue to be very low and probably not sufficient for the sustainability of its agriculture (see below), even if it doubled (to 21 kg/ha) from the current extremely low levels.

The projections imply that total fertilizer consumption in the developing countries will grow at a lower rate compared with the past (Table 4.14). This would be a continuation of the longer term trend for consumption to grow at a decelerating rate. In part, this is explained by the relatively high levels of per hectare fertilizer use already attained, particularly in Asia and in the Near East/North Africa, where dressings appear to be reaching their ceilings in some countries and crops. In addition, the growth rate of agricultural production will be lower than in the past. Only sub-Saharan Africa may experience a sharp acceleration in the growth of fertilizer use, given its very low levels per hectare and the need to increase production faster than in the past. Still, fertilizer consumption in the developing countries will probably continue to grow faster than in the developed countries and therefore their share in world consumption will continue to grow from the level of 43 percent of 1990 (it was only 20 percent in 1970 and 32 percent in 1980).

Many countries with high levels of fertilizer use are experiencing the *environmental problems* associated with intensive use of fertilizers. Sub-Saharan Africa however suffers from the opposite problem, namely too little use. Only few countries in sub-Saharan Africa have average dressings of more than 20 kg/ha (FAO/IFA/IFDC, 1992). Such low levels of fertilizer use when fallow periods become shorter represent a serious threat to sustainability. Crops "mine" the soil of its nutrients unless they are replaced with plant residues, manures or fertilizers. Soil nutrient deficiencies limit potential yields in many developing countries, especially in sub-Saharan Africa and Latin America/ Caribbean (see earlier discussion on the soil constraints in these regions). Smaling (1993) holds that soil nutrient mining occurs in almost all countries in sub-Saharan Africa. Soil mining, if left unchecked, will lead to significant soil degradation and declining crop yields (see Chapter 11 for a further discussion of soil nutrient mining).

Some developing countries are encountering difficulties in increasing yields in spite of increasing doses of fertilizers. The efficiency of fertilizer use is often quite low as a result of incorrect timing and poor application methods or failure to maintain the balance between the main nutrients (nitrogen, phosphate and potassium), secondary nutrients and micronutrients. Soil toxicity caused by salinity, alkalinity, strong acidity, iron toxicity and excess organic matter, also prevents the full benefits of fertilizer from being expressed. Evidence suggests that in the long term manufactured fertilizers need to be complemented by organic matter in the soil. Crop and nutrient management at plot, farm and village level will have to become increasingly sophisticated to ensure that the lack of one component does not invalidate the use of the entire package of all nutrients. The optimal use of all possible sources of plant nutrients is advocated under the Integrated Plant Nutrition Systems (IPNS– discussed in Chapter 12) approach to fertilization. IPNS uses sources of organic manure, biological fixation as well as mineral fertilizers in an integrated manner to improve or maintain soil fertility in cropping systems.

Plant protection and pesticides

The increasing intensification of agriculture has led to an increased susceptibility of crop production to pests.[13] Agricultural practices such as multiple crops per growing season, shortening fallow periods and monocultures all have helped to create conditions for pest outbreaks and to reduce natural checks on such events. Also some of the earlier modern varieties of the green revolution were often more susceptible to pest damage than the traditional varieties. Byerlee (1994), however, states that the later editions of such varieties are often superior in pest resistance compared with traditional varieties.

It is well known that pests may become resistant to pesticides, resulting in increased but less effective usage of pesticides. New pesticides therefore need to

Table 4.13   Fertilizer use per hectare in developing countries (excluding China)

	Fertilizer in kg per ha of harvested land*		Annual growth (%)
	1988/90	2010	1988/90–2010
Developing countries, excl. China	62	110	2.8
Africa (sub-Sahara)	11	21	3.3
Near East/North Africa	89	175	3.3
East Asia, excl. China	79	128	2.3
South Asia	69	138	3.4
Latin America/Caribbean	71	117	2.4

* Manufactured fertilizer in kg of nutrient content (N, P_2O_5, K_2O).

be developed continuously. Inappropriate choices of insecticides may disrupt the pest–natural enemy balance by being more deadly for the natural enemy than for the pests themselves, thus upsetting natural control mechanisms. The perception by farmers and extension services of yield losses due to pests are often higher than actual losses. This, together with the desire to reduce risks, induces the farmer to use large quantities of pesticides that have only marginal or no benefits in terms of yield gains or may even induce pest outbreaks. Also, in many countries, the overuse of pesticides was and still is encouraged by pesticide subsidies.

Pesticide use in the developing countries grew rapidly in the late 1960s and 1970s in the wake of rapidly spreading agricultural modernization. There is, however, substantial variation in the type of pesticides used and in the intensity of use, depending on the farming system and crop. In general, the demand for chemical pestidices increases with increasing land scarcity and market access (see Pingali and Rola, 1994). Traditional plant protection methods however remain important. Practices such as tillage (ploughing and hoeing), flooding and burning contribute to reduce all types of pests. Cultural control measures such as crop rotations and disposal of plant material also help to reduce losses due to pests. As noted above, with increasing agricultural intensification, pest pressure also increases due to increased spatial and temporal carryover. Together with increased market orientation, often the opportunity costs of labour increase also, leading to higher use of herbicides. Pesticide consumption varies widely by crop. Pesticide use is high on deciduous fruits, vegetables, cotton and cereals, and more moderate on citrus fruits, tropical fruits, cocoa, coffee and tea.

In the mid-1980s, developing countries accounted for about one-fifth of global consumption of pesticides. Their share in world use of insecticides is relatively high at 50 percent, while this share is 20 percent for fungicides and 10 percent for herbicides. East Asia (including China) accounts for 38 percent of

Table 4.14   Total fertilizer consumption in the developing countries (excluding China)

	Consumption (million tonnes)				Annual growth (%)		
	1969/71	1979/81	1988/90	2010	1970–80	1980–90	1988/90–2010
Developing countries	9.3	22.6	36.8	79.8	9.6	5.6	3.8
Africa (sub-Sahara)	0.4	0.9	1.2	3.3	6.2	2.8	4.8
Near East/ North Africa	1.3	3.5	5.6	13.1	10.6	4.8	4.1
East Asia	1.9	4.1	7.0	13.8	7.7	6.2	3.3
South Asia	2.9	7.3	14.7	32.8	10.3	7.9	3.9
Latin America/ Caribbean	2.8	6.8	8.2	16.9	10.1	3.0	3.5

developing countries' use of pesticides, Latin America for 30 percent, Near East/North Africa for 15 percent, South Asia for 13 percent, and sub-Saharan Africa for only 4 percent. About half of the pesticides used in developing countries are insecticides, with herbicides accounting for a minor part of total consumption. The opposite pattern is seen in the developed countries. This difference can be explained by both ecological and economic factors. In humid tropical countries, pest generations may follow each other without being reduced by low temperatures or aridity. Under these conditions damage from insect pests can be particularly severe and also the pressure from fungal infections is strong. Insect pests are also a severe threat in semi-arid areas and insecticides are widely used to control migratory pests such as locusts. Fungal diseases are often less important in such areas. On the other hand, low labour costs prevailing in many developing countries can make manual weed control more economical than herbicide use.

Pesticide consumption in the developing countries in 1985 was about 530 000 tonnes (in terms of active ingredients), down from some 620 000 tonnes in 1980. Pesticide consumption increased again by about 1 percent per year in the second half of the decade. This is about the same growth rate as in developed countries. Since 1990, however, worldwide use of pesticides has been declining. Future growth of pesticide consumption is likely to be influenced by a great number of factors such as economic profitability (which in turn is determined by such things as pesticide price, product price, price of alternative plant protection means, opportunity cost of labour, etc.), concern about environmental damage and direct damage to the health of persons handling pesticides, the effectiveness of pesticides, pest resistance of plants, the development of substitutes for chemical pesticides, etc. Overall and for the reasons set out below, a reasonable estimate seems to be that pesticide consumption in developing countries will continue to increase but at a slower pace than in the past. Most of the growth is likely to be in South and East Asia and in Latin America.

The damaging effects of indiscriminate pesticide use on the environment and human health (through exposure to pesticides and indirectly through residues in food commodities and drinking water) are well documented and have led to increasing concern and consequently to a number of national and international regulations regarding the production, trade and use of pesticides (see Chapter 11 for a discussion of the environmental issues related to pesticides use). One of the consequences has been that the development costs of new pesticides have strongly increased and that the number of companies willing and able to invest in product research and development has been reduced significantly.

Older compounds (off-patent pesticides) dominate the market in developing countries as they are far lower priced than new ones. However, strict regulations, bans and severe restrictions in developed countries will gradually also reduce their availability in developing countries, while new, less toxic and more environmentally friendly compounds may remain too expensive for many applications.

The increased awareness of the negative aspects of pesticides has resulted in new emphasis given to reduce dependence on chemical control and to develop and employ non-chemical means of pest control. Integrated Pest Management (IPM, see Chapter 12), which considers both crops and pests as part of an ecological system and combines natural factors that limit pest outbreaks while using pesticides as a last resort, is now considered as the preferred method of pest control.

Public and private research on biological control and biotechnological solutions to plant protection problems are increasing substantially. This includes the development of microbiological pesticides, the mass production of natural enemies and the development of transgenic plants with pest resistances. A particular example is the mass production of sterile male Mediterranean fruitflies to eradicate or suppress fruitfly populations. There is also increased attention to other non-chemical mechanisms like cultural control. Results of such programmes become gradually available and these techniques will gain in importance in future plant protection programmes especially as use of pesticides may become less acceptable. A drawback remains that in general the use of biological control tends to be information- and knowledge-intensive and poses high demands on management skills of farmers.

Pest control methodologies are expected to change substantially in developed countries but pesticides will remain a major tool. Pesticide reduction programmes are in place or considered in a number of countries, to decrease excessive use of and reliance on pesticides. In developing countries, there is substantial scope for a similar decrease of pesticide use in certain crops like cotton, vegetables and rice. However, as noted before, there may be an increase in herbicide use in countries where labour is becoming scarce, while intensification may make insecticide and fungicide use economically viable. On the other hand, older, cheap, off-patent pesticides will gradually disappear, increasing the price of pesticides but reducing their hazard. In the longer term, non-chemical control methods are expected to gain strongly in importance in both developed and developing countries (Zadoks, 1992).

4.6 THE LIVESTOCK PRODUCTION

The projections of production and consumption of livestock products were presented and discussed in Chapter 3, together with the implications for the growth of demand for cereals and oilseed proteins for feed. This section provides some further discussion of the likely evolution of the main parameters and other issues underlying the projections of production. In the first place, the trend for the share of the pig and poultry sectors in total meat production is likely to continue (Table 4.15). In particular, the growth of the poultry sector would continue at high rates, though lower than in the past, while that of the pig sector may be in the future much less rapid than in the past, because of the expected slowdown in East Asia, the region which accounts for 90 percent of total pigmeat production. Possible developments in China account for much of this slowdown, as discussed in Chapter 3.

Concerning the main production parameters, the increases in livestock numbers and in the offtake rates (the percentage of animals slaughtered each year) have been the dominant sources of growth in meat production in the past in the developing countries. This will continue to be so in the future. However, it must be noted that the historical data on animal numbers are often not sufficiently reliable to document developments in the sector. For example, some recent research indicates that the existing data may underestimate significantly actual livestock populations, particularly of small ruminants (Wint and Bourn, 1994, see below).

In many countries, there is only limited potential to increase livestock numbers in extensive systems. Overall, there has been a trend towards more intensive production systems in the very diverse production systems prevailing in the developing countries. This trend will continue and in the future much of the increased output of pigmeat, poultrymeat, eggs and, to a lesser extent, dairy would come from the further expansion of intensive and semi-intensive production systems with the use of supplementary feeds. This would make it possible for increases in yields per animal (carcass weight, milk, eggs) to be a more important source of growth than in the past.

Livestock production systems differ in their ability to respond to changing market conditions, reflecting primarily differences in the biological characteristics of the production process. With industrial-type systems being increasingly employed in the poultry sector of developing countries, production can respond relatively quickly to changing market conditions because of fast reproduction cycles and proximity of operations to urban markets. A wide range of commodities in elastic supply can be used to provide the required feed. Thus, compared with ruminants, poultry has a flexible feed resource base, and feed conversion efficiencies are high. However, the production systems of developing countries are usually highly dependent on imported technology and inputs. Eggs, pork and, to a lesser extent, dairy production systems tend also to be relatively responsive to changing market conditions. Due to the technological characteristics of intensive production, poultry, pig and most dairy systems cannot be gradually transformed from traditional to intensive production. Thus, at the level of the individual production unit, the growth process tends to be discontinuous rather than evolutionary.

In contrast, the production of ruminant meat and, to a smaller degree, dairy tends to be much less responsive to changes in demand, because of the long reproduction cycles, low feed conversion efficiencies and low degree of specialization. Thus, movements away from traditional systems toward more intensive methods tend to be slower and of an evolutionary character. Mixed farming systems are also unlikely to be highly responsive to growing demand for ruminant livestock products, simply because of the other functions that livestock have to fulfil at the farm. The possibility of expanding production through a gradual transformation of traditional farming systems is usually insufficient to respond effectively to growing consumer demand. Consequently, modern production systems, similar to those in the developed countries, have emerged in almost all developing countries alongside traditional systems. As the latter systems prove increasingly unable to meet the rising demand, in the future a larger share of total supply is likely to come from more intensive systems.

Some countries will probably find it difficult to move along the path of significant intensification and resulting productivity gains in the foreseeable future. For example, the low feed resource base and the import requirements for intensification of production would limit progress in this direction in most Sahelian countries. The difficulties in the transition from extensive to more intensive livestock production enhance the risk of environmental degradation. One of the main threats comes from overgrazing. It is believed that in many countries, especially in semi-arid ones, livestock numbers already exceed the carrying capacity of unimproved grazing land. There are major institutional and economic constraints on the path to achieving a sustainable balance between livestock numbers and forage and other feed resources. These problems will be difficult to overcome in the short to medium term and are likely to grow rather than diminish in scope and gravity.

The above-mentioned recent study (Wint and Bourn, 1994) provides interesting new insights in the evolutionary processes taking place in African livestock systems. The study is based on extensive surveys of livestock populations in Mali, Niger, Nigeria, Chad and the Sudan, and covers both pastoral and village livestock systems as well as arid, semi-arid, sub-humid and humid agroecological zones. The general conclusion is that in most situations the intensity of livestock production activities is closely correlated with the intensity of human activities (as measured by habitation density and share of land cultivated) and not, or only weakly, related to the distribution of natural grazing resources. This holds both for village and pastoral livestock activities. These findings suggest that a trend exists for livestock systems to become less dependent on the availability of extensive rangelands and for livestock production to be more closely related to the more secure feed resources associated with proximity to human settlements. They tally with the knowledge that feed requirements for ruminant and non-ruminant animals have traditionally been provided by a mix of grazing, crop by-products and, to a lesser extent, cultivated fodder crops. Ruminants depend the most on these feed resources. Reduced communal grazing resources due to increasing population, arable land expansion and degradation of pasture are making livestock increasingly dependent on crop residues and marginal feed resources.

Though pasture and forage sources remain the most important animal feedstuffs in developing countries, their supply has increased only slowly and has been inadequate to meet demand for livestock products. Concentrate feeds, mainly feed grains, have been increasingly used to supplement other fodder. Grain output has been growing much faster than pastures and fodder and its use as feed has increased considerably in the past 30 years (see Chapter 3). Higher proportions of intensive dairying, poultry and pigs will increase the use of cereals as feed. Its share in total livestock feed is expected to increase further as natural grazing resources become scarcer and the institutional changes required to control overgrazing and eventually reverse its consequences will take many years. Pasture improvement and the growing of forage have yet to be widely accepted by pastoralists or settled farmers, except in parts of North Africa, the Near East and China. Expected crop production in some countries will not provide sufficient by-products to meet protein and, in some cases, the metabolizable energy needs. Some nutritional deficits can be met by the use of feed additives, notably urea and molasses, but most of the shortfall in roughages will have to be met by the use of concentrates.

Concern is often expressed whether the developing countries will be able to increase cereal supplies by as much as needed to support the growth of their livestock sector. For example, Nordblom and Shomo (1993) consider that foreign exchange shortages in the Near East/North Africa region will make it difficult for the feed deficits to be met by cereal imports. Livestock production in the Near East/North Africa region is indeed fairly intensive in the use of concentrate feedstuffs and other regions are projected to follow on this path, though they will continue to use much smaller quantities of cereals per unit of livestock output than the Near East/North Africa region. The cereals projections underlying the projection of the livestock sector were presented in Chapter 3. They assume that for the Near East/North Africa region the cereal intensity of livestock production (amount of cereals used per unit of product) will not continue to grow as in the past following gains in the efficiency of livestock production.

Table 4.15 Meat production by species in the developing countries (including China)

| | Meat production | | | | | | Livestock numbers | | | | |
| | million tonnes | | | | annual growth (%) | | million | | | annual growth (%) | |
	1969/71	1988/90*	1991/92*	2010	1970–90	1988/90–2010	1969/71	1988/90	2010	1970–90	1988/90–2010
(93) Developing countries											
cattle and buffalo	12.1	18.6	19.3	20.5	32.3	2.2	2.7	798	1005	1369	1.3	1.5
sheep and goat	3.0	4.9	5.0	5.6	9.5	2.8	3.1	869	1129	1578	1.5	1.6
pig	9.7	28.3	28.6	33.5	64.0	6.1	4.0	291	486	860	2.2	2.8
poultry	3.7	12.9	13.3	16.7	36.9	7.0	5.1	2504	6469	12318	5.4	3.1
total meat	28.5	64.7	66.1	76.3	142.7	4.6	3.8					
Sub-Saharan Africa											
cattle and buffalo	1.7	2.3	2.2	2.3	4.2	1.6	2.9	129	159	200	1.2	1.1
sheep and goat	0.7	0.9	0.8	0.9	1.8	1.5	3.3	203	259	344	1.4	1.4
pig	0.2	0.3	0.4	0.5	0.8	3.3	4.3	6	11	16	3.8	1.8
poultry	0.3	0.9	0.8	0.9	2.2	5.0	4.6	339	630	1097	3.5	2.7
total meat	2.9	4.4	4.2	4.6	9.0	2.2	3.5					
Near East/North Africa											
cattle and buffalo	0.8	1.4	1.3	1.4	2.4	3.1	2.6	37	37	52	−0.1	1.7
sheep and goat	1.0	1.4	1.5	1.5	2.7	2.2	3.0	203	240	326	1.2	1.5
poultry	0.4	1.6	1.8	1.9	4.6	8.1	5.1	206	677	1125	6.9	2.5
total meat	2.2	4.4	4.6	4.9	9.7	4.1	3.8					

East Asia (incl. China)												
cattle and buffalo	0.9	2.3	3.2	3.6	6.4	4.7	5.0	118	153	332	1.5	3.8
sheep and goat	0.3	1.1	1.1	1.3	2.0	7.4	3.0	157	220	371	1.5	2.5
pig	7.3	24.6	24.8	29.5	57.2	7.0	4.1	216	388	727	2.4	3.0
poultry	1.6	5.1	5.2	7.1	17.3	6.3	1.0	1178	3335	7415	6.1	3.9
total meat	10.1	33.1	34.3	41.6	82.9	6.7	4.5					
South Asia												
cattle and buffalo	1.8	2.6	3.1	3.4	4.5	2.2	2.6	293	335	419	0.8	1.1
sheep and goat	0.5	1.1	1.2	1.4	2.3	4.0	3.5	148	247	337	2.9	1.5
pig	0.2	0.4	0.4	0.4	0.6	2.9	2.6	7	11	14	2.8	1.1
poultry	0.2	0.5	0.5	0.6	2.0	6.5	6.4	235	571	857	5.0	2.0
total meat	2.7	4.6	5.3	5.9	9.4	3.0	3.4					
Latin America/Caribbean												
cattle and buffalo	6.9	10.0	9.4	9.7	14.8	1.9	1.9	218	319	364	1.9	0.6
sheep and goat	0.5	0.4	0.4	0.4	0.7	−0.6	2.7	152	153	187	0.2	1.0
pig	1.9	3.0	3.0	3.1	5.3	2.3	2.8	63	75	103	0.8	1.5
poultry	1.2	4.8	4.9	6.1	10.8	7.8	3.9	546	1256	1822	4.5	1.8
total meat	10.5	18.2	17.7	19.4	31.6	3.0	2.7					

*Revised data as known in May 1994, but not used in this study.

Only in very few developing countries have meat and dairy industries developed to the stage where they can provide safe and regular supplies to their rapidly expanding urban populations. Over 90 percent of the livestock in developing countries is owned by rural smallholders with inadequate links to urban markets. The failure of existing structures and organizational patterns to cope with present and future demand for livestock products is evident in a number of aspects. First, demographic expansion has increased technical and infrastructural difficulties in meeting the effective demand, which are sometimes reflected in high price differentials between rural and urban areas. Second, environmental pollution, mainly waste from industrialized livestock production units and processing (in particular, slaughterhouses) is becoming ever more serious because of insufficient structures and the absence of adequate regulations or the lack of their enforcement. Lastly, insufficient food safety standards, because of technical and institutional constraints, are a continuous and growing human health hazard.

Animal genetic diversity

While only a few species of livestock are used to produce livestock products such as meat, milk, skins, fibre and draught power, those species have each been developed to produce specific products in a wide array of production environments, giving rise to a large number of unique breeds each with their own gene pool. It is the range of genetic diversity formed by this array of breeds which provides the key to future increases in efficiency and sustainability of livestock production.

Indigenous breeds have been developed to produce within their own specific environments and often possess attributes which are not immediately apparent, such as, for example, the ability to withstand local stress conditions which may not occur each year. In many cases, improved breeds are introduced which, under different conditions, have greater output and the resulting crossbreeds may well be better than the local pure breeds. Subsequent backcrosses to the improved breeds may, however, result in lower overall productivity due to lower rates of reproduction and lower survival chances, greater disease susceptibility, and the inability to cope with a high share of roughages in their feeds.

Crossbreeding can be a very valuable strategy as it exploits the benefits of hybrid vigour. However, sustainable systems of crossing are difficult to achieve in certain species, particularly in those with low rates of reproduction such as horses, cattle, buffalo and some sheep and goats. Practical bottlenecks are the more complicated logistics and reliable provision of sufficient crossbred replacement stock.

Nevertheless, new technologies such as embryo cloning combined with other modern techniques such as artificial insemination, *in vitro* fertilization and semen or embryo sexing, offer some potential for the better utilization of heterosis. For example, the continued production of first cross (F_1) females for milk production, exploits, to the maximum extent possible, the advantages from both indigenous and exotic genes. Such use of F_1 animals is commonplace in species with higher reproduction rates (pigs, poultry, some sheep).

The ability to provide now and in the future animals best suited to the various specific environments, depends on the maintenance of a wide spectrum of diversity within each domestic species. The diverse strains can then be used as required to cope with the inevitable change in production environments which occurs with development. The maintenance of genetic diversity is crucial for the full exploitation of all genetic means to provide in the most efficient way animal products and the most efficient means of sustaining domestic animal diversity is through well designed breeding operations.

Animal health

Meat, milk and egg production are still severely limited by pests and diseases. Some estimates show that at least 5 percent of cattle, 10 percent of sheep and goats and 15 percent of pigs die annually due to diseases. Apart from the direct loss of animals, indirect losses are incurred as a consequence of poor reproduction efficiency, retarded rates of growth and low levels of production. Growth in meat production between now and 2010 is expected to be derived for the greater part from increases in animal numbers with the remainder coming from improved productivity. Together with improved management, veterinary measures to control major epizootics and various disease vectors (such as ticks and tsetse fly) and preventive medicine will play an important part in increasing yields from both single animals and herds.

Major infectious animal diseases are those that are of significant economic importance, or have public health implications (such as rabies or brucellosis), or have recently been introduced and threaten to disrupt the sector (such as African swine fever (ASF) in Latin America in the 1980s). Disease eradication in developing countries has been fraught with difficulties but the successes are notable. ASF has been eradicated from Cuba, Brazil and the Dominican Republic, and babesiosis from large areas of Argentina and Mexico. Contagious bovine pleuropneumonia has been eliminated from the Central African Republic. Foot and mouth disease has been eradicated from all Central American countries and also from Chile, though it is still present in all other countries in Latin America.

There is a large group of chronic diseases that have more insidious effects than the major infectious diseases. Their importance is often overlooked and seriously underestimated. Though less obvious, they often have a serious economic impact through their effects on production or reproductive performance. Examples are helminth infestations, enzootic pneumonia of pigs, mastitis in dairy cattle, and chronic respiratory diseases in poultry. While managerial procedures and prophylactic animal health measures are facilitated when stock is raised under intensive production methods, the higher stocking rates and heightened stress can increase the risk of catching diseases. The development of more intensive production systems causes also a change in the disease spectrum. While in cattle rinderpest and pleuropneumonia decline, other diseases such as brucellosis, leptospirosis and mastitis usually become more important. For poultry, the conversion from extensive village poultry practice to intensive commercial systems shifts importance from, for example, Newcastle disease towards chronic respiratory disease, Marek's disease and Gumboro disease.

In Africa, existing and planned production facilities should make the continent largely self-sufficient in key vaccines which will be a base for substantial improvements in livestock health. The pan-African rinderpest campaign will require around 100 million doses annually, and protecting the cattle population at risk from bovine pleuropneumonia requires some 60 million doses per year. The major disease constraint in Africa however remains trypanosomiasis, which is virtually the only disease which, without preventive measures, entirely precludes the introduction of cattle, though some cattle strains have developed a degree of resistance.[14] Many sub-humid tsetse-infested areas with good agricultural potential could be utilized in a more sustainable manner with mixed farming systems using draught animals. Recent research has provided new control techniques which require low technology inputs and, unlike older methods, do not rely on large applications of insecticides. In many parts of sub-Saharan Africa, however, attempts to improve livestock breeds and dairy production will continue to be hampered unless efforts are made to contain trypanosomiasis. As in other developing regions, the likely expansion of intensive pig and poultry production in Africa will depend on regular measures to control diseases such as avian encephalitis and infectious bronchitis.

Unlike Africa, Latin America is free from diseases such as bovine pleuropneumonia, rinderpest and peste de petits ruminants. Foot and mouth disease however continues to be a problem. After the successful eradication of the New World screwworm from the United States and Mexico as well as North Africa, efforts are concentrated on screwworm control in the Caribbean and Central America.

Asia is also largely free of major infectious animal disease problems in cattle and buffalo. A programme for rinderpest control has been implemented in the Near East and another one is planned for South Asia, both having substantive vaccine requirements. Control of foot and mouth disease and brucellosis in the Near East is of particular concern. Developing preparedness to deal with emergency animal diseases has become an important activity in the region. In many Asian countries, notably the Philippines, Thailand, Indonesia and Bangladesh, the poultry sector is very important and control of Newcastle disease crucial, particularly at the village level.

Where intensive pig industries have been established, classical swine fever is one of the main threats facing production and trade in pigs and pig products. Diagnosis needs sophisticated laboratory testing and trained personnel, while facilities are necessary for control and eventual eradication. African swine fever would have devastating effects on the large pig population in Asia and careful monitoring is needed to prevent it from entering the region.

Further development of good quality low-cost vaccines against most bacterial and viral diseases is expected in the near future. The failure of some international eradication campaigns (e.g. rinderpest) was however not due to the lack of a suitable vaccine, but to poor veterinary infrastructure. One of the preconditions for improved veterinary health will be further investment in expansion of diagnostic facilities and in training of veterinary personnel.

NOTES

1. There are a number of indications that data on cultivated area in China under-report land actually in use (e.g. see USDA, 1991, 1993b). Some sources indicate that under-reporting amounts to about 30 percent, i.e. instead of a cultivated area of about 96 million ha (official data), actual cultivated area would be 125 million ha. Data on average cropping intensity (151 percent) seem to be fairly accurate as are data on production. If correct, this would imply a harvested area of 189 million ha instead of 145 million ha and yields and inputs per harvested hectare would be lower than official data show. Until such uncertainties about the actual data on area and yield by crop are resolved, it is difficult to make meaningful projections for such variables.

2. See, for example, Delgado and Pinstrup-Andersen (1993) for a critique of the land expansion projections to 2000 in the 1987 edition of this study (Alexandratos, 1988). Yet the actual data, halfway into the projection period, show that expansion of harvested area for the major crops in the developing countries (see below) had been taking place largely on the projected path.

3. These crops are: millet, sorghum, maize, spring wheat, winter wheat, barley, bunded rice, upland rice, sweet potato, cassava, white potato, phaseolus bean, groundnut, soybean, cowpea, chickpea, oil palm, sugarcane, banana, olive and cotton.

4. For example, for maize in the moderately cool tropics, the maximum constraint-free yield is 10.9 tonnes/ha at the high technology level and 2.7 tonnes/ha at the low technology level while in the warm tropics these yields are 7.1 and 1.8 tonnes/ha respectively (see FAO, 1978–81, Vol. 3: 124–31). Thus, land which could produce less than 20 percent of these yields (2.2 tonnes/ha of maize under high technology, 0.54 tonnes/ha under the low one in the moderately cool tropics) was classified as not suitable for maize though it was included in the agricultural land if it met the minimum yield criterion (20 percent of MCFY) for one or more of the other crops.

5. For a more extensive explanation of the methodology, see FAO (1978–81, 1982). The estimate of total potential (2573 million ha) presented in Table 4.1 (and in Appendix 3 tables) is well above the 2142 million ha that had been estimated for the 1987 edition of this study (Alexandratos, 1988) for a number of reasons. Although also the 1987 estimates were based on soil data of the FAO–UNESCO *Soil Map of the World* (SMW, FAO, 1971–81), measurements now were made with the aid of a Geographical Information System while the 1987 estimate was based on a manual grid count. This improved reading of data led to a number of changes. The number

of crops for which suitability was tested was increased from 11 in 1987 to 21 in the present study. Likewise, suitability was evaluated on the basis of the yield results under any one of the three technology levels, while the 1987 estimate was based on the yield results under the low technology level only. Furthermore, a number of rules applied in the procedure have been changed to reflect refinements in the methodology developed since 1987. Finally, a number of ex-post adjustments to the mechanically derived results were made in 1987 to take into account man-made changes to land resources (for example, terracing of slopes deemed unsuitable for agricultural production in the mechanical procedure). Such *ad hoc* expert judgement-based changes were not made in the present study.

6. The data underlying the maps are somewhat different from those in the tables because for map-drawing each map unit was allocated *in toto* to the agroecological cell and land class which occupied 50 percent or more of its area. For this reason, the term "dominant land class" is used in the legends of the maps.

7. Human settlement areas were estimated as follows: the only country with systematic data is China, for which there are data on both population density and non-agricultural land use per person (residence and infrastructure areas) for about 2000 counties. Based on these data a function was estimated linking non-agricultural land use per person to population density (the higher this density, the lower area per person used for non agricultural purposes). This function was subsequently used for all countries and all agroecological zones in each country. Estimates of population densities by agroecological zones were derived from the data used in the study. "Land Resources for Populations of the Future" (FAO/UNFPA, 1980). All these estimates are tentative and probably subject to large margins of error. The most recent world data-set for land has data on built-up areas for 17 developed countries and none for the developing countries (World Resources Institute, 1994).

8. Environment Canada (1989). The editor is grateful to Dr J. Dumanski of Agriculture Canada for bringing this document to his attention.

9. The land classified as not suitable for rainfed crop production (class NS) was tested for its suitability for forest tree species using the same methodology employed for the evaluation of suitability for rainfed crop production, except that the crop growth requirements were substituted by those for trees.

10. Classes I, II, III, IV of the Categories for Conservation Management of the International Union for the Conservation of Nature (IUCN) (IUCN, 1990).

11. Maps and inventories of National Parks, Conservation Forest and Wildlife Reserves were made available by the World Conservation Monitoring Centre, Cambridge, United Kingdom.

12. The data presented earlier for the land with rainfed crop production potential by LGP classifications can be used to draw inferences about water supplies for rainfed agriculture, as discussed in Chapter 2.

13. The term "pest" here is used to describe organisms that cause crop damage, such as animal pests including insects, mites, nematodes, and pathogens such as fungi and bacteria, and weeds. The term "pesticides" refers to all chemical means to control pests. "Active ingredients" refers to the biologically active part of pesticides.

14. Some cattle, sheep and goat strains in Western Africa have developed tolerance to trypanosomiasis through natural selection. Limited information (from the International Livestock Centre for Africa) on the productivity of trypanotolerant cattle suggests that under light tsetse pressure their loss of performance is negligible and comparable with breeds outside trypanosomiasis risk areas. However, under medium and high tsetse pressure the productivity index falls around one-quarter and one-half respectively. No evidence suggests that the trypanotolerant sheep and goats have a lower level of productivity than other sheep and goats in Africa.

Africa

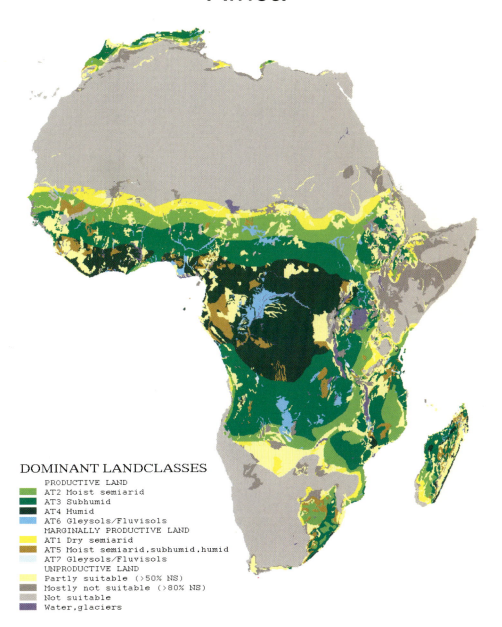

DOMINANT LANDCLASSES

PRODUCTIVE LAND
- AT2 Moist semiarid
- AT3 Subhumid
- AT4 Humid
- AT6 Gleysols/Fluvisols

MARGINALLY PRODUCTIVE LAND
- AT1 Dry semiarid
- AT5 Moist semiarid,subhumid,humid
- AT7 Gleysols/Fluvisols

UNPRODUCTIVE LAND
- Partly suitable (>50% NS)
- Mostly not suitable (>80% NS)
- Not suitable
- Water,glaciers

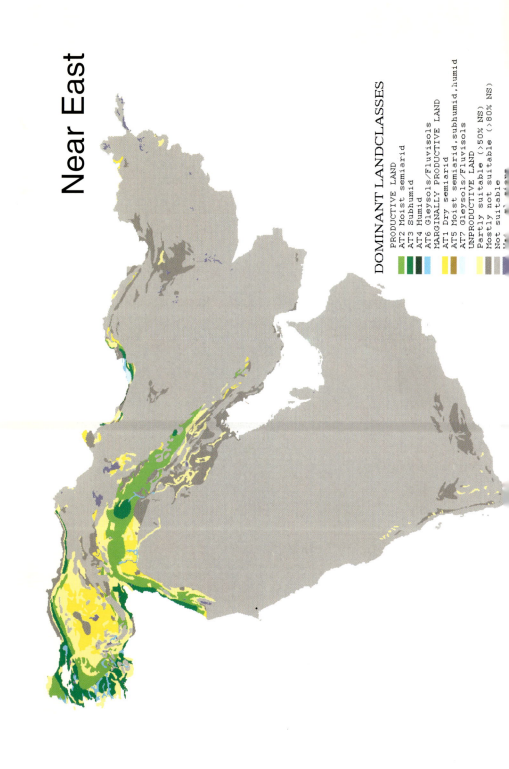

Near East

DOMINANT LANDCLASSES

PRODUCTIVE LAND
AT2 Moist semiarid
AT3 Subhumid
AT4 Humid
AT6 Gleysols/Fluvisols
MARGINALLY PRODUCTIVE LAND
AT1 Dry semiarid
AT5 Moist semiarid,subhumid,humid
AT7 Gleysols/Fluvisols
UNPRODUCTIVE LAND
Partly suitable (>50% NS)
Mostly not suitable (>80% NS)
Not suitable

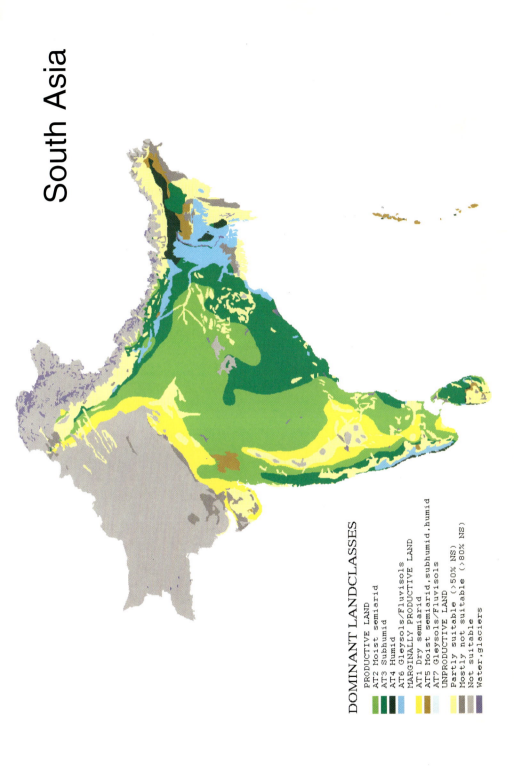

South Asia

DOMINANT LANDCLASSES

PRODUCTIVE LAND
AT2 Moist semiarid
AT3 Subhumid
AT4 Humid
AT6 Gleysols/Fluvisols
MARGINALLY PRODUCTIVE LAND
AT1 Dry semiarid
AT5 Moist semiarid,subhumid,humid
AT7 Gleysols/Fluvisols
UNPRODUCTIVE LAND
Partly suitable (>50% NS)
Mostly not suitable (>80% NS)
Not suitable
Water,glaciers

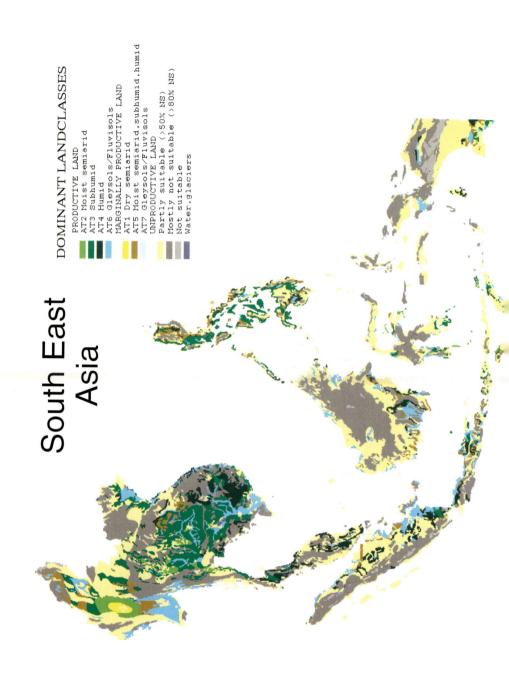

South East
Asia

DOMINANT LANDCLASSES

PRODUCTIVE LAND
AT2 Moist semiarid
AT3 Subhumid
AT4 Humid
AT6 Gleysols/Fluvisols
MARGINALLY PRODUCTIVE LAND
AT1 Dry semiarid
AT5 Moist semiarid,subhumid,humid
AT7 Gleysols/Fluvisols
UNPRODUCTIVE LAND
Partly suitable (>50% NS)
Mostly not suitable (>80% NS)
Not suitable
Water,glaciers

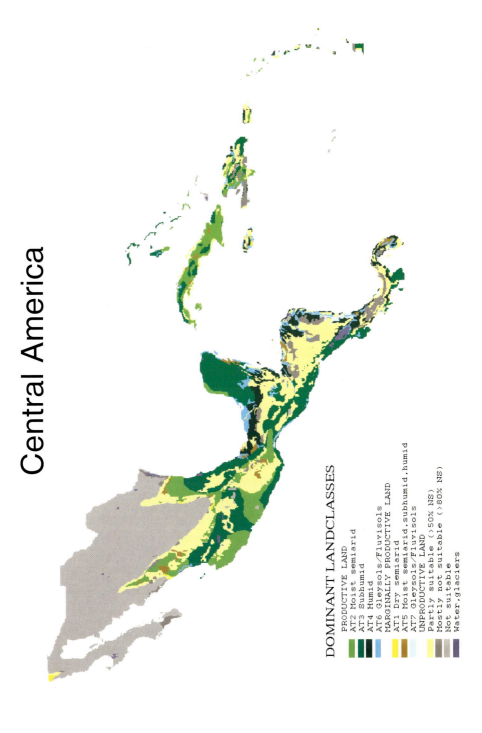

Central America

DOMINANT LANDCLASSES

PRODUCTIVE LAND
AT2 Moist semiarid
AT3 Subhumid
AT4 Humid
AT6 Gleysols/Fluvisols
MARGINALLY PRODUCTIVE LAND
AT1 Dry semiarid
AT5 Moist semiarid,subhumid,humid
AT7 Gleysols/Fluvisols
UNPRODUCTIVE LAND
Partly suitable (>50% NS)
Mostly not suitable (>80% NS)
Not suitable
Water,glaciers

South America

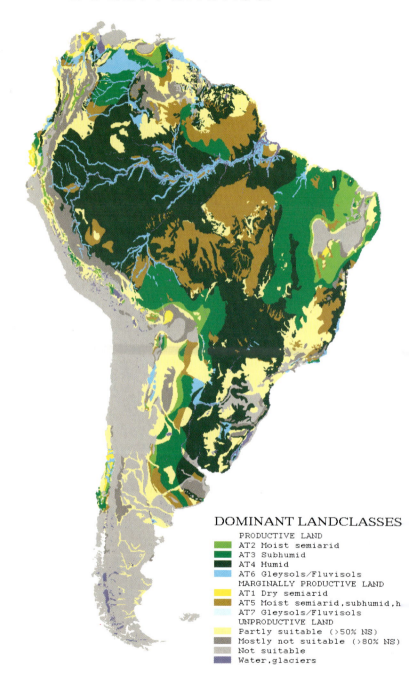

DOMINANT LANDCLASSES

PRODUCTIVE LAND
- AT2 Moist semiarid
- AT3 Subhumid
- AT4 Humid
- AT6 Gleysols/Fluvisols

MARGINALLY PRODUCTIVE LAND
- AT1 Dry semiarid
- AT5 Moist semiarid,subhumid,h
- AT7 Gleysols/Fluvisols

UNPRODUCTIVE LAND
- Partly suitable (>50% NS)
- Mostly not suitable (>80% NS)
- Not suitable
- Water,glaciers

ANNEX: MAPS OF DOMINANT RAINFED LAND CLASSES

Key

Dominantly productive land

AT2	Moist semi-arid	LGP 120–179 days	> 50% of area VS + S land
AT3	Sub-humid	LGP 180–269 days	> 50% of area VS + S land
AT4	Humid	LGP 270–365 days	> 50% of area VS + S land
AT6	Fluvisols and gleysols	Naturally flooded land (NFL)	> 50% of area VS + S land

Dominantly marginally productive land

AT1	Dry semi-arid	LGP 75–119 days	> 50% of area VS + S + MS land
AT5	Moist semi-arid, sub-humid and humid	LGP 120-365 days	> 50% of area MS land
AT7	Fluvisols and gleysols	Naturally flooded land (NFL)	> 50% of area MS land

Dominantly unproductive land

Partly suitable	LGP 75–365 or NFL	20–50% of area VS + S + MS land
Mostly not suitable	LGP 75–365 or NFL	0–20% of area VS + S + MS land
Not suitable	n.a.	
Water, glaciers	n.a.	

CHAPTER 5

Forestry

5.1 INTRODUCTION

Forests constitute a major form of land use and have a number of highly important functions. They provide services which contribute to the well-being of people; they have a major role in the environment; and their products are vital in the economy and daily living of people. Forests constitute a renewable resource capable, with sound management, of producing valuable products. They are capable of regrowth and regeneration, and at the same time, they fulfil environmental functions of soil and water conservation and the conservation of biodiversity. The products of the forest and forest industry are generally environmentally beneficent; they have potential for being recycled either in production or in energy generation and the industries themselves have potential for high energy efficiency and low negative impact on air and water quality.

Through history, forests, as the major natural occupants of land, have been subject to clearance for agriculture, pasture and human settlements. This clearance and cutting wood for fuel, construction and industry have been in the first instance without consideration of the need to ensure sustained delivery of products and services of the forests for the future. With the growth of population and wealth, the demand for land for agriculture as well as the demand for forest products and services increase. At the same time, the side effects of other activities influence the production and service functions of the forest. Thus, the production of more and more of the goods and services of forests are in competition with one another and with other uses of the forest land.

The future development of the sector has to confront the increased demand for its products, for its services and for the conservation of increasingly scarce ecosystems and biodiversity, as well as for providing sustainable livelihoods to forest-dwelling and forest-dependent communities. At the same time (as discussed in Chapter 4), the forest area, and thereby its supply capacity, will continue to be subject to increasing pressures for the transfer of forest land to agriculture, infrastructure and urban uses. Meeting these intricately interrelated demands and resolving the conflicts among them will be a major challenge for the future. Responding to this challenge will require more efficient

Table 5.1   Forest area in 1990

| | Forest area | | Other wooded land |
	(billion ha)	(% of total land area)	(billion ha)
World	3.4	26	1.6
Developed	1.4	26	0.5
Developing tropical	1.7	37	1.0
Developing other	0.3	13	0.1

and environment-friendly technologies for producing forest products and safeguarding the service functions of the forest; and the recognition of forestry's important role for ensuring sustainable livelihoods in reconciling the very diverse interests in making land use decisions.

The following sections consider, in turn, the state of forests in land use and changes therein, the present and projected future role of forest products in the economy and the interrelationships between the forest and the environment. The chapter concludes with a review of the future perspectives for sustainable development of the forest.

5.2 FOREST IN LAND USE

The world forest area was estimated in the FAO 1990 Forest Resources Assessment (FRA 1990; see FAO, 1993f) to be 3.4 billion ha – an average of some 0.7 ha per head of population. In this estimate, forests are defined as ecological systems with a minimum of 10 percent crown coverage of trees. In addition, there were some 1.6 billion ha of other wooded land with some woody vegetation.

Forests in the tropical zone

Natural forest

The 1990 area of the natural tropical forest, i.e. not including forest plantations, was 1.76 billion ha and accounted for 37 percent of the total land area of the tropical countries included (Tables 5.2 and 5.3). The area of forest per caput in the tropical zone was 0.72 ha. About three-quarters of the tropical forest is in the tropical rainforest zone and the moist deciduous forest zone. Dry lowland formations and upland formations each constitute about 12 to 13 percent of the total area. Deforestation is a major issue for the tropical forest. For the period 1980–90 it was estimated that the gross loss of tropical forest area (i.e. before accounting for the area added to forest by reforestation and afforestation) amounted to 15.4 million ha, or 0.8 percent of the forest area

Table 5.2 Tropical forest, estimates of forest cover area and deforestation by geographical sub-region

Geographic sub-region/region	Number of countries	Land area (million ha)	Forest cover		Annual deforestation 1980–90	
			1980 (million ha)	1990 (million ha)	(million ha)	(% p.a.)
Africa	40	2236.1	568.6	527.6	4.1	0.7
West Sahelian Africa	6	528.0	43.7	40.8	0.3	0.7
East Sahelian Africa	9	489.7	71.4	65.5	0.6	0.9
West Africa	8	203.8	61.5	55.6	0.6	1.0
Central Africa	6	398.3	215.5	204.1	1.1	0.5
Trop. Southern Africa	10	558.1	159.3	145.9	1.3	0.9
Insular Africa	1	58.2	17.1	15.8	0.1	0.8
Asia and Pacific	17	892.1	349.6	310.6	3.9	1.2
South Asia	6	412.2	69.4	63.9	0.6	0.8
Continental SE Asia	5	190.2	88.4	75.2	1.3	1.6
Insular SE Asia	5	244.4	154.7	135.4	1.9	1.3
Pacific	1	45.3	37.1	36.0	0.1	0.3
Latin America and Caribbean	33	1650.1	992.2	918.1	7.4	0.8
C America and Mexico	7	239.6	79.2	68.1	1.1	1.5
Caribbean	19	69.0	48.3	47.1	0.1	0.3
Tropical S America	7	1341.6	864.6	802.9	6.2	0.7
Total	90	4778.3	1910.4	1756.3	15.4	0.8

Note: The 90 tropical countries of this table do not include the countries of the region Near East/North Africa used in other chapters of this report, not the temperate countries of Asia (China, among them) and South America (Argentina, Chile, Uruguay). Of the 93 developing countries of this study, the FRA1990 data are available for 69 countries (see country list in Appendix 1). These 69 countries account for 1690 million ha of the 1756 million ha of forest area for 1990 shown in this table. The difference (66 million ha) is the forest area of 21 developing countries not covered individually in this study.

Table 5.3 Tropical forest, estimates of forest cover area and deforestation by ecological zone

Ecological zone	Forest formations	Land area (million ha)	Population density 1990 (persons/km²)	Pop. growth 1981–90 (% p.a.)	Forest cover 1990 (million ha)	(% land area)	Annual deforestation 1981–90 (million ha)	(% p.a.)
	FOREST ZONE	4186.4	57	2.4	1748.2	42	15.3	0.8
	Lowland formations	3485.6	57	2.3	1543.9	44	12.8	0.8
1	Rainforest	947.2	41	2.2	718.3	76	4.6	0.6
2	Moist deciduous	1289.2	55	2.4	587.3	46	6.1	1.0
3	Dry and very dry	1249.2	70	2.3	238.3	19	2.2	0.9
4	Upland formations (hill and mountain forest)	700.9	56	2.6	204.3	29	2.5	1.1
5	NON-FOREST ZONE (Alpine areas, deserts)	591.9	15	3.1	8.1	1	0.1	1.0
	TOTAL TROPICS	4778.3	52	2.4	1756.3	37	15.4	0.8

annually. This indicates a higher rate of deforestation than was estimated in the 1980 assessment.

In the assessment of the tropical forest resources, a high correlation has been found between the change in the forest area and the change in population density. The exact nature of the relationship varies between ecological zones. According to the model used, the process of population/forest interaction resembles a biological growth process where deforestation is observed to increase relatively slowly at initial stages of increases in the population density, much faster at intermediate stages and slowly in the final stages. This model has been used to estimate the change in the forest area where survey information on change was not available (for further discussion see FAO, 1993f).

Forest plantations in the tropical zone

Forest plantations have been established for the renewal of forests after they have been harvested and to replace forests that have been cleared for both the production of timber and fuelwood. The establishment and efficient management of forest plantations will contribute to secure both the productive and protection functions of forests, though biodiversity may be locally reduced. The total plantation area reported in 81 countries of the tropical zone was 43.9 million ha as at the end of 1990. It is estimated, taking account of imperfect stocking, that this is equivalent to an effective net area of 30.7 million ha. About one-third of the area is primarily for industrial production. The annual rate of afforestation and reforestation of 2.6 million ha is small, less than 20 percent of annual gross deforestation. The established forest plantations have a potential for wood production which is already at a level equivalent to or exceeding the developing countries' consumption of industrial wood of some 300 million cubic metres (m^3) per annum.

The effective extent of tree planting is greater in that considerable numbers of trees are planted outside the forest, around the farm or household or on boundaries, roadsides and embankments. Trees outside the forest make a major contribution to fuelwood, fodder and timber supply. In addition to forestry plantations, agricultural plantations of tree crops, such as coconut, palm and rubber, have potential as a source of wood. The total estimated area (for Asia alone) is 14 million ha, made up of rubber (7.2 million ha), coconut (4.2 million ha) and oilpalm (2.7 million ha). Several million cubic metres of rubber wood and coconut stems are being utilized in sawnwood production.

Protected areas in the tropical zone

In assessing the forest resources of both the tropical and temperate regions, special consideration has been given to the degree to which measures have been put in place to conserve the wide variety of species and habitats. One such measure is the establishment of protected forest areas. Currently protected forest areas conforming with categories I–V of the International Union for the Conservation of Nature (IUCN) cover 266 million ha or 5.4 percent of the total land area in the tropics, coverage being about 1 percent higher in Latin America than in either Africa or Asia and the Pacific.

Forests in the temperate zone

Forests in temperate countries cover some 1.64 billion ha, of which some 1.4 billion ha are in the developed countries and the balance in the temperate forests of some developing countries (including China). The main areas are in the former USSR (0.75 billion ha), North America (0.46 billion ha) and Europe (0.15 billion ha). Forests in the developed countries cover 26 percent of the land area and the area of forest per inhabitant averages 1.13 ha. In the developing countries situated in the temperate zone the percentage of land area under forest is 13 percent and the average forest area per caput averages 0.15 ha. The land area under forest in the developed countries is largely stable with slight changes due to removal of forest for urban use or addition to forest through afforestation of surplus and unused agricultural land and pasture. Forest which is cut for timber is usually reforested by planting and natural regeneration. In Europe, the forested and wooded land area increased by 2 million ha between 1980 and 1990 (FAO, 1994b). Problems for the forests in temperate countries include damage to the forest attributed to air pollution and damage from pests, diseases and fire.

Some 300 million ha of forests are designated as protected areas in the temperate zone countries, of which some 250 million ha are in the developed countries (4.5 percent of their land area). Two percent of the land area of China and 7 percent of other temperate developing countries is designated as protected areas.

Issues of forest in land use

The state of the world's forests and the trends in changes are described above. The future development of these trends will determine the ability of forests to meet the demand for their products and services. In this discussion various factors which play a role in the changes in forests are considered with a view to identifying better the areas where policy may play a role in the future development of land use.

The forests of the developing countries are under pressure from population growth and the extension of agriculture and pasture use of land. The situation in the developed countries indicates stability in land use (see also Table 4.6), and increase in the stock and yield of forests in response to management, but there is some evidence of forest decline due to fire and environmental factors.

Institutional issues in temperate zone forests

In the developed countries forests may be in public or private ownership, or in various forms of common ownership. In most developed countries the ownership of forests is effectively demarcated and regulated under the law. The management of forest lands is in most countries subject to constraints aimed at conservation of the soil and land stability in upland regions and in a number of countries harvesting is regulated to ensure sustained yield of timber. In recent years, issues relating to the impact of harvesting and tree planting on the environment and biodiversity have emerged and have led to new policies and new legal constraints, particularly affecting public forests and policy instruments such as subsidization of private forestry. In some countries, the rights of original people to forest assets have been prominent in public debate. The issues of ownership, management and privatization of forests are matters of current discussion in many of the countries in transition in the ex-CPEs.

Tropical deforestation – causes and institutional factors

The FAO 1990 Forest Resources Assessment shows that deforestation has assumed important dimensions in the tropical zone. This section discusses possible causes and factors affecting these changes. It must be recognized, however, that data inadequacy and the complex interactions between the different factors contributing to deforestation make it difficult to establish quantitatively the extent of deforestation attributable to any particular cause. Therefore, the indications suggested here are necessarily qualitative inferences.

As noted already in Chapter 4, land use statistics are unfortunately not so precise that they provide reliable information for monitoring the movement of land from one use to another. The FAO land use data indicate net increases in arable land and pasture of respectively 32 million ha (Table 4.6) and 13 million ha between 1980 and 1990 in the developing countries (excluding China). In the same period it is estimated that there has been a net reduction in the tropical natural forest area of some 150 million ha (Tables 5.2 and 5.3). Only part of this reduction is the recorded transfer from forest to agriculture and pasture. In addition to conversion to recorded agricultural uses and pasture, the main causes of deforestation are the following:

1. Conversion to subsistence agriculture and rough grazing not recognized by official agricultural land use statistics as conversion to agriculture or pasture.
2. Persistent over-cutting for fuelwood and charcoal production which reduces the forest to "other wooded land" or completely eliminates woody growth.
3. Commercial felling of timber. In the absence of further intervention such areas will naturally revert to forest. However, the construction of logging roads for commercial felling provides access and frequently facilitates conversion of forest land to other uses.

The pattern of deforestation and degradation indicates that the most intensive extension of forest clearing and forest degradation radiates out from centres of population and established agriculture. A large part comes from the informal extension of marginal agriculture and pasture by small farmers and landless people, a process which occurs without support of selection of land or crops according to productive capability, without any extension service to support the establishment of productive and sustainable agriculture and frequently without recovery of wood and woody biomass for productive use as timber or for fuel. Frequently this informal settlement of forest land involves land with poor agricultural potential and is often in upland and hilly areas. The settlers tend to be the least privileged and have the least potential for adopting the required technology and inputs for sustainable agriculture.

Population

In many developing countries the high population growth rates, in combination with limited employment opportunities, persistent poverty, inequality of access to land and insecurity of food supply, mean that the only option for subsistence is migration, often to forest areas, to find land for agriculture or pasture and shifting cultivation. Thus, population growth occurring in such unfavourable conditions stimulates this kind of out-migration and consequent deforestation. While on average some 60 percent of the net population increase in the developing countries is absorbed by migration to the urban areas, the rest is the net increase of rural and agricultural population – more in sub-Saharan Africa and South Asia, less in Latin America and the Near East/North Africa.

Security, resource control and people's participation

In the developing countries forest may have been assigned to public ownership but this has not always been followed up with effective demarcation. The assignment may be in conflict with traditional and communal ownership by local people. Common rights of access to and use of the forest by local people may be exercised but may or may not be legally recognized. Whether or not the forest is formally vested in public ownership, it may be regarded as available to open access by the people. National policy may encourage or permit settlement on forest land, or, perhaps more frequently, condone encroachment. The settlers may, however, have no security as regards their future supply of forest products or use of the land and therefore may have no incentive to adopt sustainable utilization practices involving investment in future production or conservative use of the forest or the land.

A consequence of population increase is a greater pressure on limited areas available to rural communities. Privatization and encroachment reduce the areas available for communal use. Traditional forms of common property resource management tend to break down. In these circumstances, the growth of population tends to lead to ungoverned and unsupported migration to areas which are less able to support an agricultural population, thus worsening the instability of an already unstable community with tenuous rights over the land which it occupies. Among the possible responses to this unsatisfactory relationship between marginalized rural communities and the use of the forest is the adoption of measures designed to increase the involvement of those communities in the management of the forest to their benefit. In these cases there is a clear need to promote people's participation in solving the problems of deforestation and low agricultural productivity.

Intersectoral policy impacts

It is clear from the foregoing discussion that the factors that have an impact on the forest, include not only the direct policies and decisions of forestry authorities but also the policies regarding crop and livestock production and the more general government policies supporting settlement and communications. They also include actions by urban authorities in search for water or energy supplies and policies to promote exports of forest products or promoting foreign investment in mining or energy generation. The pressures emanating from many directions impinge not only on the forest but also on the welfare of traditional communities dependent on the forest and on many aspects of environmental conservation.

In many countries, forest policy and legislation aim at the conservation of forest area and sustained production of timber. However, in the absence of effective regulations, institutions and incentives to secure sustained production, the long production cycle of the forest tends to lead to an exploitation aimed at obtaining immediate returns. Thus, concession agreements, if not appropriately formulated and implemented, may provide incentives for the immediate utilization of existing stocks of mature timber with little regard to the need for sustainability. Few such agreements have adequate enforcement mechanisms or provide incentives to the short-term concessionaire to secure sustained management. Thus, the practice of harvesting combined with the pressure to use the land for agriculture can contribute to deforestation, conservation policies notwithstanding.

5.3 FOREST PRODUCTION IN THE ECONOMY

A major function of the forest is to supply wood and other conventional products. Forests and the forest industry contribute to the economy through the production of and trade in wood for energy, sawnwood, panels and paper. In 1990, the value of forest products was estimated to be somewhat over US$400 billion, of which about one-quarter was in the form of fuelwood and the balance as the value of forest products in industrial use. Products of the forest other than wood, such as cork, resin, mushrooms, wild fruits and gum can be important, in particular in the subsistence sector, but no estimate of their value is available. The value of exports of forest products was $97 billion in 1990, just over 3 percent of world merchandise trade.

Wood in energy supply

Table 5.4 shows the data and projections for the uses of wood and its products. The quantity of wood used directly as fuelwood and in the generation of energy is about 1.8 billion m^3, just over half of all the wood consumed. In addition, some 300 million m^3 of residues from the manufacture of wood products are recovered for energy production, making a total equivalent to 0.52 billion tonnes of oil. This is approximately 5 percent of world energy consumption. The bulk of world fuelwood consumption is in the developing countries, where it represents 80 percent of their annual wood production. This volume, equivalent to 0.4 billion tonnes of oil, constitutes 15 percent of developing country energy consumption. However, in 40 of the world's poorest countries, wood is the source of more than 70 percent of national energy consumption. In these countries consumption of energy from wood ranges from 0.1 to 0.5 tonnes of oil equivalent (toe) per caput, averaging 0.25 toe. It is noteworthy that the average use of wood in energy supply in developed countries is 0.2 toe per caput, where wood in all forms contributes only 1 percent of all energy consumed (FAO, 1994b).

Wood remains the main fuel for rural communities in many developing countries and for urban communities too, where the people do not have access to, or cannot afford, alternative fuels. In remote rural areas, especially in poorer countries, modern fuels are virtually unobtainable in any substantial quantities. Among communities dependent on traditional fuels, wood is the preferred fuel, but in regions where wood supply is scarce, twigs and leaves may be used as fuel and in some countries other biomass such as crop wastes and cowdung are used as fuel. This is particularly the case in the Indian subcontinent, where 50 percent of biomass energy is from crop wastes and dung, and in certain areas of Africa.

The predominant use of fuelwood is for household requirements, namely cooking and space heating. The efficiency of use of fuelwood in traditional fireplaces is low, with a useful energy recovery of only 10 percent. Programmes to introduce improved stoves using low cost local materials, are being carried out in many developing countries in order to reduce fuel used for cooking and heating and at the same time to reduce smoke in the kitchen and to improve kitchen hygiene. This is especially important for women and children who are often responsibile for cooking and fuelwood supply. Toxic fumes from traditional burning of wood can cause respiratory diseases.

The consumption of fuelwood by commercial and industrial enterprises is substantial in many developing countries in particular by rural industries such

Table 5.4 Current and projected consumption of forest products, 1990–2010

	1990			Growth 1990–2010			2010		
	World	Developed countries	Developing countries	World	Developed countries	Developing countries	World	Developed countries	Developing countries
	(million units)			(% per annum)			(million units)		
Fuelwood and charcoal (m^3)	1800	240	1560	1.4	0.8	1.6	2400	280	2120
Industrial roundwood (m^3)	1650	1270	380	2.5	2.0	3.8	2700	1900	800
Sawnwood (m^3)	485	373	112	2.5	1.5	4.1	790	500	250
Panels (m^3)	125	108	17	4.6	4.3	6.5	310	250	60
Paper (tonnes)	238	196	42	3.1	2.3	5.8	440	310	130

as fish, tea, coffee and tobacco drying and curing, commercial food preparation, baking and brewing, textiles, laundries, metal workshops and industries such as cement, ceramics and brickmaking. Brazil, for example, uses 6 million tonnes of charcoal per year for iron and steel production.

The supply of fuelwood from the forest tends to diminish with the clearance of forest in the areas of settlement. There is substitution of supply from trees planted around the farm, on roadsides, boundaries and on wasteland and from tree crops. The overall effect is a diminution in available supply and increase in cost because of increased competition for the available supply or because of the increased distance to the point of collection. In the neighbourhood of urban areas, in arid zones and areas of dense but poor rural populations without access to alternative energy, remaining forests are destroyed by overuse. Thus, the diminishing fuelwood supply potential of trees and forests in densely populated low-income regions will tend to become an increasingly important constraint to meeting the energy needs of the population.

Projected consumption of wood in energy

World consumption of energy from all sources may continue to expand at between 1 and 2 percent per year but the growth rate of energy consumption in the developing countries will be substantially higher. Given the above-mentioned supply constraints, the use of wood in energy supply in the developing countries would grow at a rate appreciably lower than the rate of growth of the economy and total consumption of energy, and probably lower than the growth of population. The trend for countries to substitute increasingly fossil and other fuels for wood would continue. Wood and biomass however will remain the main source of fuel for remote and poor rural populations. In developed countries the earlier trend towards decline in the use of fuelwood had been reversed into modest growth since the 1970s. This trend is likely to continue. Growing interest in some developing countries in the rehabilitation of degraded land through energy plantations and in developed countries in using set-aside land for energy trees and crops, could increase further the contribution of wood as a modern energy carrier. It is expected that the recycling of residues in energy production and particularly the use of waste paper no longer suitable for reuse in paper manufacturing, will increase. Based on the above considerations, world consumption of wood in energy use is projected to grow at 1.4 percent p.a. to some 2.4 billion m^3 in 2010 (Table 5.4).

Forest products in industrial use

Wood is used as raw material in the manufacture of sawnwood, wood-based panels are used mainly in construction, housing and furniture, and paper is used mainly in communications and packaging. It is also used in unprocessed form in the construction of housing, in agricultural fencing, posts and stakes, as the raw material for artisanal products, as transmission poles and for piling. The predominant use in the rural areas of the developing countries is in the form of unprocessed roundwood. The forest industry has grown substantially in the last 30 years: its output doubled in the developed countries, but increased five-fold in the developing ones. In the developed countries, the growth of the forest industry was slower than the growth of the overall economy, while in the developing countries the growth of the industry exceeded that of the overall economy.

Industrial roundwood

World consumption of industrial roundwood is over 1.6 billion m^3, with the bulk of it, some three-quarters, concentrated in the developed countries. Of this total, nearly 1 billion m^3 are utilized in the production of sawnwood and plywood, 0.4 billion m^3 are used directly to manufacture pulp for paper and 0.2 billion m^3 are utilized in unprocessed form. In addition, some 0.2 billion m^3 of residues from sawmilling are recycled to pulp manufacture.

Sawnwood, wood-based panels and paper

World consumption of sawnwood is close to 500 million m^3, of which some three-quarters is in the developed countries. The average consumption in the developed and the developing countries is respectively 300 m^3 and 30 m^3 per 1000 persons. Consumption of wood-based panels totals 125 million m^3 of which only 17 million m^3 are consumed in the developing countries (Table 5.4).

The world consumption of paper and paperboard of 238 million tonnes is made up of 100 million tonnes of newsprint and printing and writing papers used mainly in communications, about 12 million tonnes of household and sanitary papers, and the remaining 126 million tonnes are used in packaging, transport and other industrial applications. Per caput consumption of paper averages 45 kg worldwide, but with wide disparities between the developed countries (150 kg) and the developing ones (10 kg). The manufacture of paper utilizes three main sources of fibre: 61 percent wood pulp, 5 percent pulp of other fibres and 34 percent is recovered paper. In the developing countries, the respective proportions are 29, 27 and 44 percent. In recent years policy measures have been adopted in the developed countries to encourage recycling of used paper with the aim of reducing the volume disposed of as waste.

As noted earlier, the consumption of industrial wood products has grown substantially over the past three decades, most rapidly in the 1960s and early 1970s, more slowly in the 1980s. The rates of growth have been generally much higher in the developing countries than in the developed ones. The world growth rate of sawnwood consumption fell from 2 percent to 1 percent p.a. over the period while that in the developing countries was maintained at

5 percent p.a.; for panels, the world growth rate fell from 10 percent to 2 percent but growth in the developing countries remained at 10 percent and world growth of paper fell from 5 percent to 3 percent, while growth in the developing countries fell from 6 percent to 4 percent p.a.

Projected consumption of industrial forest products

The consumption projections are based on the estimated relationships between the rates of economic growth and of population growth and those of consumption of forest products. The projected growth of consumption of industrial forest products is shown in Table 5.4. In the developing countries the consumption growth is projected to be approximately equal to projected economic growth for these countries (about 5 percent per annum). This implies a nearly tripling of consumption in developing countries over the next two decades. To cope with this projected growth in consumption, an equivalent expansion in the production of industrial wood is needed. In the developed countries, the growth in consumption of forest products is projected to be lower than the growth of their economies, with consumption somewhat less than doubling over the next 20 years.

At the global level, the consumption of industrial roundwood is projected to be about 2.5 percent p.a., leading to a consumption of 2.7 billion m^3 by 2010. This growth rate is somewhat below that of the consumption of the industry's products. This would result from a continuing trend towards a more efficient use of wood raw material through more complete utilization of smaller size wood, recovery of residues as input to panels and paper and the increased recovery and recycling of used paper in paper manufacture. These trends are well established in the developed countries and are an area of considerable potential growth in the developing countries.

The supply of industrial wood in the developing countries depends at present mainly on supplies from natural forests. This is also the main source of supply of tropical timber entering international trade. In certain countries the current rate of harvesting is not sustainable over the long term. In some regions of high population density, forests have been cleared for agriculture following initial harvesting and thereby the potential for sustainable timber production has declined. The extension of management to secure continuing sustainable supply is an essential component of policy responses to this problem.

Trees planted individually or in plantations may be expected to play an increasing role in meeting developing country requirements for industrial wood. The plantation area so far established in the tropics has a yield potential nearly equal to current consumption of industrial wood in the developing countries. They are however not all oriented toward industrial use. An additional area of 50 to 100 million ha in appropriate locations would be needed to meet projected developing country requirements for industrial wood by the year 2010.

Developed countries' demand for industrial wood is projected to increase by nearly 50 percent over the period to the year 2010. It is well within the potential of existing forest areas to meet this demand, with sustainable management, stock improvements and improved efficiency in harvesting. The area is likely to be further increased by afforestation of land set aside from agriculture. High costs of operations, however, may lead to a continuation of the trend to reduce intensity of silviculture and to accumulate stock, particularly in less accessible areas.

Forest products other than wood

In addition to timber and fuelwood, forests also generate a wide variety of other products which make an important contribution to both the national and local economy and are significant sources of materials and food for local communities. Well-known industrial materials and commercial commodities, which also enter international trade, include cork, gum arabic and rattan, together with a wide range of gums and resins, bamboos, various oils, resin and turpentine, tanning materials, honey, seeds and spices, edible fungi, wildlife and wildlife products, barks and tree leaves and medicinal plants. The non-wood forest materials are essential inputs in artisanal activity, house construction and furniture manufacturing. They are often the basis of economic activities at the household level, mainly carried out by women. The products are traded in both rural and urban markets providing an additional source of cash income. They also provide an opportunity for productive employment in periods of reduced agricultural workloads.

Rural communities benefit directly from these products, as they provide food, fuel, medicinal herbs and extractives, building materials, materials for handicrafts, animal fodder, perfumes and dyes. The wildlife of the forest often provides the main source of animal protein for rural communities. Foods available from the forest enrich diets by providing vitamins and protein-rich components. They contribute to food security by their availability when agricultural crops are out of season or deficient due to drought.

In many countries the collection of non-wood forest products is the subject of established common rights of local people. In other cases, the collection is regulated through a system of licensing. Change of use of the forest can result in conflict with these customary activities. Clearing, felling or restriction of access may result in severe hardship to communities which depend on non-wood products for their livelihood. Governments therefore should take special action in such cases to protect the interests of local communities, for example, by entering into long-term usufruct agreements with people who agree to live in harmony with the forest, taking from it only what is necessary for their livelihoods, and ensuring the rejuvenation of valuable species.

The role of traditional forest products as materials and food for rural communities will continue to be important, and could even increase when these

Table 5.5   Trade in forest products, 1961 and 1990 (1990 dollars)

	1961			1990		
	Total $ billion	Round- wood %	Pulp and paper %	Total $ billion	Round- wood %	Pulp and paper %
Imports						
World	30	13	31	109	12	60
Developed	27	15	31	90	11	61
Developing	3	5	56	19	15	56
Exports						
World	26	12	35	97	9	62
Developed	24	7	38	84	7	67
Developing	2	60	4	13	20	28

communities gain access to markets. The continuing discovery of new products and new uses for the myriad plant and animal materials in forests, will further enhance this role. These products are of particular socioeconomic importance in providing a basis for small-scale industry, generating employment opportunities for women and men, particularly in remote rural areas.

Forest products in trade

The value of world exports of forest products in 1990 was US$97 billion, accounting for 3.3 percent of world merchandise trade and 23 percent of world exports of agricultural, fisheries and forestry products (Table 5.5). Imports and exports of the developed countries accounted for about 85 percent of this trade, with Europe alone accounting for about half of all trade. The largest importers are the USA, Japan, Germany and the UK, all with imports exceeding US$10 billion per year. The largest exporters are Canada, the USA, Finland and Sweden, all with exports exceeding US$9 billion per year. Among the developed countries, 13 are important net exporters and 16 significant net importers. Trade in forest products is particularly important to the economies of some developed countries. For three countries, forest products exceed 10 percent of their total exports and for a further five countries they exceed 5 percent of total exports.

Developing countries account for about 15 percent of world trade in forest products. The largest importers among developing countries are China, Korea Republic and Egypt, each with forest products imports exceeding US$2 billion per year. The largest exporters are Indonesia and Malaysia with exports exceeding US$3 billion, Brazil with exports of US$1.75 billion and Chile with US$0.8 billion. Regionally, China and neighbouring countries of East Asia and the countries of the Near East and North Africa are substantial net importers accounting together for half of all developing country imports of forest products. Other developing regions are in near balance or are net exporters. About 50 developing countries depend on net imports for their forest products consumption. This number does not include some very small countries which are totally dependent on imports for their consumption. In 11 developing countries, forest product exports exceed 10 percent of their total exports and in another 7 countries they exceed 5 percent of total exports.

In the period since 1961, the world trade in forest products has increased more than three-fold in real terms. Developing country exports have increased six times, bringing their share in world total from 8 percent to 13 percent. Over this period, the structure of the forest products trade has changed. In 1961, unprocessed roundwood accounted for 60 percent of developing country exports. By 1990, the real value of these exports had more than doubled but constituted only 20 percent of the total. In 1961, developed and developing country exports of unprocessed roundwood were about equal. In 1990, with the inclusion of chips and particles, the export of unprocessed industrial roundwood of developed countries was more than double the developing country export. In 1990, the trade in pulp and paper dominated the forest products trade and accounted for more than 60 percent, up from the 30 percent of the early 1960s. Developed country exports of pulp and paper have increased from 4 percent of total exports of forest products in 1961 to 28 percent of the much larger exports in 1990.

Although subject to considerable annual fluctuations, the real price of forest products has tended to be roughly constant over the last three decades. Products departing from this broad tendency have been tropical logs and sawnwood, the prices of which have shown a slight upward tendency of about 0.5 percent per year, and wood-based panels and paper which had a declining tendency in prices exceeding 1 percent per year over the period 1961–80, but have experienced constant real prices over the last decade.

Outlook for trade in forest products

Total trade in forest products is expected to grow in proportion to the expansion of aggregate consumption. The expansion of forest product exports of the developing countries may see a lower rate than the expansion of consumption as priority is given to meeting domestic demand. Likewise, imports of the developing countries may grow less rapidly than consumption as priority is given to increase self-sufficiency in forest products for which there is comparative advantage in domestic production. Expansion of trade will be greatest in manufactured products, reflecting the strong tendency to concentrate on products with higher value added, with a concomitant decline in the trade in unprocessed wood raw material.

Net importing regions of East Asia, particularly China and Japan, and the Near East and Europe may be expected to continue to increase their import demand. The main net exporting regions will continue to be North America, Scandinavia, Insular South East Asia, the Russian Federation and South America.

The trade environment for forest products

Generally speaking, imports of unprocessed roundwood are free from tariffs. Producing countries frequently impose restrictions and bans or discriminatory taxes on export of unprocessed wood with the objective of stimulating local processing, securing raw material supply to local industry or discouraging forest depletion. Concerning manufactured wood products, high tariffs are in place in some countries particularly on wood-based panels and paper. The objectives may be general protection of the industry or protection of an "infant" industry.

In recent years developed country environmental groups, concerned about tropical forests, were convinced that by stopping trade in tropical timber, damage and destruction of tropical forests would be reduced. They have pressed for embargoes and boycotts on imports of tropical timber and some companies and local government authorities in developed countries have excluded the use of tropical timber in their products and contracts. Others have introduced the idea of labelling timber to prove that it originates from sustainably managed forests. The International Tropical Timber Organization has approved best practice guidelines for the sustainable management of tropical forests and set the year 2000 as a target date by which all exports of tropical timber should come from sustainably managed forests.

FAO (1994b) reports that recent empirical studies contradict the view that logging for international timber trade is a major cause of deforestation and environmental degradation. Nearly all logging in the tropics is for domestic consumption, and only about 6 percent of the wood cut in the tropics enters the international timber trade. Country case studies show that bans on trade in tropical timber, meant to protect tropical forests, are mostly ineffective and can even be counterproductive, resulting in higher environmental costs. They are ineffective mainly because of the already noted minor importance of exports in total tropical timber production decisions and because of the opportunities for trade diversion to countries without import restrictions. They can be counterproductive where diminished prospects for export earnings lead to reduced incentives to manage forests in a sustainable way or even to conversion of forest areas to alternative uses such as agriculture.

The provisions of the Uruguay Round Agreement (described in Chapter 8) imply reductions of tariffs by the developed countries on wood, pulp and furniture from an average rate of 3.5 percent to 1.1 percent, and those for wood-based panels from 9.5 percent to 6.5 percent. The result would be to reduce further the prevailing degree of tariff escalation faced by forest products in developed country markets. A number of developing countries will also reduce tariffs but levels will remain relatively high. For example, plywood tariffs after the reductions will still be 20 percent in Brazil, 35 percent in China and 40 percent in Indonesia.

An important impact of the decline in tariff rates for forest products in developed country markets is that the tariff differential between "Most Favoured Nation" (MFN) and "Generalized System of Preferences" (GSP) rates has been reduced significantly. Most tariff reductions have led to a general decline in the MFN rates, while the GSP rates have been left largely unchanged. This suggests that exporters facing the full MFN rates may gain more from falling forest products tariff rates than developing countries that previously benefited from GSP and other preferential schemes.

People in forestry and employment

Employment generated in forestry and forest industries is considerable. Part of it is the formal employment by enterprises and part is informal employment of members of households to meet their own consumption needs. The informal sector work includes the collection and harvesting of wood for fuel and charcoal making, as well as the collection of foods, medicinal and artisanal materials and hunting. People are also involved in the cultivation of forest land for food and cash crops as well as in the collection of fodder and grazing of livestock in the forest.

Formal employment by enterprises includes employment in forest management, silviculture and transport of wood. There is also the employment in research, education and training and extension services. In the forest industry sector there is the employment in management, production and marketing of sawnwood, wood-based panels, pulp and paper, together with the considerable employment in further manufactures, joinery, furniture, packaging and paper products.

A broad and crude estimate of employment in terms of work-years may be derived from the estimates of the value of the sector's output. This suggests the equivalent of 60 million work-years globally. Of these, some 12 million are in the developed countries, more than 90 percent occupied in industry-related activities, and 48 million are the work-years in the developing countries, half in fuelwood gathering and charcoal production and half in industry-related activities. These estimates are indicative of much higher average levels of labour productivity in the developed country industry, dominated by the very high output per unit of employment in capital intensive industries. The estimate of some 20–25 million work-years in fuelwood and charcoal production in the developing countries refers to production and does not include work related to the delivery of fuelwood from tree to hearth. An estimated 3 billion people in the developing countries depend mainly on wood for fuel (FAO, 1994b). Fuelwood gathering is mainly carried out by members of the household, and therefore this type of work forms part of the daily tasks of the members of some 650 million households.

5.4 FOREST AND THE ENVIRONMENT

As forests occupy some 26 percent of the land area of the globe, they are an important part of the environment, they provide environmental services and are in turn influenced by the environment. Forests provide the habitat for a large proportion of the world's plant and animal species, are the home and living environment of indigenous people and constitute a resource from which people derive sustenance. In their service function, forests contribute to the conservation of mountain watersheds, soil and water and provide shelter from wind and help prevent desertification and conserve biological diversity. Forests and trees have a role in modulating the microclimates and the local climates of regions. Forests, comprising a major component of the terrestrial biomass, enter significantly into the carbon cycle and play a part in determining the level of carbon dioxide in the atmosphere and thereby have an impact on the global climate change attributed to changes in the levels of CO_2 and other "greenhouse" gases in the atmosphere.

Changes in the forest influence its performance of environmental services. Thus, the use of the forest for production functions or the change in use is inextricably interrelated with its performance of environmental functions. Forest ecosystems are subject to change due to natural causes such as volcanic eruptions, cyclones or lightning fire. Considerable areas of forests in the tropical and temperate regions are more or less undisturbed as natural ecosystems and habitats of their indigenous flora and fauna. However, human intervention is a major determinant of the course of change.

In many countries, protected areas have been established to secure the conservation of ecosystems, species and their heritable variation that are endangered and vulnerable. The management of mountain forests in vulnerable watersheds and catchments aims at the conservation of soil and water and the control of erosion and siltation to alleviate flooding downstream and to modulate water flows for sustained supply. Forests have been planted for protection against wind and to control desertification. Programmes of fire control are instituted to reduce damage to the environment. Specific programmes of afforestation have been initiated with the explicit purpose of sequestering CO_2. Forests are managed and trees are planted to improve the atmosphere and landscape for urban populations. Trees are planted in conjunction with agriculture and livestock for shelter, fertilization of soil and soil conservation as well as for their complementary products. In short, the management and conservation of forests is a multipurpose activity.

In the case of necessary transfer of forest land to agriculture and the ensuing deforestation, the key issue is how to manage the process to ensure sustainability of development. It has to be recognized that after the change of land use, the forest is no longer there to perform its productive or service functions. Some functions, such as ecosystem conservation, will be totally lost; others, such as soil and water conservation, will require alternative approaches for the sustainable management of the land resources. In the uncontrolled degradation and destruction of forests, both the productive and the service contributions of the forests may be lost, unless specific action is taken to prevent this from happening.

Forestry, forest industries and the environment: compatibility and conflict

The use of the forest and its products in the economy is often compatible with environmental objectives but there are also areas of competition and conflict. The production of wood is a renewable process, to a high degree compatible with and complementary to the functions of forest in conserving soil, water and biodiversity. Because of renewability, the use of wood and its derivatives is benign in respect of the carbon cycle. In the best case, CO_2 eventually released in burning wood for energy production is sequestered in the growth of wood that replaces it. The use of wood in energy production substitutes for fossil fuel and, given the possibilities for regrowth, can reduce the net release of CO_2. Mechanical wood products require in general low energy inputs in their manufacture and they substitute for high energy structural materials such as steel, aluminium and cement. Although paper production requires a relatively high energy consumption, its production process has become rather energy efficient, particularly where it uses spent liquors in energy generation combined with the process of chemical recovery.

Forestry and forest industries may involve conflict with environmental objectives. Cutting wood from the forest disturbs the ecological balance and, if extensive, may have a significant impact on ecosystems. Disturbance of the forest, road construction and logging activity may have a significant impact on soil and water relations. The development of access roads may facilitate colonization and clearance of remaining forest for agriculture and pasture, reinforcing the impact on the ecological balance and soil and water relations. Destruction of forest may be associated with burning and release of CO_2. Forest management and reforestation for wood production may locally reduce biodiversity. Forest and plantation management may involve pesticides, herbicides and fertilizers which, if misused, may have adverse effects on the environment. Forest industries involve use of energy, water and inputs in the production process and the generation of residues, effluent and emissions resulting in pollution of land, water and the atmosphere. The products of industry may be damaging to the environment due to leaching of components such as preservative materials or the noxious gaseous emissions from adhesives. The inputs and additions in manufacture may render the wood or paper material less easy to recycle.

To minimize the harmful impacts, countries take action to secure sustainable development of the forestry sector through management of forests and public land and regulation of cutting and regeneration of private forests. Some countries have policies and subsidy programmes supporting private forest management and reforestation. The European Community has recently introduced incentives to favour afforestation as a means of "setting aside" land from agricultural use. In the USA cutting has been restricted on public forest lands that provide the habitat for species classified as endangered.

Forest industries are subject to regulations controlling the permitted levels of noxious waste and chemicals in the water, effluent and noxious gases in smoke stack emissions. Specific regulations have been introduced in a number of countries requiring use of recycled fibre in paper products and requiring the collection of used paper packaging. The reuse of waste paper is subsidized in some countries.

A particular matter of international concern is the high rate of deforestation in tropical regions and particularly the threat that this poses to the conservation of biodiversity and the natural resource base. The Tropical Forests Action Programme (FAO, 1985, 1991f) was initiated jointly by FAO, the World Bank, UNDP and the World Resources Institute (WRI) as an international initiative to assist tropical countries to confront this issue and particularly to increase their commitment, capability and resources towards the sustainable management of their forests.

Several environmental factors influence the state of the forest and its role in sustainable development. Atmospheric pollution has significant impacts on the health and growth of forests in some regions. Significant climatic change, affecting seasonal temperatures and rainfall, as well as increased CO_2 levels in the atmosphere would have an impact on the growth and eventually on the distribution of species. International agreements on measures to control emissions of noxious chemicals in the atmosphere and eventually to control the level of "greenhouse" gases would contribute to contain these effects on the forest.

The above summary of positive and negative relationships between the use of wood in the economy and the environment and the review of institutional approaches and policies, provides a partial picture of the complex set of options. For sound decisions, the trade-offs between the relative benefits of wood as an industrial material and its potential for environmental harm, the issues of complementarity and competition between use of wood in the economy *vis-à-vis* alternative materials and the pros and cons of the exclusive use of forest for environmental benefits have to be assessed.

5.5 FORESTS IN SUSTAINABLE DEVELOPMENT: FUTURE PERSPECTIVES

Consideration of the role of the forest in sustainable development to 2010 and beyond involves three main areas. The forests have a productive role contributing directly to the economy and the material well-being of people; there are competing demands for land occupied by forests for growing food and for human settlements and infrastructure development; and the forests and the use of their products relate intricately to the environment. Those considerations are dealt with in turn.

Demand for forest products in the economy

The demand for the products of the forest will continue to increase. Wood and biomass have a significant role in energy supply both for rural communities and as a renewable energy source in total energy supply. Forest and tree biomass have advantages in supplying these energy needs because of their location near rural populations and because they compensate for the CO_2 released in burning through its sequestration in the regrowth process. Growth of consumption of forest products is an essential component in the expansion of economies and particularly in increasing the material well-being of people in developing countries. Significant expansion of demand is projected to the year 2010. Thus the future supply of wood and forest products other than wood, the development of an industry to manufacture forest products and the development of trade in these products are essential components of strategies to promote sustainable development.

Conversion of forest land

The necessary expansion of agriculture in the developing countries to meet the demand for food and to provide employment and incomes to the rural population will involve a net increase in the area of land in agricultural use (see Chapter 4). Part of this extension will be met by conversion of land from forestry use. Conversion of land to other uses such as reservoirs may involve an additional million hectares of forest land per year.

The current rate of tropical zone deforestation of about 15 million ha per year is much higher than what would have been required under sound management of the agricultural expansion process. For the future, much greater efforts are required to increase the efficiency of converting land to other uses, e.g. the undertaking of adequate assessments of the productive potential of land, identification of appropriate technology for agriculture and support to the communities involved to ensure access to the essential inputs for efficient land use. In addition, measures to ensure continued adequate soil and water conservation, including the appropriate use of tree planting in agro-silvipastoral activities must play an important role in managing the process of forest land transfer to other uses.

The efficient use of limited forest and non-forest land resources requires a sound understanding of the productive potential of the land and of the technical options for forestry and agriculture. This information must be available and usable by the communities that effectively make the decisions on land use. These communities must be involved in the planning involving changes in land use and must have security that the benefits will be shared equitably.

Environmental demands on forests

At the same time as demands for the products of the forest and for alternative uses of forest land are increasing, also demands for a more secure environment with respect to soil conservation, the water supply, the protection from flooding, and the conservation of the remaining heritage of biological diversity, are increasing. There is an overall demand for stabilization of, or increase in, living biomass through conservation, renewal and extension of forest.

Meeting these demands requires an effective identification of forests that are more important for soil and water conservation and of forests more important for the conservation of species diversity and ecosystems. Expected action includes the expansion of protected areas, *in situ* and *ex situ* conservation of genetic resources, extension of sustainable management of existing forest and watersheds, reforestation and afforestation in both developed and developing countries and greening of wasteland, degraded forest and surplus land.

The impact on forests of atmospheric pollution may be contained by measures to control the level of noxious industrial emissions in the atmosphere. However, the absolute consumption of chemicals and fuels is expected to increase, and therefore also the risk of atmospheric pollution in established or new forms. Projections of world energy consumption leave no doubt that the emission of greenhouse gases will be difficult to contain at current levels and that such emissions are most likely to increase. The impact of atmospheric pollution on the climate and in turn on forests requires further research.

Summing up

Concluding, the perspective for forestry development to the year 2010 is one of intensifying competition for the goods and services of the forest and for the use of forest land. The demand for the products of the forest will continue to grow with growing populations and economies. In the developing countries, part of the forest land must be transferred to agriculture. The increasing scarcity of undisturbed forests makes the need to conserve forests in their service functions relating to soil, water, ecosystems, genetic diversity and the composition of the atmosphere more urgent.

As noted earlier, a major part of the tropical deforestation is due to the pressures for expansion of agriculture, grazing and fuelwood gathering, many of them originating in the growth of poor rural populations. Reducing such pressures depends above all on more general economic and social development that would provide alternative income-earning opportunities as well as contribute to reduce the rate of population growth. However, if development can reduce the pressures emanating from rural poverty, it also generates increased demands for both the products of the forest and food, in particular livestock products. These demands should be met by adequate technological progress to prevent further unsustainable harvesting and expansion of farming. Thus, efficient forest management and provision of incentives for conservation are an essential part of policies to check deforestation even when, and perhaps particularly when, poverty-reducing development occurs.

On the positive side, increasing incomes tend to upgrade concerns regarding nature conservation in people's priorities and preferences. They also provide the means to pursue this objective. However, this stage is likely to be reached at advanced levels of per caput income rather than at the early phase of income growth and poverty reduction.

The urgent areas for action to contain the adverse effects on the forest include the adoption of improved technology to secure high productivity from both agricultural and forest use of land combined with a careful assessment of land potential to permit allocation to the best use. Investments in research, training and dissemination of the necessary technology are required to secure its optimum use as are the adjustments of policy and planning to support its implementation. A fundamental requirement is the awareness, commitment and full participation of the *de facto* decision-makers, namely the populations and communities involved in forests and neighbouring agriculture.

There is increasing awareness and international commitment to address these issues, which finds its expression in the Tropical Forests Action Programme, the adoption of Agenda 21 and the Forestry Principles by the United Nations Conference on Environment and Development and the guidelines for sustainable management of tropical forests of the International Tropical Timber Organization. It has to be recognized that an effective implementation of these plans and principles requires a major effort to reach the localities where the forests are and where effective decisions are made.

CHAPTER 6

Fisheries

6.1 INTRODUCTION

The world's fishery resources are an important source of protein as well as employment and economic revenue. The historical developments as well as the future prospects of the sector are conditioned, to a significant extent, by the wild characteristic of the resource and the fact that, for most species, the levels of production are limited by nature. This has three important consequences. First, beyond certain levels, additional investment in fishing effort does not produce additional yields and, in many cases, actually leads to declines in total catch as well as to economic waste. Such an increase in fishing effort is inevitable in those, almost universal, situations where there is ineffective fisheries management. Second, with growing demand and limited supplies, the real prices of fish products inevitably increase. This has important and damaging consequences for low income consumers, particularly those from the developing countries. The third major, and more positive, result is that limited natural supplies and high prices serve to stimulate increased production through the cultivation of those species that allow it.

The potential for increasing total production much above present levels is rather limited. Attaining and maintaining these somewhat higher levels will depend on greatly improved management of the fishery resources. Without such improvements, there is a risk that even present levels may not be maintained. Under the circumstances, the prospects are that the real price of many species of fish, particularly those used for direct human consumption, will continue to rise. Better management would, therefore, contribute to containing such price increases in the longer term, and to reducing or eliminating economic inefficiencies believed to be widespread in fishing operations at present.[1] Improved management of capture fisheries may, at least in the short term, lead to a lower volume of production, albeit of better quality. Eventually, the overall production constraints may be somewhat relaxed by further growth of aquaculture and stock enhancement programmes such as culture-based fisheries.

6.2 HISTORICAL DEVELOPMENTS AND PRESENT SITUATION

Since the 1600s, the principle of the freedom of the seas dominated the use of the oceans and their resources. Beyond narrow limits of national jurisdiction (3–12 miles), the resources were open to all comers. With declining catches per vessel in the traditional grounds, the fishermen either moved to new areas or adopted more intensive techniques. In more recent years, the pace of exploration and exploitation was expedited by the development of automotive power, synthetic fibres in nets and refrigeration equipment.

This pattern had three consequences: one was the generalized depletion of conventional stocks; a second was the global extension of fishing effort to new, less conventional species as well as to far distant waters; the third was the increase in conflict between the local fishermen of the coastal states and the distant water fishermen from foreign states fishing close to shore. This stimulated increasing claims by coastal states to extended jurisdiction.

The major maritime powers generally succeeded in maintaining the principle of the Freedom of the Seas, which benefited their military and fishery interests, during the First and Second United Nations Conferences on the Law of the Sea (in 1958 and 1960). But the pressure for extended jurisdiction was inexorable and, while the discussions at the Third UN Conference were still under way in the 1970s, a regime of a 200 mile extended fisheries zone was established, resulting in a redistribution of the seas' wealth.

The choice of 200 miles, however, has no relevance to the habits of fish. Some species (e.g. oyster, clams) are sedentary while others (e.g. tuna, salmon) swim vast distances and are found both inside and outside a 200 mile limit. Given the wide diversity of the resource, there is also no direct connection between the size of a fisheries zone and wealth of resources. Among the most fertile areas are the continental shelves rich in demersal stocks (groundfish such as cod and haddock) and the upwelling currents inhabited by pelagic species (those feeding on the surface such as herrings and sardines). Temperate zone waters tend to contain relatively large populations of few individual species; while outside upwelling areas (e.g. Peru), tropical waters have large numbers of species and small populations of each. In the open ocean, the stocks are diffused. Some high seas species have schooling habits but require high search costs for their location. Others seldom aggregate and can only be taken by gear that filters great quantities of water.

Production

Volume and species composition

Since the 1950s total world production of fish increased at a rate of about 6 percent per year until the collapse of the Peruvian anchoveta fishery in the early 1970s. After that setback, with some minor fluctuations, production continued to grow reaching the peak of 100 million tonnes in 1989. However, the overall growth rate declined to 2.5 percent per annum (Figure 6.1). World production fell to 97 million tonnes in 1990 and has remained at that level for both 1991 and 1992. It has been characteristic of these developments that the contribution of some (notably) traditionally high value species to the total catch and that of other less traditional species has declined, while the catch of predominantly high volume, low unit value species has been subject to wide fluctuations. For example, a significant part of the growth during the past two decades was due to the increase in the catch of a single species, Alaska pollack, whose catch accounts for some 5 percent of world fish production. Another large part of the increase came from a few species of small shoaling pelagic fish. After the collapse of the Peruvian anchoveta, the total catch of this group fell to 6.3 million tonnes in 1973, down from 16.7 million tonnes in 1970. By 1980, the small pelagic global catch was back up to 13.2 million tonnes and peaked at 21.3 million tonnes in 1989, making up about 20 percent of total world production. Most notable were the increases in Japanese pilchard, South American pilchard and Chilean jack mackerel. These kinds of species are subject to very wide fluctuations in abundance which are generally cyclical in nature and result from natural environmental changes. Experience has shown that heavy fishing can significantly impede their recovery. The composition of current world production by major species is shown in Figure 6.2.

During the past two decades, the catch of a large number of demersal stocks (e.g. Atlantic cod, Cape hakes, saithe, haddock, Atlantic redfishes) has declined significantly, due largely to continued, heavy overfishing. Although there are instances of stock rehabilitation through the adoption of conservation measures, these are relatively scarce in most areas of the world. In contrast, production of oceanic pelagics (e.g. tunas), cephalopods and other shellfish has shown a steady increase.

While marine catch has successively decreased from the peak year in 1989 (86.4 million tonnes), the production of inland species rose dramatically during the 1980s, to 15 million tonnes in 1991 (15 percent of total production). Most of this is accounted for by nine major species whose catch was less than 500 000 tonnes in 1970 but over 5.5 million tonnes in 1990. As noted later, these species are produced almost entirely by aquaculture and most of the growth occurred in China.

A significant aspect of these developments is the change in the value of catch. Except for the tunas, the species whose catch has been growing are relatively low priced. Most of the shoaling pelagics, for example, are caught for reduction into fishmeal. On the contrary, the species whose catch has been falling are mostly high valued. The net result is that the increase in total quantity of catch has not been matched by a commensurate increase in economic value. Overfishing of the high-valued stocks has led to their depletion and, with decreased supplies, to price increases. This pattern will eventually occur for most species which are not readily cultivable. Table 6.1 presents a qualitative overview of the main characteristics of the different market segments and species.

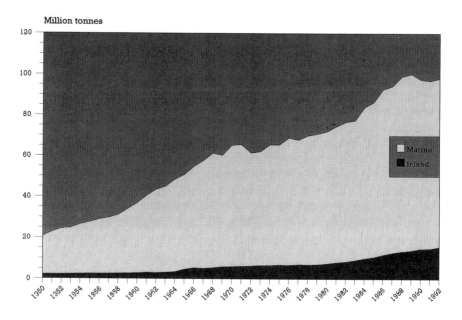

Figure 6.1   World fish total catch (marine and inland waters) 1950–92

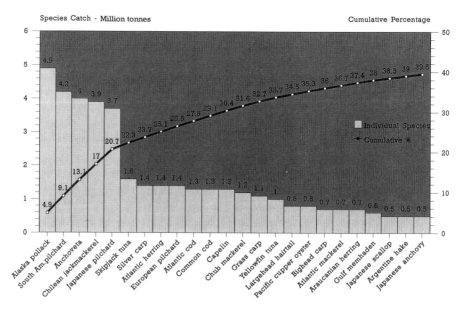

Figure 6.2   Total fish catch (major species) 1991

The dominance of a few countries

It is significant that a very few countries have an extraordinary influence on total world production. The effectiveness (or lack of effectiveness) of management regimes to which the fishery operations of these countries are subjected can have a major impact on global production. Twenty countries accounted for 80 percent of total world output in 1991 and six of them made up more than 50 percent (Figure 6.3). The concentration by countries and that by species are related to each other. For the three major developed countries, most of the increases were due to two species: Alaska pollack and Japanese pilchard. There has been an even greater dominance of individual species in the catch of two of the three major developing countries. For Peru, 90 percent of the catch in 1990 was from anchoveta and South American pilchard. For Chile, 81 percent was from those two species and Chilean jack mackerel. These are all low valued species whose abundance tends to fluctuate widely.

The development of aquaculture

During the 1980s, there was a rapid growth in the development of aquaculture. Between 1984 (when FAO began recording aquaculture production) and 1990, total production from aquaculture roughly doubled reaching about 12 million tonnes (excluding seaweeds). This has had two main thrusts: one in China and the other, more globally, for certain high-valued species.

China's aquaculture output increased about 2.5 times from 1984 to 1990, accounting for roughly 45 percent of total world aquaculture production in the latter year. To a large extent this is made up of various species of carp often raised in conjunction with agriculture since a long time ago. There have also been large increases in the culture of shrimp and mussels; to the point where China produces about 27 percent of global production of shrimp from aquaculture and about 38 percent of global production of mussels.

A separate development has been the rapid increase in output of salmon, shrimp and various shellfish to meet the demand of the luxury market in developed countries. Currently, farmed salmon makes up about 25 percent of total world salmon production from all sources and farmed shrimp about 24 percent of total shrimp production. In both cases, aquaculture output has been sufficient to significantly affect world prices. Oysters have always been cultivated and there has not been much growth in production in recent years. But there has been a large increase in farming of other shellfish since 1984. Both mussels and clams have increased about 60 percent while scallops have grown by over 300 percent.

Patterns of consumption and trade

Over the past few decades, production of fish for human use has grown more

Table 6.1   Characteristics of major market segments for fishery products

Market	Types of species	Sources	Prices (ex-vessel)	Implications
Luxury	Salmon, shrimp, sea bream, etc.	Capture and culture	$3–4/kg. Tending to decline with increased cultivation	Increased trade from culture countries. Increased demand for fishmeal. Conflicts over space and water use
	Flatfish (flounder, sole, plaice, etc.)	Capture	$3–4/kg. Increasing due to depletion of stocks	Most stocks heavily over-fished. Incentive for culture
	Tuna	Capture	$1.50–2.00/kg. Reaching limit because of substitutes (e.g. chicken)	High consumption in developed countries. Increased processing in, and exports from, developing countries
	Crabs and lobsters	Capture and production of substitutes from low priced fish through surimi process	$3–12/kg. Tending to decline with production of substitutes	
	Molluscs (oysters, clams, cockles, mussels)	Mostly culture, some capture	$1–5/kg. May decline for cockles and mussels with increased culture	Opportunites for increased production and consumption in developing countries. Sanitation problems
	Cephalopods (squid, octopus and cuttlefish)	Capture	$1–4/kg. Likely increases over the long run	Opportunities for increased capture by some developing states and for increased exports. Healthy food merits
Standard	Most finfish species making up the bulk of the market (cod, hake, haddock, jack, mackerel, grouper, croaker, etc.)	Capture	$0.50–3.00/kg. Increasing due to depletion of stocks.	Generally heavily overfished with declining total catches and decline in size of animals

Low income	Carp, catfish, milkfish, etc.	Culture	$0.20–1.00/kg	Heavy production in Asia, mostly China. Very little in Africa and Latin America
	Artisanal-caught marine and lake fish (sardines, mullet, scad, tilapia, chub, mackerel, etc.)	Capture from canoes, rafts and other small craft, generally non-powered	$0.20–1.00/kg. Rising prices due to depletion	Generally heavily overfished with declining total catches and decline in size of animals
	Frozen blocks of low quality fish of miscellaneous species	Capture by industrial vessels of former USSR	Under $1/kg	Sold to local African coastal states for various reasons. Not likely to continue for long
	Trawler by-catch (small individuals including juveniles of high-valued species)	Discards from shrimp trawling operations	$0.05–0.50/kg. Prices increasing as discards sought for feed to use in aquaculture	Locally an important source of protein for low-income consumers
Non-food markets	Small shoaling pelagics (anchovetas, pilchards, sardinellas, etc.) reduced to fishmeal and oil, mostly for feed	Capture mostly by large-scale operations	$0.10–0.40/kg. Price increases presently limited by price of substitutes for feed (e.g. soybeans)	Conversion to food use possible in future but stocks not found in Asian waters where future need will be greatest

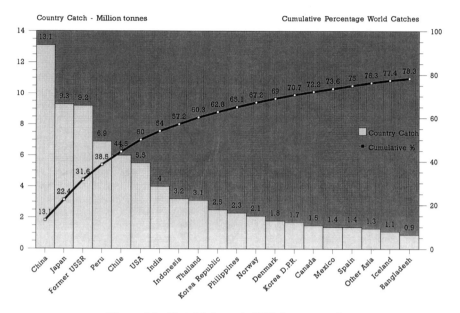

Figure 6.3   Total fish catch 1991 (by country)

rapidly than population with a resulting increase in per caput consumption. The relevant data are shown in Table 6.2. While fish provide a relatively minor source of dietary energy supplies (accounting on average for 17 calories out of 2475 in the developing countries), they are an important source of essential fatty acids, vitamins and minerals. However, although a handful of countries tend to dominate global production of fish, many countries depend heavily upon fish as a major source of protein. For the developing countries as a whole, fish currently make up about 19 percent of total animal protein consumption, or just over 4 percent of protein from both animal and plant origin. However, this latter share is very high in some countries, both developed (e.g. Japan, Norway, Portugal) and developing ones (mostly in East Asia, e.g. Philippines, Thailand, and Africa, e.g. Congo, Angola, Ghana, etc.); and it is very low in others (e.g. Argentina, many land-locked countries in Africa).

Trends in trade in fishery products closely reflect changes in production and technology development. The expansion of international trade in fish and fishery products has exceeded the growth in world fish production. World fish trade, estimated at 32 percent of world production (24 million tonnes) in 1980, increased to 38 percent in 1990 (37 million tonnes). A significant part of this increase was from increased exports of fresh and frozen products (3.3 million tonnes to 6.0 million tonnes product weight). Other large increases in exports included those of shrimp, which increased from 0.4 million tonnes to almost 1.0 million tonnes, and of fishmeal which rose from 2.0 million tonnes to

Table 6.2   Fish: historical data of food and non-food use

	1969/71	1979/81	1989/91
A. Food (liveweight equiv.)			
Per caput (kg)			
World	11.0	11.8	13.3
Developing countries	6.4	7.6	9.3
Africa (sub-Saharan)	7.7	9.1	8.0
Near East/North Africa	2.7	4.5	5.3
East Asia	8.1	10.0	14.1
South Asia	4.0	3.9	4.1
Latin America + Carib.	6.6	9.0	8.5
Developed countries	22.3	23.2	26.4
Western Europe	18.4	17.3	21.2
East Europe + FSU	19.3	20.8	21.0
North America	14.3	16.4	21.7
Japan	67.5	69.9	72.0
Others	11.1	12.2	14.4
Total (mill. tonnes)			
World	40.4	51.8	68.5
Developing countries	16.4	24.7	36.1
Africa (sub-Saharan)	2.1	3.2	3.8
Near East/North Africa	0.5	1.0	1.6
East Asia	9.0	13.5	22.4
South Asia	3.0	3.6	4.7
Latin America + Carib.	1.9	3.2	3.7
Others	0.9	0.2	0.4
Developed countries	24.0	27.1	33.3
Western Europe	6.8	6.7	8.5
East Europe + FSU	6.4	7.5	8.2
North America	3.2	4.1	5.9
Japan	7.0	8.2	9.0
Others	0.5	0.6	0.9
B. Fishmeal (liveweight equiv.)			
World totals (mill. tonnes)	19.7	19.5	27.8
Total fish use	60.1	71.3	98.2

3.2 million tonnes, although with significant fluctuations in volume during the period.

The increase in trade has been particularly marked in value terms, exports reaching a total of US$38 billion in 1991 compared with US$15 billion in 1980. The share of the developing countries in total trade has grown slowly but steadily to some 47 percent of total exports in 1989. Exports of shrimp, cephalopods, tuna (both frozen and canned) and fishmeal have been on an upward trend and contributing positively to the balance of payments in many developing countries in Asia and Latin America. Particularly marked expansions in export trade have been achieved by Thailand and China. The developed countries, especially those of the European Community, Japan and the USA, remain the principal importers, accounting for some 88 percent of total world fish imports. The USA has also become the world's major exporter of fish and fishery products, reflecting gains through the extension of jurisdiction and its increased catch in the North Pacific.

Many major fishing countries, both developed and developing, are both large importers and exporters. Most developing countries in Asia, for instance, carry out simultaneously import and export trade in fish, exporting high-value species not consumed domestically and importing fish and fish products of lower prices and quality. Export-oriented fisheries may involve the use of imported inputs (e.g. frozen tuna in Thailand). With the growth of domestic economies and per caput incomes, some exporting countries may find it difficult to maintain exportable supplies in the face of expanding domestic consumption.

The rise in prices for some fish products is being constrained by market-place competition from alternative foods. On the other hand, the culture of certain species, notably salmon and shrimp, is now producing sufficiently large quantities to put pressure on market prices. The development of new products, particularly those involving the processing of low-valued species (e.g. Alaska pollack) into reconstructed high value products, may also have some impact on prices.

Demand and taste preference for fish and fishery products and supply patterns are continually undergoing changes. As prices for the more highly preferred species rise, middle-class consumers tend to turn to species traditionally consumed by the poorer sections of the community. At the same time, although the catch of less favoured, unconventional species is increasing due to the increasing production constraints of commercially preferred species, their availability, particularly for low-income consumers in Asia, is being reduced as a result of competition from aquaculture feed demands and the degradation of coastal areas.

6.3 PROSPECTS FOR THE FUTURE: PRODUCTION

As noted, there are severe constraints to increasing aggregate fish production. The little further growth that can be achieved in sustainable ways is unlikely to be sufficient to maintain per caput supplies. Unlike other sectors of agriculture, it is more difficult to relax the production constraints in fisheries by bringing new resources into exploitation, technological change and investment. Often efforts to improve management of fishery resources will be primarily aimed at preventing current production levels from declining.

Table 6.3   Present and possible future production levels of fish
(million tonnes, liveweight)

Type	1989/91	Possible 2010
Capture	*86*	90–110
Marine	79	
Inland	7	
Culture	*12*	15–20
Marine	4.5	
Inland	7.5	
TOTAL	98	

Note: Details on production by major fishing areas are given in the Annex.

Much of the increase in catch during the 1980s came from pelagic species which are subject to wide natural fluctuations and appear to have recently passed their peaks. Some stocks are now facing significant declines. If the catches of these kinds of fish cannot increase further, the growth of the fishmeal-dependent aquaculture sector will be accordingly constrained unless technological advances are made in providing alternative sources of feed. Moreover, the major demersal species throughout the world have been overfished and they provide little opportunity for increased catch except through improved management, which inevitably will require some reduction of fishing over the medium term to allow stock rebuilding.

Marine catches can only be increased marginally over current levels. This would still entail a real risk of changing the nature of the marine food chain and affecting other species dependent on it. Total marine catches are not likely to greatly exceed 100 million tonnes and could be considerably lower. Aquaculture production may range from some 15 to 20 million tonnes from the present level of 12 million tonnes assuming that the growth rate prevailing in the last few years will be maintained. Increases in inland capture fisheries may be insignificant unless improved management of resources is combined with a better environment.

The above considerations make it clear that estimating future production levels is an operation subject to many uncertainties. With this caveat in mind, indicative estimates are given in Table 6.3. These hypothetical estimates could be feasible with better management and other interventions (e.g. sea ranching) which would favour recovery of fish stocks. The rest of this section discusses the factors that are likely to shape the future in the different country groups and species. A schematic presentation of the status and likely future potential by major fishing areas is provided in the Annex at the end of this chapter.

The developed countries

The developed countries (aside from Australia, New Zealand and South Africa) derive most of their supplies from capture fisheries in the northern temperate zones in both the Atlantic and the Pacific Oceans and from inland waters; tuna stocks in all oceans; and culture of certain species (e.g. salmon, catfish, shellfish). They also, of course, derive some supplies from imports. In developed countries technological innovation and high demand for fish have led to heavy exploitation of stocks in adjacent waters.

The northern Atlantic, the northern Pacific and Mediterranean/Black Sea where developed countries are mainly engaged in fishing, are generally considered to be overfished. At present there are very few instances of successful management, either in the multinational fisheries of the European Community or within the large exclusive zone of the USA, and there is a serious need for catch restraints to allow stock rebuilding. Although, in theory, further increases in supplies could come from the development of underutilized stocks, these are few in number and low in value and are often food fish for higher value species. The largest opportunity appears to be in the further development of the cephalopod fishery in the north Atlantic and the northern Pacific, if this can be done without affecting ecologically important by-catch species.

The situation is more serious for a few developed countries who have continued to acquire supplies from coastal countries because of high royalties and increasing costs of operation (e.g. Japan, Spain). The former USSR, an important distant water fishing nation, may play a reduced role as the heavy subsidies are being removed from the industry. Catches from the inland waters of the developed countries are relatively insignificant, except in the case of the former USSR. For many developed countries, aquaculture is becoming of increasing importance in meeting the demand for certain high value products. The most significant, at present, is from the culture of salmon. Other species include catfish in the USA, carp in the former USSR and seabream and amberjack in Japan.

The developing countries

The situation is relatively more favourable in some developing countries where resources are less overexploited than in the developed countries. For *Africa and the Near East*, the major sources of supply are the capture fisheries in the east-central and southeast Atlantic, the west Indian Ocean, and the inland lakes and rivers. Aquaculture is not, at present, an important source of fish. The west African coastal countries have significant opportunities to increase their share of the existing harvest of their coasts. However, much of the catch is of low-valued species and is now harvested by large-scale foreign industrial vessels. As industrial fisheries are a capital-intensive and high-risk sector, some of the coastal countries may still have to depend on extracting rents from foreigners while others expand the development of inshore fleets through joint ventures. There are very large unexploited resources of lantern fish (a mesopelagic species) which are found in relatively deep waters and are suitable for fishmeal. However, it is questionable whether their harvest will become economically feasible before 2010. Inland fishery resources are of importance as sources of food in many countries. However, significant increases are not likely to occur.

For *Latin America and the Caribbean*, most of the species fished in the Gulf of Mexico and Caribbean are heavily overexploited and the major opportunity for an increase in catch is from squid and octopus. The stocks of cephalopods on the Patagonian shelf appear to be fully exploited though there may be opportunities for increased catch of other cephalopod stocks in the northern coastal areas. The coastal countries may have the opportunity to increase their share of the catch presently taken by the distant water vessels. The northern areas of the west coast of Latin America contain large stocks of small pelagics which are subject to the influence of the current "El Niño" and fluctuate widely in abundance. The catch of Chilean jack mackerel has grown from negligible amounts in the mid-1970s to 3.8 million tonnes in 1990. The stock is considered to be only moderately exploited and may allow for some continued increase in catch. Aquaculture is not of great significance in Latin America with the notable exception of Ecuador and Chile which produce shrimp and salmon respectively, mainly for export. The prospect for the latter is dependent upon international market demand which appears to be presently saturated.

For *Asia*, the demersal stocks of the East and South China Seas, the Yellow Sea and the Gulf of Thailand are heavily fished. It is unlikely that recovery of these stocks, if effective management measures were to be implemented, would add more than 1 million tonnes to total catch. The pelagic stocks are fully utilized and no significant increases would be possible. There may be some possibilities for an increase in catch of cephalopods along the Chinese coast. The great majority of China's aquaculture is of herbivorous carp so that production would not be constrained by shortages of fishmeal. Increases may be feasible in eastern Indonesia where distance from markets has restrained development. Fisheries for large pelagic species and small tunas have doubled their catch in the last two decades and although there may be moderate potential for increase in the future, restrictions on dolphin by-catch may affect use of purse-seines in the future. There may be opportunities to increase the total catch of skipjack. The cephalopods are moderately exploited and may have increased in abundance as a result of the heavy exploitation of the predator demersal fish.

Brackish water and marine culture of shrimp for export to the developed countries has increased dramatically, but shrimp culture is beginning to encounter and cause some difficult problems due to disease and organic pollution from shrimp farming and limited space for expansion. Overall, improved management could lead to the largest increase in supplies to the region. Additional supplies might come from increased harvest of cephalopods and further development of resources of eastern Indonesia and the eastern Indian Ocean.

Coastal countries of the Indian subcontinent may be able to increase their catch of tunas, both the large market species and the smaller tunas. Estimates of potential yields are not available, but it is likely that the catch of skipjack in the western Indian Ocean as a whole could be substantially increased. Increase in the other large tunas may be possible but is not likely to be significant.

Inland fisheries, particularly in India and Bangladesh, provide important sources of protein for local markets. Significant improvements could be made in both countries in culture practices and in the use of water bodies where problems of ownership impede effective production.

Production prospects by species

In most of the major fishing areas production has reached a plateau and in some areas fish yields are on the decline. In some cases production levels are being maintained by catches of younger fish, and certain localized areas are particularly vulnerable. Below, supply prospects are reviewed by species.

Crustaceans are generally heavily exploited and wild penaeid shrimp production is close to a ceiling. Landings are unlikely to decline as the industry is currently overcapitalized. Penaeid pond shrimp production may also be approaching a saturation. Problems surrounding shrimp culture are the availability of fry, space, feed and environmental degradation, and, in particular, the heavy dependence on wild fish catches for feed. The main determinant of increased output will be the cost of inputs in relation to the price of shrimp. A significant potential for expanded catch can be expected from minor crustaceans (e.g. crabs and small shrimps). Molluscan shellfish, which do not depend on fishmeal, may offer good opportunities in both developed and developing countries depending on markets but face deteriorated environmental conditions. In the tropics there are potential resources of bivalve stocks.

For *cephalopods* the possibility of increased production, particularly of oceanic squids, is generally good and future levels of output are likely to depend as much on market considerations as on resource availability, as well as on technological developments in capture and processing. The banning of large driftnets in the northern Pacific could even result in a decline in catches unless new, efficient and environmentally-friendly gears are devised.

The present level of exploitation of *demersal fish* has reached the maximum and there is no potential for increased production. Stocks of cod and other groundfish have declined. Although some potential for increasing supplies exists from use of discards of lower-value species and small flatfish, in both the north Pacific and Atlantic the fishing effort has already shifted to lower-value species (e.g. blue whiting). The tightening of controls in certain countries (e.g. Morocco, Chile, Namibia) could reduce fishing operations and the total allowable catch (TAC) in order to allow rebuilding of groundfish stocks. This is likely to have a negative role on landings in the short to medium term. With good management, long-term potential would be no more than 20 to 30 percent above the present levels.

The immediate situation for *small pelagics* seems to be pessimistic. The largest global single species (i.e. the Japanese pilchard) has declined again in recent years and its recovery is likely to depend on climatic factors. The principal small pelagic fisheries of the southeast Pacific are also declining. The retrenchment of the long-range fisheries of the former socialist countries is unlikely to be offset by increased interest by other countries in establishing joint ventures with coastal states to exploit small pelagic species for food. It is unlikely that significant increases in krill production will be realized for a combination of economic and biological reasons. The current annual krill harvest of 300 000 tonnes compares with a possible yield of 3.9–6.5 million tonnes as determined by the Commission for Conservation of Antarctic Marine Living Resources. The limited production has been caused mainly by economic and marketing factors and substantial increases in catches are not anticipated for the short to medium term. The future for these fisheries may be tied to improved technology which can convert them to higher valued products and suitable promotion.

Given the spectacular success of *salmon farming*, and the possibility of its extension, the potential for increased production in the medium and long term is good. By overcoming the problems relating to a market glut through the reduction of input costs, increased production can be expected from improvements in wild stocks and ranching in the medium term. The Russian Federation has promising prospects for increasing salmon production through improved ranching.

Although some significant potential exists for the *smaller tunas, tuna-like fish and skipjack* in the medium and long term, its materalization will greatly depend on fuel costs and markets. *Conventional tuna* fisheries will also be affected by the marine mammal factor which may restrict further expansion of the use of purse seines and large-scale gillnets.

Although some growth can be expected from *inland fisheries*, the most interesting developments in this sector in the medium and long term are likely to depend on stocking programmes to enhance the stock in floodplains, rivers, reservoirs and irrigation ponds and in intensive management of small water bodies. Inland aquaculture may also have some potential for development in Latin America and Africa.

Overall assessment of production prospects

Past estimates of the annual potential supply of fish from all sources have ranged from 100 to 120 million tonnes. It is now evident that the marine capture fisheries are adversely affected at extraction levels beyond about 80 million tonnes. Inland capture fisheries' yield is about 6.0 million tonnes, of which Asia produces half and the overwhelming constraints to further increases in production rest with the appropriate allocation of water rights and water quality.

Aquaculture, although recording a remarkable growth over the period 1984 to 1990, has experienced significant problems in saturation of markets and consequent price reductions as well as environmental and disease setbacks. These, however, are expected to be only temporary and characteristic of a growing industry. The major constraint would appear to be the restricted knowledge of the requirements for growing only a relatively few species. Most notable is that finfish farming has mostly occurred for freshwater herbivore species (7.4 million tonnes) with only a modest contribution from marine species (1.0 million tonnes). The further expansion of finfish supplies from freshwater aquaculture is likely to be constrained by the freshwater environment. By contrast, the marine environment offers far better prospects for production growth, provided technology can overcome difficulties in locating pens and cages further offshore. With the exception of molluscs, carps and tilapia, the growing of fish in captivity is still in its infancy and can be compared with early attempts at animal husbandry and the domestication of wild animals.

The greatest prospects for increasing fish supplies for food are to be found in the use of small shoaling pelagics for direct human consumption. Presently these species are used for producing fishmeal, for pig and poultry production as well as for aquaculture. The salmon and shrimp aquaculture increasingly use higher-quality fishmeal as feed. This requires an upgrading in the method of harvest and handling of the low-value fish used in the production of fishmeal with consequent increases in the unit cost ex-vessel. Such increases in the price of low-value species can have adverse effects on the poorer segments of population who consume them as food, mainly in Asia. In contrast, in Latin America the supplies of this type of fish for human consumption may actually increase as the upgrading of their quality for the production of aquaculture feed would also make them more acceptable for human consumption.

The remaining option for increases in fish supplies would be that the present condition of overfishing prevails and that the majority of the marine catch increases would come from fishing further and further down the "food chain". The end limit, as has occurred in one or two areas, is a fishery that is almost entirely a "trash" fishery of mixed juveniles and other small-sized species for direct feed to grow several preferred species. That is to say, the wild production from the marine areas could end up being nearly all utilized to grow two or three species in captivity. The impact would be the loss of the present wide spectrum of food items that the existing 1000 commercial species now provide, to be replaced by very few species providing only differences in flesh, colour and texture.

Estimates of potential from mariculture are precarious to make at this juncture with only a short time-series of production showing a spectacular increase. However, there are projections of 500 000 tonnes of farmed salmon production by the year 2000 and these may be projected to 2010 by a further doubling. Shrimp farming produced 700 000 tonnes in 1991 and has been projected to expand at a reduced rate, estimated at 1 million tonnes by 2000 as a result of constraints in expertise, pollution, disease, infrastructure and market fluctuations. Any projections for farmed fish will be determined by demand pull and the price relationship between capture fish and other food products.

6.4 IMPLICATIONS FOR CONSUMPTION

The constraints limiting increases in production of fish will put severe strains on the nutritional situation of the countries and population groups with high dependence on fish for their protein supplies and on small-scale fisheries for employment and income. Such impacts would be the greatest in East and South Asia where an additional 8.5 million tonnes will be required by the year 2010 to maintain the present levels of consumption. This region contains a number of countries where fish plays a vital role in the diet (e.g. Bangladesh, China, Indonesia, Myanmar, Philippines, Sri Lanka, Thailand), in all of which fish accounts for one-half or more of the animal protein supplies.

The consequence of a shortfall in supply will be increases in the price of fish. Such increases by themselves will further stimulate aquaculture production and provide the incentive for further technological advances. Increased prices will also mean that consumer demand will switch to lower priced substitutes. Many of the presently preferred species will move to "the luxury food" class but it is expected that the broad range in fish products that has been the characteristic of fisheries will remain, thereby providing fish at an array of prices. This has been the experience of the past whereby lobsters, shrimps, crabs, salmon, flatfish and cephalopods have had a relatively inelastic demand. By contrast, demand for cod, hake, haddock, tuna, mackerel, redfish, jack, mullet and Alaska pollack is generally much more responsive to price changes.

Projections would indicate that a number of the species with elastic demand would shift to those groups having an inelastic demand. The substitution effect would draw on these less preferred species. The overall consequence would be that the existing supplies of low value fish that are important to the poorer sections of the population would be removed from within their purchasing power.

Currently, almost 30 percent of total world catch of fish is used for *non-food purposes*. Most of this is reduced to fishmeal which is used in combination with other ingredients such as soybean meal and skimmed milk powder, in the preparation of protein feeds for animals, in particular poultry, pigs and, increasingly, high-value cultivated fish such as salmon and shrimp. While in the past the most important consumers of fishmeal have been the developed countries, their apparent consumption has increased only moderately or declined. Several developing countries have become increasingly important in the 1980s. In all cases, the rapid growth in fishmeal use has been associated with rapid expansion of aquaculture.

Although fishmeal is also derived from sources other than the small shoaling pelagic species (trash fish as by-catch of trawling, refuse from processed food fish, and even, in the case of China, mussel culture), the bulk comes from fish caught specifically for the purpose of reduction to meal. The demand for fishmeal is dependent in part on demand for protein feeds, although the share of fishmeal in these feeds can vary considerably with its price relative to the prices of fishmeal substitutes (e.g. soybean, coarse grains). In recent years fishmeal has tended to establish itself as an essential element in feed compounds, due to growth factors such as its immunological effects, and the price competition of substitutes has become less critical. A continuation is foreseen of the high demand for special quality fishmeal for aquaculture purposes. Anticipated demand for shrimps and salmon will have to be met by the aquaculture output because their wild stocks are currently fished at maximum levels. New opportunities may be generated for special quality fishmeals. The industry forecasts indicate that in the next decade the share of prime fishmeal will increase from the present 8 percent to 25 percent of the total fishmeal production. However, there are serious limitations to the expansion of the world reduction industry because most stocks of fish used as raw material are highly variable and the resource situation seems likely to prevent substantial increases in production.

6.5 MAJOR POLICY ISSUES FOR THE FUTURE

Beyond the issues related to the food and nutrition problems likely to emerge from the supply constraints of the fisheries sector, those related to management of the resources and the environmental dimensions also require urgent and adequate policy responses.

The essential need for management

As described earlier, the most important impact of the likely supply–demand gap and the consequent projected increases in the real price of fish is the stimulation such price effects will have in maintaining the excessive levels of fishing intensity and the continuation of overfishing. It is clear that, without directed government intervention to protect and manage fisheries, the resource base will continue to degenerate at a rate corresponding to the increases in real prices of fish. This will continue to occur until governments can establish effective controls over the rate and type of exploitation of the fishery resources. It will become increasingly difficult to manage fisheries with successive price increases stimulating further pressure for greater exploitation levels, particularly since rebuilding depleted stocks will require periods (up to a decade for long-lived species) of reduced catches.

The major contribution that fishing countries could make to solving the problems relating to overfishing is by controlling better and, in some cases, reducing their fishing effort. Another important contribution would be the design and introduction of more selective and efficient fishing gear and practices, thus helping to reduce the wasteful incidental catch of non-target species, not only those of commercial value but also the endangered ones.

The concept of "responsible fishing" embraces not only the impacts of fishing gear and methods upon the overall sustainability of the fisheries but also many other aspects of policies and practices to maintain the quality, quantity, biological diversity and economic availability of fishery resources and to protect their environment. The International Code of Conduct on Responsible Fishing, which is currently under preparation by the FAO, is intended *inter alia* to bring marine living resources under improved management and ensure better prospects for the fisheries sector.

Small-scale fisheries and environmental degradation

Small-scale fisheries are critically important in many countries as sources of both employment and protein. They are, however, being seriously affected by two developments: conflict with large-scale operations in the inshore waters and degradation of the coastal environment. Often the small-scale fishermen have been displaced from agricultural or other natural resources employment and have turned to fisheries because the access to resources is free. There is high mobility into small-scale fisheries but very few opportunities to move out of them. Small-scale fishermen have limited range for their activities. Medium- and large-scale vessels, particularly shrimp trawlers and, more recently, those using large purse seines, often find it advantageous to fish in the inshore waters. This creates conflict over the resources as well as space and is damaging to the vulnerable small-scale operators.

The problems of excessive pressures on the inshore stocks and damaging competition among different gears is compounded by environmental degradation. The coastal zone receives large amounts of pollutants including: organic wastes from municipalities, chemical wastes from industries, pesticides and herbicides from agriculture and siltation from forest land clearing and road building. In addition, activities within the coastal zone also affect the environment. These include mining of coral reefs and destruction of mangrove swamps. Fishermen themselves contribute to these kinds of damage by converting mangrove swamps to mariculture ponds for shrimp; by excessive use of feed and antibiotics in cage culture; and by using dynamite, poison and other kinds of techniques that destroy coral reefs.

The effects of these alterations of the coastal environment on fish production are not easily measurable. Some changes may actually be positive: the production of pelagic stocks in parts of the Mediterranean is increasing, possibly as a result of nutrient discharges in these semi-closed seas. But more often, the effects are negative. Pollution can lead to eutrophication (reduction of dissolved oxygen) which causes mass mortalities of stocks. Changes in the marine environment have also apparently led to an increase in red tides, with toxic effects on both fish and man. The destruction of mangrove swamps is quite likely to have diminished nursery areas for many species of fish. Inland fisheries suffer from dams and diversions which affect fish migrations as well as aquatic productivity.

Although these kinds of damage can affect all fishing operations, they are particularly acute for the small-scale fishermen in developing countries, in particular those of Asia where demand is high and the resources are of such vital importance as a source of food as well as employment. Adoption of effective coastal area management is critically important and could facilitate the rehabilitation of stocks and increased yields, as well as alleviate the hardships of the small-scale fishermen.

Environmentally sound management for sustainable development should therefore be based on the integration of all components of sectoral development. Multi-sectoral integration should occur at the conceptualization stage of policies, plans and programmes; components of an integrated plan should then be implemented by different ministries under the technical leadership and coordination of a single agency. Integrated coastal area management (ICAM) has been initiated in a number of countries, both developed and developing, as a means of providing for the rational management of coastal resources. Integrated coastal fisheries management is being developed by the FAO with the financial support of the UNDP as part of ICAM where the fisheries sector plays an important role in the management of coastal resources.

6.6 CONCLUSIONS

There will be a significant global shortage of supply of fish in the future. Although the severity of the shortage will differ among countries, the overall effect will be a major rise in the real price of fish, which will have critically important consequences in several regards.

A basic problem is that with the absence of efficient systems to control access to the resources under open access conditions, the rise in prices will stimulate even greater investment in fishing effort than already exists. A vicious circle is established whereby stock depletion reduces supplies, leading to additional price increases.

This vicious circle can partly be broken by the establishment of systems of exclusive use rights which provide the fishermen with a stake in the resource and an interest in future returns. However, as many governments have found, this is difficult to achieve. The creation of exclusive use rights, by definition, awards benefits to some at costs to others and thereby redistributes wealth. At national levels, fishery administrators generally do not have the mandate to make such decisions. In international areas or areas where stocks are shared by countries (e.g. the northeast Atlantic), negotiators cannot readily agree to controls which limit the rights of their own fishermen.

But as the problems become increasingly severe, the issues are raised to higher political levels and, eventually, will force the necessary decisions. Several countries have already taken the basic steps to create exclusive use rights and have achieved significant benefits. Although the systems still contain many imperfections, the improvements that have been produced provide valuable lessons for other countries.

There is some hope, therefore, that the management of fisheries will eventually improve. However, although the benefits will be significant in reducing biological and economic waste, they will still not be sufficient to overcome the limits to supply. There will be continued increases in real prices with severe effects on low-income consumers, particularly those of the developing countries of Asia and Africa for whom fish is a critically important source of animal protein. Alleviation of these hardships will require major efforts by governments to adopt policies that ensure the most effective use of the scarce resources.

NOTE

1. A special FAO study concludes that the industry is overcapitalized, mainly as a result of the still largely prevailing open access regimes to ocean fisheries resources and of heavy subsidies provided by major fishing countries, e.g. up to quite recently the former USSR and countries of Eastern Europe, but also the EC and Japan. The study documents the extent to which the value of the global catch falls well short of covering the fleet's operating costs, when these are valued without subsidies (see FAO, 1993e). With the reforms under way in the ex-centrally planned economies of Europe, a substantial part of their subsidized operations has become openly uneconomic. The consequent reductions in the fleets of these countries is leading to significant structural change in the world fishing industry.

ANNEX: SUPPLY PROSPECTS BY MAJOR FISHING AREAS

		Production (million tonnes)	
		Historical maximum (year)	Present 1989/91
West		4.2 (1970)	3.1
East		12.3 (1975)	9.6
			12.7
of which culture			(0.8)
West		2.2 (1975)	1.8
East		4.1 (1990)	3.9
			5.7
of which culture			(0.1)

Marine fisheries

Northern Atlantic. Generally overfished; catches of most important high-value species (Atlantic cod, capelin, Atl. herring) in decline; consequent pressures on low-value stock (pollack, silver hake), already fully exploited; few instances of successful management; even successful management could add only 2–3 million tonnes to total catch; possibility of increasing cephalopod catch by 1 million tonnes. Over the last decade notable is the increase in landings of invertebrates, making up 32% of the catch in 1990, and high proportion of the total value. The coastal states curtailed or eliminated distant water efforts in their EEZs. Management of shelf areas seaward of 200 miles has been conducted by the Northwest Atlantic Fisheries Organization (NAFO) to maintain stocks levels. In the Northeast Atlantic TAC (total allowable catch) systems have been used as the standard tool in managing stocks. Agreed TACs have exceeded the TACs recommended by the International Commission for Exploitation of the Seas, and actual catches have exceeded the agreed TACs.

Central Atlantic. Current catch some 5.8 million tonnes; generally fully exploited. In Western Atlantic marine fisheries include small and large pelagics, reef fish, coastal demersal fish, crustaceans and molluscs. Some underutilized resources like cephalopods. Many of the resources are shared by several countries. In Eastern Atlantic catches are recorded by 21 coastal countries and more than 18 non-coastal countries giving a markedly international character. The share of landings by non-African long-range fleets remains high, being at 58% in 1989/90.

Southern Atlantic. Important fisheries include hake and blue whiting fishing. The latter is considered moderately exploited. Several management measures (e.g. licensing, mesh size regulation, etc.) have been in force for some years. Rapid development of offshore fishing in the Southern Patagonian shelf and slope by long-range fleet is source of concern in the Southeast Atlantic, southern Angola and Namibia have an exceptionally high biological productivity due to the Benguela current, but currently Namibian policy is low catches and stock rebuilding. However, the total fishing potential of the Benguela region is not well known

West 2.4 (1987) 2.2
East 2.8 (1975) 1.6
3.8

Mediterranean and Black Seas. Stocks are fully exploited with possible exception of mackerel, horse mackerel and sardines. The recent decline has been caused by the environmentally-induced collapse of fish catches from the Black Sea; overfishing of most of the highly valued demersal species such as hake, red mullet and clams. Few stock assessment recommendations have been made. The complex of problems faced in many Mediterranean countries with fisheries management, are focused on the coastal zone, where critical habitats for fisheries are often encountered, and there is uncontrolled expansion in the use of the same areas for other human activities

2.1 (1988) 1.5

of which culture (0.16)

Indian Ocean. Catches of most important high-value species (shrimp) are fully or overexploited. Main opportunities increase of catches 4.3 million tonnes for small pelagic stocks off Mozambique and Somalia, for demersal fish off Mozambique, Madagascar and Tanzania and increased utilization of increased by-catch from shrimp trawlers. Potential development for small pelagics (anchovies, scads, Indian and Jap. mackerel, round herring) in the southwestern Indian Ocean, needs to be assessed. Some form of management is required for trawl fishing in Somalia and shrimp fishing in Mozambique and Madagascar. Stock assessment work is needed in the southwestern Indian Ocean. Underexploited deep-sea resources may exist in the shelves of the Amadaman and Nicobar Archipelago, and Myanmar, but generally fisheries of the northern Indian Ocean are heavily exploited to overfished. Fishing by foreign vessels in the eastern Indian Ocean has grown less intense. Scientific management of fishery resources is not yet well established although development of fisheries has passed the point where this is becoming urgently needed

West 3.4 (1989, 1990) 3.4
East 2.8 (1990) 2.8
6.2

(continued)

ANNEX *(continued)*

	Production (million tonnes)	
	Historical maximum (year)	Present 1989/91
West	26.7 (1988)	25.5
East	3.4 (1987, 1990)	3.2
		28.7
of which culture		(2.7)

Northern Pacific. The total landings remain among the highest in the world: 31% of the total world marine catch. The recent decrease was primarily from the decline in landings of Alaska pollack and the Jap. pilchard. In Northwest Pacific the stock of pollack is fully exploited and there has been a significant increase in the proportion of undifferentiated fish in the catch. Most cod stocks are fully exploited. The demersal stocks in East China Sea and Yellow Sea seriously depleted and estimated 1/5 to 1/10 of their highest levels. Little sign of recovery. The Jap. pilchard exceeded 5.4 million tonnes in 1988, making it one of the largest single species catches but subject to wide fluctuation. Probable future declines possibly counterbalanced by other shoaling pelagic species. The total catch of pelagic fish ranged from 6 to 8 million tonnes. Due to major changes in relative abundance of different species, it is difficult to establish long-term sustainable yields for individual species. Cephalopod stocks around Japan fully exploited but oceanic squid in the northern region and neritic cephalopods in the southern region may provide certain opportunities if bans on certain current fishing gears are not extended (i.e. large driftnets). Salmon stocks appear to have stabilized due to improved artificial breeding and releasing techniques. Shrimp stocks fully exploited. There is no functional multilateral organization to assess and manage shared fish stocks. In Northeast Pacific the marine environment linked to "El Niño". The total catch ranged from 3.2 to 3.5 million tonnes in recent years. Alaska pollack consistently accounts for slightly less than half of the total catch. The pollack stocks tend toward a gradual decline. Major decline in salmon stocks attributed to degradation of the freshwater habitats resulting from urbanization and long standing drought conditions. Canada initiated individual vessel quotas (IVQs) for habitat on a two-year trial basis. Stocks of cods, hake and sablefish seem to be declining. Pressures are now growing for the institution of individual transferable quotas (ITQs) or alternatively, community development quotas (ACDs)

West	7.9 (1991)	7.5
East	1.8 (1989)	1.6
		9.1
	of which culture	(0.3)
West	1.1 (1989, 1991)	1.1
East	15.3 (1989)	14.5
		15.6

Central Pacific. Western Central Pacific generally overexploited; small-scale fisheries contribute most of the total catch. Demersal and small pelagic fish comprise most of the total catch. Shrimp and tuna are the major export fisheries. Total catch has continuously increased over the last 20 years primarily due to the extension of fishing on to new grounds, but the rate of increase has slowed. Many stocks of coastal shrimp in Asian waters and in Northern Australia are fully exploited. Few countries have formulated fisheries management plans. Some of the excess fleet have had to compensate fishing in the waters of neighbouring countries through various bilateral agreements. In Eastern Central Pacific fisheries are strongly influenced by the California current system, and dominated by pelagic species and shrimps. Sardine stocks are subjected to wide natural fluctuations. Shrimp stocks are fully exploited except the stock off Nicaragua. Catches of tuna are relatively stable

Southern Pacific. The Southwest Pacific generally overexploited both in pelagic and demersal species. Cephalopods show large fluctuations. The ITQ systems adopted by New Zealand and Australia have managed to stabilize a number of fisheries at more appropriate economic levels: in the Southeast Pacific the fisheries, particularly anchoveta and squids, are greatly affected by "El Niño". The current level of production (14 million tonnes in 1990) exceeds the record catches of 13.8 million tonnes in 1970. Small pelagics are dominant, representing 90% of total landings. Potential annual yield is estimated to be 2 to 5 million tonnes and the stock is fully exploited. The potential yield of squid is believed to be much higher than the current level of production. Chilean jack mackerel is considered to be moderately exploited although catches are relatively high

(continued)

ANNEX (continued)

	Production (million tonnes)	
	Historical maximum (year)	Present 1989/91
Southern Oceans. Most of the catch is krill, accounting for about 90% of the total. Finfish are also taken but their resources are very limited. Most of the catch taken by the former USSR fleet and Japan. Because the annual net production of krill is low compared to the available biomass, the resource could be vulnerable to overfishing, and supplies the food requirements of Antarctic mammals and birds. There is a need for precautionary management measures	0.5 (1989)	0.4
Marine total:	of which culture	83.6 (4.06)
Inland Fisheries. They represent 15% of the total supply from all sources. The production of fish, crustacean and molluscs has increased steadily during the last decade; inland *capture fisheries* have shown sluggishness, whereas *inland culture* made a remarkable growth reaching about 55% of total inland production. Inland fisheries are becoming increasingly conditioned by degradation of environment. Management tends to centre around mitigation of adverse environmental impacts. Formulation of adequate legislation for the protection of stocks in lakes, reservoirs and rivers is needed	15.2 (1991) Capture Culture	14.5 6.8 7.7
Aquaculture. (All figures quoted are already included in above table as appropriate). Current production some 12 million tonnes. Aquaculture development has been rapid, registering a mean growth of 10.9% per year; coastal aquaculture growth has been less than that of inland aquaculture. Important increases in yield have been achieved through culture of shrimps in the tropics and salmon in the temperate zones. Aquaculture has expanded over most of the world and is responsible for the increasing contribution of inland waters to the world fisheries production. The intensification of production of export-oriented communities in fish such as salmon and shrimp is largely industry driven. Severe strains are on the environment and allocation of sites, disease control and feeds. The general failure of rural aquaculture can be largely traced to managerial problems	12.0 (1990) Mariculture Inland waters	12.0 4.5 7.5
World production	100.3 (1989)	98.2

CHAPTER 7

Agricultural Development in the Economy-wide Context: Approaches to Policies and Strategies

7.1 INTRODUCTION

A careful reading of the academic and policy debate in the last decade reveals a shift in the formerly prevailing paradigm in economic development, with profound implications for policy making. The fundamental characteristic of the new approach to economic development is its emphasis on an expanded role of markets and of private initiative, and on a reduction and/or reorientation of the role of the state in the economy. The general thrust of this approach calls for the withdrawal of the state from those tasks that the market can perform efficiently, and the concentration of its activities on those tasks in which markets fail. The new approach is the result of several decades of accumulated development experience, and theoretical developments.

The disappointing performance of most developing countries during the 1980s (for some a "lost decade" for economic development) provided an impetus for a re-examination of the dominant development paradigm that emphasized market failures and advocated heavy government intervention. The idea of a different economic theory to apply to developing country problems lost its popularity, in favour of a "standard neoclassical" framework emphasizing maximizing behaviour on the part of economic agents (Fishlow, 1991).

In terms of developing country policies, the trend towards more market-oriented economic environment implies a shift away from inward-oriented import-substituting development strategies to more outward-oriented ones, and from the use of shadow prices derived from planning exercises to prices based on international opportunity cost of resources as guides for resource allocation.

As far as agriculture is concerned, the prevailing view in the 1950s and 1960s, i.e. that agriculture should be artificially "squeezed" on behalf of more dynamic sectors, is now largely rejected. The role of agriculture has been elevated from a sector destined to provide resources to foster industrialization, to a crucial sector for increasing export earnings, generating employment and improving food security.

Agricultural development presupposes integration of agriculture in the macroeconomy requiring a framework for overall economic policy guaranteeing: (a) stability of rules and signals for a decentralized calculation of gain and loss – a clear regime of property rights is quite essential in this regard; and (b) a government capable of identifying and providing socially productive physical and intangible capital that private markets cannot provide or can only do so inadequately, and catering for those social issues that markets either cannot solve by themselves or do so with long delays.

In what follows, a number of policy issues are examined which have as a common theme the interaction between agriculture and overall development mainly in developing countries, with particular reference to how the macroeconomic and institutional environments condition both agricultural performance and the effectiveness of sectoral (agricultural) policies. Within this context, attention is paid to the role of the public sector in shaping the economic environment of agriculture.

Section 7.2 discusses the evolution of theoretical perceptions on economic development and the exogenous (international) economic environment facing developing countries and how they influenced thinking about economic policies for agriculture. The objective is to learn from the analysis of past experience how policies for agricultural development and those for the overall economy can be made more mutually supportive.

Section 7.3 discusses the economic crisis of the 1980s and the economic reforms that several developing countries had to initiate as a result of the crisis. The fiscal, monetary and agriculture-specific policies are examined from the standpoint of their impact on both overall and agricultural growth during the period of economic crisis and adjustment. The methodological difficulties in evaluating the effects of policy reform programmes are discussed.

Section 7.4 examines the role of price, non-price and institutional factors for stimulating agricultural supply and growth. The question is to what extent and under what circumstances price reforms alone can promote private investment and carry the major burden of reviving agricultural growth. The types of interventions that are likely to promote or retard agricultural development are examined, with due regard to their effects on other objectives. Particular attention is given to the institutional and infrastructural problems prevailing in Africa, which may limit the effectiveness of price incentives in stimulating agricultural supply. Drawing on the experience of past policies, the role of the public sector is discussed with respect to several aspects of economic activity related to agriculture: provision of public goods and of a framework to facilitate market activity and increase private sector participation.

Section 7.5 draws some general conclusions for short- and medium-term policies on the basis of the analysis in the preceding sections. The discussion then shifts from short-term policies to long-term growth strategies. The realization that neglect for the agricultural sector has had negative effects on overall economic growth has given new impetus to agriculture-led development strategies. Two approaches to such strategies are critically reviewed. In turn, the implications of recent theories of economic growth for agriculture and agricultural policies are examined, emphasizing the importance of non-physical capital. The chapter ends with some general conclusions.

7.2 EVOLUTION OF THINKING ON AGRICULTURE AND DEVELOPMENT

The 1950s and the drive for industrialization

During the period immediately following the Second World War, no readily available conceptual/theoretical framework existed to analyse largely peasant agrarian societies. Thus the background of the major intellectual contributions in development economics reflected the post-war economic policy thinking in developed countries. In turn, this was influenced by economic conditions and concerns in the developed countries, particularly by the need to mobilize and reallocate large amounts of resources for the rebuilding of Europe and to avoid 1929-like experiences. It was considered that these major tasks could not be entrusted to markets and price signals. Market adjustments were considered to be slow, and marketing institutions for inputs and outputs inflexible. Thus, direct participation of governments and the planning of the production/ resource allocation system were considered essential (Stern, 1989).

A stronger case in favour of government intervention and resource planning was made for developing countries where inflexibility in input and output markets (such as price rigidity and labour immobility) was assumed to be even more pronounced. At the same time, the perceived economic successes of the industrialization policy in the USSR and the simple observation that the then developed countries had undergone a transformation from mainly agricultural to mainly industrial economies, fuelled the drive towards industrialization. Empirical evidence demonstrated that *inter alia* the process of development had been characterized by a movement of labour from low-productivity agriculture to high-productivity manufacturing (Kuznets, 1955).

Capital and labour were perceived to be more productive in industry where economies of scale and external economies prevailed, rather than in agriculture which was subject to diminishing returns. Thus, it was suggested that policies to turn the terms of trade against agriculture could increase accumulation for investment, by shifting resources towards industry. As agriculture was thought to be relatively unresponsive to prices, transfers of agricultural surplus through this type of "taxation" could be made without sacrifices in agricultural output and in food availability to the urban sector.[1]

The perceptions of the relative roles of agriculture and industry in economic development were strongly influenced by a number of theoretical and empirical contributions to the economic analysis of development. The existence of "surplus labour"[2] in agriculture was well accepted during the 1950s. The presumed existence of surplus labour in agriculture meant that labour could be attracted from agriculture to industry without loss of agricultural output, and at low wages. The resulting profits in the capitalist industrial sector could be reinvested to increase capital and promote growth.[3]

Another important reason for underestimating the role of agriculture as an "independent" source of economic development was the lower income elasticity of the demand for food compared with that for other products. If true, in the process of development, demand for food would grow more slowly than demand for non-agricultural goods. The implication was that agriculture would naturally decline relative to non-agriculture. It was furthermore asserted that there was a secular decline in the international terms of trade of primary commodities *vis-à-vis* manufactures, casting doubts on the potential role of agriculture as a foreign exchange earner. Therefore, a successful development strategy should be characterized by a move away from primary commodities and towards industrial products. As the demand for imported manufactures was considered to be price inelastic, the development of a domestic industrial sector implied a move from tariff-based measures to a regime of physical import controls of industrial goods to protect industry.[4]

A number of contributions underlining the merits of a development strategy based on *government-led industrial growth* appeared in the 1950s. They emphasized the positive external economies (spillover effects) associated with industrial investment (as learning by doing, etc.). They also stressed the large number of linkages between industry and other sectors and the concomitant multiplier effects industrial development could have on the overall economy. As individual entrepreneurs could not take into account those secondary "external" effects in their calculations of returns on investment, simultaneous government-coordinated industrial investment planning needed to be undertaken by the state.

Thus, during the 1950s, development was tantamount to the transformation, with heavy involvement of the state, of low-technology/agriculture-based economies into modern-technology/industrial ones, and the role of development economics was to study ways to transfer surplus resources (labour, savings) and surplus production from agriculture to industry. The emphasis was on agriculture's role as a resource reservoir. The "agricultural transformation" process was interpreted as a process of resource transfers (mainly labour), to be achieved through changes in the terms of trade against agriculture and in favour of industry.

Restoration of agriculture as a key sector
in the quest for economic development

As more data on developing countries became available in the 1960s, empirical evidence showed that freer markets, more liberal trade regimes and a growing agricultural sector were conducive to overall economic growth. It was demonstrated that export pessimism was largely unfounded, and that agricultural production and exports did respond to incentives (and disincentives). This "neoclassical resurgence" in development economics, particularly in its approach to international trade, agricultural policies and development planning, coincided with the decline of "enthusiasm" for planning in a number of countries. In parallel, the demographic projections of the 1960s showing rapidly increasing population in the developing countries gave rise to preoccupations with food availability in those countries, and generated a lot of discussion on the "World Food Problem" (Little, 1982). Agriculture's important role in overall development started to be recognized as it became evident that agricultural stagnation could choke off industrial development and cause food shortages and hunger. Hence the perceived need for more attention to the development of the agricultural sector (Jorgenson, 1961; Ranis and Fei, 1963).

Furthermore, the "false dichotomy of agricultural versus industrial development" was attacked, and agriculture was "elevated" by many analysts to an active source of economic growth. The important ways in which an expanding agricultural sector could contribute to economic growth included the provision of food and foreign exchange, transfer of labour and savings to the non-agricultural sectors and provision of an expanding market for the products of domestic industry (Johnston and Mellor, 1961).

The important new element in such a perception of agriculture was that resource transfers from agriculture occurring during the process of transformation should be effected as a result of agriculture's declining dependence on resources, i.e. as a result of growing productivity rather than as forceful extraction of agricultural surplus. This approach emphasizes a simultaneous effort to develop agriculture and industry through their mutual links.

During the 1960s the idea that peasants would not respond to economic incentives was tested using empirical data. The findings demonstrated the contrary, i.e. that farmers in developing countries do respond to economic incentives, allocate resources efficiently under existing technologies, and respond to profitable innovations provided that they are not too risky (Schultz, 1965). By the end of the 1960s it had become widely accepted that policies should primarily endeavour to enhance the opportunities for the above-mentioned rational behaviour of farmers to manifest itself and lead to increased production and improved productivity, e.g. by investing in research and extension, providing infrastructure, etc.

The 1970s: external shocks and the developing countries

The economic environment facing developing countries in the 1970s was shaped by a number of exogenous factors: (a) the two oil crises of 1972–73 and

1979; (b) the increases in commodity prices including the doubling of the world prices of cereals in 1973–74; (c) the shift to floating exchange rates after the breakdown of the Bretton Woods system and the ensuing volatility in world financial markets; and (d) the drastic increase in world liquidity, as oil exporting countries recycled oil revenues through commercial banks, and as developed countries in the early and mid 1970s attempted to "inflate away" oil price increases through money creation.

The combined effects of the oil shocks and financial market instability contributed to an economic slowdown and increased protectionism by developed countries after 1973, that subsequently led to a reduction in the growth of trade. Developing countries attempted to minimize the negative external effects on their economies by borrowing heavily and at favourable terms in the world financial markets to maintain high levels of domestic expenditure. Thus, even several oil importing countries were able to grow at relatively high rates despite an adverse international environment.

The "pro-agricultural shift" of the 1960s continued and was further strengthened in the 1970s. The strains posed by higher food and oil prices on the ability of developing countries to import food and a series of poor harvests brought about the prospect of an agricultural sector unable to feed and provide a living for the growing population. FAO's "Freedom From Hunger Campaign" in 1963 and the publication in 1970 of FAO's "Provisional Indicative World Plan for Agricultural Development" had increased the awareness of the problems facing agriculture and the rural sector and underlined the need to accord a higher priority to increasing food production. Increasing emphasis on poverty alleviation and social equity naturally directed attention to agriculture, as the majority of the poor live in the rural areas. During the 1970s, it was realized that the rapid development of the 1960s had made only a small contribution to poverty alleviation and that industrialization had not produced the expected results with respect to employment. The employment question was particularly acute in South Asia, Latin America and some African countries where landlessness was on the rise.[5]

Although the linkages between agriculture and industry were recognized, the approach to agricultural development was driven by the urgency to increase food production and, as a result, was narrow, i.e. it was not placed within a framework of macroeconomic equilibrium and overall efficiency in resource allocation. The pro-agriculture development strategy of the 1970s focused on relaxation of supply constraints to food production, investment in agricultural infrastructure, extension and agricultural research and integrated rural development schemes to combat poverty.

In parallel, the drive for industrial growth through import substitution was on the decline. As world trade was expanding twice as fast as world production, countries could neglect export opportunities only at a high cost. In addition, evidence from research on trade and industrialization showed that countries with higher rates of export expansion showed higher GDP growth rates (Myint, 1987). The views that attributed export difficulties to failure of domestic policies to exploit comparative advantage started gaining ground.

In the industrial countries, the dominant thinking of the time emphasizing a "trade-off" between inflation and unemployment was seriously questioned. Industrial countries tried to counter the negative effects of the first oil shock on their economies by easing their monetary policies (monetary expansion). Such policies, while failing to prevent economic recession and unemployment, gave rise to severe inflation.

7.3 MACROECONOMIC ENVIRONMENT AND AGRICULTURAL GROWTH

**Macroeconomic policies and economic performance:
the experience of the 1980s**

In the 1980s, there was a drastic change in the international economic environment facing developing countries. In the wake of the second oil shock, most developed countries opted for tighter monetary policies that resulted in higher real interest rates and a severe contraction in their economies, the negative effects of which were felt by developing countries in the form of declines in their aggregate exports. In addition, international commodity prices fell sharply for a number of reasons: (a) decrease in demand due to economic slowdown; (b) contraction in world liquidity itself;[6] and (c) increases in supplies following the earlier surge in agricultural commodity prices and the support policies in some developed countries. High interest rates, lower export receipts and the increasing interest burden on the accumulated debt, worsened the creditworthiness of a number of developing countries and limited their ability to borrow from private credit institutions. Credit to developing countries from private sources essentially dried up. The declaration of a moratorium on debt repayment by Mexico in 1982 had spillover effects to countries which under "normal" conditions would have been able to borrow.

While external negative shocks were catalytic for the economic crisis in the developing countries, the severity and duration of the crisis varied significantly among countries. Countries which modified their macroeconomic policies during the crisis years suffered less than those which did not and were able to take advantage of the economic recovery of the developed countries and of international trade in the latter part of the 1980s. Among the countries that faced serious economic difficulties were also those that had experienced (at least temporary) increases in their terms of trade either as oil exporters or as exporters of other commodities. The latter observation demonstrates that lower international terms of trade is probably only one of the many factors, and perhaps not the most important one, responsible for the economic decline of many developing countries in the late 1970s and early 1980s.

Country studies searching for causes of poor growth performance could not identify a single macroeconomic or trade characteristic that could explain this performance in all countries. Thus, while most poor performers were characterized by inward-looking economic policies, some could be found with relatively open economies. Similarly, while there is a strong correlation between "acute" inflation and growth collapses, one could find countries with moderate inflation that experienced growth collapses and countries that did not. Instead, studies at the country level seem to indicate that at the root of the growth crises were policy "packages", the main element of which was the unsustainable large fiscal expansion associated with the substantial capital inflows to developing countries during the 1970s. The cessation of those inflows resulted in severe economic crisis (Condos, 1990; Corden, 1990; Lal, 1990).

The sources of capital inflows varied from country to country. Borrowing in the international capital markets was a significant source of foreign exchange inflows (total long-term debt of developing countries increased from US$63 billion to US$562 billion between 1970 and 1980) as were international aid and favourable swings in the international commodity terms of trade (minerals, oil, coffee, etc.). Some countries experienced increased foreign exchange inflows from a combination of sources, i.e. from both export earnings and borrowing against expected future earnings.

The growth implications associated with (cessation of) foreign capital inflows are not uniform across sources. Borrowing in international capital markets (including borrowing against future earnings) could have the most serious implications. If invested in non-productive projects, or consumed, repayment obligations could make the countries, all other factors constant, worse off than if the inflow had not occurred. Temporary export booms become a problem to the extent that the inflows are spent on consumption and/ or unproductive investment and treated as if the foreign exchange boom is permanent. In such cases the additional resources from these inflows can be completely wasted. A comparison between the experiences of Indonesia, Nigeria and Botswana in handling export windfalls is instructive in that respect (Pinto, 1987; Hill, 1991).

Although it would be an unwarranted conclusion to say that foreign resource inflows are bad for development, their mismanagement can be outright detrimental. It happened in a number of developing countries when such inflows were translated into high budget deficits, high inflation, overvalued real exchange rates and a worsening of the current account balance. This seems to have worked as follows.

Often the fiscal expansions associated with such inflows were directed towards domestically produced non-tradable goods and services (including expansion of services of the public sector itself, expansion of services of agricultural parastatals, urban construction, large and possibly hasty infrastructural projects), as well as towards imports. The resulting excess demand caused increases in prices and wages which, combined with a fixed or slowly adjusting nominal exchange rate, resulted in an overvalued real exchange rate. The combination of higher domestic costs, and an overvalued real exchange rate, shifted incentives away from tradables towards non-tradables[7] and imports, and worsened the trade balance, exacerbating the effects of increased direct spending on imports. The inability to match increases in domestic fiscal expansion with increases in fiscal revenues resulted in large budget deficits financed mainly by money creation and the inflation tax. The magnitude, timing and duration of domestic and external imbalances varied with the distribution of spending between consumption and investment, the efficiency of investments undertaken, the flexibility of exchange rates, the response of output to increased spending and the ability of governments to collect revenues.

Domestic and foreign imbalances could be sustained only to the extent that foreign exchange inflows continued to be available (ability to borrow, autonomous inflows of aid and investment, a lasting export earnings windfall). As this was not the case in the 1980s, those imbalances and the underlying spending were unsustainable and had to be reversed. Immediate policy responses by some countries to deteriorating budget deficits and current account imbalances included the imposition of import restrictions and cuts in domestic spending. Which sectors were affected most from such cuts was often a matter of relative political leverage rather than of economic efficiency considerations. Available evidence shows that in a sample of 24 countries, expenditure cuts from 1974 to 1984 resulted in capital expenditure declines of about 28 percent, while the declines in subsidies and transfers to parastatals were 11 percent and in the public sector wage bill 14 percent. Expenditure in infrastructure experienced a steep decline (25 percent), while defence and social sector spending declined by 7 percent and 11 percent respectively. Within the capital budget, infrastructure declined by 41 percent (Knudsen *et al.*, 1990).

The inability of several countries to meet their debt repayment obligations and the incipient crisis raises some questions as to the investment selection criteria used to allocate inflows generated during the "boom periods". Some rough estimates indicate that for the low- and middle-income countries as a group, productivity of investment fell by one-third between the 1960s and the 1970s, although part of the decline should be attributed to the drastically changing international economic environment (Fardoust, 1990). The important conclusion of the analysis above is that a distinction should be made between sustainable, productivity-based growth spells, and "demand driven" ones generated through increased public spending during periods of increased capital inflows. Several of the growth experiences of the 1970s belonged to the second category, could not be sustained, and they have had detrimental effects on the agricultural sector and overall development, a topic taken up in the next section.

The above scenario describes the policy-related aspects of the crisis, and should not be taken to mean that external factors were unimportant. Both the severe decline in the (barter) terms of trade of developing countries and the increase in world real interest rates would have had negative effects on growth even in the absence of highly expansionary fiscal and monetary policies. Among the external factors should be counted the eagerness of private lenders to provide loans to the developing countries. The world banking system was inundated with liquidity and tended to "recycle" funds through bank consortia without adequate "creditworthiness" controls, sometimes in the conviction that banks would be bailed out by their own governments in case of default. In conclusion, although unpredictable exogenous factors have been important, the analysis above shows that, given the severity of such factors, unsustainable public spending expansions (often under donor blessing) exacerbated the situation transforming it (for a number of countries) from an economic slowdown to a growth collapse.[8]

Asia was the only region with no declines in per caput incomes during the 1980s. Countries in Asia followed diverse development strategies and experienced diverse patterns and rates of growth. The extent and modalities of government intervention in the economies of Asian countries span a wide spectrum ranging from heavy interventionist to highly liberal (World Bank, 1993b). In terms of the macroeconomic framework discussed above, the large economies that make up most of the region have not experienced growth collapses due to a combination of several factors: (a) low debt ratios; (b) maintenance of creditworthiness that permitted them to finance the negative effects of the external shocks; (c) prudent fiscal policies; and (d) quick responses to the initial shocks.

For instance China, and to a smaller extent, up to quite recently also India, are clearly among the countries with low external debt. Thailand, by following prudent fiscal policies, is a classic example of a country that has managed to maintain creditworthiness, as has Indonesia. Indonesia, Korea Republic and Thailand responded quickly to external shocks although the time paths of adjustment were different with Thailand following a more "gradualist" adjustment path (Corden, 1990).

Macroeconomic imbalances, sectoral policies and agricultural incentives

The preceding discussion on the emergence of fundamental macroeconomic disequilibria provides the background for examining their impacts on agriculture. The importance of macroeconomic policies in affecting agricultural incentives through the four major "macro-prices" (interest rate, exchange rate, the general price level and the wage rate) has been the focus of attention in the 1980s. The detrimental effects of adverse developments in the international terms of trade were compounded by macroeconomic and trade policies that caused declines in overall economic growth. Spending increases directed towards domestic non-tradable goods and services caused decreases in the relative prices of agricultural commodities, mainly exportables or import substitutes, *vis-à-vis* those of non-tradable commodities (including those of services and domestically produced capital goods). Those adverse effects on the agricultural terms of trade were reinforced by overvalued exchange rates maintained through capital controls and foreign exchange rationing. Although overvalued exchange rates constitute (in principle) disincentives for both agricultural and non-agricultural tradeables, the latter were protected through tariffs and quantitative restrictions. Thus, macroeconomic and trade policies had detrimental (indirect) effects on the terms of trade of agriculture *vis-à-vis* both non-tradables and non-agricultural tradables.[9]

The negative effects of macroeconomic policies, especially on exportables, were often compounded by agricultural-sector-specific pricing policies, such as border taxes on agricultural exports, price controls, and the wedge between border prices and those at the farm level created by the monopsonistic behaviour of government parastatals and state marketing boards. Such direct interventions in agriculture have been uneven across commodities and, in some cases, they have benefited some agricultural commodities.[10]

The effects of pricing policies on agricultural price incentives were attenuated by the increases in aggregate absorption (i.e. total expenditure by domestic residents) generated by increased public spending, a share of which was spent on agricultural commodities, and by spending on public infrastructure investments in agriculture.[11] Input (mainly fertilizer) subsidies also tended to attenuate the negative effects of macroeconomic policies although their impact varied across classes of farmers as often subsidized inputs had to be rationed. While such counterbalancing forces may have been operating in the 1970s in some countries, in many cases they were reversed in the 1980s in the wake of the economic recession and associated reductions in aggregate demand.

The combined effects of macroeconomic and sector-specific policies for agriculture were largely negative. In the World Bank study on the direct and indirect sources of protection[12] of major tradable agricultural commodities, the authors reached a number of important conclusions:

1. The indirect effects were much stronger than the direct effects. Namely, policies directly affecting prices resulted in a positive protection (on the average) of imported food products at an approximate rate of 20 percent and the taxation of exported commodities resulted in a negative of 11 percent. Incorporation of the effects of macroeconomic and trade policies resulted in a *negative total* protection of 7 percent for imported food crops and 35–40 percent for exported agricultural products.[13]
2. The degree of total negative protection of exported commodities was in most countries higher than for imported ones (mostly food staples). The widespread drive for food self-sufficiency accounted for the lower negative (and sometimes positive) protection of food staples. A summary of the findings by region is shown in Table 7.1.

The study showed that for the 18 countries examined, the income losses of

Table 7.1   Direct, indirect and total nominal protection rates, by region, 1960–84 (%)

Region	Indirect protection	Direct protection	Total	Direct protection of importables	Direct protection of exportables
Asia*	−22.9[†]	−2.5	−25.2	22.4	−14.6
Latin America[‡]	−21.3	−6.4	−27.8	13.2	−6.4
Mediterranean[§]	−18.9	−6.4	−25.2	3.2	−11.8
Sub-Saharan Africa[¶]	−28.6	−23.0	−51.6	17.6	−20.5

Source: Reproduced by permission from Krueger *et al.* (1991).
Note: The period covered is generally from 1960 to 1984, but it varies somewhat in a number of countries.
*Republic of Korea, Malaysia, Pakistan, Philippines, Sri Lanka and Thailand.
[†]In South Asia (Pakistan, Sri Lanka), the indirect nominal protection rate was −32.1 percent, while in East Asia (Korea Rep., Malaysia, Philippines, Thailand) it was −18.1 percent.
[‡]Argentina, Brazil, Chile, Colombia and Dominican Republic.
[§]Egypt, Morocco, Portugal and Turkey.
[¶]Côte d'Ivoire, Ghana and Zambia.

the agricultural sector due to taxation were substantial. Net average *direct* taxation (i.e. after the subtraction of subsidies) amounted to 4 percent of agricultural GDP between 1960 and 1984. Including *indirect* taxation brings income losses to 46 percent of agricultural GDP, the percentage varying between 37 percent for the "average taxers" to 140 percent for the "heavy taxers" (in the sense that an agricultural GDP of 100 with the interventions would have been 240 without them). In only two countries it was found that price-based negative protection was compensated by increased infrastructural investment in agriculture. The major beneficiaries from income transfers out of agriculture were the government sector, urban consumers and industry.

The taxation of agriculture, and the subsequent decline in the performance of the sector, had feedback effects on the rest of the sectors and the macroeconomic system. Direct and indirect taxation of export commodities resulted in a shift by producers away from export crops, which, combined with lower world commodity prices, resulted in a collapse in export earnings and reduced the ability of countries to import in support of industrialization. Thus, industrialization strategies based on agricultural taxation often turned out to be self-defeating. Likewise, losses of parastatals constituted a substantial burden on the government budget in a number of countries. The problem was more pronounced in Africa where marketing boards established by colonial powers were expanded by governments and were assigned the task of regulation and control of most aspects of agricultural marketing activities. Parastatals were often assigned functions going beyond performing marketing activities, e.g. management of commodity stocks, subsidization activities, etc.

They were also used (along with other parts of the public sector) as employers of last resort. Provision of those services by parastatal organizations contributed to their big deficits which were financed by increasing marketing margins or from the public budget.[14] The combination of parastatals' marketing with their other functions contributed to the lack of transparency, made controls difficult and increased the costs of providing such services. Whatever welfare gains they generated for some parts of the population (not always the most needy parts) were bought at a heavy cost and proved unsustainable.

Agriculture under economic adjustment policies

As the resource inflows that sustained excess spending dried up by the early 1980s, several developing countries found themselves with low foreign exchange reserves, and unable to borrow from the private sector. Thus they were forced to turn to the international lending agencies (mainly the World Bank and the IMF) for their financing needs. Loans from those agencies were made conditional upon acceptance of comprehensive policy reform packages of macroeconomic stabilization and structural adjustment. Stabilization aims at reductions in domestic budget and current account deficits through reductions in public spending and credit ceilings especially to the public sector. As wages and prices of non-tradables were relatively inflexible, exchange rate devaluations were required to restore relative prices and switch incentives and the pattern of production towards tradables. Structural adjustment was to be achieved through medium-term supply-side policies to enhance efficiency and remove bottlenecks. The major thrust of structural adjustment reforms was the enhancement of the role of the markets in guiding resource allocation. Thus, a number of measures were proposed to liberalize markets for inputs and outputs, including reductions or elimination of subsidies, reductions in agricultural export taxation, elimination of marketing and transportation controls, etc.

   The agricultural focus of structural adjustment programmes varies extensively across developing regions. For sub-Saharan Africa from 13 loans between 1980 and 1987, 77 percent included conditions for agricultural policy. This percentage was matched only by trade policy conditions (77 percent). In other developing countries during the same period, 38 percent of loans (out of a total of 16 loans) contained agricultural policy conditions (World Bank, 1988). Price policy reforms included the reduction of the gap between border prices and those received by the producers for both inputs and outputs. Their effects in terms of incentives are diverse across product categories: while for exportable commodities it meant a reduction in taxes on prices levied by parastatals, for imports it meant the abolition of quantitative restrictions and/ or lowering of tariff levels.[15] On the input side, it usually meant the abolition of subsidies with negative effects on all producers having access to them. In return for the implementation of import liberalization, donors undertook the financing of the importation of a list of essential items which often included fertilizers and agricultural chemicals.

Agricultural parastatals and marketing boards were to be abolished or substantially reformed to increase their financial accountability and improve management structure. In most cases, parastatal support from the public budget ceased and the policy of employer of last resort was abandoned, often worsening the immediate problem of unemployment before the hoped-for positive effects from the policy reforms were generated. As a result of the reform of agricultural parastatals and marketing boards, barriers to inter-regional movement of commodities were abolished, as were pan-territorial and pan-seasonal pricing schemes.

Evaluating the effects of structural adjustment programmes undertaken by the countries themselves or imposed as part of conditionalities on the agricultural sector is difficult for several reasons:

1. The degree and consistency to which reforms were implemented and sustained varies widely among countries (inconsistencies include nominal devaluations not matched by supporting monetary policies, cuts in aggregate demand through reductions in government investment rather than consumption, etc.).

2. The occurrence of external shocks such as changes in the international prices of commodities, and other economic developments in the rest of the world complicate the isolation of the effects due to policies.

3. For regions such as Africa, the volatile political and ecological conditions and dependence on rainfed agriculture make difficult the apportionment of credit or blame for agricultural performance to policies, political instability or weather conditions when comparisons are made over short time periods.

4. Conclusions may vary with the evaluation method chosen: should performance be evaluated in a "before and after adjustment" manner or in a "with versus without adjustment" framework? The latter would require constructing a counterfactual scenario describing both the feasibility of continuing pre-adjustment policies and their effects on the sector.

5. Initial conditions in the adjusting countries vary, with countries implementing structural adjustment programmes "entering" the 1980s with weaker economies (see discussion below on price versus non-price incentives).

6. Large discrepancies exist between the performance of countries in different regions undergoing similar adjustment programmes. Evidence shows that middle-income countries have had a greater rate of success in resuming growth than low-income countries, especially the ones in sub-Saharan Africa.

7. There is often a confusion between policy adjustment and adjustment lending. Thus, what is often tested are the effects of adjustment lending

rather than the effects of policy reforms. The two may be different since there are countries that undertook drastic reforms without adjustment lending and others that received adjustment lending but abandoned reforms (Summers and Pritchett, 1993).

A number of studies looking at raw data or using statistical techniques, have concluded that structural adjustment programmes halted the decline of, or even increased, agricultural production and exports, or increased agricultural growth (Lele, 1992; Faini, 1992). Data analysis presented in FAO's 1990 "State of Food and Agriculture" provided indirect support for this hypothesis by looking at 65 developing countries. Namely, it was found that, in general, countries which had gone through "healthy adjustment" processes and were meeting the basic objectives of stabilization programmes (reduction of budget and current account deficits, etc.) had also better agricultural performance.[16] In another study, the overall agricultural performance of sub-Saharan African countries implementing policy reforms (including exchange rate adjustments and price and fiscal reforms) was compared to that of non-adjusting countries. The results show that among countries "starting out" with similar agricultural growth rates before adjustment, growth rates were increasingly higher for those that adjusted (Binswanger, 1989). Studies can also be found that dispute the proposition of an improvement in agricultural performance of countries undertaking structural adjustment reforms.[17]

A recent World Bank study (World Bank 1994b) compared the performance of adjusting and non-adjusting countries in sub-Saharan Africa, taking into account the "degree of adjustment". It concludes that the overall growth performance was superior for the adjusting countries. However, the record is mixed for agriculture. There was no clear pattern of differences in agricultural growth between the countries which improved their macroeconomic situation and those which did not. But countries which experienced improved prices for their agricultural exportables (e.g. through devaluation) had a better agricultural growth record than those which did not.

Results derived from cross-country comparisons may be indicative but are not very informative or conclusive, as such comparisons mask wide differences in the experiences of individual countries, not only in terms of the effects of policy reform but also on the initial conditions at the initiation of reforms (extent of economic crisis, economic position, etc.), and modality of application of the reforms (timing, consistency, etc.). Given the methods used, what is often being tested is not only the correctness of the policies themselves (or of the principles underlying such policies) but also the effectiveness of implementation.[18]

In addition to the issue of the effectiveness of policies in improving agricultural performance, serious concerns have been raised as to the effects of policy reforms on the more vulnerable parts of the population. Demand contraction, the abolition of parastatals or the reduction in their role, and the withdrawal of the state from certain activities, combined with rigidities in resource mobility and slow response by the private sector, may result in increases in unemployment due to delays in redeployment of labour. Given the fixity of the capital stock in the short run, an increase in the supply of labour due to the reduction in the public payroll coupled with an inelastic demand for labour may cause short-term reductions in both employment and the real wage. Reductions in public investment outlays may not be matched by increased private ones (at least in the short run), while expenditures for infrastructure, health services and social programmes may be affected by the overall budget reductions. Likewise, increases in food prices may affect adversely the food security of the more vulnerable parts of the population, while cuts in food and other subsidies have more pronounced effects on the poor and the unemployed (FAO, 1989b).

It is difficult to obtain definitive answers as to the exact causes of observed changes in social indicators by making aggregate cross-country comparisons. Policy reforms affect individual population groups in different ways and it is difficult (and dangerous) to generalize conclusions drawn from individual country analyses. Changes in the social indicators do not point always in the same direction and vary also among regions. While it is difficult to identify with precision the effects of policy reforms on social indicators from those that follow as a consequence of the full-scale economic crisis of the early 1980s, it is now well accepted that a new class of poor has been created by adjustment, although some of the "old" poor may have benefited from policy reform. Policy reforms that reduce macroeconomic imbalances, restore relative prices and halt economic decline do not necessarily reduce poverty by themselves. Often such policies must be supplemented by special programmes and interventions targeted at the poor (see Chapter 9 for developments in the incidence of poverty and discussion of anti-poverty policy interventions).

Reforms should also be analysed from the point of view of the alternatives that were available to the countries that faced the crisis of the 1980s. Policy reform programmes implemented through the initiatives of the adjusting countries themselves or under the pressure of international lending agencies, should be viewed as a response to the unsustainable policies characterizing development strategies for a long period of time. Thus, the issue of the sustainability of the previous policies should be considered. The countries that were facing very serious distortions and imbalances were the ones that sought to reform first, but also the ones in which the reforms were slow to produce results given the magnitude of distortions, suppression of private sector activities, etc.

In conclusion, while differences in opinion may exist on implementation issues (timing and sequencing of reforms, modalities of protection of weaker groups of the population, etc.) regarding policy reforms, there are very few arguments against their principal directions: (a) on the fiscal side, prudence should be exercised to avoid large and persistent deviations of spending from their sustainable means of financing; (b) careful screening of public investments is required to evaluate their economic and social returns; (c) conditions should be created for markets to work more efficiently and for prices to play their role as the major signals for resource allocation; (d) non-economic objectives should be pursued as much as possible by interventions aimed directly at the problem rather than by policies that distort economic incentives; (e) exchange rate reforms should be supported by proper macroeconomic policies.

7.4 POLICY REFORMS AND AGRICULTURE: THE ROLE OF PRICE AND NON-PRICE FACTORS

Price and non-price factors: complementarities and conflicts

On the issue of the role of the state in economic activity in general and in agriculture in particular, the debate on policy reforms which emphasize efficiency and private initiative had – at its inception – more to say about what governments should not do than what exactly they should do. The approach to policy reforms has since undergone changes as a result of accumulation and analysis of experiences. Thus, in the late 1980s and early 1990s, loan conditionality emphasizing demand contraction and overall efficiency through restoration of "proper" relative prices has been supplemented with policies (and funding) emphasizing sectoral aspects and the supply side to stimulate economic growth. More emphasis is currently given to helping countries improve agricultural infrastructure, implement interventions for poverty alleviation, improvements in education and health, and deal with environmental and natural resource degradation. With the passage of time, policy approaches are modified and fine-tuned as more lessons are learnt as to which policies work and which do not, and why. For agriculture, a major issue concerns the additional policies or interventions required to supplement policies that "get prices right", which, most often, result in conditions favourable to agriculture.

Although the responsiveness of production of individual agricultural commodities with respect to price incentives may be high, for agriculture as a whole it is likely to be low due to aggregate resource constraints, at least in the short to medium term. More importantly, in countries or regions with poor resource endowments, price incentives under traditional technologies and poor infrastructure may not be sufficient to increase production significantly, stimulate investment and overall agricultural growth. Thus, although price reforms are necessary for agricultural growth, the pessimism about the extent to which output of the sector as a whole could respond to price incentives alone over the short to medium term, calls for emphasis on "non-price measures", in particular improved research, input delivery and infrastructure services.

If structural impediments predominate in agriculture, it will be difficult to achieve sustained growth in production by price incentives alone unless

Box 7.1   Policy reform and agricultural development in sub-Saharan Africa: the
crucial role of infrastructural deficiencies*

In evaluating the efficacy of reform programmes to spur overall agricultural
growth, the issue of supply response of agriculture to price incentives is critical to
the extent that almost all reform programmes concentrate on "getting the prices
right" which, for the case of agriculture, often means an increase of relative
agricultural prices.

There is a great deal of controversy as to the actual magnitude of and the
proper method to evaluate supply response. Agreement exists that: (a) the short-
term price elasticity of the agricultural sector as a whole (as opposed to the one
for individual commodities) is very small, hardly significantly different than zero
(in the statistical sense); and (b) that the long-run elasticity can be quite
substantial as intersectoral resource movements (capital, labour) take place and
as new technologies are introduced.

For sub-Saharan Africa, very little empirical evidence exists on the supply
response of agriculture to prices. From what is known (Bond, 1983) short-run
elasticities are not lower than the ones in other countries (about 0.18 for the
period 1968–81), whie the long-run elasticities seem to be significantly lower than
those that have been reported for other areas.

While the magnitude of the price elasticities is a disputed issue, little
disagreement exists on the substantial response of agricultural output to non-
price variables especially infrastructure facilities and services. Empirical evidence
demonstrates clearly large effects of *changes in* the levels and/or quality of
infrastructure on agricultural production. Furthermore, such effects tend to be
higher the lower the level of existing infrastructure in the country. Evidence also
exists that the estimates of the price elasticity of supply tend to fall as
infrastructural variables are included in the estimation, i.e. there is an interaction
between supply response to price and availability of infrastructure.

There is abundant, though scattered and largely anecdotal, evidence to both the
serious deficiencies on one hand, and the importance, on the other, of agricultural
and rural infrastructure in sub-Saharan Africa. Particularly acute is the situation
concerning transport and communications infrastructure, the lack or bad state of
which inhibits the transmission of price signals to farmers and causes large
disparities between prices at the producer level and prices at the consumer centres
or ports. Collapsing transport infrastructure inhibits market integration of remote
rural communities and discourages the production of market surpluses even when
relative prices turn out to be favourable.

Table 7.2 demonstrates some significant differences in the infrastructural
situation of African countries compared to the situation in some Asian ones.
Those differences may be obscuring quality differences, and also the specific
problems that may exist in the quantity and quality of rural transportation
services. World Bank estimates show that the restoration and maintenance of the
present rural road network in Africa will require an expenditure of about 0.5
percent of the region's GDP for the next 10 years. Expansion of the network to
handle the increases in marketing of agricultural commodities will require an
additional 0.6 percent of GDP.

The major cause of the communications/transportation problems in sub-
Saharan Africa is the low population densities in the continent compared with
other regions. Thus while on one hand Africa has the greatest need for a good

(continued)

Box 7.1   *(continued)*

transport and communications network given the spatial distribution of its population, on the other hand not only is it under-equipped relative to its needs, but the little infrastructure that exists cannot be properly maintained let alone expanded due to the low per caput incomes prevailing in the region.

The causes of the deficient transportation system in Africa are rooted both in the colonial past of the region as well as in some ill-focused interventionist policies of post-independence governments. Namely, during colonial periods the transport infrastructure was heavily skewed in favour of the colonial enclaves or to connect ports to mines, as massive railway building in the sparsely populated Africa was considered uneconomic. The little rural road capacity that was built either during or after the colonial period is heavily under-utilized (traffic consists mainly of pedestrians and head-loaders) due to shortage of foreign exchange necessary to buy spare parts for trucks and vehicles. In addition, a nexus of transport regulations, trading limitations, interregional transport restrictions, reliance of parastatal truck fleets, and the stagnation of the agricultural sector that limited transport volumes, discouraged private investment in transport capacity and vehicle servicing. Such policies stifled local attempts to develop transport means suited to the region's technological capabilities (means of transport in between head-loading and motorized vehicles such as bicycle trailers, handcarts, motorized rickshaws, etc.). The system of pricing and marketing of fuel created serious chronic shortages further exacerbating the situation. Such deficiencies in the transport systems, coupled with poor information flows, inhibit the production of marketed surplus (especially food items) and increase the tendency of farmers towards subsistence production. They also increase the interregional variability of prices and impair food security in food-deficit areas. Such deficiencies prevent the integration of small rural communities in the market system, the creation of markets and create a set of "semi-tradable" commodities with relative prices being determined by local conditions rather than by international price movements (Delgado, 1992).

*Based on Platteau (1993).

---

extension services, marketing, transport and storage facilities, etc., are improved (issues of marketing and rural credit are discussed in Chapter 9, those of extension services in Chapter 10). This view recognizes a significant role for governments in providing public goods and assisting structural change. Examples of positive reaction of agricultural output to public goods provision abound (Delgado and Mellor, 1987; Binswanger, 1989), especially in countries with a low "stock" of infrastructural facilities and a low level of infrastructure services (see Box 7.1).

Structural considerations notwithstanding, improved price incentives are a necessary component of policies to alleviate input shortages in agricultural production and to stimulate investment of private capital. The responsiveness of total agricultural output to such incentives may be low in the short run, but is considerably higher in the longer term. Often the responsiveness of agricultural supply to price and non-price factors is compared. Such

Table 7.2   Selected indicators of surface transport for Africa and Asia, 1990

Region/country	Rail and road mileage per 1000 persons	Rail and road mileage per 1000 ha of cultivated land	Number of motorized vehicles per mile of paved road
Africa			
Benin	0.17	0.36	9.0
Kenya	0.30	1.09	19.2
Malawi	0.28	1.55	17.0
Senegal	0.44	0.60	12.2
Tanzania	0.15	0.09	14.2
Togo	0.37	1.26	13.5
Zimbabwe	1.09	4.11	35.7
Asia			
Bangladesh	0.07	0.75	47.5
India	0.68	3.58	49.0
Pakistan	0.73	3.71	42.5
Philippines	1.65	7.38	51.8
Korea, Rep.	0.58	11.41	67.2

Source: Platteau (1993), with data from Ahmed and Donovan (1992).

comparisons have no meaning to the extent that price and non-price incentives are complementary. In one case, though, they may be substitutes. Namely, modalities of financing the provision of public goods for agriculture may be (at least partly) related to the price incentives to agricultural producers. Given the limited capacities of developing countries in levying income taxes, a part of their fiscal receipts necessary for increasing the provision of public goods to agriculture will have to come from border taxation of agricultural exports and imports as well as imports of agricultural inputs, thus affecting agricultural incentives.

With respect to price and non-price incentives, some issues need to be considered:

1. From an efficiency standpoint, the prices used to calculate project priorities are critical. The implication for policy sequencing is that restoration of "proper" relative prices should precede planning for public goods, though it must be recognized that the "proper" level of relative prices is itself a function of the level of public goods.
2. The long-run effects of improved price incentives will differ between countries and regions depending on the availability and quality of land and water resources and the "stock" of public goods already in place. In situations characterized by the existence of land rents, price increases may induce private investment, even with less government provision of public goods or where public goods are already in place, than in cases where land rents are absent (Delgado and Mellor, 1987). This implies that a careful

analysis of long-run private sector response to price incentives (in terms of capital accumulation, reversal of migration to urban areas, etc.) should be made before the appropriate levels of public goods are decided.

3. Often, low aggregate agricultural production response to price incentives may be due to a number of institutional reasons other than the lack of public spending on infrastructure.[19] Some of them are examined in later sections. It is now well recognized that the fact that similar policies seem to work in some countries and not in others is, to a large extent, due to differences in the quantity and quality of existing physical but also institutional infrastructure, and of political consensus to carry out often unpopular policy measures.

Provision of public goods: what role for the state?

Ensuring that the fruits from an increased provision of public goods "trickle down" to all producers (especially small farmers) is an important task of the public sector. The cost-reducing effects of infrastructural improvements in the system of agricultural marketing and distribution will not be shared by farmers if the retreat of monopsonistic government parastatals is replaced by monopsonistic traders or transporters, a situation more likely to occur under conditions of incomplete credit markets. An active role of government is needed in such cases to prevent such situations from occurring and to rectify them when they do occur.

Although agreement may exist "in principle" on the need for governments to provide public goods, the modalities of provision, institutional organization, and especially the relative roles of the public and private sectors may be different for different types of public goods. Firms investing in the discovery of new technologies may only capture a small part of the net income gains resulting from it. This reduces their incentives to invest in research, with the consequence that "too little" research may be produced by relying on the private sector only (Timmer, 1991). In such situations, public-sector-supported research is a necesary component of the "non-price" policy interventions (for broader discussion of the case for publicly supported agricultural research see Schultz, 1990).

The case of other types of public goods is less clear-cut. One example of a "mixed case" is irrigation. The complexity of management and large financing requirements of irrigation projects make them inaccessible to private investors in developing countries. Inclusion of externality costs in those projects (water-borne diseases, depletion of underground aquifers) may make the intervention of the public sector necessary for social efficiency (Ellis, 1992). On the other hand, the private sector or local communities, cooperatives, etc., may have an advantage in the construction, design and operation of such projects. For rural roads, while a central authority is more competent to undertake construction, experience shows that such projects are more successful when users are involved in the planning stage, and when local resources are mobilized for the maintenance and rehabilitation of the rural roads network. The setting of clear rules allocating benefits and responsibilities among the central authority and user groups is critical for the success of decentralization of decision making. In general, while the responsibility for planning and overall execution and supervision of infrastructural facilities rests with governments, the provision of infrastructure services may very well be the responsibility of the private sector and local communities.

Pricing issues are also important in public goods (such as electrification, education, irrigation, etc.). Experience shows that few countries charge "fair" or cost recovery prices for water use. This increases the burden on public budgets, cuts into budgets for maintenance and rehabilitation, and results in the overuse of the resource with secondary environmental effects. Public investments should primarily strive at attracting private sector resources in agriculture, but should also consider wider non-strictly economic objectives, such as employment generation and poverty alleviation, through the execution of rural public works programmes. Public investment in rural infrastructure, research and extension are of great importance as a motive force for a broad-based rural development.

Policy reforms: the role of institutional capacities and political constraints

As no clear-cut general conclusions can be drawn on how to choose among alternative options for providing public goods, careful analysis of needs, means available, and policy impacts in the individual countries and contexts is required. The analytical skills and institutional capacities of the public sector for performing such tasks are currently lacking in a number of countries. Large public projects, despite their usefulness a priori, can turn into white elephants without the proper administrative and implementation capacity. Such capacities should be carefully considered at the planning stage.

Often the lack of developed private sector institutions to take over the role of the retreating agricultural parastatals in marketing, storage, imports and exports of agricultural inputs and products, complicates and frustrates liberalization efforts. For countries with a history of limited private sector involvement in marketing and distribution activities (other than parallel market activity), the private sector may not "step forward" to cover the vacuum left from the withdrawal of the state. There are also cases where rapid privatization entails short-term problems associated with staff redeployment and increases in social tensions. In such cases, reform of parastatal enterprises and marketing boards needs to be gradual, with emphasis on increasing accountability, streamlining operations and reducing costs. Steps are also needed to rid such organizations of multiple developmental tasks that they are often called upon by governments to perform (price stabilization, provision of guaranteed marketing outlets, food security, employer of last resort). Mitigating the adverse effects of employment should be the object of alternative targeted interventions. Experiences with privatization in developed countries show that introducing competition from the private sector can be an important first step towards successful liberalization, although recent evidence from sub-Saharan Africa is mixed.

In many cases, liberalization efforts are not accompanied by a clear legal framework to sanction the activity of the private sector. The lack of such a framework creates risks that discourage private participation in productive and marketing activities. Examples include continued or erratically applied subsidies to parastatal agencies competing with the private sector and quantitative restrictions to cross-regional marketing.

In relation to the origins of the policy reforms, existing evidence suggests that "home-grown" adjustment programmes have a better chance of being consistently implemented and produce results. It is not always possible for countries to come up with the necessary political consensus needed for implementing a wide-ranging credible reform programme. To the extent that price distortions and the level and composition of public spending reflect a structure of political power, their removal or prospect of removal, will be met with differing degrees of resistance. If the beneficiaries of reforms (private merchants, traders, informal sector) are able to capture the opportunities created by economic liberalization and replace the state in the productive and distributive functions it (the state) relinquishes, then the reforms have a better chance of being effective. For a pre-reform consensus to be achieved, the potential beneficiaries have to have some political leverage before the reforms are initiated. The importance of institutional and political issues in conditioning success of reforms is presented in Box 7.2.

7.5 LESSONS AND POLICY IMPLICATIONS: SHORT-RUN ADJUSTMENTS AND LONG-TERM STRATEGIES

Policies for the short and medium term

The implication of the analysis in the previous sections is that a "proper" macroeconomic environment (price stability, a competitive exchange rate, an interest rate that reflects the balance between credit demand and savings) can greatly contribute to improving the conditions for agricultural growth. The main conclusion from reviewing past experience is that piecemeal solutions cannot work as crises are usually the result of not one but many policies, often working at cross-purposes. A consistent set of policies is necessary to avoid crises or reverse them. For instance, demand contraction without devaluation in the presence of inflexible prices and wages, may cause a crisis, in countries suffering from reversals in foreign capital inflows. On the other hand, real

---

Box 7.2 Determinants of success of policy reforms: the case of South Asia and
sub-Saharan Africa

The variety of experiences with (apparently similar) policy reforms warrants a
closer look at the socioeconomic and/or political factors determining consistency
in reform implementation and success in achieving growth. In a paper written for
FAO (Subramaniam *et al.*, 1994) a comparison was made between a "typical"
poor Asian and a "typical" poor African country (hereafter referred to as Asia
and Africa) to show how differences in the economic and institutional structures
may determine the nature of adjustments needed as well as reform outcomes. The
major results are summarized below:

1. In Africa reforms were more often initiated not by the countries themselves but
   imposed by lending agencies. They were general and uniform rather than
   country specific, and often they ran contrary to the interests of the ruling
   coalitions. Reforms in Asia were often initiated by the countries themselves,
   were more sensitive to the local needs and conditions, invited less opposition,
   and were politically more sustainable. Therefore their implementation was
   more consistent.
2. Short-run political sustainability was hampered by the frequency and
   magnitude of the external shocks to which the African economies were
   subjected, and the ensuing economic hardships that prevented the
   compensation of the "losers" from the reforms. The inability to provide
   compensation hindered consensus.
3. Despite the fact that the "post adjustment" or "redefined" role of the state was
   similar for regions (providing public goods, dealing with externalities, reducing
   transactions costs, assuming social functions) several differences exist: The
   required reduction in the role of the state was smaller in Asia than in Africa,
   since in the latter region the pre-reform state intervention was more extensive.
   As a result of these differences, agencies supplying public goods were less
   disrupted by the reforms in Asia than in Africa.
4. In agriculture, private marketing networks and indigenous rural institutions
   were weaker and more suppressed in Africa where the mistrust for the private
   sector was greater than in Asia. The implication is that in the "typical" Asian
   country the private sector was able to step in and assume functions previously
   in the domain of the state.
5. Growth in the typical African country is hampered by the undeveloped state of
   financial markets. The weaker fiscal system makes the African economy more
   dependent on commodity taxes for collection of fiscal revenues. The relative
   inability of the African state to collect revenues contrasts with the finding that
   the marginal impact of additional infrastructure in terms of higher agricultural
   productivity is larger in Africa than in Asia.

---

devaluations are not sustainable without proper monetary and fiscal policies.
The case for a wide flexibility in the exchange rate policy is gaining ground.
However, devaluation, even when necessary, cannot substitute for the need to
achieve, through productivity increases, long-term improvement in
competitiveness. Despite the fact that devaluation may give a boost to

agriculture by restoring the price incentives of tradables, and may provide short-term relief in the balance of payments, long-term performance of the agricultural sector and the economy is contingent upon real productivity gains.

The question whether, in principle, a system of fixed or flexible exchange rates is superior is not the right one to ask. The real issue is whether monetary and fiscal policies are supportive to the exchange rate system chosen. The reason why flexible exchange rates may be preferable derives from the fact that fiscal and monetary policies to support a fixed exchange rate are almost untenable in practice, especially in the presence of external shocks. For countries constrained in terms of their ability to adjust the value of their currencies, a fiscal policy that avoids overvaluation and prices and wages that are flexible to respond to disequilibria constitute the cornerstones of a supporting macroeconomic stability. The recent decision to devalue the CFA and CF francs (the currencies of the 14 African countries of the franc zone – described in *IMF Survey*, 24 January 1994) demonstrates the difficulties in adhering to the "iron laws" of fiscal and monetary policies needed to prevent real exchange rate overvaluation under a "nominal anchor" system, and in the face of external shocks. Namely, while the fixed exchange rate system was a source of stability, low inflation and growth until the mid-1980s, the decision to maintain a fixed parity in the face of reductions in the terms of trade and an appreciating French franc meant that international competitiveness needed to be restored through containment of production costs, and/or with the use of trade measures. Failure to contain costs through internal adjustments caused a collapse of exports, investment and growth and a financial crisis in a number of these countries.

Experiences with adjustments undertaken in the wake of the economic crisis show that several countries used trade restrictions and foreign exchange rationing to deal with balance of payments crises. As allocation of restricted imports and rationed foreign exchange often reflects the ability of groups or sectors to exercise political power, agriculture has mostly lost out in the process.

Within a framework of macroeconomic reforms as described above, funding for agricultural infrastructure will be conditioned by the constraints imposed on fiscal expansion under the reforms. As was mentioned previously, in the short run and until more efficient income tax collection mechanisms are developed, taxation, *inter alia,* of agricultural exports and imports, may be unavoidable. If such direct taxation of agriculture is to be imposed (or continued) then it is more important that indirect taxation of agriculture in the forms of overvalued exchange rates, etc., be removed. Uniform taxation or border protection although sub-optimal from an efficiency standpoint may be the best way given the implementation difficulties and the possibilities for rent-seeking associated with non-uniform measures. Converting quantitative restrictions to tariffs will add to revenue collection. Several developing countries have already made such shifts.

Other sources of funding also need to be sought. Savings from phasing out various unproductive/uneconomic subsidies and from the reform or privatization of state enterprises and parastatals can be used to finance public goods as well. In several countries additional funding can be provided by reallocating overall expenditures among expenditure lines, e.g. from recurrent expenditures to capital spending as the efficiency of the public sector improves.

In addition to an "enabling" macroeconomic framework, liberalization of the agricultural sector will be ineffective unless balanced reforms are undertaken across economic sectors. Promoting efficiency reforms in agriculture by lowering import controls, tariffs or subsidies, but maintaining protection in other sectors, puts agriculture at a distinct disadvantage. Lowering half of the distortions in the economy is different from lowering all distortions by half.

The international prices of some commodities (e.g. cereals, dairy products, sugar) are lower and often more volatile than would be the case in the absence of trade-distorting policies of developed countries. However, in some cases these very policies contributed to enhance security of supplies in world markets, particularly of cereals. This is because they caused stocks held by these same countries to be higher than they would have been in the absence of such policies (O'Brien, 1994; see also Chapter 2).

International market prices represent the actual opportunity cost of resources facing developing countries and could be used, in principle, as the basis for calculating relative protection or taxation of commodities, i.e. as an anchor for calculating distortions. However, such prices are sometimes abnormally low and opaque as a result of heavy, irregular and often not transparent export subsidies. The issue, therefore, is whether such prices should be allowed to penetrate unhindered the agricultural economies of the importing countries because they can disrupt the orderly development of domestic production.[20] The gains from the lower import prices can be offset in varying degrees, or more than offset, if producer incentives are subject to such erratic and artificial signals from the international markets. It is for this reason that several importing developing countries have adopted border measures to filter such signals, e.g. through the adoption of "price bands" for import prices. Such price bands are meant to allow longer term price trends to penetrate the domestic economy, but not the erratic and temporary price movements caused by certain types of export subsidization, both open and hidden ones.

In conclusion, in considering the role of international market prices as opportunity costs influencing incentives to domestic producers, allowance should be made for the higher risks involved in such use of distorted international prices. They are not only lower but also more volatile than they would be without these policy distortions. They are also less transparent (e.g. when export subsidies are not fully reflected in quoted prices) and they may be less durable than price trends generated by market forces, because the policies that lowered them in the first place may be changed at short notice.

Arguments have been advanced in favour of positive protection of (especially food) crops, mainly for improving food security (often wrongly associated with food self-sufficiency) as well as for supporting rural employment and incomes. Some of those arguments in favour of increased food self-sufficiency can be valid also on efficiency grounds if they correct market failures or other distortions. Likewise, if food production has multiplier effects that private producers fail or are unable to take into account in their decision-making process, or are unable to act upon due to, e.g. lack of credit, some public intervention may become necessary to maximize net social welfare (Matthews, 1989; this issue is discussed below). When such policy objectives are pursued, the instruments by which food production is to be promoted need careful consideration, so that they attack the root of the problem. Thus, interventions to increase productivity in the smallholder sector through non-price-based measures (such as supportive credit organizations and cooperatives, facilitating technology adoption, etc.) may be more efficient or more cost effective than interventions which affect the structure of price incentives. The possible broader implications of those productivity-increasing interventions for agricultural and overall growth are examined in the next section.

Increased volatility of both import bills and export receipts may lead to occasional national food security crises in low-income, food-deficit countries. In these conditions a greater degree of food self-sufficiency than would be suggested by pure efficiency considerations may be justified to the extent that it reduces the risk element associated with world market price volatility and to the extent that it is difficult to implement alternative policies to cope with this risk, e.g. maintenance of foreign exchange reserves targeted for food imports, use of variable tariffs to compensate for the effects of artificially and erratically volatile world prices due, *inter alia*, to trade-distorting policies, or resort to world futures or option markets.

As was mentioned previously, deviations of domestic from international prices may be justified to the extent that international prices are the result of unsustainable policies of food exporters (rather than technological improvements), implying a high risk that they will be eventually reversed. This is important, especially if decisions are to be taken on investment projects in the food sector. The rate of return on infrastructural projects of long duration should take into account "sustainable" levels of world prices. Several other arguments may be advanced in favour of increased self-sufficiency over and above what would be under strict efficiency criteria. Such arguments may reflect social and political values and realities in different countries, their validity should be examined on a case by case basis, keeping in mind the income losses or gains associated with such choices as well as whether or not they can be sustained.

From short-term policies to long-term strategies: the role of agriculture in economic development

The trend towards fiscal prudence and market-oriented policy reforms in developing (but also developed) countries is likely to continue. The particulars of reform (timing, sequencing, speed, etc.) may change as lessons are learnt and more data are gathered on performance of countries undergoing reforms. It is now clear that more time has to be allowed than was initially thought, for the reforms to produce results, especially in low-income countries where institutional and infrastructural deficiencies prevail. In such countries the response of agriculture to macroeconomic reforms has been mixed.

In the future, reform programmes will likely continue to pay more attention to alleviating infrastructural bottlenecks, promoting research and extension, providing funding for environmental conservation and for direct measures to alleviate poverty in its different aspects (food security, health, etc.), in addition to restoring equilibria in external and internal balances and in relative prices.

Most of what has been discussed so far relates to short- and medium-term policies. The purpose of such policies is to create the conditions for efficient allocation of productive resources among sectors and within agriculture. Short-term policy reforms aimed at correcting economic imbalances and alleviating associated crises, while necessary, may not be sufficient for the resumption of growth. Long-term growth strategies involve going beyond short-term policies, and are based on the identification of those sectors where public investment funding will be given priority.

The severe economic costs incurred as a result of the neglect of agriculture in developing countries in the context of import substitution/industrialization, prompted a closer look at agriculture as a driving force for overall economic development, especially for countries with large rural populations, and a high share of agriculture in GDP, exports and employment. Thus, agriculture-based development strategies have gained new momentum.

Agriculture-based development strategies go beyond the adoption of short-term measures that eliminate discriminatory policies against agriculture. They suggest an active support of the agricultural sector in terms of massive public investment allocations, pointing out the broad benefits of such a strategy in terms of improvements in food security and nutrition, an egalitarian income distribution, employment generation and, more importantly, stimulation of overall growth through production (supply) and income (demand) linkages between agriculture and the rest of the economy.

Mellor and Johnston (1984) proposed an "interrelated" strategy which emphasizes the benefits resulting from increases in the productivity of smallholder (mainly food) producers through the development of infrastructure and the dissemination of appropriate labour intensive technologies. Such a strategy would improve the nutritional status of the population by both increasing food production and (more importantly) improving access to food through the secondary effects of increased agricultural production on employment and incomes.[21] Increased smallholder production and incomes have spillover effects on the rural non-farm and urban manufacturing sectors through increased demand for agricultural implements, and consumer goods that are income elastic. Both of those non-agricultural sub-sectors constitute presumably labour-intensive activities and their growth further increases employment. Increased non-farm employment and incomes absorb the increased food production without a collapse in prices. Surplus to industry is transferred through cheaper food and savings. Increasing productivity and incomes of small farms, and the resulting increased farm and non-farm employment, result in improved income distribution. Thus the strategy can generate overall economic and social gains.[22]

A similar development strategy, emphasizing the primacy of the agricultural sector in the development process, has been proposed by Singer (1979) and empirically tested by Adelman (1984) and especially by Vogel (1994). As in the case of the "interrelated" strategy, the proposed agricultural-demand-led industrialization (ADLI) strategy is based on the strong demand linkages between agriculture (especially food production by small- and medium-level farmers) and the non-agricultural sectors. Emphasizing smallholder food production results in increased derived demand for (mainly traditional) intermediate inputs which can be satisfied domestically, in contrast to export agriculture's demand for modern intermediate inputs a large part of which "leaks" into the international economy.

Furthermore, increases in incomes of smallholder producers result in increased demand for domestically produced basic consumer goods, thus providing a stimulus to domestic light manufacturing and further increasing employment. The ADLI strategy is proposed as an alternative to past trade-based strategies emphasizing import substitution or export promotion policies. Empirical testing of such strategies for their superiority in enhancing overall growth is subject to many caveats, including those emanating from the correct, or otherwise, representation of the specificities of the agricultural sector in the relevant models, e.g. resource and technology constraints or the functioning of international commodity markets (see Alexandratos, 1992).

Under the "linkages" criterion, the allocation of public investment priority to agriculture is justified when it increases GDP (in the sector itself and, through the demand linkages, in the rest of the economy – principally in the sectors producing inputs and consumer goods for agriculture) by an amount exceeding that obtainable from alternative allocations of public investment. The most comprehensive empirical test of this question to date has been performed by Vogel (1994), using social accounting matrices from 27 countries to derive agricultural and non-agricultural multipliers.[23] The results of the study confirm agriculture's strong backward links to non-agricultural production activities at low levels of development. Also, at low levels of development the rural household income multiplier (i.e. the effect of increases in rural household income on the demand for non-agricultural consumption goods) is the dominant one, while during the development process, the backward agricultural input–output multiplier becomes dominant, i.e. the demand for agricultural inputs, including agricultural services and credit, is expanding. The findings indicate that the rural household multiplier declines with development as one would expect.

Whether for correcting years of neglect and discrimination or as a result of an "active" strategy that places agriculture at the leading sector in the development process, agriculture is and will probably continue to be the focus of the development efforts in developing countries, especially low-income ones. The case for an agriculture-led development strategy seems strong for several countries, but care should be exercised not to make sweeping generalizations. The different "models" or strategies giving priority to agriculture were developed in the context of more-or-less specific regional environments, stages of development, and/or global economic climates. For instance the interrelated strategy is based largely on evidence from the successes of the green revolution in Asia. Several countries for which an agriculture-based development strategy seems to have worked (for example Taiwan (province of China), Thailand, Malaysia) are countries that were also promoting an export-led growth strategy. Therefore, it is difficult to determine which element (pro-agriculture or pro-export) was at the basis of the growth successes in these countries.[24]

ADLI strategies, looking at demand linkages, assume that supply elasticities in all sectors are infinite, so no disruptive large price increases take place. This means that success of such demand-based strategies hinges on large supply elasticities which in many countries and/or sectors may not exist and may have to be "created" in the long run through the elimination of various bottlenecks. In a situation where supply elasticities are low, the objectives of the strategy are not met no matter how large the magnitude of the multipliers.

Problems will also be created where there is an asymmetry in the response of the agricultural and manufacturing sectors to demands from the other sector. For instance the stimulation of too rapid an increase in food production without a concomitant increase in employment and incomes in the rural economy (and given inadequate export structures or high transfer costs) may result in collapse of food prices in producing areas unless the excess is removed from the market or area. Alternatively, government stockholding may be needed to keep prices from crashing (Mellor, 1986, reports such a case in India). Such delicate intersectoral equilibria are difficult for most governments to manage and it is almost impossible to plan them ahead of time. In the same fashion, if the supply elasticity of non-agricultural goods is small, then an increase in demand for those goods will cause inflationary pressures or "leakages into imports". ADLI "strategists" suggest measures, including infant industry protection, to counter the problem. As already noted, such policies may result in chronic protection and inefficiencies. The above-mentioned "interrelated" strategy supports a liberal open trading regime even if that means that part of the increased incomes would be spent on imports.

A strategy (e.g. a scheme to allocate public investment resources or recurring development expenditure among sectors) should consider several options taking into account the long-term comparative advantage of the country. Although the "linkages" criterion may be considered, this should not be the only one. Public investment should also be directed where it can attract private investment activity. The distinction between "export promoting" and "domestic market expanding" strategies may be misleading. Even for countries that experience declines or stagnation in their markets for traditional exports, an inward-looking strategy may not be advisable. The potential for non-traditional agricultural exports, for agriculture-based products with higher level of processing, etc., should also be explored.

In general, increasing smallholder food production although desirable due to its possible multiple growth and social effects, should not be pursued at any cost to overall social and economic efficiency, i.e. even when strong comparative advantage exists in other agricultural (or non-agricultural) activities. If such sectors are neglected, the overall social welfare can be reduced substantially. In case an alternative strategy is followed, the "second-round" effects of an agriculture/food-based development strategy (such as improvements in nutrition or rural poverty alleviation) may be more efficiently achieved by direct interventions. Although there is little doubt that in most low-income developing countries agriculture can be the most important sector in terms of potential returns (both social and economic) to public investment, caution should be exercised in general in picking "winners". The successful experience of East Asian countries in picking winner industries may not be feasible for a number of countries.

Agricultural growth and the non-agricultural rural sector

Another dimension of the agriculture-based development strategy has to do with the implications of such a strategy for growth through agriculture's links to rural non-farm activities. Non-farm employment constitutes an important share of total rural employment especially where communities are organized in small rural towns. The share varies among developing countries and country regions depending on the concentration mode of the population. Thus, shares are higher in Latin America and Asia where the rural population is more concentrated in rural towns, than in Africa where rural populations are more dispersed in small rural settlements[25] (for more extensive discussion see Haggblade et al., 1989, 1991; Hazell and Haggblade, 1993).

Non-farm rural activities offer employment and income opportunities to the rural poor – landless or near landless – and women, the latter being generally a significant share of total rural non-farm employment (food preparation and processing, tailoring, etc.). As in the case of links between the farm and overall economy, strong production and consumption linkages exist between the farm and non-farm rural economies.[26] Growth in the rural non-farm economy will have positive effects on poverty reduction and income distribution, and will stem the rural to urban migration. According to the available empirical results, the average agricultural income multiplier is around 1.6, i.e. each additional unit of agricultural income generates additional income of 0.6 units in the non-farm rural sector. The multiplier seems to vary by developing country and region, being in general lower in Africa than in Asia and Latin America.[27]

Research has demonstrated the importance of rural towns in strengthening the linkages between the farm and rural non-farm sectors and for increasing the multiplier effects of agricultural growth (Hazell and Haggblade, 1993). Thus, for Africa, the lower multiplier effects can be partly explained by the very low concentration of population mainly in small rural settlements and the difficulties in connecting population concentration points. This result is in accordance with what has been presented in Box 7.1, i.e. the importance of rehabilitating and expanding the transport network in Africa for increasing agricultural growth and facilitating the realization of its "secondary" effects.

From a policy standpoint, the implication is that apart from policies that promote growth in the agricultural sector, policies to strengthen the ability of the rural non-farm sector in responding to the demands of a growing farm economy have to be considered. In addition to increasing infrastructural facilities and human resource capacities in the rural areas and small towns, a clear, enabling legislative framework for private small business activity is needed including legislation concerning labour markets. Donor support should consider promoting activities such as training in small business skills (accounting, etc.), technical assistance in workshop facilities and the creation of an efficient credit system in the rural areas.

Support of non-rural farm activities also addresses the important issue of export diversification of developing countries to products with a higher share of value added. A move away from their dependence on raw agricultural commodities, in addition to the growth implications mentioned above, reduces the vulnerability to the wide swings of international commodity prices. Developed countries can contribute to this effort by removing discrimination to imports from developing countries with a high value added share, e.g. through tariffs escalating with the degree of processing.

The above results are relevant in rethinking the process of economic transformation of developing countries from basically agrarian to basically industrial economies. The growth potential of other sectors such as rural commerce and rural services has to be considered along with (or instead of) manufacturing.

Concluding remarks: agriculture-based strategies and the role of the state

In considering the role of agriculture in development, a strategy cannot and should not ignore the fact that economic development leads ultimately to a reduced role of agriculture in the overall economy in the long run. The debate on agriculture-based development strategies does recognize the need for such transformation, but rejects the idea of agriculture as simply a resource *reservoir*. Instead, these strategies view the transfer of resources to non-agriculture as a result of *surplus generation* in agriculture as opposed to *surplus extraction*. In other words, this transfer is effected as a result of growth in agricultural productivity which, in combination with the fact that demand for agricultural products grows less fast than that for other products, makes it economically advantageous for resources to be increasingly directed to other sectors as development advances. The role of the state is crucial in these strategies in promoting technological improvements in agriculture (including human capital) to increase productivity, and promote institutions that facilitate the smooth functioning of markets for inputs and outputs.

In this connection, it is noted that more recent thinking on economic growth tends to underplay the importance of physical capital in economic development in favour of a broader definition of capital, including the stock of knowledge and human capital. As knowledge generated in a sector may have positive spillover effects on other sectors, public investment in research and development is crucial, as is the promotion of industries that produce knowledge.[28] For agriculture, the positive effects of human capital, learning, and research and development have long been demonstrated (see Chapters 4 and 10) and have been embodied in the literature on agricultural development since the 1960s, so this evolution in thinking contributes little that is new to the analysis of the determinants of agricultural growth.

Other reasons for which the new growth theories are of limited use for the understanding of agriculture in economic development are: (a) they deal with one sector models (at least in their recent form); and (b) they have been largely concerned with long-run steady states. Descriptions of steady states may be of limited usefulness when describing agricultural growth and the changing role of agriculture in economic development, especially during the agricultural transformation.[29]

In summary, although there are strong arguments in favour of priority to agriculture in development strategies in well-identified situations, there are no ready-made universal answers of general applicability. Research results on the effectiveness of public investment at the country or regional level are difficult to generalize into a single strategic approach to agricultural development. The dynamic long-term effects of government interventions and public investment activities have not been assessed due also to the lack of detailed data of public expenditure going to agriculture as opposed to rural areas in general, and on the amount of private investment that they can generate, including investment financed from farm savings and own-labour investment not giving rise to observable financial flows (Timmer, 1991).

It is also risky to make generalizations regarding the appropriate "mix" of public and private participation. The tasks generally accepted to be appropriate for the public sector, as discussed in this chapter, are neither well determined for each country nor may they be considered to be immutable over time. The view that development depends to a large part on organization, good governance, structure of institutions (including people's participation in the decision-making process) and political consensus is gaining more and more ground. These characteristics differ among countries and over time within the same country as institutions and organizational structures evolve. While the general proposition that governments are not generally well equipped for direct interventions in the production and distribution systems is now widely accepted, it is equally well recognized that they must become more effective in the provision of public goods, etc. The extent to which they can do so will go hand in hand with improvements in their organizational/managerial abilities.

Development failures in the past were to a large extent due to attempts by governments, sometimes encouraged by the "donor" community, to do too much, in their effort to balance the (often contrasting) interests of too many groups, by circumventing or supplanting market activity. The relevant experience indicates that greater reliance than in the past should be placed on markets and private agents, and that the conditions that cause markets to fail (information deficiencies, lack of clear property rights, lack of a clear and stable legal framework) should be corrected. Within a market-oriented framework, there is plenty for governments to do: establish and enforce rules and regulations for the smooth functioning of markets; establish and enforce property rights, especially land tenure rights; provide public goods and correct externalities; establish quality standards for food; carry out specific interventions to alleviate poverty and improve food security and nutrition.

NOTES

1. Such arguments were the cornerstone of writings by Soviet authors such as Preobrazhensky (Sah and Stiglitz, 1984).
2. Labour is in "surplus" in a particular sector (e.g. a traditional sector) when it can be transferred out of that sector and into another sector (e.g. the modern sector) *at the prevailing wage* without loss of output in the former. The concept presupposes that no additional capital is used to substitute for labour in the traditional sector. This concept, developed by A. Lewis, became one of the most influential contributions in development economics.
3. Lewis himself did not identify the backward sector with agriculture, and the capitalist sector with industry. Plantation agriculture, for instance, was part of the advanced sector. Interestingly enough, the Lewis model on surplus labour was used later to support a "pro-agriculture" development strategy (see Johnston and Mellor, 1961). In fact, as early as 1953, in his advice to the government of Ghana, Lewis emphasized stagnant agricultural productivity as the main obstacle to industrialization.
4. The "secular decline" argument developed simultaneously by Prebisch and Singer

was based on a combination of assertions and data analysis results. The question of a *secular* decline in primary commodity prices, although generally accepted today, has a number of different dimensions. Research undertaken by FAO shows that ". . . on a purely qualitative basis the study confirms the original observation by Prebisch and others that the international barter terms of trade of primary commodities have a tendency to deteriorate". But it cautions that the measured tendency to decline is: (a) small in size; (b) statistically significant at the lowest confidence level; (c) in most cases reversing itself given a sufficiently large number of years; and (d) erratic in sign and size if considered over a small number of years. Stronger tendencies to decline can be detected though over smaller periods (Scandizzo and Diakosavvas, 1987). The possibility that the estimated trend in real commodity prices is biased downwards is raised in a recent World Bank study (World Bank, 1994a: 14). Namely, if the manufactures price index is adjusted for quality improvements, then the real commodity price index will be increased by about half a percentage point per year between 1900 and 1992. The same index would be increased by a further 0.1 points p.a. if the deflator is only for the price of manufactures exported from the developed to the developing countries; and by another 0.4 points if adjustments are made in this latter index for quality changes "within product categories". The sum of the three components means that the long-term trend in real commodity prices may have been underestimated by as much as one percentage point a year.

5. For a detailed discussion and review on the reasons behind the rise in interest for "pro-agricultural" strategies see Staatz and Eicher (1990).

6. On the role of liquidity, see Rausser *et al.* (1986).

7. Tradable commodities are those whose prices are, by and large, determined in international markets.

8. The above analysis is not meant to assign blame to lenders or borrowers. The interest is to extract useful lessons on the policy dimensions of the crisis, i.e. the extent that it depended on policies under the control of countries.

9. Detailed data on a number of countries are provided in a five-volume World Bank study (Krueger *et al.*, 1991). Results are summarized in Krueger *et al.* (1988), and in Schiff and Valdés (1992). It should be noted that the study contains only a small sample of developing countries from all regions, and thus the results should be interpreted bearing this in mind. See also Norton (1992).

10. A number of such schemes are presented in FAO's study on agricultural price policies (FAO, 1987).

11. Public investment projects expanded rapidly in the 1970s although many questions have been raised *ex post* about their economic rates of return. An ex-post evaluation of the World Bank's agricultural investments for six African countries showed that 36 percent (in value terms) had negative economic rates of return while 18 percent had rates of return lower than 10 percent (Lele and Myers, 1987).

12. For definitions and methods of measuring the rates of direct and indirect protection of agriculture see Krueger *et al.* 1988, 1991.

13. For further evidence on the intersectoral consequences of macroeconomic and trade policies see World Bank (1986).

14. The reverse could also be true. Governments used the profits made by parastatals or stabilization funds during favourable years to finance budget deficits. For evidence see Claassen and Salin (1991).

15. The extent to which these measures result in a net disincentive to producers of competing products depends on the total package of policies. The latter may include compensating influences from devaluation and/or adjustments of border measures (e.g. price bands) to filter the penetration of signals from non-permanent and

volatile distortions in world market prices and conditions of trade (subsidized credit, etc.)

16. The countries considered to be "healthy adjusters" were able to reduce domestic imbalances by increasing savings rather than reducing investment, and external imbalances by increasing exports rather than by reducing imports.

17. Platteau (1993) performed statistical analysis of the growth performance differences reported in Binswanger (1989). He could not find statistically significant differences of growth rates in agricultural production between countries providing a "favourable price environment" to agriculture and those that did not.

18. In the quest for evaluation of the effects of adjustment on economic performance, several criteria have been used to classify countries as "adjusting" or "non-adjusting" according to timing, intensity and consistency of adjustment efforts. Countries have even been put into "finer" categories (early intensive adjusters, early non-intensive, late intensive, etc.). The criteria used have been the number of structural adjustment loans received, or the period between agreement on and disbursement of a structural adjustment loan. It is unlikely that either measure constitutes a credible indicator. The "Third Report on Adjustment Lending" of the World Bank suggested a closer look at the record of reforms to evaluate the performance of adjustment programmes rather than relying on number and timing on adjustment loans. The reason is that (according to the World Bank): (a) mere existence of the programme does not guarantee reforms taking place; (b) reforms agreed are not fully implemented; and (c) reforms executed are often reversed (see World Bank, 1994b). Structural adjustment programmes have often been criticized as being too rigid, i.e. as ignoring the political realities and dynamics of the implementing countries. Thus, failure of implementation may at times be an endogenous feature of some of the programmes. It is shown elsewhere in this chapter that when governments have to cut expenditure, less "visible" (soft) items in the budgets are cut first. Thus, public investment programmes are reduced before cuts are made in wage bills of the civil service, or in military expenditure.

19. Lack of imported consumer goods in the rural areas has been identified as an impediment to price response by farmers. For instance, this phenomenon is believed to have been present in Tanzania in the 1970s. Non-availability of consumer goods, in either the official or black markets, pushes the marginal utility of additional income earnings close to zero. Thus, with higher prices, the same level of income can be achieved with more leisure, and the supply response may actually be negative (Bevan et al., 1987).

20. The lack of transparency in the formation of international prices under various open and hidden forms of export subsidies can be deduced from the following examples: quoted export (Fob) prices of wheat were in early 1994 $130–150/tonne (US Soft Red Winter No. 2) or around $120 (Argentina Trigo Pan). At the same time, Kenyan producers complained that "wheat is pouring into Mombasa at $90 and $100 a tonne — we can't compete with those prices" (Financial Times, 28 June 1994). China is reported to have purchased in January 1994, 815000 tonnes of US SRW wheat with an EEP (US Export Enhancement Program) bonus of US$65.6 per tonne, resulting in an implicit discounted price of US$93.

21. The strategy considers the "dualism" in capital allocations (i.e. unduly large allocations of capital to industry and the unproductive elements of the private sector) as a major obstacle to the efforts of countries to increase employment, combat rural poverty and malnutrition. It is thus defined as a development strategy with a central role given to food production by a unimodal smallholder-based agricultural sector.

22. The observed positive correlation between food production and food imports in

many developing countries is used as a major argument in support of the strategy. The conclusion is that an increase in food production causes an increase in food imports due to its general positive effects on overall income, but this conclusion applies in varying degrees to the different countries and may not apply at all to some countries (see de Janvry *et al.*, 1989).

23. The social accounting matrix (SAM) is an accounting framework that maps out the aggregate structural interrelationships between different sets of economic "agents" and traces the circular flows of incomes and expenditures on goods and services.

24. According to Mellor, it was the critical importance of trade expansion in an agriculture-based strategy that created the perception of those countries as examples of export-led growth rather than as examples of successful agriculture-based strategies (see Mellor, 1986).

25. Data show that rural non-farm employment (excluding earnings from seasonal and part-time activity) accounts for 19, 36 and 47 percent of rural employment for Africa, Asia and Latin America respectively, when rural towns are included. Respective income shares range from 25–30 percent for rural Africa to 30–40 percent for rural Asia and Latin America (including part-time and seasonal employment but excluding rural towns) (Hazell and Haggblade, 1993).

26. A third linkage (labour market linkage) between the rural farm and non-farm activities has also been investigated. Growth in agriculture may increase the rural wage rate and cause substitution of rural non-farm economic activity to shift from low skill, labour intensive to higher skill more capital intensive, high return ones (see Hossain, 1988).

27. Results from different regions vary around the average according to the type of agricultural activity but also the methodology used. Most studies use demand-driven multipliers, and input–output models which assume a fixed proportions technology and a perfectly elastic supply of non-tradable commodities. Thus the derived multipliers are probably overestimated. The study by Haggblade *et al.* (1991) constitutes an exception in which prices are endogenous, input substitution is allowed and the supply of non-tradables is not perfectly elastic. Using this method, consistently lower multipliers are obtained compared to the fixed-price models.

28. For a comprehensive presentation of the evolution of new growth theories, see Stamoulis (1993).

29. For a critical review of the role of agriculture in economic development in the context of various growth theories, see Stern (1994).

CHAPTER 8

International Trade Issues and Policies

8.1 INTRODUCTION

This chapter discusses current and future trade policy issues in the context of the likely developments in international trade of the major commodities presented in Chapter 3. It was noted there that these developments could lead to the agricultural (crop and livestock products) trade balance of the developing countries turning from positive to negative, i.e. the developing countries as a whole are likely to become growing net agricultural importers.

The reasons for these expected developments may be summarized as follows. First, the developing countries as a whole import food products with relatively high income elasticities in their markets and, with a few exceptions (e.g. fruit and vegetables), their exports to the mostly saturated markets of developed countries comprise products with low income and, for some of them, also low price elasticities. Increasing competition among developing countries for agricultural export markets in developed countries often leads to small increases in volumes exported and price declines with the net result that real export earnings from some commodities fall rather than increase.

Second, policy interventions through international commodity agreements (ICAs) have tended to fail in their objective of preventing price declines. Resort to other instruments, e.g. futures markets, use of options, etc., may help to cope in part with the problems of price fluctuations but these instruments are not designed to counteract the more fundamental factors determining the long-term evolution of prices. Third, the consumption by the developing countries of their export commodities tends to absorb an increasing share of their production (e.g. tobacco, rubber, cotton) and reduces the extent to which supplies are available for exports to developed countries.

Fourth, the expected trade deficit on agricultural account is likely to be offset at least in part by a growing net surplus of manufactures based on agricultural raw materials, e.g. cotton. Increasing net imports of raw cotton or hides and skins is a desirable development because it serves the input requirements of their rapidly growing and increasingly export-oriented textile, clothing and leather goods industries.

Fifth, developing countries are likely to continue to face significant market access barriers in the developed countries for some of their produce, e.g. sugar, as well as for processed agricultural products and manufactures. The policy reforms in prospect under the Uruguay Round Agreement on Agriculture (AOA) provide some scope for reduction in protection and other trade-distorting practices, but will not eliminate such practices (see below). The rest of this chapter is devoted to discussing the policy issues and the policy reforms currently under way or in prospect which have a bearing on the agricultural trade environment of interest to both developing and developed countries.

8.2 POLICY ADJUSTMENTS AFFECTING AGRICULTURAL TRADE: MAJOR COUNTRY GROUPS

For many *developing countries* the typical policy reforms under way are those in the general category of structural adjustment programmes (SAPs), as discussed in Chapter 7. Of interest to this chapter is the orientation of these policies towards deregulation, corrections of exchange rate misalignments, opening of the economies to foreign competition and promoting export expansion. They favour sectors producing tradable commodities. As such they could contribute to create a more favourable environment for the development of both agricultural and non-agricultural trade. As discussed later, the AOA offers considerable flexibility to the developing countries for adopting border measures in their agricultural trade, including the possibility, of which many have taken advantage, of setting bound tariffs for the future at relatively high levels, sometimes above the tariff equivalents prevailing in the base period.

Systemic reforms in the *ex-CPEs* are having profound influences on the environment and prospective evolution of agricultural trade. A significant change in this direction has been the shift from semi-barter or clearing arrangements to trade in "convertible currencies and at world market prices", stimulated by the dissolution of the Council for Mutual Economic Assistance (CMEA) and an increasing trade orientation towards the market economies, especially towards Western Europe. Hungary, Poland and the former Czech and Slovak Federal Republic signed Association Agreements with the EC, which reinforced the new trade relationships (FAO, 1993c; Rollo and Smith, 1993). In parallel, export credits and assistance programmes became more common in the agricultural trade of the former USSR. With the prospect of some countries of Eastern and Central Europe joining eventually the European Union, there will be enhanced scope for further agricultural policy changes in both Western and Eastern Europe to create conditions for future convergence of policies (Nallet and Van Stolk, 1994).

The longer term prospects for the agricultural trade of Eastern Europe and the former USSR were discussed in Chapter 3. Such prospects point to the direction of Eastern Europe becoming a modest net exporter of cereals and dairy products and the former USSR becoming a much smaller net importer or fully self-sufficient in cereals. But the materialization of this prospect is not for the immediate future. Some studies make a case that a combination of reduced domestic consumption and the potential of productivity gains in the former USSR could ultimately turn it into a net exporter of cereals (Mitchell and Ingco, 1993; Johnson, D. G., 1993).

For *the OECD countries* there have been, or are in the making, important reforms in policies for agriculture which, in combination with those required by the Uruguay Round Agreement on Agriculture (AOA), will have significant impacts on agricultural trade. The main thrust of such reforms is to shift provision of support to agriculture from market measures (essentially maintenance of producer prices at levels considered necessary for farm income support) to alternative means of supporting farm and rural incomes, mainly payments per hectare. It is not the object of this chapter to describe in any detail the intricate combinations of the many policy instruments through which the objective of reforms is pursued. This objective may be generally defined as: (a) allowing an enhanced role for market forces to determine production, consumption and trade outcomes; (b) containing the budget and wider economic costs characteristic of the agricultural support and protection policies; (c) relaxing trade frictions, particularly those conflicts waged with export subsidies; and (d) reaffirming the commitments of societies to pursue rural development, though by means other than market support to increase production. Whether the new policy measures are largely or only partly neutral as incentives to production is an open question. It is fair to say that not many of them can be considered as completely decoupled, i.e. as not influencing at all the production decisions of farmers, or even the decision to continue to be a farmer.

The first signs of the reforms are already visible in most countries in the form of significantly reduced real producer prices for major commodities. But the wedges between effective producer prices and world prices resulting from the agricultural policies persist at high levels in most OECD countries. The relevant estimates are shown in Table 8.1. At the same time, the total support, as measured by the producer subsidy equivalent (PSE) and the consumer subsidy equivalent (CSE),[1] has continued to increase in recent years (measured in current dollars). The relevant data are also shown in Table 8.1. The most recent OECD evaluation brings this out clearly when it states that "an estimate of total transfers, based mainly on the PSE and CSE calculations, indicates that the combined transfers from consumers and taxpayers in the OECD as a whole (excluding Iceland) declined in 1993 by less than one percent, to just over $335 billion when compared to 1992" (OECD, 1994: 15); and this after increases of about 2 percent in 1992, 10 percent in 1991 and 15 percent in 1990 (OECD, 1994: 122). Some of the countries with the highest PSEs will be joining the European Union. This will likely lead to their rates of protection being reduced, at least the ones making for high divergence between domestic producer and world market prices.

Table 8.1 Developments in agricultural support and real producer prices, OECD countries

| | Producer subsidy equivalent (PSE) | | | | Consumer subsidy equivalent (CSE) | | | | Average nominal assistance coefficients for producers‡ | | Real producer prices in 1993 | | |
| | $ billion (net PSE) | | %* | | $ billion | | %† | | | | Index (1988 = 100) | | |
	1988	1993	1988	1993	1988	1993	1988	1993	1988	1993	Wheat	Beef	Milk
Australia	1.0	1.0	8	9	−0.3	−0.3	−6	−6	1.08	1.10	67	99	100
Austria	2.1	3.0	46	56	−2.0	−2.4	−47	−53	1.90	2.34	79	85	93
Canada	5.5	4.8	38	32	−2.3	−2.2	−21	−21	1.51	1.40	65	99	96
EC-12	69.1	79.6	46	48	−56.3	−57.0	−40	−39	1.84	1.93	73	80	82
Finland	3.9	2.7	70	67	−3.0	−2.1	−67	−66	3.78	3.39	82	76	96
Japan	35.6	35.0	72	70	−37.9	−42.5	−55	−51	3.10	2.93	84	81	93
New Zealand	0.3	0.1	7	3	−0.1	0	−6	−3	1.07	1.03	88	109	92
Norway	2.6	2.7	74	76	−1.3	−1.3	−59	−60	4.57	4.49	79	95	129
Sweden	2.7	1.9	56	52	−2.7	−1.5	−57	−45	2.39	2.03	56	68	69
Switzerland	4.7	4.5	77	77	−4.0	−3.2	−65	−56	4.21	4.10	83	64	83
USA	22.3	24.2	24	23	−8.1	−10.9	−8	−12	1.29	1.29	70	94	87
OECD	130.2	139.3	42	42	−99.9	−106.6	−34	−34	1.68	1.69			

Source: OECD (1994); 1993 data are provisional.

*Total PSE as percentage of the total value of production (valued at domestic prices) adjusted to include direct payments and to exclude levies.

†Total CSE as percentage of the total value of consumption (valued at producer prices), including transfers, such as consumer subsidies.

‡Indicators of the wedge between world prices and effective producer prices created by agricultural policies (ratio of border price plus PSE to border price).

8.3 GLOBAL ADJUSTMENTS: THE AGREEMENT ON AGRICULTURE OF THE URUGUAY ROUND

The conclusion of the Uruguay Round of the Multilateral Trade Negotiations in December 1993 is the major policy development that affects the rules governing the conduct of agricultural trade. The provisions of the Agreement on Agriculture (AOA) and the related one on Sanitary and Phytosanitary Measures are reviewed briefly below.

The agreement on agriculture

The AOA prescribes rules for "permitted" policies that have effects on agricultural trade. They may be classified in three broad categories. First, there are policies having an impact on *market access*, i.e. those that determine the rules under which domestic buyers may provision themselves in world markets. These are for the most part "border" measures, e.g. tariffs, variable levies, quotas, etc. However, other measures not in the "border" category, e.g. intervention purchases, determine the price at which domestic producers sell and compete with imports. It is for this reason that the AOA also disciplines the policies of *domestic support* to agriculture as expressed in an *aggregate measurement of support* (AMS). The third broad category concerns the extent to which exports may be subsidized (*export competition* rules), defined in terms of limits to the total amount of subsidies (monetary terms) as well as in terms of quantities that may be exported with subsidies.

The major AOA provisions in these three broad categories are summarized in Table 8.2. The following comments may be made by way of explanation. Non-tariff barriers (NTBs), including variable levies, have to be converted into their tariff equivalents for the base period specified in the AOA. These equivalents are to be roughly equal to the difference between reference domestic and world market prices, expressed either in absolute money amounts per unit of product (specific tariff) or as percentage of the world market price (*ad valorem* tariff). The resulting tariff equivalents have then to be reduced by the percentages and over the periods indicated in Table 8.2 and the resulting level is to be "bound", i.e. it should not be increased. Developing countries that had unbound tariffs were allowed, and indeed have done so, to offer "ceiling bindings" on their tariffs without necessarily setting these bound tariffs equal to the tariff equivalents of their NTBs in the base period.

Two particular features are worth noting:

1. The average reduction of 36 percent is to be understood as the simple arithmetic average of the individual "tariff lines", but no tariff line may be reduced by less than 15 percent. This offers considerable flexibility to countries to choose which tariff lines to reduce by how much, provided they

respect the two overall constraints of 15 percent minimum reduction per tariff line and a simple average for all tariff lines of 36 percent.
2. The resulting bound tariffs are the maximum permitted. Countries may apply lower tariffs at any time. In principle, this creates the flexibility for countries to vary tariff levels in pursuit of domestic policy objectives, e.g. to mitigate the transmission to the domestic economy of world market price fluctuations. In practice, and subject to the constraint that tariffs may not exceed the bound levels, countries can vary the level of tariffs up and down.

The other aspect of the *market access* category of measures concerns the obligation for countries to maintain existing access conditions including existing preferential access arrangements granted to specific exporting countries and, in their absence, create conditions for minimum access of imports to their markets, 3 percent of their domestic consumption initially, rising to 5 percent later. For these quantities countries are to open "tariff quotas", i.e. defined quantities of imports on which minimal tariff rates will be applied, and in any case at rates lower than the "bound" levels. The idea is that the minimal or lower tariffs will induce importers to buy in world markets the quantities in the quotas, but there is no implication that countries should oblige their importers to do so. However, it is reasonable to expect that if the general bound tariff is high, the wedge between domestic and world market prices will also be high. For example if the former is 100 percent, domestic prices will be roughly double those in world markets. The opportunity to import up to 3 percent of domestic consumption at a tariff of, say, 30 percent will probably provide sufficient incentive for importers to do so.

The other two categories of AOA provisions (*export competition and domestic support*) are fairly straightforward in their interpretation from the information provided in Table 8.2. It is noted that domestic support as measured (in absolute money amounts) by the AMS need not include support provided under: (a) measures which have "no or minimal trade-distorting effects on production" such as resource retirement schemes, domestic food aid, safety net programmes, advisory services and the like (the so-called "green box" measures); (b) direct payments under production-limiting programmes, provided they are based on fixed areas and yields, are made on 85 percent or less of the base level production or, for livestock, are made on fixed number of head ("blue box" measures); and (c) measures which result in support not exceeding 5 percent (10 percent for developing countries) of the total value of a product (*de minimis* clause). For export competition the reductions in export subsidies and in the quantities exported shown in Table 8.2 are to be applied to each commodity exported with subsidies in the base period. Commodities exported without subsidies in the base period may not receive export subsidies in the future.

The Final Act incorporates two Agreements with the potential to lower substantially technical or non-tariff barriers to trade. The Agreement on the

Table 8.2 Agreement on agriculture: summary of major provisions

	General	Developing countries*
Implementation period	1995–2000	1995–2004
Export subsidy reductions		
Base period	1986–90†	1986–90†
Expenditure (for each commodity)	36%	24%
Quantities (for each commodity)	21%	14%
Domestic support reductions		
Base period	1986–88	1986–88
Aggregate measurement of support (AMS)	20%	13$^{1}/_{3}$%
Credits starting from:	1986	1986
Exemptions	• "green and blue" box support policies • if product-specific support does not exceed 5% of the total value of a product (or product group), this support need not be included in the AMS nor be reduced (*de minimis* percentage) • the same as above for non-product-specific support which does not exceed 5% of the value of total agricultural production	• "green and blue" box support policies • if product-specific support does not exceed 10% of the total value of a product (or product group), this support need not be included in the AMS nor be reduced (*de minimis* percentage) • the same as above for non-product-specific support which does not exceed 10% of the value of total agricultural production
Market access *A. Tariffs*		
(a) ordinary customs duties	• reduction commitments to be implemented on the duty level as in 1986–88	• reduction commitments to be implemented on the duty level as in 1986–88

(b) Other border measures (including non-tariff barriers (NTBs))	• to be converted into ordinary customs duties in their tariff equivalent of the base period ("tariffication")	• to be converted into ordinary customs duties. Countries with unbound tariffs have the option to offer "ceiling bindings" not necessarily equal to the tariff equivalents of the base period NTB or the level of unbound tariffs
(c) Tariff reductions	• the resulting duties from (a) and (b) are to be reduced on average by 36% (simple average), with a minimum of 15% for each tariff line	• the resulting duties from (a) and (b) are to be reduced on average by 24% (simple average), with a minimum of 10% for each tariff line

B. *Minimum access (for importers)*

Base period	1986–88	1986–88
Minimum access (for each commodity)‡	3% of base period consumption increasing to 5% in 2000	1% of base period consumption in 1995 increasing to 2% in 1999 and 4% in 2004

*The least developed countries are exempt from reduction commitments, but should tariffy all NTBs and may not increase their support to agriculture beyond the 1986–88 level.

†If 1991–92 subsidized exports exceeded 1986–90 subsidized exports, 1991–92 may be used as starting point. Volume and budgetary commitments to be reached at the end of the implementation period (2000) are however based on the 1986–90 situation.

‡Countries seeking special treatment regarding tariffication can in certain circumstances opt not to tariffy but should offer minimum access of 4% rising to 8% of domestic consumption over the period. For developing countries a similar special clause is applicable and access opportunities of 2% rising to 4% should be created.

Application of Sanitary and Phytosanitary Measures will provide a uniform interpretation of measures relating to food safety and animal and plant quarantine. It establishes a framework for mutual recognition of food control and quarantine regulations and inspection procedures on the basis of equivalence of performance, taking into account the assessment of risk associated with the application, or non-application, of each measure. The Agreement provides for the use of international standards developed by or under the FAO/WHO Codex Alimentarius Commission (food safety), the Office International des Epizoöties (animal health) and International Plant Protection Convention (plant health and quarantine). The other Agreement, on Technical Barriers to Trade, covers other aspects of governmental and non-governmental technical regulations and requirements. The implementation of the two Agreements will require the development of national skills and infrastructures in a number of developing countries if they are to reap the trade benefit from them.

Implementing the AOA and extent of movement towards freer trade

Countries have submitted their reduction commitments under the AOA in the form of "schedules", e.g. showing what is base period tariff equivalent and what will be the bound level in the future, the domestic support reduction commitments and the level of subsidized exports in the base period together with the target level to be achieved by the terminal year of the implementation period. At the moment of writing the "schedules" had not been published, but some information from various sources gives useful indications of the magnitudes of changes involved. Thus, an UNCTAD document (UNCTAD, 1994b) gives the scheduled reductions in the AMS of the USA, the EC, Japan and Canada. Their aggregate AMS in the base year is given as US$143 billion and it will be reduced to US$117 billion by 2000. Likewise, *Agra Europe* (1993) estimated the EC tariff equivalent in the base period for common wheat to be ECU 149/tonne when the reference world price was ECU 93/tonne. If this tariff equivalent were reduced by the full 36 percent of the AOA, the future bound tariff would be ECU 95/tonne or, under a special provision, a duty that essentially did not exceed 55 percent of the domestic intervention price (see *Agra Europe*, 17 June 1994). In the same source, the base period subsidized exports of wheat and wheat flour are given as 17 million tonnes; 79 percent of it, or 13.4 million tonnes, would be the permitted level in 2000.

The above examples give an idea of the extent to which the AOA may be taken as representing a movement towards freer agricultural trade. It is fair to say that strong protection will continue to prevail. But the movement towards agreed principles to discipline the conduct of trade (in particular as regards export subsidies) represents decisive progress towards a more certain and transparent trading environment, less subject to erratic interventions and attendant distortions.

Naturally, there is no guarantee that the different provisions of the AOA will prove to be compatible with each other in all countries. The situation is one resembling a problem subject to multiple constraints. In principle only one of these constraints will prove to be the binding one in any particular country situation. For example, reduction of subsidized export quantities by 21 percent may not imply reduction of the value of the relevant subsidies by 36 percent or more. In this case subsidized quantities may have to be reduced by more than 21 percent and down to the level required to reduce the corresponding value of subsidies by at least 36 percent. In the same vein is the issue of the compatibility of the impacts of the AOA concerning imports and exports and those concerning reductions of domestic support, in particular when the latter are viewed together with the other measures foreseen in domestic agricultural policy reforms, e.g. shift of support from price to direct payments. In brief, the issue is whether the changes in domestic support would generate production, consumption and trade outcomes that would be compatible with the application of the trade policy measures of the AOA.

Some of these issues had been investigated for the EC on the basis of what was known about the CAP reform, the provision for the AOA contained in the Draft Final Act of the Uruguay Round of December 1991 and the subsequent "Blair House Accord" between the EC and the USA. The conclusions of the relevant studies (e.g. Tangermann, 1992; Frohberg, 1993) seemed at that time to indicate that these reforms would be largely compatible with each other, in the sense that their impacts on net trade of the EC in the major temperate zone commodities would be fairly similar. Naturally, the results of these studies differ from each other and many of their assumptions, methods and degree of coverage of agriculture can be questioned. For example, uncertainty persists as to the future growth of cereal yields in the EC. If such growth were to be stronger than assumed in these studies, the surpluses generated by the CAP reform may exceed those that would be compatible with the trade policy provisions of the AOA.

The AOA and the policy environment for the developing countries

Under the AOA, the developing countries have received "Special and Differential Treatment" (SDT). It provides for less demanding policy changes on the part of the developing countries, e.g. smaller reductions in the AMS, longer implementation periods, etc. The essence of SDT provisions is shown in Table 8.2. It is noted in particular that the least developed countries, of which there are 41 in total, are exempt from reduction commitments even though they are supposed to change their border measures to tariffs and may not increase price-distorting support to agriculture beyond the 1986–88 level.

Of particular importance may prove to be the above-mentioned option offered to the developing countries in the SDT clause for the products for which the tariffs they applied in the base period were "unbound". In such cases, the developing countries can declare future "bound" tariffs ("offer ceiling bindings") at levels they consider appropriate for their policy objectives. Where other countries have not objected to such offers, the developing countries have been able to set high enough tariff ceilings to allow themselves the flexibility of applying in the future any tariff below the ceiling they consider necessary, e.g. in cases of temporarily abnormally low prices, as discussed in Chapter 7, or abnormally high ones. Another aspect of the AOA worth noting is that it does not address the issue of policy measures implying negative protection for producers and exporters, a phenomenon not uncommon in a number of developing countries, particularly as regards taxation of export crops. Such policies are being changed as part of the wider domestic policy reforms. In the first place, the widespread introduction of structural adjustment programmes has already, in many cases, reversed previous trends towards taxing agriculture. Thus, the current emphasis on promoting primary sector activity, particularly where it has export potential, may involve a tendency to introduce measures to support production. In addition, even where agricultural production in aggregate has been taxed in the past, such measures which involve the transfer of resources out of agriculture have often been accompanied by interventions to subsidize specific groups of producers, often for social rather than economic reasons. In these circumstances, an understanding of the extent to which individual policy interventions are likely to be acceptable in the new international environment is a standard requirement for policy makers.

In conclusion, therefore, the implications of the AOA for the developing countries are significant mainly in the way in which agricultural policies will be formulated in the future. Whether the pressure for change comes from the new disciplines of the Final Act or from those deriving from structural adjustment policies, both point in a rather similar direction, one where influencing prices is no longer the main instrument of agricultural policy. Whether, however, it will always be feasible for developing countries to adopt suitable non-price distorting policies is a matter that requires case-by-case examination. A review of selected policy alternatives to developing countries in the post-Uruguay Round policy environment is given in Box 8.1.

Developing countries and world market changes

In addition to the changed policy environment, the implications for the developing countries of the Uruguay Round agreements derive from changes in world markets – the size, stability and prices fetched.

As regards the effects on world agricultural markets, it was already noted earlier that although the Agreement on Agriculture is comprehensive it only represents a partial liberalization agreement. Overall, a large degree of distortion in the world market of agricultural commodities will still remain even after the complete implementation of the reduction commitments.

Box 8.1. Review of policy alternatives to developing countries*

Policy	Efficiency concerns	Compliance with GATT	Comments
Output price support	Inefficient targeting, resource misallocation, can be high cost	Poor: subject to limitations outside of which distorts prices and increases AMS	May be a case for price stabilization involving limited support. Generally regressive in effect. Difficult to target
Input subsidies	Resource misallocation, can be high cost	Moderate: may be used under certain conditions. Otherwise contributes to AMS and price distortion	Offers a degree of targeting: marginally preferred to output price support. Distributionally regressive
Credit subsidy	Efficient targeting, relatively efficient resource allocation	Moderate/good: less distorting effect, possibility of exemption	More favoured form of intervention, and potentially easy to target
Food security stocks	Minimum distorting effect when objective of stocks is to eliminate extreme market fluctuations only and not to maintain a narrow market price band	Moderate/good: purchases and sales can be at administered prices, but subsidy to producers must be included in AMS. Such stocks must be integral part of national food security programme	Process of stock accumulation and disposal need to be financially transparent
Subsidized food distribution	Market distortion is minimized when subsidized transfers are well targeted and, in the case of general subsidies, the market is not crowded out by too low and static subsidized prices	Good: eligibility to receive food and/or money to buy food at market or subsidized prices subject to clearly defined criteria. Subsidization of prices on a regular basis also permitted	Food purchases by government to support subsidized programmes shall be at market prices; required financial and administrative transparency

(continued)

Box 8.1   *(continued)*

Non-tariff barriers	Inefficient resource allocation; tariffs preferred	Poor: distorts prices and increases AMS, tariffs should replace non-tariff barriers	May need to phase out tariffs slowly
Direct income payments	If feasible might involve excessive cost	Good: no distorting effects, no increase in AMS provided meet criteria	Not feasible in most developing country contexts
Public investment (extension, research, infrastructure, marketing and storage facilities)	Efficient resource allocation with minimum distortion of market activity	Good: in general no distortionary effects or increases in AMS	Results may be too long term, particularly infrastructure. Investment in marketing and storage most beneficial. Difficult to target

*Reproduced from FAO (1994c).

In general, according to most studies and compared with the situation without the effect of the Uruguay Round, moderate increases can be expected in the prices of temperate zone products (5 to 10 percent on average) but smaller increases or even slight declines in the prices of the principal tropical products. The price changes of both temperate and tropical products are of concern to the developing countries. Moreover, the expansion in world trade in these commodities, which has been projected to be slower in the future than during the 1970s and 1980s, will only be stimulated to a limited extent by the Uruguay Round agreements. Thus in general it may be expected that there would be no major changes in the global volumes of trade although there will certainly be some changes in the patterns of trade and scope for the more competitive exporters. Beyond agriculture *per se*, important changes are expected from the expansion in trade under the liberalized Multifibre Agreement. A large rise in exports of textiles to the developed countries is expected while the upward pressure on price could curtail demand somewhat in the developing countries, where the bulk of textiles consumption takes place. On balance the demand for textile fibres could be stimulated, which could be of considerable interest to a number of fibre-exporting developing countries. At the same time a beneficial effect on the expansion of world agricultural markets could come from the boost to world income from the Uruguay Round. This boost to income, mainly in the developed countries, would presumably increase the demand for higher valued products as well as for niche market products like exotic fruit and vegetables, cut-flowers and horticultural products.

Food import costs

The increases that are likely in the prices of the main temperate zone food products together with the reduction in export subsidies could imply rather significant increases in the import prices paid by the net food importing developing countries, which are the large majority of developing countries. In this context, the *Decision on Measures Concerning the Possible Negative Effects of the Reform Programme on Least Developed and Net Food Importing Developing Countries* could, in principle, help this group of countries in the event of higher world food prices and import bills. This is an important text even though it is still rather weak on concrete action. The idea behind the *Decision* is that agricultural trade liberalization is likely to lead to higher world prices for food while a reduction in export subsidies will also raise the effective price paid by importers.

Food aid

There is also some concern that the volume of food aid, which historically has been closely linked to the level of surplus stocks, could be more limited in future as the surplus stocks are run down. The *Decision* recognizes these issues and provides for some redress. First it promises action to improve food aid via: (a) reviewing the level of food aid; and (b) providing an increasing share on grant terms. The *Decision* also promises that full consideration will be given to requests for technical and financial assistance to improve agricultural productivity and infrastructure. It goes on to promise that any agreement on export credits would make "appropriate provision" for differential treatment in favour of these countries. Finally, the *Decision* makes some provision for short-term assistance in financing normal commercial imports from international financial institutions under "existing facilities, or such facilities as may be established, in the context of adjustment programmes". This concession appears to be rather weak. The loans (probably not grants) would be from existing facilities at the World Bank or IMF and would be subject to conditionalities of such loans, including "in the context of adjustment programmes". Finally, there is no commitment to new facilities set up specifically to compensate food importers for the higher import bills arising from a reform process, only a reference to "or such facilities as *may* be established".

Food stocks

While it is likely that world agricultural prices may be slightly more stable as a result of the Uruguay Round, there is a question mark over food stocks. The general move towards liberalization and a reduced role of the government in price support activities could lead to a fall in government stockholding of agricultural commodities. The reduction may not be large but there is a question as to whether the private sector would step in to fill the gap. If not, as seems likely, then global food stocks are likely to be reduced. Fortunately, however, support to food security stocks undertaken in a prescribed fashion has been excluded in the Final Act from reduction targets. As called for by FAO's Intergovernmental Group on Grains at its 25th Session in 1993, it is to be hoped that countries would take advantage of this exemption and build up adequate food security reserves, but developing countries may not be able to make major efforts on this score as holding stocks is an expensive undertaking. The costs and benefits of building and utilizing food reserves or relying on the world market for food supplies need to be carefully weighed.

8.4 BEYOND THE URUGUAY ROUND

Regional trade blocs

One of the main developments that characterize the world agricultural trading system is the formation or expansion of existing *regional trade blocs*. A number of regional trade agreements are under active discussion or have been agreed upon. Examples are the completion of the Single European Market in January 1993, the Protocol between the EC and the European Free Trade Agreement (EFTA) to form a European Economic Area, and the launching in January 1994 of the North American Free Trade Agreement (NAFTA), which extends the Canada–USA Free Trade Agreement to include Mexico. A number of other countries in Latin America are at present negotiating free trade arrangements with the NAFTA countries, or are strengthening and amplifying existing arrangements between themselves. The United States in addition has proposed an Enterprise of the Americas Initiative which would liberalize trade and investment flows between the countries of North, Central and South America. Discussions have also been held on trade cooperation in the Asian and Pacific Rim area, though it is less clear at present what form such arrangements might take. There are many regional agreements among developing countries but not many of them, so far, cover agricultural products.

Large trade blocs do not necessarily imply a movement away from trade liberalization since they could *in principle* maintain or enhance open trading arrangements with non-members (and other blocs). But the temptation to write rules for the blocs which have the effect of discriminating against outside competitors might be hard to resist. A strong GATT will be needed to make sure that regional trade blocs create trade rather than merely divert trade to their own membership at the expense of outsiders. Under the GATT discipline, regional arrangements are subject to certain conditions one of which is that such arrangements on the whole should not raise trade barriers against other GATT members. Developing countries could probably align themselves with developed country's trade blocs if they so wished, but they would presumably have only limited influence in shaping the rules for such regional trading arrangements. The role of developing countries that might not wish to join such blocs is problematic in the absence of a strong GATT, as they would be dependent upon the blocs providing them with preferential status so as not to be excluded from the major developed markets. The prospect of developing countries forming their own strong trading blocs is limited both by their own economic structures and by the lack of countervailing power that such blocs would possess. Finally, trading blocs seem unlikely to solve agricultural trade problems of a global nature. Existing trade blocs have not found the answer to such issues.

World market prices

The problem of *low and highly volatile prices* for some major agricultural export commodities of the developing countries will continue to loom large. As noted before, efforts to cope with this problem through international commodity agreements (ICAs) have proven to be of limited value and will probably continue to be so. For the longer term, the main hope for improvements or avoidance of further deterioration in real prices lies in the potential growth of consumption and imports in countries with still relatively low consumption levels of some of these products, namely the ex-CPEs and the developing countries themselves, and of course, in a change in the fundamentals of the producing and exporting countries, i.e. the low opportunity costs for labour due to the low level of productivity in other sectors, both agricultural and non-agricultural (see Chapter 3). But as noted in Chapter 3, these developments are not for the immediate future, not even for the medium-term one.

In order to cope with these problems, developing exporting countries could attempt to stop the decline in agricultural prices and to reduce fluctuations, or alternatively they could take steps to offset, or compensate for, their consequences. The main way that countries sought to stem the decline and stabilize international agricultural prices has been through the Uruguay Round of Multilateral Trade Negotiations. These have had the objective of boosting agricultural trade and prices via substantial reductions in protectionism, mainly in the developed countries. The other main approach to arresting price decreases and to stabilize them, given the failure of international commodity agreements, is to develop market demand via export promotion, commodity development policies and diversification. The Second Account of the Common Fund for Commodities represents an attempt at the international level to assist in this process although it has only recently started to be operational.

The alternative approach to international action on commodity markets is to take steps to offset the effects of price declines and price instability. The effects of world market price declines can be offset at the national level by measures to boost productivity and to cut export taxation or provide protection to agricultural producers. Many countries have followed these paths with the aggregate result of exacerbating the international problem. Similarly, domestic price stabilization measures adopted by many countries, both developed and developing, aggravate the international problem by reducing the extent to which these countries absorb fluctuations in world markets. In adopting a national policy response to declining and unstable world prices, the developing countries are at a distinct disadvantage. They cannot afford to provide extensive protection to their agriculture or engage in massive research and development expenditures to boost productivity. Other ways of offsetting the effects of declining and unstable prices have therefore been developed in the form of compensatory financial transfers and the use of other financial instruments. These include the compensatory financing facilities of the IMF and the EC's Stabex scheme, both meant to deal with fluctuations in export earnings. Finally, there has been the development in trading techniques that offer exporting countries new ways to counter the fluctuations in their commodity prices. These include long-term contracts with fixed prices, forward contracts, the use of options or hedge prices through commodity exchanges, over-the-counter markets and the use of swaps and commodity-linked bonds. As noted before, however, despite the usefulness of these various instruments to lessen the risk deriving from price fluctuations, they are unlikely to address the more fundamental factors underlying the long-term decline of prices of some agricultural commodities.

Trade in relation to environment and sustainability

The other main issue that is expected to characterize the world agricultural trading system in the near future is the growing concern with *environmental and sustainability questions* and how these relate to a more liberalized trading system. In practice for most agricultural commodities trade only represents a small proportion of agricultural output. For a few, however, such as tropical beverages, trade is much more important and it is always important for some countries. Trade, the environment and sustainable agricultural development are related, mainly because trade permits the location of production to be separated from the point of consumption. Trade may therefore affect the environment if it causes agricultural production to shift from places where it is less sustainable to places where it is more sustainable or vice versa. The two are also related indirectly because trade can help economic development and through it the perceptions of people about the value to be attached to environmental conservation.

Reference is made to Chapters 2 and 3 where it was indicated that the future is one of growing population and higher per caput consumption. Pressures on environmental resources will continue to grow with attendant threats to sustainability, as discussed in Chapter 11. It remains to note here that agricultural trade flows and policies affecting them determine in part not only the location of production but also the extent to which sustainable or otherwise technologies and practices are used to obtain any given amount of agricultural produce. For example, the high support and protection policies in some developed countries have encouraged excessive use of agrochemicals or intensive livestock units with consequent adverse environmental impacts. With more liberal trade and expanded market access, part of the produce thus obtained could have been produced in, and exported from, other countries with more environmentally benign technologies, e.g. meat, wheat and other cereals in South America rather than in Europe.

It follows that more trade can reduce overall pressures on the environment as measured at the global level, if it shifts production to locations where resource endowments are more appropriate to produce any given amount of output in more sustainable, or less unsustainable, ways. At the same time, there is no guarantee that this will be always so, if countries ignore or do not account in an appropriate manner for the environmental externalities generated by production for export or import substitution. In the end, any given amount of output will be produced with the least demands on the environment if both importing and exporting countries have policies that embody the environmental externalities into the costs of production and the prices of the traded goods. For the world as a whole, when all countries have such policies, production would occur in the areas where the sum of production costs and environmental costs is smallest (Anderson and Blackhurst, 1992; FAO, 1993j). Naturally, any given environmental cost defined "objectively" in physical units (e.g. tonnes of soil lost to production caused erosion in the same type of land, i.e. other physical facts being equal) will be valued at different prices by different societies, just as any other factor of production is valued. It follows, therefore, that determination of production location patterns through trade (with all countries having environmental policies reflecting their own valuations) will minimize total costs including environmental costs as perceived by the different countries, but not necessarily as measured by some "objective" physical yardstick. This is as it should be because, as discussed in Chapter 3, sustainability and the value of the environment are anthropocentric concepts.

Thus, provided that this way of viewing the value of environmental resources is accepted and that appropriate environmental policies can be introduced, agricultural trade and environment can be made compatible but this may need multilateral support to make it effective. This will be one of the main challenges of international agricultural policy making in the years ahead, in particular for the Sub-Committee on Trade and Environment which has been established pending the negotiation on the successor to the GATT, namely the new World Trade Organization.

If national environmental policies are adopted by only a few countries, this does not represent a satisfactory approach to the pursuit of sustainable agricultural development. The imposition of taxes which make producers in one country face the environmental costs of particular production systems may simply lead to production being concentrated in countries where such costs are ignored or only partly taken into account. To avoid this, some countries could be tempted to adopt a unilateral approach, with policies which tend to equalize the treatment of domestic and foreign producers via the imposition of domestic charges on both local and imported goods or via import measures. Such approaches can, in many cases, be fully compatible with GATT rules but not all environmental policies would qualify. Thus a strong case can be made for a multilateral approach to reinforce the feasibility of adopting national agricultural environmental policies, through for instance international commodity-related environmental agreements. Multilateral approaches are also needed to avoid the misuse of environmental policies as a new form of disguised protectionism. For example, developed countries should not go so far as to deny poorer countries the advantages of profitable trade by demanding that they meet strict environmental standards reflecting values of much wealthier societies. At the same time, low-income developing countries will continue to face constraints in adopting "high" environmental standards. Assistance to enhance their development prospects as well as specific assistance in the area of sustainable development and the environment would contribute to relax these constraints.

In addition to national environmental issues, there are a number of transboundary environmental issues, which may or may not directly involve trade. In cases where appropriate national policies are not introduced and where damage is caused to other countries, resort may be made to multilateral action to encourage "good practice" via, for example, international environmental agreements (IEAs). The growth in the number and scope of these IEAs has been rapid and may be expected to continue in the future. An important issue for international policy will be to ensure that such action is based on objective, scientific criteria and to recognize the authenticity of differences among countries in the valuation of environmental goods. Action should be non-discriminatory, proportionate to the damage caused and result in minimal economic damage. Increasingly what is at stake is where the environmental problem does not derive from the production of the commodity itself but where it is the production and processing methods that are damaging and where trade measures are used to attempt to influence the choice of technologies used in other countries. These subjects are likely to be an important feature of the agricultural trade policy agenda for a long time to come and it will be important that agreements are reached which discipline the use of trade measures under the IEAs to make sure that these measures are not used as a new form of protectionism.

In conclusion, the widespread existence of environmental externalities means that markets alone cannot ensure sustainable agricultural development, since private market values normally do not take account of social costs and benefits. As a result if sustainable agricultural development is to be attained, governments should have appropriate policies which modify the behaviour of markets, otherwise resource allocation is likely to be sub-optimal from the point of view of the society as a whole. The range of policy options is very wide. Some imply the use of corresponding trade policies and others affect trade indirectly via their effects on production or consumption. However, it must be noted that inappropriate policies, what is called intervention failure, can also cause sub-optimal allocation of resources and need reform just as much as market failure requires action. It seems likely that countries will increasingly adopt environmental policies and that this will be a major aspect of agricultural policies by 2010.

NOTE

1. The PSE measures the value of the monetary transfers to producers from consumers of agricultural products and from taxpayers resulting from a given set of agricultural policies in a given year. The CSE measures the value of monetary transfers from domestic consumers to producers and from taxpayers to consumers of agricultural products resulting from a given set of agricultural policies in a given year. In other words, the CSE measures the implicit tax imposed on consumers by agricultural policies. For this reason the estimates have a negative sign (OECD, 1994: 251–3). The PSE and CSE measures were initially developed in FAO (FAO, 1973).

CHAPTER 9

Agriculture and Rural Poverty

9.1 INTRODUCTION

Over 1 billion people in the developing countries are poor, with a substantial majority of them living in rural areas. The development of agriculture can play a direct role in rural poverty alleviation, since the majority of rural poor depend on agricultural activity for providing the main source of their income and employment. It also contributes indirectly to alleviate rural poverty because the state of agriculture influences that of the non-farm rural economy. It can also play an important role in alleviating general poverty, since the agricultural sector makes a significant contribution to overall economic growth through its linkages with other sectors of the economy (Chapter 7). Furthermore, the increasing recognition of poverty as the root cause of problems of hunger and undernutrition assigns agricultural development a pivotal role in efforts to improve nutrition through increasing the quantity, quality and variety of food supplies, and through creating employment and income earning opportunities for the poor.

In this chapter, the focus is on a review of: (a) the empirical evidence concerning the effects of agricultural and more general economic growth on the incidence of poverty; (b) the role of interventions to improve access to land and develop the rural finance and marketing systems; and (c) the scope for direct interventions to create employment, or facilitate access to food. The related topic of the non-farm rural sector is covered in Chapter 7 and that of future developments in agricultural technology and research requirements in Chapters 4 and 12. This chapter was prepared in parallel with, and drew upon, a more comprehensive study of issues in rural poverty alleviation. This latter study (Gaiha, 1993) can be usefully consulted for more in-depth discussion of the topics in this chapter and in Chapter 10.

9.2 INCIDENCE OF POVERTY

The only available comprehensive estimates of poverty are those provided by the World Bank. The relevant table from the World Bank is reproduced here (Table 9.1). These estimates indicate that over 1.1 billion people in the developing countries were poor in 1990. This is approximately 30 percent of

Table 9.1 Estimates of the Magnitude and Depth of Poverty in the Developing World, 1985–90

Region	Number of poor (millions)		Headcount index %		Poverty gap index %	
	1985	1990	1985	1990	1985	1990
Aggregate	1051	1133	30.5	29.7	9.9	9.5
East Asia and the Pacific	182	169	13.2	11.3	3.3	2.8
Eastern Europe	5	5	7.1	7.1	2.4	1.9
Latin America and the Caribbean	87	108	22.4	25.2	8.7	10.3
Middle East and North Africa	60	73	30.6	33.1	13.2	14.3
South Asia	532	562	51.8	49.0	16.2	13.7
Sub-Saharan Africa	184	216	47.6	47.8	18.1	19.1

Source: Reproduced by permission from World Bank (1933d).
Note: The poverty estimates are for 86 countries, representing about 90 percent of the population of developing countries. They have been updated from those used in *The World Development Report 1990* and are based on national household sample surveys from 31 countries, representing roughly 80 percent of the population of developing countries, and on an econometric model to extrapolate poverty estimates to the remaining 55 countries. The estimates do not incude the countries of Indochina or of the former Soviet Union. The poverty line is $31.23 per person per month at 1985 prices. It is derived from an international survey of poverty lines and represents the typical consumption standard of a number of low-income countries. The poverty line in local currency is chosen to have constant purchasing power parity across countries based on 1985 purchasing power parity exchange rates. The headcount index is the percentage of the population below the poverty line. The poverty gap index is the distance of the average income of the poor below the poverty line (zero for the non-poor) expressed as a percentage of the poverty line.

their total population. Apparently the incidence of poverty declined in the two decades to the mid-1980s, but no significant progress has been made since then and the absolute numbers of the poor increased (Lipton and Ravallion, 1993). The highest incidence of poverty is encountered in South Asia and sub-Saharan Africa, where nearly 50 percent of the population is estimated to be below the poverty line, with the former accounting for the bulk of the world's poor because of its large population. The data and methods used to estimate the incidence of poverty and to construct poverty profiles are subject to many limitations. Some methods are better suited than others as aids in the design of policies for alleviation of poverty or mitigation of its effects (for a review see Ravallion and Bidani, 1994).

Comprehensive estimates of *rural poverty* are not available. The available data for a number of developing countries indicate that the incidence of poverty is highest in the rural areas (Table 9.2). High concentrations of poverty also occur in urban areas. In Latin America and the Caribbean, much of the increase in total poverty during the crisis decade of the 1980s occurred in urban

Table 9.2   Rural poverty in the 1980s

Region and country	Rural population as percentage of total	Rural poor as percentage of total poor
Sub-Saharan Africa		
Côte d'Ivoire	57	86
Ghana	65	80
Kenya	80	96
Asia		
India	77	79
Indonesia	73	91
Malaysia	62	80
Philippines	60	67
Thailand	70	80
Latin America		
Guatemala	59	66
Mexico	31	37
Panama	50	59
Peru	44	52
Venezuala	15	20

Source: Reproduced by permission from World Bank (1990).

areas (World Bank, 1993d). Rural and urban poverty are linked through rural–urban migration flows, as well as other factors.

Evidence concerning access to education and health services, housing and sanitation also indicates that the incidence of poverty is comparatively more severe in rural areas. While much of the information comes from case studies, some comparative estimates on urban–rural differences in access to safe water and health services in the developing countries are given in Table 9.3.

From a policy perspective it is necessary to know the characteristics of the population groups making up the rural poor. A recent study on rural poverty identifies the small farmers, the landless, the women, the nomadic pastoralists, the artisanal fishermen, the indigenous ethnic groups and the displaced persons as the functionally vulnerable groups in the rural sector and indicates that alleviating poverty in each may require different policy instruments and approaches (Jazairy et al., 1992). Although few indicators are disaggregated by gender, the available data on literacy indicate wide gender disparities. Thus the female adult illiteracy rate in 1990 was 46 percent compared with a total rate of 36 percent. In 1987 there were only 81 females per 100 males in primary education and 75 females per 100 males in secondary education. Women's inferior educational endowments, combined with reduced access to assets, cultural constraints on engaging in certain types of employment and their continuing daily responsibilities for children, make it even harder for them to escape the poverty trap.

Table 9.3   Rural–urban disparities in access to selected services

| Countries | Percent of population with access to: | | | | | |
| | Health service | | Safe water | | Adequate sanitation | |
	Rural	Urban	Rural	Urban	Rural	Urban
With UNDP human development index:						
High	n/a	n/a	56	84	n/a	99
Medium	67	97	69	91	58	89
Low	41	81	53	69	14	47
All developing countries	49	90	60	82	21	69
Least developed countries	39	85	42	57	16	48
Sub-Saharan Africa	36	80	28	65	17	56

Source: UNDP (1992).

The regional patterns of poverty presented here are fairly similar to those presented earlier for the incidence of undernutrition (Chapter 2). Data on related indicators such as the prevalence of underweight children or the incidence of micronutrient malnutrition further confirm the close association of deprivation in the nutrition/health nexus with that of poverty (FAO/WHO, 1992; WHO, 1992). It can be expected that progress in poverty alleviation would be largely reflected in amelioration of these poverty-related indicators of deprivation in the nutrition, health and other areas. However, public policy which targets directly the areas of nutrition, health, education and housing will also be essential. The fact that countries with similar per caput incomes have widely differing rates of undernutrition, morbidity and illiteracy, proves that there is considerable scope for such public policy. Moreover, the increasing recognition of the role of human resources in overall development strengthens the case for such public policies.

9.3   RURAL POVERTY AND AGRICULTURAL GROWTH

A general case can be made from several studies that when the economy grows the incidence of poverty (percent of population below the poverty line) tends to decline. Whether such a relationship holds in all cases, depends essentially on what happens to income distribution when the economy grows. It is possible that income distribution could become less equitable, thus offsetting partially or completely the potential benefits of such growth to the poor. Whether this happens is an empirical question. It is difficult to establish whether systematic relationships exist between overall economic growth and changes in the distribution of income. Comparative studies seem to indicate that economic growth is as likely to be associated with increases as with decreases in inequality.[1] A recent study illustrates such diverging effects by comparing the experiences of rural India and Brazil (Datt and Ravallion, 1992). According to this study, in India the positive growth effects on poverty were reinforced by improvements in the distribution of income. In Brazil, on the other hand, a worsening of the income distribution led to one-half of the potential poverty reduction effects of growth being cancelled.

With regard to the relationship between agricultural growth and rural poverty, the empirical evidence seems to lend support to the common-sense proposition that the distribution of benefits from increased agricultural production will approximately reflect the initial distribution of productive assets and of access to inputs and services, as well as changes in the distribution of such assets brought about by the process of agricultural growth itself. It is, therefore, possible for agricultural growth to be associated with worsening of the distribution of income. If, moreover, the deterioration is sufficiently severe, it can even cause parts of the rural population to become poorer in absolute terms. This seems to have been the case in Latin America in the 1980s when the incomes of the rural poor declined despite considerable increases in aggregate agricultural production. In this region, the most severely affected category – landless labourers – suffered a 23 percent fall in real wages between 1980 and 1987 (UN, 1992).

The specific impacts of agricultural growth on different socioeconomic categories of rural producers and labourers, as well as the mechanisms through which these impacts are mediated, depend on the nature of the growth processes and the structural factors underlying the social organization in rural areas. Much of the empirical knowledge on these matters comes from studies (mostly from India) of situations in which rapid agricultural growth occurred as a result of the spread of the green revolution.

The evidence in these studies indicates that the advent of the new technology, in the form of biochemical innovations, was associated with reductions in rural poverty, e.g. the proportion of the poor among the cultivating households declined and so did the severity of the poverty of those households which remained poor. At the same time, however, some of the poor became poorer and some of the non-poor were pushed below the poverty line. Much of the adverse impact on selected groups of the rural population was the consequence of the initial inequality in access to land. Since smallholders were impeded by restricted credit markets, input supply problems, limited access to extension services, risk aversion and tenure insecurity, the benefits of the new technology tended to accrue mainly to large landholders. The latter expanded acreage through resumption of land for personal cultivation (by evicting tenants) and/ or through leasing or purchase of land from small landowners. As a result, the distribution of gross cropped area became more unequal and the percentage of landless households increased from 25 percent to 35 percent (data and analysis for rural India for 1968–70 from Gaiha, 1987).

On the other hand, the longer period (1974–84) evidence from North Arcot (a small region in South India) indicates that where small-scale, owner-occupied farms dominated, the effects were more favourable to the poor, largely because there was a conducive institutional setting, in which state and local governments made credit and modern inputs available to small farmers, and invested heavily in infrastructure (Hazell and Ramasamy, 1991).[2] Even then, the early adopters of the modern varieties (MVs) were typically large farmers. But eventually, over 90 percent of the paddy area was planted with the MVs, with no systematic differences by farm size, and similar yields were obtained on large- and small-scale farms alike. There was no evidence of a general increase in land concentration or of loss of land by smallholders and there were sizeable absolute gains in income of all household categories and a decline in the incidence of absolute poverty (for more general discussion of technical change and rural poverty see Lipton and Longhurst, 1989).

In addition to the fact that agricultural growth may be associated with at least some parts of the rural population becoming worse-off economically, concern with rural poverty must also encompass those sections which are chronically poor and too marginal to be affected by agricultural growth, one way or the other. These sections comprise people living in remote, resource-poor regions, without any infrastructure; backward sections of society, debarred from owning assets, denied access to education and condemned to menial occupations; and the disabled and the aged, incapable of augmenting their incomes above a bare subsistence level.

Over a longer term perspective, the sustainability issues related to agricultural growth assume increasing importance for the rural poverty problem. Continued population growth in the context of rural poverty, and in particular when it occurs in conditions of inequality of access to land, tends to push the rural poor to expand agriculture into ecologically fragile areas. This causes deforestation and exploitation of the land in ways which damage its productive potential. This process sets the stage for continued poverty of the part of the population concerned. Examples of this process abound, from the Himalayan and Andean ecosystems, large parts of Africa and the colonization experiences (spontaneous or officially sponsored) of the tropical rainforest (e.g. Brazil, Indonesia). To the extent that this process takes place in parallel with production increases in the higher potential areas, it provides another example of agricultural growth bypassing, and often making worse-off, part of the rural poor, e.g. by lowering the prices they receive for their own production. The sustainability issue and its relation to rural poverty is also relevant for the better off agricultural lands, in so far as more intensive agriculture, if not carefully managed, can reduce the productive potential of the land and water resources and threaten the sustainability of the poverty-reduction effects obtained initially.

It can be concluded that, on balance, agricultural growth can be expected to bring about reductions in rural poverty (Lipton and Ravallion, 1993).

However, some parts of the rural population, including the rural poor, may become worse-off economically in the process of agricultural growth. The structural characteristics of the rural economy at the inception of agricultural growth play a predominant role in the distribution of benefits from higher production. Scale-neutral technical change can benefit large and small farmers alike, if institutional rigidities do not stand in the way. With regard to the latter, an activist public policy in the area of institutions, research, credit, etc., can be instrumental in ensuring the wider spread of benefits.

9.4 INTERVENTIONS TO IMPROVE ACCESS TO LAND

It was noted above that the structural characteristics of the rural economy – particularly land ownership and land tenure systems – play a decisive role in determining the distribution of benefits and the rural poverty effects of agricultural growth. The pros and cons of interventions and the lessons of experience from efforts to enhance the access of the poor to land are discussed below with regard to: (a) redistribution of ownership rights; (b) regulation of tenancy contracts; and (c) the role of land titling.

Land redistribution

The most recent attempt to take stock of progress in redistributive land reform was undertaken in 1991 for the quadrennial FAO report on progress under the Programme of Action of the World Conference on Agrarian Reform and Rural Development – WCARRD (FAO, 1991b). The report concludes that progress has been limited, mainly because the implementation of land distribution programmes was strongly affected by political realities.

The *equity case* for land redistribution from large landowners to the landless and/or small owners of land rests on at least three considerations: (a) the landless/small owners are usually poorer than large landowners; (b) in general, but with important exceptions, total employment and production per hectare increases as farm size decreases; and (c) inequality in the distribution of land conditions the poverty effects of agricultural growth not only because of the resulting unequal distribution of the income attributed to land but also because it breeds social stratification patterns inimical to the poor in many other areas, e.g. the distribution of political power or access to credit.

The *efficiency case* requires that land redistribution increases, or at least does not reduce, farm output and the potential for growth. Since often an *inverse relationship* between farm size and output per hectare is observed, land redistribution has the potential of increasing output.[3] In most cases, the inverse relationship is due to a higher cropping intensity, and a more labour-intensive and higher-value crop mix, on smaller farms. Land quality differentials may account for part of this inverse relationship, e.g. when large farms contain a higher proportion of inferior quality land compared with the small farms.

Controlling for these differences in land quality can attenuate the inverse relationship, but it does not cancel it. The inverse relationship, however, can be modified with the onset of green revolution where land augmenting technology tends to equalize the yields attained by small and large farms. But differences in labour use per hectare tend to remain, and this is an argument in support of the continued relevance of the efficiency case for land reform even under the new technology. This is because in the context of rural market imperfections, particularly of labour markets, large farms tend to use factor proportions skewed in favour of capital (mechanization) and against labour beyond the proportions dictated by social allocative efficiency considerations, even in the absence of policies distorting prices in favour of capital, i.e. even when governments have implemented policies to "get prices right". This is a rather important consideration in the debate on what is the appropriate mix of policies for getting prices right and those for bringing about changes in basic structural characteristics of the rural economies (see Platteau, 1992).

The case for a more equal land distribution is further strengthened when linkages with non-agricultural activities in the rural sector are considered. Some South Asian evidence suggests that in villages with relatively equal land (and farm income) distributions, the share of locally produced labour-intensive non-farm goods in total consumption is higher compared with that in villages characterized by higher inequality. Thus, it can be expected that a more equal land distribution contributes to rural poverty alleviation indirectly via its effects on rural non-farm employment.

The extent to which the changes brought about by policy interventions to redistribute land prove durable and the persistence of the unavoidable upheavals in the production structures are important elements in the land reform process. Chile's experience with land reform is instructive. Initiated during 1964 to 1970, land redistribution was extended during 1970 to 1973 and reversed during 1973 to 1976. Expropriation of land was stopped. Land worked by the campesinos in the transition period *(asentamientos)* was partly confirmed to them, partly restored to the previous owners and the balance was sold at public auction. New farm enterprises with strong financial resources entered the sector. The modern sector became strongly export oriented. In parallel, the lack of technical and credit support induced many small farmers to retreat into more traditional patterns of subsistence production. Producer cooperatives were dismantled. But more than half of the land expropriated was ultimately confirmed to the beneficiaries (Jarvis, 1989; Gomez and Echenique, 1991). The crisis of 1982–83 resulted in a notable reversal of policy. From 1987 the Government started financing a considerable part of the costs of technical assistance to smallholder beneficiaries of agrarian reform (Meller, 1988).

Policy interventions also led to structural changes in production in the Philippines where land reform was limited to rice and maize lands which were managed predominantly through tenancy arrangements. This limitation induced landlords to divert their land to other crops, often at the expense of both efficiency and equity. For instance, rice lands with higher income-earning and labour absorptive capacity were converted to less labour-intensive crops such as coconuts.

Success is more likely to be achieved when distributing state-owned land, where resistance is less than when trying to redistribute land away from large landholders. In the Philippines, between 1987 and 1990, two-thirds of the total land distribution targets of state-owned land were fulfilled. However, only 2 percent of the target for private land redistribution was achieved, due to disputes with owners over appropriate compensation (information from country working papers for the preparation of FAO, 1991b).

Another important issue is the extent to which the state of modernization of agriculture is related to the process and chances of success of redistributive reforms. It was noted above that under the new technology, the inverse relationship between yields and farm size tends to weaken. This can weaken the efficiency case for reform. The experience also shows that the threat of reform can be instrumental in prompting larger farmers to promote modernization as a defensive action. For example, in some Latin American countries, the threat of expropriation and incentive policies (input subsidies, tax breaks) were successful in inducing large farms to modernize, hence increasing agricultural output. One outcome of this modernization, however, was to render expropriation with compensation very costly. Further, as was recently found in Colombia, larger farmers often successfully used their influence to extract promises from the government that their land would not be expropriated if they modernized. As a consequence, redistribution of land to the poor was negligible. Interestingly, modernization of agriculture had the opposite effect on land redistribution in the Philippines. There, compensation had been fixed at pre-green revolution land values and the economic gains associated with modern seed and fertilizer rice technology allowed the beneficiaries to capture significant economic surpluses.

Government support to the beneficiaries of land reform is an essential component of the whole operation. The case of Mexico illustrates the pitfalls of unsupported land reform. There, the reform was not accompanied by significant productivity gains as most of the small farmers of the agrarian reform sector were left with non-irrigated land, and government support policies were not always effective. Even in the case of small farmers who had some irrigated lands, the government-supported cooperatives did not provide the needed services. By contrast, most of the irrigated land was left in the medium and large holdings and government support policies were heavily skewed in their favour.

The preceding discussion is more relevant to cases where the reform aims at creating more equal distribution of landholdings to be owned and operated as individual units. But there have been experiences with alternative ownership and exploitation structures in the post-reform periods. The creation of producer cooperatives is a case in point. Here the general conclusion is that the experiments with producer cooperatives led to disappointing results – particularly in some Latin American countries. In Peru, for example, earlier reforms had led to some two-thirds of agricultural land being controlled by producer cooperatives in 1979. However, these cooperatives suffered from serious diseconomies of scale and work incentive problems and many were broken up in the early 1980s and the land was distributed as individual holdings.

In Nicaragua, producer cooperatives were initially thought to be better suited for the large-scale production of export products such as coffee, cotton and beef. Subsequently, the emphasis on land distribution moved away from the establishment of producer cooperatives towards direct distribution to individuals. This followed the realization that dividing large farms into smaller holdings would not necessarily cause a reduction in output if adequate credit and other support were provided.

Land reform will continue to be a relevant issue in the future in the quest for poverty alleviation and more equity in the rural areas. However, it may cease to be the burning issue it once was, especially in those countries where the non-agricultural sector will be increasingly the main source of additional employment and income-earning opportunities and land will lose its primacy as the main form of wealth. As indicated in Chapter 3, a number of developing countries are expected to have over the next 20 years economic growth rates high enough to imply that the bulk of additional wealth will be generated in the non-agricultural sector. It can be expected that trends for increasing farm size will emerge in such situations, just as they did in the developed countries. This is because pressures build up for the incomes of people in farming to follow (though not necessarily become equal to) those that can be potentially earned in the non-agricultural sector. A combination of more land per person and higher income earned per unit of land is normally the result of these pressures, brought about by technological change and the flow of labour from agriculture to other activities, though not necessarily always urban-based ones.

Many developing countries may not, however, enter this phase of transition in the foreseeable future. In many low-income countries with unfavourable overall growth prospects, high incidence of rural poverty and continued high population growth rates, the number of people seeking to make a living in agriculture will continue to grow. In these circumstances, the distribution of land ownership and the potential role of interventions to change it towards patterns more conducive to poverty alleviation and enhanced equity will continue to be live issues.

It is to be noted, however, that a more equitable distribution of a growing agricultural income can only contribute in limited ways to make significant dents in rural poverty directly, so long as the population dependent on agriculture continues to grow. This is because even an optimistic agricultural growth assumption (e.g. around 3.5 percent p.a. in gross value terms) will probably mean a growth in average per caput incomes of the growing agricultural population of under 2.0 percent p.a. Welcome as such an outcome is for its potential for reducing rural poverty, it cannot be compared with the long-term benefits obtainable from a combination of vigorous non-agricultural growth and declining agricultural population.

Tenancy reforms

Tenancy is the term commonly used to refer to those land tenure arrangements (legal and customary) which regulate access to land in forms other than acquisition of ownership rights. It refers to all those situations where a person's access to land is through some arrangement with another person or entity who enjoys superior land rights. Policies to reform tenancy arrangements are often predicated on grounds of both efficiency and equity or poverty alleviation. It must, however, be noted that regulation of tenancy contracts could result in a contraction of the supply of land for tenancy and may thus lead to an increase in the number of the landless in rural areas as tenants are evicted (as happened in the Philippines, India and Sri Lanka in South Asia, and numerous Latin American countries; Osmani, 1988).

One of the major issues concerns the relative merits of alternative arrangements for renting land, e.g. fixed payment for a definite period of time (fixed rent), sharecropping, labour service or mixtures thereof. A major thrust of tenancy reform policies has been to restrict or prohibit sharecropping contracts. The concern with sharecropping was also associated with the feudal conditions existing in agrarian societies. In retrospect, however, it has been found that such policies can have unintended negative effects on the poor. There is now increasing realization that under special circumstances, sharecropping contracts can be an efficient vehicle for *risk-sharing*[4] with positive impacts on both efficiency and equity. For example, evidence shows that there is a higher implicit rent on sharecropped rented land (which may reflect a risk premium) and a higher frequency of share tenancy in areas with variable weather. Cost-sharing arrangements in sharecropping can allow poor farmers to have access to certain inputs which they would not otherwise obtain because of their limited access to financial resources. For example, because fixed rents on leased lands generally must be paid in advance, poor farmers without access to credit may be prevented from leasing land. This constraint is overcome under share tenancy as payments are made only at harvest.

The experiences of tenancy reform in China, Laos and Vietnam indicate that changing from forms of socialized farming systems to ones based on the household economy, where allocative decisions, ownership of other productive means, and longer land-use rights are given to the individual households, can bring substantial efficiency and equity gains. In China, for example, the increases in agricultural production, and the development of the non-farm rural economy were spectacular and led to strong broad-based rural economic growth and significant reductions in the incidence of rural poverty. This process has slowed down in recent years. Vietnam, after the implementation of land tenure reforms, became self-sufficient in food grains for the first time and subsequently a net exporter of rice.

Concerning Africa, recent evidence suggests that most indigenous land tenure systems are adapting efficiently to changes in resource availability. As a policy option, it may therefore be better to concentrate on providing an appropriate legal and institutional environment for more efficient transactions than to restrict land sales and rental markets with tenancy legislation.

Land titling

Three arguments are usually put forward in support of land titling:

1. Titling is assumed to increase tenure security with a view to promoting investment in land and water conservation and capital inputs, and adoption, where appropriate, of permanent crops
2. By providing collateral, titles may increase access to institutional credit
3. Land titles are considered necessary for the development of land markets which are essential for promoting commercial development of agriculture.

Recent evidence from Africa suggests that the expected positive relationship between tenure security or the extent of land rights, particularly inheritance, and long-term investment in the form of land improvements hold in some areas, but not in others. Further, formal ownership titles are not necessary for tenure security, since under most communal tenure systems, a farmer has usage rights to specific plots which are often hereditary. In other places where such hereditary rights are not available, however, lack of title does bias production decisions in favour of short-cycle crops.[5] Although some evidence from Africa shows that titling may not have a significant effect on access to credit, evidence from some countries in Asia and Latin America is rather more favourable, with significant increases in access to institutional credit after land titling.

Land titling often aggravates inequality since wealthier and more influential individuals can obtain greater rights than they formerly enjoyed. In these conditions the poor are exposed to increased risk of landlessness and loss of common property resources after implementation. The position of women in land titling requires a special focus. Land titling programmes, for example, tend to concentrate on the land parcel as the relevant target unit, paying little attention to distribution of land rights within the household. Granting land titles to male household heads tends to diminish women's control over land usage and transfers. Land titling may also cause the loss of secondary rights in land, such as the right to gather fuelwood, which are of particular importance to women. There is thus a strong case for designing land legislation and reforms to target women as direct beneficiaries.

Advances in electronic data processing have opened up new horizons for traditional land titling and cadastre systems. There is no inherent reason why land registration and cadastre cannot be made sensitive to special cultural conditions on the one hand, and to equity and efficiency criteria of sustainable rural development, on the other. With the evolution currently taking place in thinking about land tenure and the shift from social to private property models, the compilation of land records (land registration and cadastre), and local community involvement in land regulation (taxation, zoning, etc.) can be expected to be major activity areas in the coming decades.

9.5 RURAL FINANCE

The nature of agricultural production imposes significant discontinuities between the time resources are committed (immobilized) in production at planting and the generation of revenues after the harvest. Farmers who do not have sufficient resources to invest in such immobilization for the required part of the year, including both for purchased inputs and for living expenditures in the waiting period, depend more than the people in other sectors on the availability of credit. Facilitation of access to credit for the rural poor has therefore a role in the panoply of policy instruments for alleviating rural poverty.

The environment for rural financial intermediation has changed significantly in recent years. The concept of privatization has been embraced by an ever increasing number of countries, and the role of markets in the determination of prices for traded agricultural products has been enhanced. Food and agricultural input subsidies, including those for agricultural credit, have been reduced or eliminated. A larger share of rural credit comes from private sources, and a declining share from the state. As subsidies on credit are reduced the cost of credit increases, and as subsidies on other inputs are reduced the credit requirements rise.

Until the early 1980s most attention was focused on formal finance, i.e. that sector of the financial system regulated by a central monetary authority; only occasionally was mention made of financial activities that were not regulated, i.e. informal finance. During the 1980s, however, research increasingly showed that informal finance plays an important role in rural development, especially for poor people: small farmers, landless people, micro entrepreneurs, and particularly women within these groups. It also became apparent in a number of countries that the informal system operated more efficiently and equitably than did the formal financial structure.

Far too often talk about financial services is limited to credit alone. Surprisingly large amounts of savings deposits can be mobilized even in low-income countries and among people who would fall in the category of the poor, when a reliable and effective system for doing so exists. There are a number of studies which confirm that a large percentage or, in some cases, the entire seasonal lending for agricultural production could be financed with locally mobilized funds (local in this context meaning rural). Again, this requires improvements and further developments in the financial system, which at village and district levels makes possible the mobilization of savings. At the national level, the financial system must also be capable of transferring such savings from surplus to deficit areas, at the same time maintaining the confidence of savers in the safety of their deposits. As an agrarian economy develops, the flow of savings generally has been from rural areas to urban centres, a flow often stimulated by the adverse terms of trade for agriculture stemming from government policies that explicitly or implicitly tax the sector (see Chapter 7).

Specialized credit institutions and commercial banks

When governments and donors alike started focusing attention on credit as a means of fostering rural development, a variety of specialized, mostly government-owned, credit institutions were created. The overall experience with these types of institutions has been quite unsatisfactory. They were directed to extend below-cost loans to target groups or activities identified either by the government or by external funding entities. Because the selection of those groups or activities was often made on criteria other than commercial ones linked to their financial performance, the loan repayment rate was poor. In most cases the lending agencies were supervised and controlled by ministries which were not equipped to deal with financial institutions. All these negative features, together with excessively high transaction costs, have led many of these institutions into great difficulties, and made them increasingly dependent on state subsidies for survival (see Besley, 1994, for a review of arguments for interventions in rural credit markets).

Simultaneously with government efforts described above, commercial banks were urged to increase their activities in rural areas, particularly by lending to the agricultural sector. Again, results were generally below expectations and the intended target group, often small farmers, benefited little from these measures. Small farmers were ignored because lending to them was expensive and they were considered to constitute a higher risk than large farmers, though there is no adequate empirical evidence to support the latter argument.

Partly on their own initiative and partly as a means of complying with the government directives, commercial banks in some countries have experimented with group lending schemes for small farmers. For example, in Ghana, such schemes were initially fairly successful, but when the numbers of groups increased the staff involved could not manage them properly. This shortcoming was reflected in loan default and consequently considerable sums were lost in these schemes. However, during the last three to four years some Nigerian commercial banks have made special efforts to build up lender–customer relationships with selected cooperatives, so that the latter can satisfy the demand for credit by their members. Initial results have been encouraging.

Except for a few recent successes, both specialized credit institutions and commercial banks often lack adequate institutional and operational arrangements at the grassroots level. In particular, they are too far removed from their clientele to make optimum lending decisions and to implement sound loan collecting procedures.

Cooperatives and other rural organizations in rural finance

To overcome these crucial problems, increasing efforts are being made to involve in the provision of financial services various rural organizations, such as cooperatives, informal groups of small farmers and other rural people, and traders dealing with agricultural inputs and produce. Cooperatives permit economies of scale for their members in access to financial services. They are particularly well suited to providing financial services to rural people as they operate at grassroots level among people who know each other well, a basic condition for trust. Often a cooperative is the only financial institution (or formal organization) in a rural area, and is therefore an obvious structure for the operation of new financial services to supplement the traditional, informal sources of credit.

Cooperatives and other less formal group arrangements have the potential of reducing the transaction costs of lending to small farmers and other segments of the rural population subject to disadvantages as well as of improving the management of risk. Successful lending programmes have shown the importance of factors such as homogeneous borrowing groups, which are jointly liable and themselves assume some managerial and supervisory responsibilities, and are bound in loyalty by a common bond other than credit. Important factors for the success of cooperatives include bottom-up institutional development, extensive training at all levels, reliance on savings mobilization and equity contribution rather than external funds, gradual expansion of activities, and strict monitoring and auditing.

Informal rural finance

For the rural people, and particularly for the rural poor, the main sources of credit have been and continue to be various types of informal arrangements. Although more common among poor people, informal loans and savings activities are known among and between all economic classes. Traditionally, informal rural finance has been viewed by outsiders as a plague for poor people, whereas in fact large numbers of the poor benefit from it. Furthermore, contrary to widely held opinions, there is surprisingly little evidence in recent studies of exploitation. Women, in particular, often have to resort to informal finance because of institutional and legal barriers to formal credit such as lack of collateral or requiring a husband's signature on loan agreements.

A great number of financial intermediaries operate in the informal financial markets. Probably friends and relatives are the most common source of credit, particularly in rural areas, in some countries accounting for more than half of all informal loans. In most cases, no interest or collateral is involved and repayment conditions are very flexible. These attributes have great merit for those without collateral, such as the landless or those without land titles, and in situations where production risks are high. Furthermore, loans are often received in kind, such as seeds and fertilizers, and may be repaid also in kind.

Rural communities of some countries may save jointly for a variety of purposes, generally not for lending but for the bulk purchase of farming inputs (e.g. in Zimbabwe) and for various social functions. Informal rotating savings groups among women have also become popular as a means of maintaining their financial independence. More sophisticated groups are the ROSCAs (rotating savings and credit associations),[6] which are found in many low-income countries and which have been extensively studied in recent years. In many areas, more individuals participate in ROSCAs than deal with formal financial institutions; recent research in the Cameroon suggests that the volume of deposits moving through ROSCAs may sometimes be larger than amounts held in banks (Schrieder, 1989).

The predominance of informal credit arrangements in rural credit markets necessitates the analysis of the complex transactions that occur in such markets. Recent developments in the analysis of informal credit institutions emphasize the role of informational deficiencies in shaping such transactions. In developing countries, incomes of rural borrowers are uncertain, collateral is often lacking and repayment, if not willingly made, is extremely difficult and costly to enforce. Thus, when a loan transaction takes place, it is very costly for the lender to determine the default risk of a borrower and monitor that her/his behaviour makes repayment likely.

Borrowers and lenders in their effort to reduce transaction costs of screening and monitoring loan performance may link the terms of the loan contract to transactions taking place between them in other markets. Such transactions may be between traders and farmers (traders making loans to farmers for purchases of inputs), landlords and labourers (landlords making advance payments to workers to secure their labour when needed in the future), etc. Such interlinkages[7] lower the transaction costs of screening the borrower's creditworthiness, provide a source of control by the lender on the borrower's earnings and income, and give the opportunity to the lender to affect the probability of loan repayment by manipulating the terms of trade in other markets. For instance, a trader who is also a lender may provide better prices for modern inputs to his borrower, since the use of such inputs reduces the probability of default on the loan (Hoff and Stiglitz, 1990). High information and transaction costs often restrict loans to members within geographical or social boundaries (a village or kinship group) where transactions are sanctioned by the community. This type of behaviour may explain the high segmentation of rural credit markets.

The analysis of informal markets shows that the scope of financial market liberalization will be limited if the basic reasons for the distortions in rural credit markets (i.e. informational asymmetries) are not sufficiently dealt with. Given the significant role of interlinked transactions in rural credit markets, actions by governments in other markets as well as risk reduction policies may have beneficial secondary effects on rural credit markets. For instance land titling, increasing market integration, improvements of rural infrastructure and other risk-reducing policies, will increase the credit receiving capacity of rural borrowers and reduce the importance of information constraints. High segmentation in rural credit markets may introduce monopolistic elements. Existing evidence points to the existence of such elements in the behaviour of lenders. In cases where highly priced rural credit is the result of monopolistic or collusive behaviour by local lenders, entry should be encouraged.

The conclusion of the preceding discussion is that highly specialized credit institutions are no longer considered the most suitable type of rural finance, particularly not for the rural poor. Intermediaries which accept savings deposits, such as local banks, cooperatives and other rural organizations, have gained popularity among the rural people themselves and have demonstrated promising results. It is commonly accepted that there should be a choice of institutions offering financial services and that they should compete with each other and thus improve services to their customers. Acceptability of institutions by prospective customers should be the main criterion in promoting different types of financial intermediaries in rural areas. For the poor people in rural areas it is extremely important to have informal financial intermediaries included and to have appropriate operational linkages between them and formal financial institutions established.

9.6 MARKETING

The structure of rural markets influences the incidence and persistence of rural poverty, because it is instrumental in determining the way in which trade takes place and the processes governing the terms of exchange. Marketing systems in developing countries are diverse in structure and often complex. They comprise arrangements for credit, storage, transport, and involve a hierarchy of intermediaries, e.g. large and small private operators, cooperatives and state agencies engaged as traders, processors, distributors, wholesalers and retailers. The one common characteristic shared by many developing countries is that transactions, especially by the poor, tend to be relatively small. The rural poor participate in those exchanges as producers with small quantities of cash crops or food surpluses to sell, as net purchasers of food and other basic necessities for own consumption, as petty traders in staple food producing areas, and as labourers in agricultural production, food processing and local distribution.

The limited scope of the market increases both transaction and production costs; the former because of informational deficiencies, and the latter because of the limited nature of specialization and division of labour. Moreover, where the formal market structure is absent or incomplete, the small trader, consumer or producer is impeded by barriers to entry, insecure property rights and poorly enforced laws. In response, there are frequently alternative "organizational forms" and "rules and conventions" that provide a structure for exchange.

State intervention

State interventions in marketing have been quite common in many developing countries. The policy instruments used in these interventions have ranged from steps taken to improve market infrastructure, to price interventions, to creating and fostering marketing parastatals and cooperatives. The pros and cons of direct government involvement in agricultural marketing are part of the wider issue of the role of the state in economic life which has come under increasing scrutiny in the context of structural adjustment policies. This wider issue is discussed in Chapter 7. Here, the focus is mainly on the significance for the rural poor of a government role in agricultural marketing.

Improving market infrastructure

A well-recognized role for policies in this area is the provision of marketing facilities and support services. These include urban and rural wholesale and retail markets, rural assembly markets, auction rooms, etc.; collection, analysis, and dissemination of marketing information on prices, quantities, qualities and crop conditions to enhance market integration and market transparency; establishment of a uniform system of grades, weights and measures; and provision of marketing extension services, e.g. advice on what to grow, how to handle it and where to market it. These public sector activities can bring significant long-term benefits to the poor as consumers and small farmers, particularly through reduced transaction costs and lower real prices of food. Whether government intervention should range beyond these functions raises more contentious issues relating to price policy and direct state involvement in marketing.

Price interventions

Price interventions can take many forms and have multiple objectives. Stabilization of prices and/or incomes (not the same thing and often incompatible with each other) can confer important benefits to the poor because they provide a more secure investment climate, easier access to credit and reduced short-term stress on household consumption. Even when such stabilization schemes are appropriate and cost effective, however, care must be taken not to introduce sustained price biases. If prices do not conform to the long-run opportunity costs represented by world market prices (see Chapter 7), the resulting misallocation of resources can seriously hinder economic efficiency, growth and poverty alleviation.[8]

Marketing agencies

In the past, rural traders were often mistakenly associated with exercising monopoly power and exploitation of small farmers and this presumption in fact was one argument for the creation of marketing boards. In practice, however, rural traders often operate with low overheads and margins, and there is little systematic evidence that small farmers have been disadvantaged. Indeed, in many cases, agricultural parastatals have not performed any better, and sometimes far worse, than private agents. Often they tax farmers indirectly by absorbing a large part of prices of commodities they handle. This has discouraged production or at times encouraged farmers to bypass the formal sector in favour of the informal sector, with the additional costs which that entails.

Despite these deficiencies, however, some marketing boards have performed well, successfully extending to small farmers the benefits of large-scale marketing organizations – especially in remote areas, e.g. the Grain Marketing Board in Zimbabwe. In other cases, marketing boards have been operationally efficient and relatively cost effective, but due to inappropriate pricing policies and multiple objectives imposed by governments they were led to accumulate large losses, with subsequent burdens on government budgets.

Cooperatives

Advantages of economies of scale similar to those for parastatals are often cited for cooperative ventures. The smallholders can benefit from the economies of scale in processing, storage or transport, where there are few private agents competing to provide these services. Vertical integration would be attractive to the small farmer producing traditional cash crops, or milk and livestock products, but is less appealing for those producing food crops, which require little processing and which are sold locally. However, timeliness and appropriateness of pricing decisions, and the coordination of processing, storage and transportation call for considerable managerial competence. Finding appropriate leaders, with the broad range of management skills required, is often difficult.

The overall lesson of experience with direct state involvement in marketing is that less of it is better than more of it, but some continuation of such a role will be necessary to ensure a smooth transition to a system based on private agents. The critical question is how to move smoothly from one organizational form to the other. Where major disruptions in services take place, it is likely that the poor will suffer first and probably most. Finding ways to avoid the adverse impact of reform on vulnerable groups, while attaining the benefits of a more efficient market structure, will be the main challenge in marketing in developing countries for some time to come.

9.7 SELECTED DIRECT ANTI-POVERTY INTERVENTIONS

Rural public works

Public employment schemes have long been used in many developing countries in emergency situations, such as during periods of drought and famine (and, more recently, during periods of macroeconomic stabilization and adjustment), and when there is large-scale, transitory unemployment and underemployment in the rural sector. However, in recent times, many developing countries have incorporated such schemes as regular elements of an anti-poverty strategy.

The experiences of Asia and sub-Saharan Africa provide useful examples of the worth of public employment schemes used for poverty alleviation during periods of drought and associated threats of famine in the 1980s (e.g. Botswana in 1983–85, India in 1987). In South Asia, rural public works (RPW) form the core of government anti-poverty strategies, creating useful employment for many of the rural unemployed and underemployed, as well as significantly reducing income variability. In Latin America many countries, such as Bolivia, Chile and Peru, have used public employment schemes to counter the temporary drops in labour demand that occurred during periods of structural adjustment or macroeconomic shocks.

Apart from providing substantial welfare benefits for the poor, RPW programmes often contribute to economic growth through the creation of assets such as roads, schools and canals. Social and economic returns to assets created through such programmes can be enhanced by ensuring that the projects are well integrated into existing rural development plans. Community participation in project design and execution can help to select highest priority projects, avoid wastage and promote labour-intensive methods. It can also help in the maintenance of the assets after they are created, even though regular financial provisions for maintenance are also necessary.[9]

While there are few estimates of the cost effectiveness of RPW, simulations for 1980–2000 based on Indian data suggest that RPW programmes can have a greater impact on the poor than investment in irrigation or schemes of public distribution of food at subsidized prices (Narayana *et al.*, 1988; Parikh and Srinivasan, 1989). However, care must be exercised in targeting the poorer households if the aim is poverty alleviation. This has been achieved, for example, in the Maharashtra Employment Guarantee Scheme and the

Bangladesh Food for Work Programme, where the programmes pay lower-than-market wages. In contrast, Bolivia's Emergency Social Fund (ESF), which largely financed local infrastructure projects executed by private contractors and permitted the hiring of construction workers at market wages, was not well targeted. Fewer than one-half of the workers employed on ESF schemes were drawn from the poorest 40 percent of Bolivian households (World Bank, 1990).

Food and nutrition interventions

Some sections of the poor (e.g. the old and the handicapped, as well as some of the groups referred to earlier) are not likely to be in a position to benefit from the direct anti-poverty interventions like those discussed above. Special interventions to raise their income levels are necessary. Unanticipated fluctuations in food prices may also have serious consequences for the "entitlements" of the poor, particularly for casual agricultural labourers. In this context, there is a strong justification for direct interventions to enhance their access to food, usually in the form of food subsidies.[10] The design and implementation of food subsidy schemes raise many contentious issues. Such subsidies take a variety of forms (e.g. general food subsidies, food rations, food stamps, etc.) and the choice of an appropriate form is often very difficult. The direct welfare effects of food subsidies may be quite different from the indirect effects operating through other markets (Timmer, 1986). The experiences with food subsidization policies are summarized below.

General price subsidy schemes

Their general characteristic is that they supply unlimited amounts of specific subsidized food products to any one who wishes to buy it. The subsidy may cover a portion of the total production, storage and marketing costs. The price wedge may be administered at a point of import, or a point of processing, storage or sale. Such schemes have been extensively used in developing countries because they are administratively convenient, especially where private market channels exist. However, they also tend to be costly, because of leakage of benefits to the non-poor.[11] For this reason some countries have limited general food subsidies applied to commodities that are consumed mainly by the poor, as in Egypt where the benefits of coarse flour subsidies accrue mainly to low-income groups, or to specific geographic regions, as in the Philippines where rice and cooking oil were made available to selected poor villages through local retailers at discounted prices. It is estimated that 84 percent of the total cost of such subsidies accrued as benefits to the target groups.

Ration schemes

An alternative to a general subsidy is to provide a quota, or ration, of subsidized food to each eligible household and let the market supply any further requirements. Ration schemes are designed to ensure access to a regular supply of basic staples at "reasonable" prices. The absolute transfer under a general ration is similar for *all* income groups. Thus, rations tend to be more progressive than general food subsidies.[12] However, as is true of general subsidies, ration schemes are often limited in coverage – especially of the poor in rural areas – because the infrastructure and distribution networks needed to implement them are lacking. In some parts of India as well as in other countries, including Bangladesh and Pakistan, the benefits of ration systems accrue disproportionately to urban consumers, despite the fact that, as noted earlier, poverty is largely a rural phenomenon in South Asia.

Food stamp schemes

A subsidy scheme that is similar to ration is food stamps, where the rations are measured in terms of nominal currency units rather than in commodity weights or volumes. There are, however, some important administrative differences in their functioning. Food stamps do not require the government to handle directly any commodities. They do, however, require that retailers accept a parallel currency and are able to redeem this currency conveniently.

The experiences with food stamps in Sri Lanka and Jamaica suggest that such schemes can function if they are well targeted. In Jamaica targeting was achieved by selecting well-defined needy groups such as pregnant and lactating women and children under five registered at health centres. However, even if the programme is broader in scope, as was the case in Sri Lanka, where stamps for food and kerosene were provided to families with self-reported incomes below a certain minimum, the financial burden of such programmes can be much smaller compared to the implementations of general ration schemes and other food subsidies. If the inflationary pressures cannot be kept under control through appropriate macro policies, the value of the food stamps can quickly erode, partially nullifying the beneficial effects on the poor.

Despite these two relatively successful experiments, the schemes in Colombia, Egypt, Peru and Venezuela ran into difficulties because of the problems in establishing the stamps as an alternative currency (Alderman, 1991). Taken together these experiences indicate that in the absence of sound macroeconomic management and the lack of a well-developed market network in rural areas, the implementation is likely to be difficult with the gains to the poor eventually eroding.

Supplementary feeding programmes

These are a form of highly targeted ration or in-kind transfer scheme. Their main objective is to reduce undernutrition. Subsidized or free food is distributed through schools, nutrition and health centres or community organizations for direct or home consumption to those deemed specifically vulnerable to nutritional and health risks. The beneficiaries usually comprise children under five, school children and pregnant and lactating women. Additional targeting on the basis of growth monitoring, location or income helps to identify the neediest members within these groups. Older children and adults are also fed in many emergency situations.

The Indian experience with two supplementary feeding programmes provides some useful lessons. Initiated in 1975, the Integrated Child Development Services (ICDS) programme aims to improve the nutrition and health of children 0–6 years of age by simultaneously providing supplementary feeding, immunization and curative medical care to children and pregnant and lactating women, and health and nutrition education to mothers. However, the primary emphasis in the ICDS programme is on providing meals. The results have been mixed, mainly because of difficulties faced in targeting the beneficiaries, strong urban bias and other reasons. By contrast, the Tamil Nadu Integrated Nutrition Project (TINP), initiated in 1980, is *area-targeted* (to the rural areas of six districts having the lowest calorie consumption in the state), *age-targeted* (concentrating exclusively on children 6–36 months of age), and *need-targeted* (depending on weight gain over a certain period). Since the children are on the supplementation programme only for the duration of time their weight gain is below standard, it is essentially a short-term intervention that endeavours to avoid fostering long-term dependence of beneficiaries on public assistance.

Home and community gardens

The production of secondary food crops in home and community gardens and small livestock and poultry rearing or fish farming can make an important contribution to improving household food security by improving food consumption, particularly during times of seasonal scarcity of food. In addition, this would also provide additional income to families from sale of surplus produce utilizing available family labour.

NOTES

1. Fields (1989). A more recent study on Latin America shows that periods of recession in the 1980s tended to be associated not only with increasing poverty but also with worsening income distributions, meaning that recessions hit the poor the hardest (Psacharopoulos *et al.*, 1992).
2. Evidence from other regions also indicates that small farmers' adoption of new technologies depend on access to credit markets, input supplies, extension, tenure security and problems of risk aversion. Rural women are particularly disadvantaged in these areas.

3. The inverse relationship refers to land productivity (physical yields or gross value of output per hectare) as opposed to total productivity and is, in fact, generally associated with a higher level of inputs per hectare on small farms, in particular labour per hectare. However, in countries where labour is abundant and land is the scarce factor of production, maximizing output per unit of land is of primary importance.

4. A landlord has the option of either cultivating the land himself with the help of hired workers or leasing it out to a tenant for fixed rent or for a fixed share of the output. Suppose first that the only kind of risk is in production. Output depends not only on the inputs, but on the weather. In the owner-operated system, the entire risk is borne by the landowner because the labourers earn a fixed wage and the owner earns the residual. In a fixed-rent system, the tenant bears the entire risk. Thus, given risk aversion, share tenancy may be preferred because it reduces the effects of the risk element in the decision-making process with respect to investment, input use, etc.

5. For instance, squatters on government land in Jamaica devoted half as much land to permanent and semi-permanent crops than did titled farmers. A third of recipients under a government titling programme moved away from short-cycle crops after the change in their status (Feder and Noronha, 1987).

6. A typical ROSCA is a group of 15–30 members who contribute a fixed sum weekly or monthly to a pool which is distributed among members in various, predetermined ways.

7. "An interlinked transaction is one in which two parties trade in at least two markets on the condition that the terms of all such trades are jointly determined" (Bell, 1988).

8. Price interventions have also been used as means of transferring resources between different sectors or interest groups, e.g. in the form of pan-territorial pricing schemes, as in several African countries, of income supports to the producers of selected products, or of taxation of, mainly, export commodities. From an efficiency point of view, alternative forms of intervention, such as lump sum transfers or asset redistributions may be more effective.

9. For an analysis of the experience in the maintenance of common property resources by villagers themselves in rural South India see Wade (1987).

10. However, other efforts to improve the ability of poor households to make better use of the resources available can also have significant nutritional benefits. These efforts include, for example, nutrition education, introduction of appropriate technologies, expansion of water and sanitation facilities, access to cooking fuels, etc.

11. Costs ranged from less than 1 percent of total public expenditure in Colombia during 1978–80 to 10–17 percent in Egypt between the mid-1970s and 1984. In Egypt, only about 20 cents out of each dollar spent reached those in the poorest quartile of the population (World Bank, 1990).

12. Experience in Sri Lanka after 1978 indicates that a targeted rice ration scheme covering the poorest half of the population benefited the poorest groups (bottom 20 percent of the population) much more than the general wheat and bread subsidy schemes implemented during the same time. A similar pattern of transfers is seen in the distribution of food grains through fair price shops in certain states in India. In Kerala in 1977, for example, the poorest 60 percent of the population received 87 percent of the foodgrains distributed.

CHAPTER 10

Human Resources Development in Agriculture: Developing Country Issues

10.1 INTRODUCTION

There is increasing evidence and recognition that what matters for development, more than natural resources and man-made physical capital, is the capability of people to be effective and productive economic agents, in short, human capital. In the particular case of agriculture, most studies on the subject establish that the education and skills of agricultural people are significant factors in explaining the inter-farm and inter-country differences in agricultural performance, along with the more conventional factors such as availabilities of land and water resources, inputs, credit, etc.

With the shrinking of per caput agricultural resources following demographic growth, with the agricultural labour force in the developing countries projected to continue at positive (though declining) growth rates and with the share of young people in the total also continuing to grow, the task of upgrading the literacy, the skills and other capabilities of the agricultural people is enormous, for coping with both the increases in numbers and the backlog inherited from the past. Moreover, the increasingly binding character of natural resource scarcities imposes severe limits on the extent to which production increases can be had through expansion of extensive agriculture. The generation and diffusion of technology and management capabilities for more intensive and modernized agriculture and supporting services become imperative. This can only be achieved through the upgrading of the quality of human resources employed in agriculture.

It is noted that many dimensions of the human resources development (HRD) issue are final end-objectives of development, e.g. literacy, better health and nutrition, etc. Although this chapter is concerned with policies to upgrade the quality of people to become more productive and more energetic economic agents, the need to make progress in literacy, health, nutrition, etc., as objectives in their own right, should not be lost sight of. This is important, since it implies that evaluation of returns to investment in these areas must take into account the value of improvements in literacy, etc., as increasing the welfare of individuals directly and not only indirectly through making them more productive economically. These considerations cannot but influence the criteria for making decisions concerning the allocation of scarce resources, e.g. between promoting basic education versus creation of more directly productive agricultural skills.

For practical purposes, this chapter does not cover the entire set of variables whose evolution determines HRD outcomes. In particular, it does not cover aspects of health, sanitation and nutrition (direct policy interventions to improve nutrition are discussed in Chapter 9). It rather focuses on those actions aimed directly at upgrading the productive potential of people making a living in agriculture. Section 10.2 presents the magnitude of the target population, now and in the future (the population economically active in agriculture). Section 10.3 focuses on basic education and agriculture. Section 10.4 discusses policies and actions to diffuse technical and management knowledge to the persons working in agriculture through the extension services. In both Sections 10.3 and 10.4 the historical developments and present situation are presented and discussed before discussing the needs and possible developments in the future. Section 10.5 highlights the important place of technical and professional education in agriculture in HRD itself and in development in general.

10.2 MAGNITUDE OF THE TASK

A first impression can be had by observing that the present population economically active in agriculture (PEA) in the developing countries of just over 1 billion is likely to continue to increase by some 13 percent in the next 20 years (Table 10.1). The growth rate is slowing down from 1.2 percent p.a. in the last 20 years to 0.6 percent in the next two decades, and indeed the PEA is about to peak in the regions of Latin America/Caribbean and East Asia. But this is not likely to happen in the two regions with the highest shares of their population in agriculture and with high incidence of rural poverty (sub-Saharan Africa and South Asia). This means that 20 years from now, these two sub-regions are still likely to have 60 percent of their labour force depending mainly on agriculture for employment and income. This contrasts with likely developments in the Latin America/Caribbean and Near East/North Africa regions which seem to be transiting towards patterns of labour force dependence on agriculture more typical of southern Europe.

Naturally, the human resource development effort has to provide for the entire agricultural, and indeed the rural, population, not only those classified as economically active. In particular, interventions in the areas of basic literacy, health and nutrition have to reach people well before they grow to become members of the PEA. The magnitude of the task can be appreciated from a few related parameters. In the first place, the numbers in Table 10.1 have to be

Table 10.1 Population economically active in agriculture (millions)*

	1970	1980	1990	2000	2010
All developing countries	790	923	1051	1130	1190
(% of total econ. active popul.)	(71)	(66)	(60)	(53)	(47)
93 Study countries	780	912	1039	1120	1180
(% of total)	(71)	(66)	(60)	(53)	(47)
Africa (sub-Sahara)	98	118	140	170	205
(% of total)	(81)	(76)	(71)	(66)	(60)
Near East/North Africa	31	32	35	38	39
(% of total)	(57)	(46)	(37)	(30)	(24)
East Asia	411	488	549	550	530
(% of total)	(76)	(71)	(63)	(55)	(47)
South Asia	203	235	275	320	365
(% of total)	(71)	(68)	(65)	(61)	(57)
Latin America/Caribbean	37	39	41	41	40
(% of total)	(41)	(32)	(26)	(21)	(17)

*The data, and in particular the projections, should be understood as indicative of broad orders of magnitude. They are, as far as possible, standardized for comparability among countries and regions. They may differ from those obtained from the routine labour force survey statistics. For discussion, see FAO (1986). Data by country are given in Appendix 3. The basis of these estimates is the historical data up to the early 1980s from ILO's work providing internationally comparable statistics. ILO is in the process of updating these data.

multiplied by a factor of 2.2 for the developing countries as a whole to obtain the estimates of the total agricultural population (a lower factor applies to East Asia, a much higher one to Near East/North Africa). Secondly, the age structure of the rural population implies that some 13 percent of the total, or some 350 million, are in the age group 15–24 years, a group commonly referred to as youth in the HRD programmes. Their numbers will be edging up towards 400 million in the future. Indeed, from the point of view of providing basic education services, these estimates will have to be more than doubled to account for children in the age agroup 6–15. Finally, the share of economically active women in the PEA is about 30 percent for the developing countries as a whole, but with wide regional variations, e.g. 56 percent in sub-Saharan Africa, 37 percent in Near East/North Africa, 31 percent in Asia, but only 12 percent in Latin America. It is obvious that the data referring to women are of great importance for focusing the HRD effort in the rural areas given the increasing recognition of the role of women-in-development in policy making in combination with the fact that past HRD policies have tended to favour men rather than women.

10.3 BASIC EDUCATION AND AGRICULTURE

Basic education, often referred to as literacy and numeracy education, is the most fundamental of HRD efforts, not only as a universal right of the

Table 10.2 Average social and private rates of return to education by region*

Region	Social			Private		
	Primary	Secondary	Higher	Primary	Secondary	Higher
Africa	27	19	14	45	28	33
Asia	18	14	12	34	15	18
Latin America	35	19	16	61	28	26

Source: Schultz (1988: 575).
*These rates of return are based on statistical associations between the market earnings and schooling of individuals, therefore they do not include other possible benefits such as the effects of education on the productivity of non-market time such as the time spent by farmers in own cultivation and the time spent by women in home production, on infant mortality and female fertility, etc. Private returns are typically the internal rate of return to investments made by individuals in their own education. The investments include both explicit (tuition fee, costs of uniform and books, etc.) and implicit (opportunity cost of time) costs of education. In calculating social returns, all costs of education, including public sector subsidies, are included on the cost side of the calculation. The social rate of return is lower because the same benefits (incremental income of the person receiving the education) are compared with total costs of providing the education, not only with those financed by the person concerned. In this case, the term "social" may be misleading because it does not include the benefits of education of a given person accruing to other persons and society at large (externalities).

individual but also as the foundation for any further initiative in human resource development in agriculture designed to improve agricultural production and, hence, incomes and welfare. Basic education can improve significantly the efficacy of training and agricultural extension work which in turn affect agricultural production through: (a) enhancing the productivity of inputs, including that of labour; (b) reducing the costs of acquiring and using information about production technology that can increase productive efficiency; and (c) facilitating entrepreneurship and responses to changing market conditions and technological developments (Schultz, 1988). The relationship between education and agricultural development cuts both ways and the two are mutually reinforcing, with demand for schooling rising as rural incomes increase.

An analysis of 37 sets of farm data from developing countries showed that farmers completing four years of elementary education had higher productivity by, on average, 8.7 percent (Jamison and Lau, 1982). The same authors estimated social returns to investment in rural education of 7–11 percent in Korea (Republic), 25–40 percent in Malaysia, and 14–25 percent in Thailand, under various assumptions. A recent review of research findings for Asian countries comes to similar conclusions (Tilak, 1993). But the most extensive studies related to rates of return to education at different levels have been conducted using data at the national level. A summary of the findings of these studies is presented in Table 10.2. The rates of return are highest for investment in primary schooling in all regions for which there was information. Since rates of return on public investment in most other sectors are commonly well below those presented in Table 10.2, there is a strong prima facie case for strengthening public provision of education, including the reallocation of funds within the total education budget in favour of primary education. But the relative emphasis to be attached to the different levels of education will vary among countries and over time, depending on the level of technology used in agriculture. The higher the actual or potential technological environment (use of MVs, agrochemicals, irrigation), the higher the required level of education, which must increasingly focus also on the need for sustainable use of resources, aspects of nutrition, health, etc.

Public expenditure on education in developing countries has climbed from 2.9 percent of GNP in 1970 to 4.1 percent in 1988 (UNESCO, 1991). However, the improvements have not been uniform across different regions, e.g. sub-Saharan Africa and Latin America and the Caribbean experienced reversals in the 1980s. Nor have figures been that impressive for the least developed countries when real expenditures on education are adjusted for population growth. Notwithstanding the reported declines in expenditure in some regions, the enrolment rates in all regions seem to have consistently improved, as evidenced by the increases in expected years of school enrolments.[1]

Nevertheless a large proportion of the adult population, both female and male, in rural areas in many developing regions continues to be classified as illiterate (Figure 10.1). Apart from the factors already noted, possible misallocations of public resources within the education sector, relatively long gestation periods required for training teachers, and relatively higher unit costs of providing educational services in the rural areas may have also contributed to the persistence of high rural adult illiteracy rates. The relevance of the first factor can be illustrated by considering the fact that despite the relatively lower social rates of return to higher education, the "developing countries as a group spend over 25 times as much per student for the 7 percent of the school-age population enrolled in higher education as for the 75 percent in primary education" (FAO, 1991e). Moreover, since the education system produces its own main input, the rate at which the system expands is limited by the capacity of the system to produce teachers. Thus in the initial phases of developing the education system, the rate at which educational services expand can be slow and costly. The fact that rural populations tend to be spatially more dispersed exacerbates these problems and increases the unit costs of providing educational services in rural areas.

The situation of access to education of rural women deserves a special mention. The importance of women's labour in planting, cultivation, weeding, harvesting and processing of food, in feeding their families and in rearing children, brings into sharp focus the urgency of improving women's access to educational services. Despite the increases in enrolment rates for women, in line with overall increases mentioned above, "in the low- and middle-income

Percentage

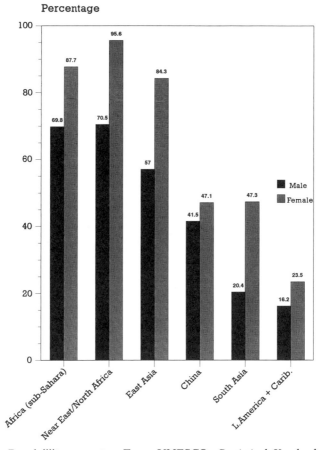

Figure 10.1   Rural illiteracy rates. From UNESCO, *Statistical Yearbook* for those countries for which data are available. Data do not necessarily correspond to the same year for each country

countries as a group, there were still only 81 females per 100 males in primary school and 75 females per 100 males in secondary school in 1987. In sub-Saharan Africa, there were only 77 and 59 females per 100 males in primary and secondary school, respectively, with the lower number of females relative to males in school reflecting both fewer female entrants and a higher dropout rate among women" (FAO, 1991e). Indeed, these disparities are even larger in the rural areas.

The reasons for the gender disparity in enrolments (and educational attainments) are both cultural and economic. Tradition often demands special concern for the privacy and social reputations of women. In cultures where female seclusion is practised, the impact of that tradition on girls' enrolment after puberty is substantial. These concerns prevent parents from sending girls to school, unless schools are located close to home, well supervised and served by female teachers. When parents themselves lack education, they are more reluctant to challenge tradition to educate their daughters. Traditional constraints tend to be far more severe in rural areas.

Among economic reasons, the high opportunity costs of sending girls to school weigh heavily in household decisions. These costs include chore time, children's forgone earnings, and – especially for girls – mothers' forgone earnings. The opportunity costs of sending girls to school are likely to be higher for poor families in rural areas, since they tend to make a greater contribution to family welfare.

10.4 AGRICULTURAL EXTENSION AND TRAINING

Agricultural extension "assists farm people, through educational procedures, in improving farming methods and techniques, increasing production efficiency and income, bettering their levels of living and lifting the social and educational standards of rural life" (FAO, 1984a). Publicly supported agricultural extension services for farm people are an innovation of the twentieth century. For example, the USA established its Cooperative Extension Service in 1914.

This HRD innovation in agriculture has been spreading in recent years. Out of 198 extension organizations in 115 countries which provided reports to FAO in 1989, only 10 percent had been established before 1920, while 50 percent had been established after 1970 (Swanson et al., 1990). The increasing adoption of organized extension services in the developing countries reflects the realization of their importance for agricultural development and the high rates of economic returns that have been demonstrated by the experience of countries where extension has been properly delivered. For example, a study reports that in the USA a $1000 increment in extension spending was associated with a $2173 increase in farm output within a two-year period (Evenson, 1982b). Comparative studies from several countries provide further support to the finding of relatively high economic returns to investment in agricultural extension (Evenson and Kislev, 1975; Evenson, 1982b; Feder et al., 1985). More recently, studies on productivity-increasing effects of agricultural extension services have been reported. A report on World Bank support to agricultural extension services in 22 sub-Saharan African countries is illustrative. An average of 40 percent increase in yields in the first year has been recorded in a 1989–90 study (World Bank, 1992b).

The findings of the recent Global Consultation on Agricultural Extension of FAO indicate that the rate of return to investment in extension is influenced by a variety of factors. These range from the economic value of the farm product, with cash and export crop farmers enjoying higher returns than food crop farmers; the general economic climate, with returns being lower in relatively poorer agricultural nations; to the nature of the extension services provided, with larger returns to those systems that embrace large numbers of farmers and have lower costs per farmer (Contado, 1990). More generally, it is recognized that an effective extension system cannot be considered on its own and that it "needs a supportive environment that includes a long-term commitment to agricultural growth expressed through the provision of adequate agricultural support services – of which extension is but one – and macro-economic policies that, at a minimum, do not disfavour agriculture" (Hayward, 1990).

Scope of extension effort worldwide

Agricultural extension services in the world have been expanding during the past three decades. Around 1959, there were approximately 68 organized extension services, with as many as 180 000 agricultural extension personnel. By the year 1980, the number of organized extension services had increased to around 150, with a total personnel of about 350 000 (Evenson, 1982a). The estimated extension workers in 1989 stood at approximately 600 000, of whom nearly two-thirds were located in developing countries. A least developed country like Mozambique, for example, had in 1989 about 350 professional/technical staff in extension following the establishment of a national extension service in 1986 with UNDP/FAO assistance. The data collected for FAO's Global Consultation indicate that agricultural extension expenditure was approximately US$4.6 billion in the 98 countries for which data were available, of which nearly 87 percent was in developing countries. If it had been possible to include all countries of the world, the estimated total expenditure on extension would probably have exceeded US$6 billion per year (FAO, 1991d).

In spite of the tremendous increase in the numbers of agricultural extension workers in the last three decades, the actual coverage of agricultural extension services in the developing countries has been limited. In the USA, Canada and Europe, one public extension agent covers about 400 economically active persons in agriculture, even before counting the services of private sector extension agents. In the developing countries an extension worker covers on average about 2500 such persons. This suggests that in actual practice, only one out of every five economically active persons in agriculture receives extension services in the developing countries. This rate is likely to be lower when one considers that about one-fourth of the extension worker's time is devoted to non-educational duties, which was equivalent to approximately 140 000 full-time years of extension workers' time in 1989 (FAO, 1990a).

Another issue is the kind of farmers served by the extension agents. Data from the Report of the FAO Global Consultation on Agricultural Extension show that in the reporting developing countries, 6 percent of extension agents' time and resources is devoted to large commercial farmers, 26 percent to smaller commercial farmers, while 24 percent is devoted to subsistence farmers and 6 percent to farm women. In a well-documented case study of the extension programme in two provinces of Turkey, however, the record indicates that 100 percent of the 5100 large-scale farmers were served by the extension service, while only 55 percent of the 62 300 small-scale farmers were receiving extension services. Of the 17 900 medium-scale farmers, 90 percent were receiving services from extension (Contado and Maalouf, 1990).

The preceding paragraph illustrates the problems associated with public extension services. The private sector has varying degrees of involvement in agricultural extension work in most countries. In developed countries, the trends towards privatization relate to budgetary problems (Legouis, 1991). In the developing countries, the need to increase coverage and contain the costs of public extension are the main driving forces behind the increasing involvement of the non-governmental organizations (NGOs) and the private sector in extension (Maalouf et al., 1991). For example, in Colombia, 35 percent of the 2315 extension agents are provided by an NGO – the National Federation of Coffee Growers. In Uganda, 7 percent of the 2040 agricultural extension agents are provided by a private company (FAO, 1991d). However, at the end of the 1980s, worldwide, NGOs and private firms providing agricultural extension services constituted only 7 percent and 5 percent, respectively, of the total number of extension agencies (data for 113 countries covering 186 extension organizations; FAO, 1990a). Furthermore, their coverage in the developing countries tends to be small and concentrated among cash crop farmers. But with the move towards structural adjustment policies and privatization of production enterprises and services, there has been an increasing involvement of the private sector in providing extension services. Even if this involvement is concentrated on large commercial farmers and commodity producers, it could result in freeing public funds which could then be channelled to other farmers.

Issues for the future

As noted, the population economically active in agriculture (PEA) will continue to grow in the developing countries, to about 1.2 billion by year 2010. There would be a need for over 2.4 million extension workers in order to provide effective extension coverage, since a ratio of 500 economically active persons in agriculture to 1 extension worker is deemed to be the upper limit for providing effective extension service. If the rate of increase from 1980 to 1989 continues, there could be 2.1 million extension workers by 2010, which would be very close to the requirements. There are, however, two factors that may hinder the attainment of this projected number: first, the considerable contribution of China to the increase of extension workers during 1980–89 might not be repeated in the next two decades; second, in Africa the trend has been for the resources devoted to extension by the Ministries of Agriculture to decline (25.6 percent of total budget of the Ministries of Agriculture in 1980,

22.3 percent in 1985 and 18.8 percent in 1988; FAO, 1990a). But there are also countervailing factors. One is the increasing allocations for extension by Ministries of Agriculture in Asia and the Pacific, Near East and Latin America and the Caribbean. Another is the trend towards expanding the role and share of the private sector (NGOs and private commercial firms) in agricultural extension services. A third factor is the expansion of agricultural development schemes where farmers pay for agricultural extension services through commodity levies such as the Rubber Industries Smallholders Development Authority (RISDA) and the Federal Land Development Authority (FELDA) in Malaysia, the National Federation of Coffee Growers in Colombia, etc. A fourth factor is the increasing number of countries adopting a partnership in funding extension services between the central government and the local government as in the case of China, Poland, etc. Finally, there will be increased availability of people trained in agriculture for extension work in most developing countries as the intermediate and higher level schools of agriculture established in the 1960s and 1970s mature and turn out more graduates in the years towards 2010.

Another critical problem in HRD in agriculture is the gender issue. In developing countries, an important proportion of farm work continues to be done by women but only 17 percent of agricultural extension workers are women. If the proportion of women extension workers remains constant over the projection period, there will be approximately 330 000 female extension workers by the year 2010. But by giving stronger recognition to the role of women in agriculture, increasing the number of female students in agricultural schools and colleges and increasing resource allocation for extension services directed to women farmers, it may be possible to raise the proportion of women extension workers to 20 percent of the total in the developing countries.

The low level of training of a large proportion of extension workers is another issue for developing countries that must be addressed in the future. Given the increasing number of middle-level and college agriculture graduates in many developing countries, it is probable that the older high-school trained extension workers would be replaced gradually by persons with higher educational qualifications. This is already happening in many countries of Asia, Latin America and the Near East, where the proportion of low-level extension workers could decline from 40 percent in 1988/89 to perhaps 20 percent by the year 2010.

10.5 TECHNICAL AND PROFESSIONAL EDUCATION IN AGRICULTURE

The amount and quality of trained technical and professional manpower in agriculture are critical factors, both in agricultural development and more general HRD. This "human capital" is relatively scarce because training takes years and is costly. However, investing in technical and professional education has a high multiplier effect when trained personnel are properly employed as extension agents, trainers, researchers, programme managers, policy makers and in the private sector.

Although many developing countries still have serious shortages of trained manpower in fields related to agriculture, considerable progress has been made during the last three decades. By 1983, for example, there were over 400 000 trained agricultural personnel in 46 countries in Africa. In 25 of these countries, moreover, the existing institutional capacity was sufficient to allow the training of the required number of agricultural personnel for the year 2000 (FAO, 1984b). Worldwide, increases in institutional capacity for training are reflected in the increased number of extension personnel mentioned earlier, as well as of agricultural research personnel. ISNAR reports that agricultural research personnel in developing countries increased at the rate of 7.1 percent annually from 19 753 to 77 737 during 1961–65 to 1981–85 (ISNAR, 1992).

However, when these numbers are compared with requirements, especially in the least developed countries, there are still considerable shortages. For example, Ethiopia would have to graduate 231 people at the professional level and 1254 at the technical level annually to reach the minimum estimated requirements for trained manpower in agriculture by the year 2000. As mentioned earlier, the problem in agricultural extension is a shortage of well-trained extension agents in many developing countries. In the case of research, UNESCO data show that there were approximately 500 scientists and engineers per million population in a sample of developing countries as compared to more than 3000 scientists and engineers per million population in developed countries (UNESCO, 1991).

For most developing countries, the supply of technical and professional manpower in agriculture for the next decades will remain problematic. In Africa, 18 of 46 countries surveyed in 1983, had reported that their technical agricultural personnel was less than 50 percent of the year 2000 minimum requirement. Even in those developing countries where, in general, numbers of technical and professional manpower met the minimum requirement, the problem is an excess of numbers in certain fields and shortages in others. For example, in Africa only 7 percent of technical and professional trained manpower in agriculture are in forestry, 5 percent in fisheries and 11 percent in livestock (FAO, 1984b).

The major problems of developing countries in the area of agricultural education and training as they face the new century include inadequate institutional capacity, relatively low level of public and private support to agricultural education, and limited resources and experience to cope with new areas of training in agriculture, i.e. environment and natural resource management, biotechnology, farming systems management and agri-business.

NOTE

1. Based on data provided in the World Bank's *World Development Report–1984*, Schultz estimated that over the period of 1960–1981 the expected years of school enrolments in low-income countries had increased from six to eight years, with greater improvements for middle-income countries. The apparent "paradox" of increased enrolment with lower public expenditures in low-income countries is explained in terms of: (a) declining quality of schooling per student; (b) declining unit costs of production of educational services of a constant quality relative to the general price level; or (c) errors in the underlying data (Schultz, 1988: 552–7). It must, however, be stated that data referring to the 1980s indicate temporary reversals also in enrolments in most regions.

CHAPTER 11

Pressures on the Environment from Agriculture

11.1 INTRODUCTION

Chapter 4 sketched out the growth prospects and the main underlying parameters for the crop and livestock sectors in the developing countries. These prospects imply further intensification of land and water use. More mineral fertilizers and, to a lesser extent, pesticides will be used in the future. The progressive introduction of more environment-friendly technologies can only moderately attenuate growth in pesticide use in the next 20 years, but will not reverse past trends for growing use (see below). The projected growth path may be challenged therefore on a number of grounds concerning its environmental impacts and sustainability. Two related issues stand out in this regard.

First, what are the technological options for putting agriculture on to a more ecologically sound pathway while at the same time not only achieving the projected output, but also laying the foundation for sustainable agricultural development in the longer term? Second, given the inevitability of some trade-offs between the environment and development in the medium-term future, what other actions are required to minimize them and ensure progress towards the objective of sustainable agriculture and rural development? These issues were at the heart of the debate at the 1992 UN Conference on Environment and Development (UNCED) and at the more technical meetings leading up to it, notably the 1991 FAO/Netherlands Conference on Agriculture and the Environment.

The first issue on technology is dealt with in Chapter 12, while Chapter 13 examines the wider, and in the main complementary, policy and institutional actions that are needed at the national and international levels to avoid or minimize the environment and development trade-offs. Examples of such actions are those for resource use planning, infrastructural development, farmer support services, and the wider ones affecting overall development and international economic relations. Several of these issues have been discussed in previous chapters. It remains for this chapter, which deals primarily with the developing countries, to describe and where possible quantify, the agricultural pressures on the environment that are implicit or explicit in the projections of the probable growth paths of the crop and livestock sectors.

11.2 PRESSURES ON LAND AND WATER RESOURCES

Competition for land and water

Competition for *land* between sectors and production systems is projected to intensify. It is expressed most accurately in the expansion of the use of land for arable and tree crops, shifting cultivation and grazing of livestock and its conservation under forest. Then there is the competition between crop and livestock production and, on a much smaller scale, between crop production or mangrove swamp preservation and aquaculture; and, as noted in Chapter 5, there will be further pressures on the forest for timber and fuelwood extraction. Finally, increasing population and economic growth will contribute to further diversion of land to human settlements and infrastructure.

Forest land loss to agriculture is primarily a tropical rather than a temperate zone problem (see Chapter 5). In the former, deforestation rates are currently about 15.4 million hectares a year, of which a large part is thought to result from extension of grazing and cultivation, particularly shifting cultivation, a considerable proportion of which will eventually revert to bush and then tree fallow. In the temperate zone, shifting cultivation is no longer a significant feature of agricultural production systems and in many countries net afforestation, either naturally or through plantations, is taking place. For example, China has a considerable programme of afforestation (People's Republic of China, 1992).

It is possible that the rate of deforestation would slow down for a number of reasons. First, the lagged effect of recent policy improvements: for example, those to remove incentives and tax distortions favouring deforestation or to strengthen the controls on logging practices. Second, the increasing scarcity of forest land in the right locations suitable for arable cultivation. Third, the projected slower growth of the population dependent on agriculture in the developing countries. Fourth, technological improvements which meet agricultural demands through land intensification rather than land expansion. Fifth, the adoption and implementation of sustainable agriculture and rural development policies and programmes. And sixth, policy changes in some financing institutions which are adopting more rigorous environmental impact assessments of investment projects.

These reasons are not, however, going to be sufficient to reduce the pressure altogether. Even if current rates of deforestation are considerably reduced, significant areas of tropical forest are likely to be converted to some form of agricultural use during the next 20 years or so. As noted in Chapter 4, some 90 million ha of additional land may come under crop production by year 2010 in the developing countries, excluding China. This is a small part of the 1.8 billion ha of land with crop production potential not used now for agriculture. However, it is estimated that at least 45 percent of the latter land overlaps with forest and the real overlap is probably much larger.

This continued pressure highlights two priority actions which are considered in Chapters 12 and 13: greater emphasis for agricultural research on sustainable alternatives to shifting cultivation; and a more holistic strategy for tropical forest conservation such as that sought by the FAO Tropical Forests Action Programme, since current forestry policies have tended to treat the deforestation problem as a forestry problem in isolation from a better understanding of the reasons for agricultural intensification and as a consequence have commonly failed. In addition, emphasis on wider rural development and agrarian reform could contribute to the slowing down of migration into marginal, environmentally fragile areas.

Crop production in dryland areas (land class 'dry semi-arid' in Table 4.4) is projected to expand by 6 million ha, a relatively small figure. However, much of this land is currently rangeland and its conversion carries the risk of either increasing grazing pressure on the remaining pastures or displacing livestock on to even more marginal land, which must be managed carefully if degradation is to be avoided.

The competition of aquaculture for land may not be very significant quantitatively, because the areas involved are generally quite small. It can, however, be highly important qualitatively, because in some countries the land being developed is relatively unique mangrove swamp, with potentially valuable biogenetic resources and an ongoing role as a breeding ground for coastal fisheries. No projections have been made of the total area at risk, but specific studies for key localities suggest that major losses could continue during the next two decades unless greater protective measures are taken.

Finally, moving on to the competition between agriculture and human settlements, including urban/industrial and infrastructural development, there are a number of unknowns that make it difficult to be emphatic on the magnitude of the pressure, yet there are clear grounds for adopting the precautionary principle, which is embodied in the Rio Declaration.

Between 1990 and 2010, the population of the developing countries is projected to increase by about 1.8 billion, but there is great uncertainty about how much land these people will occupy. Economic growth, industrialization and continuing urbanization will further increase pressures for expansion of land under human settlements. The tentative and very speculative projections presented in Chapter 4 place the additional land to be occupied by human settlements in the period to 2010 in developing countries (excluding China) at about 34 million ha, of which nearly 20 million ha would be land with agricultural potential (Table 4.3). This is only a small fraction of the 1.8 billion ha of land with agricultural potential.

Thus, for the developing countries as a whole, land loss to human settlement appears not to be a substantial threat – a modest improvement in the productivity of the existing land or a small addition from the land not yet used for agriculture could compensate for the loss. But the aggregate picture is misleading (Norse *et al.*, 1992): first, because some countries have little or no additional land; second, much of the urbanization tends to be in areas with high quality soils, while the land not in agricultural use tends to have poorer quality soils; third, the population of the developing countries will continue to grow well into the second half of the twenty-first century, causing the possible loss of further substantial areas, which once built on would be lost for ever. The situation in coastal areas of many small island countries is particularly critical because of the pressure on land from tourism. And of course the problem does not end here, because expansion of human settlements is not the only factor causing land loss. Degradation, for example, is causing both land loss and reduced productivity over a far wider area (see below). Long-term global warming and climate change also could threaten as much as one-half of the high quality land resources of some countries through sea level rise or deterioration in agroecological conditions, e.g. in Bangladesh or the Gambia. These facts are put forward as justification for the above conclusion that the precautionary principle should be applied so as to minimize the loss of land to human settlement, and this notwithstanding the contrary possibility that some developing countries will be able to follow the pattern now exhibited by many developed countries, namely the reduction in arable area as a consequence of successful intensification. The zoning and other land use planning policies required are discussed in Chapter 13.

The combined effect of population and economic growth will exert even greater pressures on *freshwater* supplies than they will on land (Falkenmark and Widstrand, 1992). Chapter 4 indicated that technological growth would generally continue to make it possible to increase agricultural production with relatively modest expansion of the land in agricultural use. This, however, has not been the experience to date with water consumption, and major improvements in water use efficiency are unlikely in the medium term. Though technological progress has raised water use efficiency in a few areas, it has been insufficient to compensate for income growth and wasteful consumption patterns which collectively can cause a manifold increase in per capita water demand for non-agricultural uses (Table 11.1). The future need not be like the past, but technological improvements and changes in consumption patterns seldom take less than 15 years to have an appreciable effect, and generally the time lag is much longer.

These facts have serious implications for the next two decades and beyond. Africa and Asia already show a worsening shortage in per capita freshwater availability, although much of Latin America is well endowed (Table 11.2). Many countries are already closer to their water supply limits than to their land limits, and the need to increase the agricultural production will accentuate pressures on the water resources. Three aspects are of particular importance.

Table 11.1   Sectoral water withdrawals by country income groups

| Income groups | Annual withdrawals per caput (m³) | Withdrawals by sector | | |
		Agriculture (%)	Industry (%)	Domestic (%)
Low income	386	91	5	4
Middle income	453	69	18	13
High income	1167	39	47	14

Source: Reproduced by permission from World Bank (1992a: 100), based on data from the World Resources Institute.

First, food supplies in the developing countries are already heavily dependent on irrigated cereals production, which accounts for about one-half of the total production of cereals. This dependence is projected to grow somewhat, in spite of the high cost of irrigation in some countries and the pressures to remove the subsidies (hidden or open) on existing irrigation, which will become increasingly difficult to justify on economic grounds.

Second, the growing water demand for irrigation and the rising industrial and domestic demand will increase competition for water and could push up its price beyond a level that is profitable for staple food production in some areas. Agriculture is the dominant water user, accounting for nearly 70 percent of total consumption of managed water resources, compared with about 21 percent and 6 percent for industrial and domestic use respectively. The latter users are commonly in the situation where they can only expand their consumption of water by "taking" it away from agriculture, and can generally afford to pay more for it than agriculture. Third, overextraction of groundwater is a growing problem in many areas. It is most acute in the Near East, where it is leading to salt-water intrusions that ultimately make the water unsuitable for crop production. But it is also a problem in large areas of South Asia, where food security is heavily dependent on irrigation. Over-pumping in these areas is causing water levels to fall beyond the reach of shallow tubewells with the risk that irrigation may eventually become too expensive or physically impractical (see Box 11.1). Finally, the supply problem will be intensified by degradation of existing irrigation systems to the point that they have to go out of use, and deterioration of water quality (see below).

Degradation of land and water resources

The effects of existing degradation, notably erosion, nutrient mining, salinization of soils and contamination of water, are included implicitly in the production analyses of this study because the base year yields and fertilizer response ratios reflect the productivity effects of such damage. Less tangible problems, like desertification and forest degradation, are not included.

Box 11.1   Groundwater mining

Much of the successful expansion of irrigation in recent decades has come from the exploitation of groundwater using tubewells. They have the advantage of small scale, low cost and rapid construction without the loss of fertile land and the destruction of human settlements commonly associated with large-scale, gravity-fed schemes based on reservoirs. The expansion has been very rapid. In India alone, the number of tubewells jumped from nearly 90 000 in 1950 to over 12 million in 1990. Behind this success, however, is the neglect of the fact that agricultural development based on groundwater is unsustainable, when it uses "fossil" water or when extraction rates exceed rates of recharge.

The rapid expansion of tubewell irrigation has put extreme pressures on what is commonly a static resource because natural rates of recharge are low. Moreover the problem has been intensified by pressures that are generally some distance from the site of extraction, largely through deforestation in upland watersheds, or overgrazing and other forms of land degradation that accelerate runoff and reduce rainfall infiltration. Consequently, water tables are dropping and causing a wide range of environmental, economic and social problems. Saline intrusions are becoming a problem in many coastal regions. Over-pumping has led to increased investment or operating costs as falling water tables have necessitated deeper wells and greater energy consumption for pumping. In some instances poor farmers without the capital to deepen their wells have had to revert to rainfed production. In others the necessary adjustments have been too late and the land has been desertified, as in some places in India.

The magnitude of the current extent and intensity of degradation has recently been estimated using a standardized methodology – the Global Assessment of Soil Degradation (GLASOD) – and is more defensible than earlier calculations that pooled the results of diverse national and regional assessments without common definitions and methodologies (ISRIC/UNEP, 1991). Soil erosion is by far the most widespread cause of degradation, with water erosion being the principal agent (Table 11.3). Of greater consequence still are the estimates of the intensity of degradation. Nearly 1 billion ha of arable land in developing countries are estimated to be so degraded that productivity is being moderately or severely affected. Some 9 million ha worldwide, of which 5 million ha are in

Table 11.2   Per caput water availability by region, 1950–2000 ('000 m^3)

Region	1950	1960	1970	1980	2000
Africa	20.6	16.5	12.7	9.4	5.1
Asia	9.6	7.9	6.1	5.1	3.3
Latin America	105.0	80.2	61.7	48.8	28.3
Europe	5.9	5.4	4.9	4.4	4.1
North America	37.2	30.2	25.2	21.3	17.5

Source: from FAO (1993d), based on Ayibotele (1992).

Table 11.3   Soil degradation by type and cause (classified as moderately to excessively affected) (million ha)

	Water erosion	Wind erosion	Chemical degradation	Physical degradation	Total (million ha)
Regions					
Africa	170	98	36	17	321
Asia	315	90	41	6	452
South America	77	16	44	1	138
North and					
Central America	90	37	7	5	139
Europe	93	39	18	8	158
Australasia	3	—	1	2	6
TOTAL	748	280	147	39	1214
Major causes	(%)	(%)	(%)	(%)	
Deforestation	43	8	26	2	384
Overgrazing	29	60	6	16	398
Mismanagement of					
arable land	24	16	58	80	339
Other	4	16	10	2	93
TOTAL	100	100	100	100	1214

Source: Adapted from ISRIC/UNEP (1991).

Africa, have had their original biotic functions fully destroyed and reached the point where rehabilitation is probably uneconomic.

More recent information suggests that the estimates of the extent of land degradation and its effects on productivity as reported in GLASOD have been exaggerated in a number of cases. At the same time, however, the effects of land degradation are often masked by the compensating effects of improvements in agricultural technology or increasing applications of plant nutrients so that yields may have been increasing even on degrading land. Also in some cases, farmers and rural communities in some areas of Africa and Asia have demonstrated that land which GLASOD classifies as unrestorable can in fact be rehabilitated. Thus, for example, farmers in Kenya (see Chapter 12) and China have rehabilitated formerly abandoned or heavily degraded land at modest cost given appropriate incentives and technologies, though the conventional approaches assumed by GLASOD would probably have been uneconomic.

The production projections assume some changes in national policies and farm practices to correct part of the soil/water degradation, though the full benefits of such changes will not be felt in the short term. Thus some degradation is assumed to continue. However, projections of future degradation and its consequences, particularly its impact on productivity, are very difficult to make as discussed below.

Soil erosion

There is widespread evidence of erosion resulting in losses greatly in excess of 50 tonnes of soil per hectare per year, losses that may be five or more times the natural rate of soil formation. The impact, however, of such losses on crop yields or production has not been well established in physical terms though there have been many attempts to do so. The relationship between erosion and productivity loss is more complex than previously thought, as is that between man-made and natural erosion. Similarly, the experimental techniques once used to quantify these losses are less effective than previously thought. The relationship between soil erosion and yield loss is not linear, i.e. it is not directly proportional to the thickness of the layer lost, nor to the type of particles lost. Moreover, the impact on soil structure, particularly the impact on the size of the air spaces, can be more important than that on its chemical composition. Mitchell and Ingco (1993) and Crosson (1992) report that a number of studies on the relation between land degradation and losses in crop yields conclude that such losses are probably of minor importance, being in the order of 2 to 4 percent over very long periods (up to 100 years). All these studies, however, refer to the generally deep, fertile soils in North America. Their conclusions cannot be simply transferred to other regions, mainly because tropical soils and biological processes in them can be very different from those in the temperate zone soils. Where soil fertility is mainly concentrated in the surface layers, soil erosion can lead to greater yield losses than those estimated for the temperate zones.

It is also possible that yield loss in one area may be compensated by gains further down the slope, valley or plain, where the soil is eventually deposited. Though here again the issues are not straightforward: first, because deposition may also have negative external consequences, e.g. where it silts up reservoirs and irrigation canals and reduces their effective life (Table 11.4); second, because man has commonly been blamed for much of the silt load of rivers, whereas it is now considered that a substantial proportion results from upward and ongoing movements in the earth's crust. In China, for example, whereas the severe erosion of the loess plateau was once attributed largely to man's activities, and is still presented in these terms by some observers, it is now thought that over 60 percent of the erosion is due to such movements.

Soil nutrient mining

This is an issue brought into prominence by the ongoing debate on environmental accounting. The shortening of fallows and prolonged crop harvesting without adequate technological responses to replace the soil nutrients taken out by crops with organic or mineral fertilizer inputs, leguminous crops, nitrogen-fixing algae and so on (see Chapter 12), is lowering the nutrient status of soils and the actual or potential crop yields. It

Table 11.4   Siltation of selected Indian resevoirs

Reservoir	Assumed rate (acre-feet p.a.)	Observed rate (acre-feet p.a.)	Expected life as % of designed life
Bhakara	23 000	33 475	68
Maithon	684	5 980	11
Mayurakshi	538	2 080	27
Nizam Sagar	530	8 725	6
Panchet	1 982	9 533	21
Ramganga	1 089	4 366	25
Tungabhadra	9 796	41 058	24

Source: Data from Central Water Commissioner, Ministry of Water Resources, Government of India.

consequently threatens the sustainability of agricultural production. Part of the problem is often masked by the gains from unbalanced fertilizer use, which although it raises yields, also introduces an additional economic cost because the lack of balance lowers the technical efficiency of the mix of fertilizer nutrients (Twyford, 1988).

The situation is serious in many countries but most acute in sub-Saharan Africa, both for major nutrients like nitrogen and phosphate, and for micronutrients like boron and manganese. In the mid-1980s all countries in the region were estimated to be suffering from nutrient mining to a greater or lesser degree (Smaling, 1993), with the most serious problems occurring in semi-arid areas where livestock manure is in short supply and the use of mineral fertilizers is seldom economic. The projected situation shows some improvement but the time horizon to 2010 is too close to remove the main gaps in technology and infrastructure. These constraints are examined in Chapters 12 and 13 respectively. In particular there is the need for better and less costly integrated plant nutrient systems and improved transport and marketing systems to lower relative mineral fertilizer prices, and provide incentives for more sustainable agricultural practices.

Salinization of soils

This is primarily a problem of irrigated areas, but also occurs in hot dry zones where strong evaporation brings salts to the surface. In irrigated areas it is usually the consequence of bad design causing poor drainage, and/or inadequate maintenance and inefficient management leading to excessive application rates, and seepage from water courses. The end result is waterlogging, salinization, depressed crop yields and eventually, if corrective action is not taken, loss of land for agriculture. This leads to physical pressures on the finite resource base if land is permanently lost, which according to some estimates may vary in the range 0.2–1.5 million ha per year worldwide, while some 10 to 15 percent of irrigated land is to some extent degraded through waterlogging and salinization.

The production projections of this study assume a net increase in irrigation of 23 million ha in the developing countries, excluding China (Table 4.5). They also assume that appropriate measures, described in the next two chapters, will be taken so as to prevent further losses of existing irrigated land through salinization. There is some validity for such an assumption in that there is now greater emphasis on improved water management and drainage, but the required investments and institutional changes will take a number of years to mobilize. As noted in Chapter 4, gross investment in irrigation would need to be much above that implied by the net addition of 23 million ha, in order to account for the rehabilitation or substitution of irrigated land subject to degradation.

Desertification

This can be broadly defined as land degradation in dryland areas.[1] Much of the attention has been on Sudano-Sahelian Africa where the deserts were once reported to be steadily advancing, though such reports are now challenged (see below). It is not, however, just an African problem. All the major continents face desertification problems and the area of cropland and rangeland prone to desertification is estimated at 30 percent of the world's land surface.

A major shift in scientific thinking on this topic has occurred in recent years, with a growing consensus for the view that the area affected by desertification has been greatly overestimated (see, for example, Nelson, 1988; Warren and Agnew, 1988; Bie, 1990). FAO has placed particular emphasis on the weakness of the methodology used to produce some of the more extreme estimates. It is now recognized that drylands are much more resilient to drought and to man's abuse than previously thought.

Although man's role in desertification is still not well understood, there is little doubt that soil nutrient mining and the overcultivation of fragile soils does lead to dryland degradation and desertification. The projections of Chapter 4 point to increasing pressures of this type. Mitigation of these pressures will be dependent on improvements in farm practices, such as soil moisture conservation, and the development through research of leguminous live mulches and other techniques (see Chapter 12).

Water contamination

The future water supply is threatened both by the quantitative constraints examined above, and qualitative problems arising within agriculture which are a threat to crop yields and to human, livestock and wildlife health. The principal threats of agricultural origin are the following: rising salt concentrations in irrigated areas; fertilizer and pesticide contamination of surface and groundwater; and discharges of organic effluents from intensive livestock units and fish farms. All of them are projected to increase, because of the long length of time required to achieve appropriate corrective actions.

The intensification of irrigation could raise the degree of water reuse, and hence the build-up of salt concentrations, with risks to crop yields and to the sustainability of irrigation if corrective actions are not taken. Fortunately the main irrigated cereals, rice and wheat, are relatively tolerant to low or moderate salinity, but are subject to yield losses of about 10 percent at high salinity levels.

Greater applications of organic and mineral fertilizers are essential to prevent soil nutrient mining and raise crop yields, but in many developing countries application will remain below that likely to cause major pollution problems. However, in some areas with either very high application rates (for example the Punjab) or thin sandy rock strata above aquifers (as in parts of Sri Lanka), the risks could be significant if corrective measures are not taken. Per hectare application rates of mineral fertilizer in some countries of Near East/ North Africa and South Asia are projected to exceed 100 kg of nitrogen, so that by 2010 heavy rates will have been applied for 20 to 30 years or more. This is the time span over which the developed countries started to face severe groundwater nitrate problems, underlining the importance of adjusting application rates to the rate of uptake by plants. Restrictions on the use of agricultural chemicals and on the disposal of wastes from agroprocessing to protect water quality for increasing human consumption, will necessarily impose costs on consumers.

Intensive dairying, pig and poultry units are projected to account for an increasing proportion of total livestock output, and so their effluents will also grow. Many developed countries have introduced measures regulating the storage and disposal of such effluents. Livestock concentrations are strictly regulated in the areas of permeable sandy soils in the Netherlands, for example. The situation in the developing countries is likely to worsen before similar measures begin to be taken there.

Pesticide use

The production projections of this study imply three changes that, in the light of the past approaches to pest control, could pose serious threats to the environment. First, the further reduction in the length of fallow periods may not only endanger soil fertility as noted above, but in the absence of suitable corrective actions, could lead to more serious and more frequent weed, insect and disease attacks because the causal agents are able to survive in greater numbers from one cropping season to the next. Second, the increase in the area of land carrying two or even three crops a year could have similar effects as the reduction in fallow periods. Finally, the rise in the demand for vegetables, and to a lesser extent fruit, could lead to greater pollution and health risks from excessive use of insecticides. These crops tend to receive excessive applications of insecticides and often too close to harvest, either as an insurance against the risk of loosing a high-value crop or in order to improve their cosmetic appearance and hence their price. Such excessive application rates can pose numerous risks to the people applying them, to consumers, to natural predators of pests and to drinking water supplies.

It is difficult to project the future rates of pesticide use in any detail. It is assumed, however, that a combination of greater emphasis on integrated pest management (IPM) and biological control methods in general, and concerns about public health risks and ecosystem protection, will continue the present trend for a slowing down in the overall growth rate of chemical pesticide use, and a reduction in both the rates of application and the toxicity of insecticides and pesticides in general. These positive trends are evident, for example, in Egypt which has banned the use of a number of toxic pesticides, and Indonesia, where the successful introduction of integrated pest management on rice has brought widespread benefits (see Chapter 12).

Consequently, it is a reasonable assumption that the future growth rate of pesticide use will be lower than that of the past two decades, and that there will be a reduction in the environmental risks per unit of pesticide use. Thus, while the concerns expressed by certain groups are important, and there are no reasons for complacency, there are no grounds for arguments based on linear extrapolations of past mistakes.

11.3 GLOBAL CHANGE ISSUES

Loss of biodiversity

The projections indicate two particular pressures that warrant attention, namely deforestation and the loss of wetlands. Forests, biodiversity and climate change (discussed below) are closely interlinked because forests have a dual role as habitats and as major carbon sinks. The many levels of forest canopy, especially those in the tropics, with their varying light intensities and moisture levels, allow a multitude of habitats to co-exist in a small area rich in biodiversity. The closed tropical forests, for example, cover only 7 percent of the earth's surface, yet contain at least 50 percent and possibly 90 percent of the world's species. Many of these species have not been described or assessed for their utility as foods, medicines or other purposes. It is impossible to say which of them may be redundant for long-term sustainable development, though it is also difficult to argue that life as we know it could not go on without some of these unassessed resources.

Although the area of tropical forest that may be taken up by agriculture during the period to 2010 would be only a relatively small percentage of the remaining global stock, there are grounds for attempting to minimize such losses. A small proportion of the developing countries' wetlands will also probably be drained and used for crop production, since they represent one of their few remaining resources suitable for permanent agriculture. The wetlands are both a source of biodiversity and an ecosystem providing environmental services such as the purification of water, flood mitigation, fisheries and wildlife habitat and amenity. Their role needs to be protected, but as noted the areas likely to be converted to agriculture will probably be a small share of the total, and will provide a vital contribution to staple food security.

Climate change: potential global and regional impacts

Agricultural activities are a major contributor to anthropogenic sources of greenhouse gases, which in turn contribute to radiative forcing and hence climate change. In turn, climate change will have an impact on agriculture.[2] Apart from the release of carbon dioxide caused by biomass burning, mainly through deforestation and savanna fires, which together account for about 30 percent of the total amount of the CO_2 emitted, agriculture's main contribution to radiative forcing is through the emission of methane (CH_4, about 70 percent of the total emitted) and nitrous oxide (N_2O, about 90 percent of the total).

Rice cultivation appears to be the largest anthropogenic source of global methane emissions, influenced by a complex set of factors that primarily affect methane-generating and methane-absorbing bacteria. These interactions are very sensitive to the physical, chemical and biological conditions of the rice paddy environment and hence can be manipulated by management practices. For example, methane emissions to the atmosphere are higher in deep water than in shallow water rice cultivation.

The harvested area for rice in the developing countries, excluding China, is expected to increase by about 11 million ha by 2010, an increase of about 10 percent. Hence the projected increase of methane emissions from this source is relatively modest also because some of the increase in rice area will arise from the conversion of wetlands which produce some methane in their natural state. The modest incremental contribution from this source will be further depressed by the expected move away from deep water rice cultivation, mainly because of increasing competition from aquaculture.

A greater increase in methane emissions to the atmosphere than from rice cultivation may arise from ruminant livestock, the numbers of which are projected to increase by more than 35 percent by 2010 in the developing countries. Methane is released as a result of incomplete decomposition of plant material by aerobic fermentation, a process favoured by the fibrous and forage feeds of varying quality widely used for ruminants in developing countries. Thus, more methane is released from poorly fed than from better fed ruminants. The overall effect of the projections on radiative forcing is difficult to assess, being the combined effect of the numbers of ruminant livestock and the nutritional composition of their feeds, which is expected to improve moderately under the pressure of the rising demand for livestock products.

The annual rise in atmospheric nitrous oxide is relatively small and precise measurements of sources of this gas are scarce. Emissions are estimated to be mainly from biotic sources, related to nitrification and denitrification processes. These take place in the sorts of natural ecosystems such as tropical rainforests, tropical/subtropical savannas and intensive fertilizer agroecosystems. Applications of nitrogenous fertilizers are estimated to account for about 20 percent of the current annual increase in atmospheric N_2O. Such applications are expected to rise little in industrialized countries, but significantly in the developing countries, particularly on the better quality arable land. However, typically these lands do not have the seasonally hydromorphic soil conditions that specially favour N_2O emissions.

There is still a great deal of uncertainty regarding the nature, regional distribution and distribution over time of the potential impacts of climate change on agriculture. In spite of the extensive research stimulated by the Intergovernmental Panel on Climate Change (IPCC) and by national concerns, it will probably take many more years before these uncertainties are resolved. Nonetheless, there is a scientific consensus that global warming is a real phenomenon that could have a range of both negative and positive impacts on agriculture (FAO, 1990d). Some of the most severe negative impacts could be in those regions already vulnerable to present-day climate variation, notably sub-Saharan Africa, and with the least ability to research or buy their way out of trouble. In broad terms, the developing countries may be more affected by the possible changes than the developed countries, even though the temperature rise from global warming is projected to be greatest at the higher latitudes where most of the developed countries are located. The potentially greater impact on developing countries is, on the physical side, because most of them are in low rainfall areas or have significant areas of arid land, which are already subject to major agricultural production problems because of rainfall variability and associated constraints, and on the economic side because of their greater dependence on agriculture (see Chapter 3).

All of these potential climate impacts probably lie outside the time frame of this study, since they are not currently envisaged to play a notable role before about 2030. There is, however, one positive impact which may already be influencing agricultural production, namely the boost to crop growth arising from the enhanced carbon dioxide levels in the atmosphere that are one of the forcing factors for global warming and climate change. The higher CO_2 levels lead to faster growth of plant biomass and better water utilization in many crops, contributing to stronger root growth and denser ground cover. Some scientists believe that this effect accounts for a proportion of crop yield gains in recent decades, possibly as much as 10 to 25 percent of the cumulative increase. Moreover, even if the developed countries succeed in stabilizing CO_2 emissions in the early part of the twenty-first century, it is unrealistic to expect the developing countries to do so before the end of the century. Hence this CO_2 fertilization will have an important impact throughout the current projection period and beyond.

Finally, although the main negative impacts of climate change are likely to be beyond 2010, there are agricultural measures which can be justified against current socioeconomic needs and yet would help to minimize the potential impacts of climate change. These are considered in Chapters 12 and 13.

Concluding comments

Two aspects of the discussion in this chapter are worthy of emphasis. First, as discussed in Chapter 4, the future rate of growth of land expansion for agriculture will likely be lower than in the past. The pressures on water, however, will grow considerably as will those on the environment arising from the intensification of land use. Secondly, the analysis has focused primarily on pressures on natural resources in the aggregate, which does not identify clearly the distribution of such pressures among countries arising through international trade. Trade contributes to transfer pressures on resources from importing countries to exporting ones, as for example happened with the production of cassava in Thailand for export to Europe. Such effects may be significant and are partly addressed in Chapter 8 in the context of the discussion of trade issues.

NOTES

1. FAO's suggested definition is "The sum of geological, climatic, biological and human factors which lead to the degradation of the physical, chemical and biological potential of lands in arid and semi-arid areas, and endanger biodiversity and the survival of human communities." (FAO, 1993g).
2. The role of forests and forestry in relation to the carbon cycle is discussed in Chapter 5.

CHAPTER 12

Laying the Technological Foundation for Sustained Agricultural Development

12.1 INTRODUCTION

In recent years, contrasting views have been expressed regarding the environmental soundness or appropriateness of low or high external input technology paths in responding to many of the pressures outlined in Chapter 11. One issue is clear: increasing agricultural production depends on replacing most of the soil nutrients removed in the harvesting of crops, otherwise nutrient mining will take place and production will not be sustainable. Low external input systems will require large inputs of labour (which are not always available) and high external input systems will need considerable inputs of fossil fuel (i.e. non-renewable) energy. Although the use of mineral fertilizers will continue to grow, it cannot, in many situations, provide all of the inputs necessary to maintain soil fertility and must be used with organic manures and other biological inputs as part of an integrated plant nutrition system. It is also noted that low external input systems are not necessarily less polluting than high input ones. Badly timed applications of manure, for example, can be a more serious source of groundwater and surface water contamination than appropriate amounts of mineral fertilizers. Requirements therefore are not just technological, but also include manpower training and regulatory instruments.

There is growing acceptance that what is required is a balanced integration of the two systems. A number of the underlying forces are particularly significant in that they stem from changing perceptions as to priorities and pathways for technological development in the developing countries, and changing opportunities in developed countries. These aspects will be considered in the next three sections. The penultimate section (Section 12.5) will discuss the technological responses required to achieve the growth in agricultural production projected in this study with the minimum risk to the environment. The final section will briefly assess the needs and opportunities for laying the technological foundation for sustainable agricultural development beyond the year 2010.

This emphasis on technology is not intended to suggest that a new technological pathway is sufficient in itself. There is a wide range of policy and institutional measures required to provide the incentives needed for farmers, forest users and fishermen to adopt sustainable technological and resource management practices. These other measures are the subject of Chapter 13.

12.2 CHANGING PERCEPTIONS OF TECHNOLOGICAL DEVELOPMENT REQUIREMENTS IN DEVELOPING COUNTRIES

Early efforts to support agricultural development in the developing countries were heavily based on transferring technologies and management practices of the developed countries for a narrow range of crops in areas with favourable soil and agroclimatic conditions. This had a number of positive benefits but also some undesirable consequences. On the positive side, and as noted in Chapter 2, agricultural growth in the developing countries exceeded that of population, with the exception of sub-Saharan Africa. Moreover, a number of countries have been able to raise agricultural export earnings and local incomes derived from them without sacrificing domestic food production. This has been achieved through major technological improvements in the high potential areas, though in some instances with the environmental penalties described in Chapter 11. Notable achievements include the uptake of high yielding wheat and rice varieties in Asia and Latin America and the increase of sugar yields in some countries as well as the rapid growth of the palm oil sector. There is no reason to believe that these benefits cannot be maintained in the medium term while production is shifted on to a more sustainable development path.

The main undesirable consequences were as follows.

1. Traditional mixed cropping and interplanting practices with high resilience to weather fluctuations and pest attacks were commonly discouraged and replaced by less stable monocropping and row planting. It is now accepted that this shift was undesirable in some situations and deleterious in others for both short-term household food security and long-term sustainability in the more marginal environments.
2. The technological needs of the arid and semi-arid areas were neglected, except where the lack of water could be overcome through formal irrigation.
3. Plant breeding focused on cash crops and on a few major staples, and until recently generally neglected grains like millet, roots and tubers, and most legumes. Moreover, earlier breeding objectives were focused on maximizing yields rather than stabilizing them, which is a primary concern of many farmers. And in some cases, for example for sorghum and millet, by aiming for higher yields breeders were implicitly selecting for varieties with long

growth cycles, which exposed the farmer to greater risks of crop failure in areas where the length of growing period is critical.

4. Tillage systems were centred on conventional ploughing, which is not well suited to some of the fragile soils of the developing countries, where the emphasis should have been on minimal tillage systems.
5. Soil nutrient replacement was dominated by mineral fertilizer use rather than the development of integrated plant nutrition systems.
6. Soil conservation techniques were drawn up around engineering and not biological approaches to soil stabilization, and with erosion control rather than soil moisture management as the primary objective.

These undesirable consequences are now broadly recognized, and are increasingly addressed by research and extension systems (CGIAR, 1992). As yet, however, this recognition has not resulted in a major shift in national and international research priorities, and where it has, the main technological and institutional findings are unlikely to be applied widely in the short to medium term. Consequently the production projections of this study are based on the assumption that the prevailing technological path will be the dominant one for the next 15 to 20 years, particularly in the high potential areas, but that there will be a gradual shift in researching technologies for the more marginal areas.

Low acceptance rates for some of the research stemming from the 'Western' style approach to agricultural development was commonly ascribed to the lack of integration of peasant farmers into the market economy, and their low susceptibility to economic incentives. Low acceptance rates were seldom attributed to the deficiencies outlined above or to the inappropriateness of the technologies to farmers' needs. A number of changes, however, have taken place or are under way, regarding the manner in which technological needs are identified and research is conducted, which should make the production levels projected for 2010 more achievable and sustainable.

There are three notable changes concerning the identification of farmers' needs. First, widespread acceptance now of the earlier contention that peasant farmers are profit maximizers provided the technologies are not too risky, and are profitable very early in the adoption process. Both aspects have been neglected in the design and evaluation of many technologies. Second, some progress has been made in understanding the links between population pressure on resources and the development and adoption of technologies, for example, the extent to which the adoption by farmers of land-augmenting, yield-enhancing technologies depends on their access to land and market incentives (see, for example, Pingali and Binswanger, 1984; Lele and Stone, 1989; Tiffen et al., 1993). A later section considers this aspect in more detail. Third, growing recognition that the decision process of farmers is weighted more by the profitability of technologies than by their environmental friendliness (Bebbington et al., 1993).

Similarly there are three important changes regarding the conduct of research. First, there is the emphasis on farming systems research with the greater involvement of farmers in the decision process, which helps to place commodity research in a more meaningful production context. Second, the introduction of on-farm client-oriented research or on-farm research, which is an approach designed to meet the needs of resource-poor farmers, and complements as well as depends on, experiment station research. Finally, the "rediscovery" of indigenous technical knowledge, with growing acceptance of the need to build on existing technologies which have been selected and refined by farmers in harmony with their own sociological and ecological conditions (Altieri, 1987; Chambers et al., 1989), but with the understanding that they are not sufficient in themselves (Richards, 1990; Bebbington et al., 1993). The use of the velvet bean as soil cover and green manure in communities in Honduras is an example of such participatory research and development (Bunch, 1990).

12.3 CHANGING OPPORTUNITIES IN THE DEVELOPED COUNTRIES

The rise in public willingness to pay for a better environment

Although a better environment which is achieved through cleaner technologies and sustainable agricultural practices may be cost saving in the long term, there are commonly short-term economic penalties through higher production costs, restrictions on resource use, and public expenditure financed through taxes. Thus organically grown foods cost 10 to 20 percent more than the "conventional" product, and calls for restrictions on resource use in the United States to promote sustainable timber extraction and protect the habitat of the spotted owl contributed to the increase in US timber prices in early 1993.[1] Consequently, public policy changes towards sustainable technologies and practices are dependent on the willingness of consumers and the public sector to pay these additional costs. This willingness has been growing since the early 1970s, and can be expected to continue to grow, albeit with some short-term declines during periods of economic uncertainty.

Environmentally oriented shifts in technology

These shifts are driven by three forces: first, the public pressures discussed above; second, in particular in the European Community (EC), the need to address the surplus production problem; third, scientific and technological progress itself. The first and second forces, for example, are combining in the EC to restrict the use of mineral and organic fertilizers in sensitive watersheds, and to ensure more sustainable land management practices. They may ultimately result in significant amounts of agricultural land being withdrawn from cultivation and placed under pastures, forests or leisure uses.

The third force, the growth in scientific knowledge and technological progress, expresses itself in two particular ways. Firstly, in the greater understanding of the risks to human and ecosystem health of certain practices. As an example, the discovery of the link between the use of chlorofluoro-carbons (CFCs) and ozone damage and the discovery of the ozone hole over the Antarctic led rapidly to the Vienna Convention and the Montreal Protocol. Secondly, in the development of cleaner and more energy-efficient technologies that are more cost effective and hence will be taken up through market forces even in the absence of regulatory pressures.

Though these forces may have their initial impact in the developed countries, they will also benefit the developing ones. They will provide new or better tools for technology development in developing countries themselves (see subsection on biotechnology in Section 12.5), and some of the technologies for developed country markets will be directly usable, such as certain biopesticides, or adaptable to developing country conditions, such as surge irrigation.

12.4 POPULATION PRESSURE AND TECHNOLOGICAL CHANGE

Analyses of the complex relationships between growth in total population and the part of it dependent on agriculture, land use (and hence environmental change) and technological change, help to put this study's projections into a wider perspective. They also provide useful insights for policies to make them guide rather than work against the natural processes determining the intensification of land use and the adoption of technologies. Two points of view tend to prevail. The first assumes that there is a negative relationship between the growth of the population dependent on agriculture in conditions of poverty and environmental quality. The second looks at the relationship in a more dynamic way and with greater regard for the economic dimension.

The first viewpoint lays great emphasis on the fact that the pressure of population bearing on a limited land base and the slow introduction of farmer-based or formal research-based innovations in response to these pressures, has had widespread negative impacts on the environment. Reduced fallow periods and tree cover on erodible soils, together with the slowness of the natural processes that restore soil fertility, have lowered land productivity through loss of nutrients or nutrient mining.

The second viewpoint is centred on the work of Ester Boserup (Boserup, 1965, 1981) who applied the concepts of factor substitution (labour for land) and technological change to hypothesize that as population density increases, technological changes occur autonomously through shortening fallow periods, increasing inputs of labour and the adoption of improved tools (planting sticks to hoes to animal-drawn ploughs). According to this hypothesis the problem of population growth, giving rise to an increased demand for food, can lead to its own solution by altering factor prices: first increasing scarcity of land compared to labour, giving rise to increased land intensification or shortening fallow and increased use of labour; and then increasing scarcity of labour at some stage through the land intensification sequence – gathering, forest fallow, bush fallow, grass fallow and annual cropping – leading to the adoption of improved tools. Such a process of "farmer-based innovation" (Pingali and Binswanger, 1984) describes the evolutionary process of adapting production technology to changes in factor scarcity. The responsiveness of "science-based innovation" to economy-wide factors, such as its endowment of land and labour, the non-agricultural demand for labour and conditions of demand for food and other agricultural products, gave rise to the closely associated concept of "induced innovation" (Binswanger and Ruttan, 1978; Hayami and Ruttan, 1985). Thus the land-scarce agricultural economy of Japan by the late 1800s gave rise to biological innovations that increased yields per unit of land, while the USA, which at that time had 100 times more land per head of agricultural labour, adopted a mechanized agricultural technology. In the USA, biological innovations were not adopted widely until several decades later, in the 1940s, in response to rising land values.

The autonomous process, however, leading to appropriate institutional and technological responses to the pressures on the environment, such as the building of terraces to control erosion, and the utilization of organic manure to restore fertility, may not occur fast enough if population growth is rapid or, conversely, is constrained by labour shortages if an easier, less labour demanding option is open in the form of migration.[2] The adoption of research-based innovations may also be constrained by deficiencies in infrastructure, extension and marketing and credit systems. Thus the end result can be a treadmill of low rates of adoption of either farmer-based or research-based innovations, low productivity agriculture, environmental degradation and poverty. As noted in Chapter 2, failure of institutions governing land tenure to adapt to changing conditions contributes to create a vicious circle between poverty and environmental degradation.

There remains the issue of poverty, which is commonly associated with these treadmill situations. It can be argued that in some situations poverty becomes part of the solution as the poor respond by migrating; by making personal sacrifices so that their children can be educated and find better employment opportunities outside agriculture; and by diversifying their agricultural and non-agricultural incomes in response to changing market opportunities. This benign process is clearly exemplified by the recent longitudinal study of the Machakos district of Kenya (see Box 12.1). In other cases, the necessary conditions are not in place and poverty and environmental degradation co-exist and become mutually reinforcing.

Box 12.1   Environmental recovery in the Machakos district of Kenya*

Large parts of this agroecologically diverse district of some 1.4 million ha are inherently marginal for the production of the preferred staple, maize. Much of it is highly sloping land with a mean annual rainfall of less than 800 mm per year which is spread over two rainy seasons, with marked inter- and intra-annual variation. Consequently, with such constraints and low population pressure in the more favoured areas, it was largely uninhabited at the beginning of this century. This changed rapidly. The best lands were settled first, followed by the more marginal ones. By the 1930s substantial areas were so badly degraded by crop production and grazing that observers of the time thought that the district was on the edge of ecological collapse. There was severe soil erosion over about 75 percent of the inhabited area, and tree cover was down to around 5 percent. By the 1940s the population carrying capacity of the district was exceeded at the prevailing low levels of technological inputs.

Yet by 1990 the picture had changed completely. The population of over 1.4 million was nearly six times that of the early 1930s, with population densities in the most marginal agroecological zones increasing nearly 30-fold between 1932 and 1989. Agricultural output was up, dependence on food imports from other districts was down, and there was less soil erosion and more tree cover.

Thus, and contrary to the prevailing view, population growth meant less degradation and a more sustainable agriculture. The factors involved, and their sequencing, have been thoroughly analysed. There was internal and external migration from the 1920s onwards in response to land shortages, with the former being to more marginal lands within the district. External migration to urban areas was followed by return flows of remittances which provided part of the subsequent capital needs for agricultural development. Then, there was intensification of land use, starting in the late 1930s on the better and more populated lands close to the urban markets, but not until the 1960s on the less populated "marginal" lands. This intensification was largely through reduction of the fallow periods, the introduction of multiple cropping, the closer integration of crop and livestock production, and heavier applications of manure, compost, or, in the case of export crops, mineral fertilizer. It was paralleled or followed by the widespread adoption of soil conservation measures to rehabilitate degraded land, notably conservation tillage, contour farming and terracing (the proportion of treated area in total area rose from about 52 percent in 1948 to 96 percent in 1978), with substantial gains from soil erosion reduction and from enhanced rainfall infiltration and soil moisture retention. This widespread adoption was encouraged by the introduction of a range of cash crops, particularly coffee, fruit and other horticultural crops that gave higher incomes than staple crops, and thus made soil conservation more profitable. Finally, and perhaps most critically, there was investment in improved roads and other infrastructure needed for ready access to urban and overseas markets and to local processing locations. Much of the incentive and the capital for this retreat from supposed ecological disaster came from the people of Machakos themselves with significant community and local NGO inputs and relatively small central government and donor inputs.

*Source: Tiffen et al. (1993).

12.5 THE TECHNOLOGICAL CHALLENGES OF
AGRICULTURAL GROWTH

Limiting land and water degradation

Chapter 11 highlighted the extent of degradation and our lack of understanding as to its full implications for crop productivity and sustainability. Two aspects of soil conservation are becoming increasingly clear. The success of conservation measures is highly dependent on farmers receiving crop yield and economic benefits in the first or second season after implementation (FAO, 1989a). In dryland areas such gains will commonly arise more from improvements in physical structure leading to enhanced soil moisture levels and retention (Shaxson, 1992) than from the reduction of soil nutrient losses, although the latter are important (Stocking, 1986).

Failure to meet these requirements, together with institutional weaknesses, accounts for the failure of many past conservation techniques and projects, which were either very labour intensive or required costly mechanical operations. Hence they were often not profitable in the short or even the medium term and, moreover, they were costly to maintain. Farmers therefore seldom adopted the conservation techniques, or did not maintain the conservation structures after the end of the project. The success stories of today, which in a sense are success stories of the past since they commonly use or build on indigenous technologies, are consistent with these conclusions (see, for example, Reid, 1989; FAO, 1991a; Kerr and Sanghi, 1992).

The above conclusions and observations carry several important messages for technology development to achieve long-term sustainability. First, soil conservation strategies, research and extension should concentrate on measures with no or low external capital requirements, so that they are more appropriate for resource-poor farmers in marginal areas that are projected to be under increasing pressure. Second, provided the appropriate institutional support is given (see Chapter 13), known techniques could help to boost or stabilize yields. Third, these techniques are not widely used and could benefit a much larger area. A large part of the drylands in sub-Saharan Africa and in Asia could gain from them through both increased and more stable yields and by more frequent cropping. Also slopelands in the high rainfall tropics could gain from techniques better adapted to their particular constraints.

There are also other implications for research. The focus should be on biological rather than mechanical approaches to soil conservation which, as with vegetative barrier techniques or with systematic crop and residue management, either retain soil particles and eventually build up natural terraces or protect the soil surface from rain impact and erosion. Attention should also be focused on techniques that combine soil erosion limitation with wider land degradation control functions, such as the use of leguminous live mulches.[3] The International Agricultural Research Centres of the CGIAR,

---

Box 12.2   Managing the "four waters"*

From 1949 to 1980, the irrigated area in China increased by 32.7 million ha and currently covers 46.7 million ha. Due to problems of scarce water resources and the limited possibility of further expansion, the rate of growth has fallen significantly in recent years. In response, Chinese engineers and agronomists have developed an innovative method of water management, known as the concept of the "four waters", which refers to a comprehensive control and supervision of groundwater, surface water, soil moisture and rainfall for agricultural production. The objective of this approach is to produce two crops per year over the largest area possible with limited use of surface water. The basic innovation of the "four waters concept" is the dynamic control of the aquifer. Whereas traditional horizontal drainage keeps the groundwater table below a certain level to avoid waterlogging and secondary salinization, dynamic groundwater management, in addition to controlling the water table, uses the aquifer as storage. Dynamic groundwater management keeps the level of the aquifer within a specified range, which is defined by hydrological and agricultural requirements, and takes into account the constraints posed by salinity hazards and the need for efficient energy use. The "four waters concept" has been tested extensively in the Nanpi experimental station in Hebei Province and in a pilot project of 23 600 ha. The results have been positive. The implementation of this water management approach showed that large areas of saline–alkaline land could be reclaimed, and land which was previously unsuitable for irrigation due to groundwater salinity, has been cultivated. Moreover, rice yields increased by 77 percent: from 3.7 tonnes/ha to 7.8 tonnes/ha. Multi-annual hydrological simulation demonstrates that with only 550 mm average annual rainfall, as much as 43 percent of the dry season irrigation requirements can be met without groundwater mining or external water imports.

*Source: Shen and Wolter (1992).

---

notably CIAT, ILCA and IITA have been supporting national research efforts by collecting and testing suitable legumes for forage or mulch purposes, but the current national and international efforts are inadequate relative to the task and their potential contribution to sustainability.

Finally, there is the problem of salinization. As noted in Chapter 11, irrigated land is being lost and substantial areas suffer reduced yields through salinization. The most common causes are inadequate drainage, rising water tables because of water seepage from distribution canals, and excessive application rates. Consequently, the standard corrective actions have been additional drainage and canal lining, both of which can be costly. In the future, however, part of the solution seems likely to lie in conjunctive use of surface water and groundwater and the parallel use of canal and tubewell systems, with the latter providing vertical drainage as well as secondary irrigation. Experience in China, for example, has shown that a more holistic approach to water management, what is called there the "four waters concept", can prevent salinization and reclaim salinized land (Box 12.2).

Promoting integrated plant nutrition systems (IPNS)

IPNS aims at maximizing the efficiency of plant nutrient supply to crops through better association of on- and off-farm sources of plant nutrients and ensuring sustainable agricultural production through improved productive capacity of the soil. Such systems may significantly reduce needs for mineral fertilizers, because they provide timely and sufficient supplies of plant nutrients, and reduce as far as possible plant nutrient losses in cropping systems. Adoption of IPNS has the potential of increasing the profitability of mineral fertilizer use (FAO, 1993a).

Progress towards achieving these broad objectives must be viewed in the various ecological and economic contexts of agriculture in developing countries. Firstly, there are situations where plant nutrients are mined from the soil because extractions by crops and losses of plant nutrients through erosion, leaching and volatilization exceed the low level of plant nutrient supply. Here IPNS will assist in achieving a better balance of nutrients, the intensification of cropping systems with limited use of external inputs, better recycling of local sources of nutrients and, above all, a drastic reduction of nutrient losses.

Secondly, there are situations where plant nutrient efficiency is low, even with significant supplies of nutrients from various sources. In these cases, IPNS could improve efficiency through the appropriate combination of plant nutrient sources and cropping techniques. In most cases, this low efficiency is due to an unbalanced supply of plant nutrients (often too much nitrogen relative to other nutrients) or to another limiting factor such as micronutrient deficiency and physical or chemical constraints of the soil. IPNS seeks to relieve such constraints depending on the availability of resources (plant nutrients, equipment, energy, adapted varieties and irrigation).

Thirdly, there are situations where losses of plant nutrients are polluting the environment because of excessive or improper management of plant nutrient supplies. Examples are nitrates in surface and groundwater, phosphates in surface water and nitrous oxides and ammonia in the atmosphere. IPNS could reduce this type of pollution by both increasing plant nutrition efficiency and reducing plant nutrient losses.

Fourthly, there are arid and semi-arid areas where the maintenance of soil organic matter is crucial for efficient plant nutrient management, for maintaining soil permeability and waterholding capacity, and for the development of deep rooting systems capable of exploiting water stored in the soil. One challenge of IPNS is to produce sufficient biomass in order to restore to the soil the organic material lost during crop cultivation. In mining the reserves of soil plant nutrients, farmers reduce the capacity of their soils to produce biomass and cause the soils to lose organic matter. However, the mineralization rate of the soil organic matter is rapid when the soil temperature is high, and the biomass production, when plant nutrition is not a limiting factor, is directly related to the availability of water. Therefore, in the semi-arid tropics it may be difficult to restore the organic matter content of a degraded soil.

In the humid tropics, leaching of plant nutrients, erosion and acidification, and immobilization of plant nutrients in the soil may hamper the efficient supply of nutrients to plants. Additionally, competition of weeds and pressure of pests are important factors decreasing such efficiency. However, crop production, biomass production, and crop diversity are higher and the effect of temperature on the mineralization of soil organic matter is generally lower, than in the semi-arid tropics. Climatic risks are also low as compared to semi-arid areas and the natural conditions are generally more favourable for agricultural intensification. Under such circumstances, IPNS will then have to address rather diversified levels of intensification. The upgrading of the soil fertility is easier than in the semi-arid or arid tropics because the production of biomass is higher. Limiting plant nutrient losses is more complex than in semi-arid tropics because the overall quantity of plant nutrients involved is higher and the pressure of the factors causing these losses is also higher.

In the absence of water constraints in irrigated areas, the potential for plant nutrient efficiency is high. However, the efficiency of the use of plant nutrients is often quite poor because of poor control of nitrogen losses in the cropping system or unbalanced fertilization. IPNS in irrigated areas faces special problems since the use of crop residues has to be carefully regulated to avoid the development of diseases and the leaching of plant nutrients. However, fixation of nitrogen is possible either directly through traditional flooding irrigation (blue algae, azolla) or when irrigation takes place through sprinklers in mixed cropping or in relay cropping systems. IPNS in irrigated areas mainly focuses on improving plant nutrient efficiency, as maintaining soil organic matter content is simpler than in rainfed areas because production of biomass is generally high.

The efficiency of nitrogen use is a major problem in IPNS. Fixation of atmospheric nitrogen may provide significant quantities of nitrogen to the cropping system if water, phosphorus and sulphur are available. However, nitrogen biofixation cannot cover all nitrogen requirements except at low levels of intensification. There is a wide range of free-living soil bacteria that extract nitrogen from the atmosphere and make it available for plant growth. There are other bacteria, notably rhizobia, which live symbiotically with plants in small nodules (swellings) on their roots, receiving sugar from plants and providing in return nitrogen that they have taken from the atmosphere. The latter have been exploited by man for many years and sustained cropping systems in Europe and other parts of the world before the discovery of mineral fertilizers in the nineteenth century. In China, Thailand, Vietnam and other Asian countries, the alga *Anabaena azollae*, which lives symbiotically with the water fern *Azolla*, has sustained rice cropping for centuries by providing much of the nitrogen required.

The challenge at present is to exploit conventional techniques in combination with methods of genetic engineering to improve nitrogen availability and widen the range of plants and environments that can benefit. Current natural or managed plant/microbe systems in good rainfall areas can provide 20–60 kg of nitrogen per hectare, sufficient to sustain cereal yields of around 1 tonne. This could be raised by 25 percent by the year 2010. Conventional plant breeding techniques may assist to obtain better efficiency through varieties using plant nutrients more efficiently, or with stronger root systems avoiding losses through leaching. Varieties tolerant to soil constraints (salinity, lack of oxygen, free aluminium) will also benefit more than traditional varieties from plant nutrient supply.

IPNS is likely to make a significant contribution to crop production growth and to the achievement of sustainable agricultural systems over the period to 2010. Nevertheless, one should not underestimate the difficulties facing IPNS in the short to medium term, and also not overestimate the gains in the long term. Lack of livestock and labour will be a major constraint in some areas, as many smallholders cannot keep sufficient livestock to generate the required amounts of manure (up to 10 tonnes or more per hectare) or cannot provide the large labour input for collecting and spreading it. Where land constraints are severe, it may be impossible with current or foreseeable technologies to achieve or sustain high yields with only recycling or biofixing techniques for plant nutrient supply (Norse, 1988).

China has been in this situation for several decades. In spite of its efficient systems for organic residue recycling and biological nitrogen utilization, since about 1950, staple food production has become increasingly dependent on mineral fertilizers, making China the largest user of mineral fertilizers in the world. In spite of the efforts in China to improve further the efficiency of organic residue use, this trend seems likely to continue, reinforced by the increasing shortage of labour for the collection and application of organic manures. The production projections of this study suggest that several other countries or areas face the same dilemma as China.

Expanding the opportunities for integrated pest management (IPM)

The past agricultural performance carries with it the burden of mistakes arising from our previous lack of knowledge regarding pesticide toxicity, their persistence in soils and water, their accumulation through food chains and their impact on both non-target and target species. Some of the costs of these mistakes are to be seen in human mortality and morbidity, in damaged ecosystems and in the increase of pesticide resistance. There are now over 450 injurious species of arthropods that have developed resistance to one or more pesticides because of repeated applications (Georghiou and Lagunes-Tejeda, 1991). Resistance is also increasing in plant pathogens and weeds.

Following research showing the growing damage to human health and ecosystems from pesticide use, FAO spearheaded efforts in the mid-1960s to develop and implement the concept of integrated pest management (IPM). Initially, progress was slow since an understanding of predator–prey systems and other key aspects of ecosystems had to be built up. But in the past 10 to 15 years the success stories have grown and the concept has become more comprehensive. IPM now brings together five mutually enhancing control approaches.

- Pest control using crop rotations, intercropping and other management methods.
- Host plant resistance.
- Biological control using natural methods or introducing new enemies of the pests.
- Selective use of pesticides, preferably biopesticides, in conjunction with pest population monitoring and the establishment, where possible, of economic thresholds for pesticide use.
- Plant health programmes, including plant quarantine.

FAO's extensive experience in Asia with the Intercountry Rice IPM Programme adds an additional dimension: namely that of farmers becoming managers and experts in their own fields. Through "Farmer Field Schools", they learn how to grow a healthier crop, to conserve natural enemies of crop pests, and to use the appropriate pesticide where needed only.

Chapter 11 outlined the increasing pest pressures envisaged for future agricultural growth, notably because of the intensification of production. It noted that although total pesticide use may continue to grow, this will occur at a lower rate as compared with the past and that application rates, environmental persistence and toxicity levels of pesticides will be lower in the future. These prospective developments are likely to come about because of the growing political, technical and farmer support in developing countries for IPM, and against the excess pesticide use of the past.

In pursuing the further development and implementation of IPM, priority should be given to the crops accounting for the bulk of pesticide use: cotton, maize, rice, soybeans, fruit and vegetables. All have a potential for wider IPM implementation though IPM cannot be effective for the full range of major pests. The FAO rice IPM programme has reached some 600 000 farmers in Asia, cutting their pesticide applications by up to two-thirds, increasing yields and lowering costs of production. The number of trained rice farmers in Asia is expected to surpass 1 million before the end of the century, but this still represents a small fraction of the approximately 90 million rice farmers who could benefit from this programme's approach.

Although prospects for the other major crops are also good, experience thus far has been less favourable. The experience with cotton is mixed, with some countries achieving substantial reductions in pesticide use but others still increasing such use. Nonetheless, a more effective sharing of international experience could achieve widespread reductions by 2010 and lower cotton's dependence on pesticides.

As indicated in Chapter 11, the relatively high value of vegetables and the agronomic conditions under which they are grown,[4] commonly lead to the heavy use of toxic pesticides. FAO is trying to stimulate action in this area through a regional vegetable IPM programme for Asia which is based on the lessons from the rice IPM programme mentioned above. However, the benefits of this programme and of various national initiatives are unlikely to have a major impact in the near future unless the problem receives much higher priority.

Progress has been made with the use of biocontrol agents, i.e. living or dead organisms (bacteria, fungi, insects, viruses, nematodes and protozoa), but mainly in the developed countries and commonly for glasshouse conditions. The most widespread biopesticide is *Bacillus thuringiensis* which, for example, is very effective against some cabbage pests, but is only used on a minor proportion of developing country production. In Brazil, however, a baculovirus is being used on about 1 million ha of soybeans to control the velvet bean caterpillar, a major pest of soybeans.

In spite of its demonstrated benefits, however, IPM is not a panacea or a complete alternative for all conventional plant protection methods. Its success depends on a number of natural, social and economic conditions. It requires appropriate, i.e. precise timing and sequencing of its various control measures. It therefore needs well-trained farmers and a well-functioning extension system with diverse pest monitoring facilities and early warning systems, etc. This places considerable demands on research and extension staff, as well as on farmer's capabilities. The high labour intensity required at both the farm and extension service level, could reduce the competitiveness of IPM *vis-à-vis* traditional high-external input systems in the near future as opportunity costs for labour are increasing in many developing countries.

Water development and water saving

Chapter 11 presented a perspective view of growing competition for water, a competition that agriculture has lost in some developed countries (for example on the Texas high plains in the USA) and seems destined to lose in some developing countries. The projected expansion in irrigation is therefore dependent on a number of technological developments[5] that increase the efficiency of water capture or water use, particularly the following:

- Steps to safeguard the existing irrigation infrastructure through the soil conservation measures discussed earlier that will slow down the siltation of

---

Box 12.3   Emerging irrigation technologies: LEPA and surge irrigation

The low energy precision application (LEPA) method of irrigation consists of a low-pressure moving irrigation system, such as a modified centre pivot system or linear-move system, where sprinkler heads are replaced by drop tubes which deliver water to the soil surface. The crop response to this system of irrigation is similar to the response to stationary drop installation with closely spaced emitters. Saline water can be used without damage to foliage under this system. It maximizes irrigation efficiency with low pressure nozzles near the ground, and the application efficiency could be as high as 90 percent (Fangmeier *et al.*, 1990). It was reported that the average amount of water per hectare by Texan farmers dropped 28 percent between 1974 and 1978 because they adopted LEPA (Postel, 1989).

Surge flow irrigation is defined as the intermittent application of water to furrows or borders creating a series of on and off periods of constant or variable time spans. Usually water is alternated (switched) between two irrigation sets until irrigation is completed. The switch is accomplished with a surge valve and an automatic controller. Surge flow greatly reduces the intake at the top of the field because the opportunity time is much less than under continuous flow method. It is reported that efficiency of irrigation could be improved to an average of 70 percent or more. In the USA, the effectiveness of surge irrigation as a water conservation measure is demonstrated by its rapid growth and acceptance by farmers.

---

canals and reservoirs, and the actions to stop salinization and to rehabilitate saline land that are described later.

- The introduction of techniques, pricing policies and institutional changes that raise water use efficiency.
- Improvements in irrigation design and technologies to raise the efficiency and lower the costs of operation and maintenance.
- Expansion of the use of marginal quality water including brackish water and municipal wastewater.
- Removal of technical constraints to conjunctive use of surface water and groundwater resources.

Irrigation is the largest user of fresh water, yet irrigation water use efficiency can be as low as 40 percent (Kandiah and Sandford, 1993). Consequently, even a 10 percent improvement in water use efficiency can release a substantial volume of water for other uses and delay the onset of competition. The substitution of marginal for high quality water can have similar benefits. Several semi-arid countries in the Near East region already use treated sewage effluents for irrigation, releasing high quality water for other purposes.

Many of the technical solutions have been produced and implemented in the developed countries, but adoption has been slow in most developing countries, mainly because of their high cost and complexity. Most of them are overhead irrigation systems involving sprinklers, and a range of micro-irrigation systems that are twice as efficient as surface irrigation. The more recent techniques such as low energy precision applications (LEPA) can be 90 percent more efficient than surface irrigation, as well as permitting the use of saline water (Box 12.3). They are being adopted in some developing countries, such as Morocco, but are unsuitable replacements for the bulk of present irrigation because the latter consists mainly of surface systems involving complete flooding for paddy rice, or furrow irrigation. The principal opportunity for the latter seems to be surge flow irrigation (Box 12.3), but this generally needs to be adapted to the farming conditions in developing countries. Such solutions could have an appreciable impact on non-flood irrigation schemes well before 2010. For the rest, the immediate task is applied research to improve flood irrigation, and to adapt sprinkler and micro-irrigation systems more closely to developing country conditions. Given the long gestation period for the completion and implementation of such research, litle benefit can be expected before 2010.

In the past, attention has been focused on the health risks that can be posed by irrigation systems, through malaria and the spread of bilharzia, for example. Conjunctive use of water, however, presents an opportunity not only to reduce the competition for it but also to help reduce the most widespread threat to health in many rural communities, namely the lack of potable water. Irrigation canals and tubewells are increasingly recognized as a relatively safe source of drinking water, yet technologies and irrigation system design seldom take this into account. Research and technology development are urgently needed to determine the design requirements for dual-purpose systems, to identify treatments to safeguard health, and to adapt existing techniques and equipment to satisfy these requirements.

Raising livestock productivity

The projections pose a number of technological challenges for the livestock sector itself and for those supporting it with feedgrains, high-protein feeds, processing services and other inputs. The challenges differ between the land-dependent commodities like ruminant meat, and those which are becoming increasingly separated from land, such as eggs, pig and poultry meat which are increasingly produced by intensive systems dependent on concentrate feeds and located in peri-urban areas or areas with good access to urban markets. The former tend to be resource constrained and supply driven, and in many cases are exerting growing pressures on the environment through overgrazing. The latter are increasingly able to side-step resource constraints because market conditions enable producers to purchase feed concentrates and the best available production and processing technologies. They also may place unsustainable burdens on the environment through poor waste disposal.

The main challenges are to compensate for the lack or poor quality of land through measures to raise pasture and rangeland output and improve management systems; to bring about a closer integration of crop and livestock production; to raise the supply and quality of supplementary feeds; to achieve genetic improvements from conventional breeding and modern biotechnical tools; and to complement these gains by cheaper and more effective animal health measures. There is much in the technological pipeline to meet these challenges, which could have their impact well before 2010.

CIAT, for example, has selected forage legumes, grasses and browse species, which fit well into pasture management and ley systems for poor acid soils. Farmers are achieving major increases in stocking rates, and animal weight gains of 100 percent or more. In Latin America, only a relatively small proportion of the total area which could benefit from such improvements has been reached, but a much wider uptake could be achieved by 2010. The technology could also be adapted for the large land areas in Africa and Asia suffering from similar constraints.

Closer integration of crop and livestock systems is partly culture dependent, partly resource-constraint driven, and partly market driven. Land pressures are forcing nomadic pastoralists to settle and become cultivators. Land and labour pressures are forcing cultivators to adopt animal traction, and to keep livestock as a vital source of manure for maintaining or increasing crop production and cash incomes (Mortimore, 1992). Market opportunities and the desire to obtain more regular income streams are favouring dairy-based mixed systems. All of these are ongoing shifts that can be expected to strengthen during the projection period and increase sustainability.

In recent decades, conventional animal breeding has allowed the developed countries to raise productivity per animal by 1 to 2 percent per year. Similar attempts in the developing countries have been far less successful, in part because of unsuitable breeding stock, poor feed and environmentally stressful conditions, particularly temperature stress. Modern biotechnical tools now provide the possibility of modifying the genome of indigenous animals and mixed breeds to cope better with such stresses or diseases, and raise milk and meat output. Moreover this potential can now be brought to the farmer more quickly through new reproduction techniques, like embryo transfer and *in vitro* fertilization which can speed up the breeding and stock multiplication process. The techniques, however, are unlikely to be widely used in most developing countries by 2010, because of institutional and structural constraints.

Biotechnical tools and processes are also starting to have a practical impact in the animal health sphere, notably in the prevention, diagnosis and control of animal diseases, particularly vector-borne diseases. They will have an increasing impact over the projection period, with growing relevance to the needs of small-scale farmers. New vaccines are already on the market for the control of bacteria causing diarrhoea in lambs, calves and pigs.

In the medium term, vaccines should become available to control trypanosomiasis and theileriosis although commercialization of the former is unlikely to happen soon. Once achieved, however, it would help to free large areas of Africa of trypanosomiasis and replace trypanocidal drugs, new types of which have not been developed for some 30 years with as a consequence an increasing threat of resistant trypanosomes. In the shorter term, new control techniques, such as low level or selective insecticide spraying, pheromone traps and sterile male release, will continue to reduce populations of the trypanosomiasis vector (the tsetse fly) and replace the older but environmentally damaging techniques of massive spraying of residual insecticides, bush clearance and wildlife control. The eradication of rinderpest also could be achieved by 2010 if sufficient resources were to be devoted to this task.

In future research on animal nutrition appropriate to the conditions of the majority of developing countries, greater advantage should be taken of the particular digestive characteristics and complementarity of the different species of livestock. Ruminants can use fibrous feed and non-protein nitrogen sources which cannot be used by monogastric animals, which convert high energy feeds more efficiently.[6] Ruminants can be regarded as two subsystems: the rumen and its microbial contents; and the animal itself which can convert nutrients produced by the microbes and those derived directly from feeds (undegraded nutrients) that usually cost more. Therefore, greater attention is now paid to improving the rumen function through manipulation of its microbial environment. A further point is that supplements of small quantities of essential nutrients to balance those absorbed from the basal diet (generally pasture and crop residues), may greatly increase productivity.

The improved understanding of the feeding process leads to two priority areas of work to enhance livestock performance: optimize protein/energy ratio in nutrients absorbed by ruminants from diets based on poor quality forage; and optimize the digestibility of the basal feed resource. Pursuing these approaches involves research in the area of rumen manipulation, development of local sources of undegradable protein supplements, feed processing to improve the digestibility of low quality forage, and genetic manipulation of plants so that their proteins resist rumen microbial attack.

Examples of innovations embodying these new approaches in animal nutrition which are likely to be increasingly applied in the future are the following. First, the use of locally manufactured multinutrient blocks (molasses–urea) which now have been successfully tested in about 60 developing countries, including India (particularly in milk production) and the Sahel in Africa. These blocks encourage an efficient rumen ecosystem by providing a source of minerals, vitamins and fermentable nitrogen in order to correct an unbalanced nutrient supply. Secondly, leguminous forages as strategic feed supplements for ruminants. Promising species of such legumes are *Leucaena leucocephala* and *Gliricida sepium* which contribute fermentable nitrogen in the rumen as well as undegradable proteins to diets based on fibrous crop residues. The third example is the fractionalization of sugarcane for pig and ruminant feeding in Latin America. With this method, which could be a breakthrough for feeding monogastrics in the humid tropics, sugarcane juice is the basis of the diet for pigs and can totally substitute for maize. In a fully integrated system, the sugarcane tops and leaves as well as the bagasse are fed to ruminants, with the leftovers being used for fuel. Sugarcane also can be grown in association with soybeans and selected species of fodder trees to provide protein-rich livestock feed.

Developing the potential of biotechnology

Biotechnology is defined as any technique that uses living organisms to make or modify a product, to improve plants or animals, or to develop microorganisms for specific uses. Biotechnology is not new and many products are the result of a simple but effective use of traditional biotechnologies, such as fermentation processes for the production of cassava-based foods, which combined with boiling lowers the cyanide content. Biotechnology here refers to both traditional and modern biotechnology which is based on the use of new tissue culture methods and recombinant DNA (rDNA) technology, often referred to as "genetic engineering". Tissue culture includes *in vitro* fertilization and embryo culture, protoplasts and the culture of isolated cells and microspores. Such methods are used to produce pathogen-free plants and for germplasm storage. The current largest use of plant biotechnology is in the clonal propagation of plants, particularly of ornamentals because of their relatively high market value. The modern technique of rDNA offers the potential of moving any cloned gene from any organism into any other organism (the transgenic host) and is much more precise and faster in achieving results as compared to conventional plant or animal breeding techniques. However, biotechnology is not a substitute for the latter and should be seen as complementary. Indeed, the strengthening of traditional biological research is an essential prerequisite for establishing a biotechnological research capability in most developing countries.

Biotechnology offers a range of applications mainly for plant and animal production. Some are likely to have an increasing impact well before 2010 while others are of a longer term nature. The former include tissue culture of virus-free stocks of cassava and other root crops, and the introduction of microbial plant growth promoters such as mycorrhiza. The latter include cereals with the ability to fix some of their own nitrogen needs, and transgenic tree crops, but the greatest expectations are in introducing disease- and drought-resistant genes.

Many of these applications will contribute to more sustainable resource use, particularly by (a) gradually raising crop yields and reducing land requirements for a given level of production, thereby lowering the pressures on marginal lands and natural forests; (b) complementing industrial with biological sources of nitrogen for plant growth; (c) raising production performance of plants and animals through growth manipulations and by producing improved vaccines and enhancing disease resistance; and (d) lowering chemical inputs needed per unit of production.

12.6 A RESEARCH AGENDA FOR A SUSTAINABLE FUTURE

UNCED's Agenda 21 provided a philosophy for sustainable development which may be interpreted to guide the direction of agricultural research. The first emphasis is on the improved management of biological systems, based on a better understanding of their dynamic processes. Examples are integrated plant nutrition systems (IPNS) and integrated pest management (IPM) to substitute for heavy reliance on chemical-based interventions. The second emphasis is on better information management, implying the need for sound data on natural resources, land use and farming systems, agrometeorology, etc., to improve environmental monitoring capability and make better use of natural resource potential. The third emphasis is on improved farm–household system management, in order to obtain a better understanding of the differing systems and hence considering in an integrated way household, farm and off-farm activities. The emphasis also must be to obtain a fully participatory approach to development.

Moving to a more operational level, two central thrusts can be identified, each having a set of priorities. One thrust is aimed at promoting sustainable increases in productivity in the higher potential areas. The second should target marginal and fragile environments where current degradation must be reversed and production stabilized or raised. These must be supplemented by two cross-cutting and highly complementary approaches: that of rehabilitation and restoration of ecology; and that of exploiting the synergism of indigenous technical knowledge and modern science. All four actions must be supported by international efforts to strengthen the national agricultural research systems, both institutionally and financially, since they will have to undertake much of the adaptive and applied research.

The priorities for making progress towards sustainable growth in productivity include: expanding on-farm biological production and recycling of inputs and lowering off-farm mineral fertilizer and pesticide needs; raising average yields and crop yield ceilings; improving irrigation water management; limiting soil acidification; using energy more efficiently and promoting renewable energy sources; and reducing labour inputs of some multiple cropping systems.

On-farm production or recycling of inputs can serve three main purposes. First, they can provide small-scale farmers with a profitable alternative to high cost external input systems, which though effective in technical terms, imply financial risks. Second, they can help to prevent soil nutrient mining and the excessive build-up of mineral fertilizer and pesticides residues in soil water, groundwater and surface water. Third, through the greater use of live leguminous mulches, green manures and other organic residues, they can improve soil structure, maintain soil fertility and enhance the soil's role as a sink for carbon dioxide. This requires a much deeper knowledge of agroecosystem functions but may also face severe labour constraints. It is difficult to imagine, for example, how some of the complex intercropping and relay planting systems currently used in China to achieve three crops a year, which require considerable labour inputs, can survive when labour opportunity cost rises. Research priorities include nutrient recycling processes and techniques, natural resources management at village level, and integrated crop/livestock management systems.

Raising yield ceilings in difficult environments has been an issue for a number of years. There is now a better understanding of the issues involved, and progress is being made with some crops like millet and legumes. In parallel, there is renewed emphasis on raising yield ceilings of staple food crops in the high potential areas, which have been at the centre of the successes of the "green revolution" (the irrigated rice and wheat areas of Asia) and where experimental yields seem to have reached a plateau and have been virtually static for the past 10 years or more (Pingali et al., 1990). Although average farm yields in these areas are still well below experimental yields, and so yield growth could continue to 2010, thereafter the decline in yield growth rates could accelerate unless research manages to break through the plateau and bring about another shift in the production frontier (Hayami and Otsuka, 1991). This challenge has become a major priority for IRRI and other research institutions, but their efforts must be backed up by national action. Recent widespread introduction of hybrid rice in China and other countries in Asia, first for temperate-zone rice but subsequently for tropical varieties, offers great promise of significantly raising the yield ceiling for rice (see Chapter 4). Priority research tasks include: (a) production of crop varieties with enhanced tolerance or resistance to moisture stress and soil nutrient constraints; (b) studies to overcome micronutrient deficiencies; and (c) investigating soil conditions under continuous intensive crop production as well as under low input conditions.

The research agenda for irrigation has three components. One is the use of cheap sources of low quality water in place of scarce high quality water. The second is raising water use efficiency so as to reduce unit costs. The third is to improve the management of irrigation systems. Existing technologies can go a long way to sustaining growth in the medium term, but accelerated research is needed to find more economic ways of preventing the further deterioration of existing water resources and to widen the technological options for the future. Key priorities are: (a) raising the efficiency of flood irrigation by practical methods of flood depth and seepage control, better land preparation, and the use of alternate wet and dry regimes; (b) adaptation of surge irrigation to developing country conditions; (c) development of simple, efficient and economical wastewater treatment methods so as to prevent or minimize adverse impacts on human health and the environment; and (d) identifying the main institutional characteristics of successful irrigation management systems and the effects of transferring management to farmers.

A common outcome of more intensive land use, whether it be through high or low external input systems, is increasing competition between crops and weeds for the available soil nutrients and water, and for light. The conventional responses are either more hand weeding or more herbicides. Labour for the former is becoming increasingly scarce and expensive, and the increasing quantities of pesticides are a growing threat to the environment and human health. Key priorities are (a) greater emphasis on weed management; (b) increased research on biological control methods and biodegradable herbicides; and (c) researching innovative ways to reduce herbicide applications.[7]

Fulfilling the energy needs of agriculture and of rural services is at the core of improving productivity. Land preparation, harvesting, irrigation and processing require different types and levels of energy inputs, both in direct (mechanical, thermal, fossil and electrical energy) and indirect (fertilizer) forms. Without these energy inputs, agricultural productivity will remain low and probably well below its full potential. At the same time, unsustainable practices based on unnecessarily high energy inputs lead to resource depletion. There is the need, therefore, to understand better the energy–agriculture links and the potential of sustainable energy systems based on renewable energy sources, mainly biomass, solar and wind energies. The potential of agriculture itself as an energy producer requires further studies and research into the use of biomass residues, energy plantations and combined energy and food production schemes. Key priorities are (a) evaluation of the energy–agriculture interrelationships for different ecosystems, and in relation to high and low potential areas; (b) better understanding of the integrated management of energy and other inputs (water, fertilizer, pesticides, mechanization); and (c) assessment of the potential of biofuels for different environmental and land-use policy situations.

Concerning the need to safeguard the marginal areas, there are two facets to this problem. The first concerns those areas that need not be permanently marginal, since with suitable investments, institutional changes and technologies they could become moderate to high potential areas (see Box 12.1). The second relates to areas that are inherently marginal because of severe aridity constraints that cannot be overcome by irrigation, or adverse soil types that cannot be exploited economically. Here priority must be given to limiting degradation while non-agricultural employment opportunities are created, so that people can afford to buy food from the better endowed areas, instead of being forced to over-exploit the land to meet basic needs.

It is increasingly acknowledged that there are many indigenous techniques for on-farm or local water conservation that can be used now or quickly adapted to complement the above actions. Maintaining progress after 2010 will, however, require more basic and applied research in the next two decades.

As with plant breeding, lead times can be 10 to 15 years or more. Some of the minimal tillage techniques, for example, that have transformed dryland crop production in parts of the USA, took some 20 years to develop and implement. And since they are low input systems, they are much more sustainable in terms of fossil energy needs and soil fertility maintenance than the farming practices they have replaced. Research priorities include: (a) development of minimal tillage systems for low-income farmers in the drylands of developing countries; and (b) methods to improve pastures in the extensive range areas, both under tropical and temperate conditions.

While there are substantial opportunities for developing sustainable low-input systems, complete independence from external inputs is not possible except in a few special circumstances such as volcanic and aeolic dust deposition or sediment-loaded flooding. More research is required on lowering the relative costs of external inputs, or achieving the same objective by raising input use efficiency, or reducing needs through innovative ways of overcoming the factors that currently cause marginality. Some of the opportunities are well illustrated by the widespread problem of phosphate-deficient soils. Phosphate is an essential chemical for plant growth but many soils are highly deficient in it, or the phosphate present is not available to plants because of other soil factors such as aluminum or iron toxicity. Organic manures are seldom the long-term solution if high yields are to be achieved, while manure made from biomass grown on phosphorus-deficient soils will itself be deficient in phosphorus.

The conventional production of phosphate fertilizers is expensive, and when long distance transport costs are added, they tend to become even more uneconomic for farmers in marginal areas. Many countries, however, have low-grade phosphate rock or other phosphate-bearing materials that could be used if cheap methods could be found for treating them so that the phosphate becomes readily available to plants. Alternatively, and in the longer term, there seem to be possibilities for genetically transferring to other crops the properties of pigeon-peas to release bound phosphates in the soil and make them available for plant growth. Without such technological breakthroughs, however, sustainable development in many marginal areas will be impossible. Research priorities therefore include: (a) development of cheap techniques for improving the effectiveness of using low-grade phosphate-bearing rocks, such as incorporating organic matter and promoting mycorrhiza activity in the soil; and (b) identifying the pigeon-peas mechanism for releasing bound soil phosphate, and possible transfer of the mechanism to other crops.

Increasing competition between crop and livestock production for land in both high potential and marginal areas, combined with the threat to crop yields from declining soil fertility, will be a positive force for integrating the two systems. This will, however, intensify the demands on research to come up with more satisfactory solutions to a number of problems and opportunities, notably: (a) reduction in the labour needs and other constraints to the adoption of livestock-oriented alley cropping and other sylvo-pastoral systems, taking into account competition for water, light and plant nutrients; (b) development of legume-based relay, intercropping, soil conservation, and other practices to raise the supply of high protein feeds and forages; and (c) introduction and refinement of ley farming systems for acid and other low-fertility soils based on legume and grass pastures.

This chapter focused on the needs and opportunities for producing the projected output in a sustainable way and for laying the foundation for sustainable agriculture in the twenty-first century. The needs, however, are not only those of research and technology development. Unless technological progress is accompanied by institutional changes and other improvements in the incentives for agricultural development, many of the research findings will not leave the laboratory or the experiment station, or be developed on farms themselves. These aspects, some of which have been already discussed in earlier chapters, will be further explored in the next, concluding chapter.

NOTES

1. In fact the peak in US timber prices lasted barely a month, March 1993, and by May they had returned to their level of late 1992. The "spotted owl" controversy added further uncertainty to an already uncertain market. There were costs nevertheless.
2. The HIV/AIDS pandemic sweeping many rural communities in affected African countries is having a similar effect on labour supplies leading to less labour-intensive but less productive farming systems (see Chapter 3, note 2).
3. Low growing crops, which fully cover the ground and protect the soil surface against rainfall impact and wind erosion, provide nitrogen to the associated crops, food for earthworms, and raise soil organic matter levels so as to increase soil porosity, rainfall infiltration and retention.
4. These are: irrigated or well-watered conditions; relay cropping with two or more harvests a year; crops at different stages of maturity in close proximity; no or limited fallows. All favour the substantial carryover of pests from one crop to the next, and the early infection of young plants or competition with them, in the case of weeds.
5. The requirements for sustainable water use are not just technological. The institutional and policy dimensions are equally important and are dealt with in Chapter 13.
6. Recent studies have shown that, among the ruminants, buffaloes and camelids have a greater efficiency in digesting fibrous feeds and recycling urea.
7. For example, in the "Cerrado" area of Brazil, soybean farmers practising minimum tillage methods, spray herbicides at night (calm, cool and damp conditions) and successfully reduce application rates to a quarter of those recommended.

Minimizing the Trade-offs between the Environment and Agricultural Development

13.1 INTRODUCTION

The preceding chapters have demonstrated that there are commonly a number of trade-offs between the environment, food security and other aspects of development. Some of the trade-offs are avoidable, but others are not. The underlying issues were clarified and brought into prominence by the Brundtland Commission. Its report emphasized the difficult task of reconciling the short-term imperative of increasing food and agricultural production as well as incomes for the current generation with the longer term, but almost unspecifiable, need of conserving natural resources for meeting the requirements of future generations (World Commission on Environment and Development, 1987). Thus, while the long-term objective may be sustainable development of agriculture and of the whole economy, the pathways or processes involved may have to breach the environmental requirements of this goal in the short to medium term. Hence the importance of minimizing the trade-offs.

Chapter 11 spelt out as far as possible the pressures on the environment associated with the possible future evolution of agriculture. Many of the pressures on the environment associated with the quest for development have, however, become a subject of contention between the developed and the developing countries and account for some of the differing priorities pursued by these two groups at the United Nations Conference on Environment and Development (UNCED) and other international fora.

13.2 THE NORTH–SOUTH DIVIDE

The developed countries tend to give priority to the environmental dimension and to measures to limit natural resource degradation, in spite of the economic and social costs that may be associated with such measures. In doing so they seldom acknowledge that a sound environment is, in some respects at least, a luxury good which they can now afford, but which in their earlier history they had largely ignored. The developing countries, of necessity, tend to argue for different priorities. They recognize the importance of shifting on to a more sustainable growth path and at UNCED gave overwhelming support to AGENDA 21 (Box 13.1), and the conventions on biodiversity and climate change. But they emphasize the need to ensure that environmental measures do not have adverse effects on their development, arguing, for example, that unless rural poverty is eliminated, many of their people will have no alternative to overexploiting natural resources for day-to-day survival.

Most developed countries have already taken measures to overcome or continue to alleviate the more serious agricultural threats to the environment. They have, for example, taken marginal land out of production; reduced or banned the use of mineral fertilizers and residual pesticides on sensitive watersheds vulnerable to groundwater contamination; tightened the controls on waste disposal from intensive livestock units, and so on. Further measures to protect the environment from agricultural pressures are largely a question of social choice since they have the economic and technical capabilities to introduce additional control measures or technologies that are environmentally more benign and sustainable. They can also bear with less economic hardship than the developing countries the eventual economic consequences of such action, e.g. higher food costs and/or higher food imports or lower exports. With the wider acceptance of, and further progress in, environment-friendly technologies, it can be expected that trade-offs between the environment and development, as the latter is conventionally defined and measured (e.g. per caput incomes, etc.), will tend to become smaller.

The situation in most developing countries is quite different. For them, improving agricultural resources management is a social imperative rather than a social choice, because degradation of these resources is both a cause and a result of poverty (see Chapter 2). But their environmental options are often severely constrained in the short to medium term at least. They will have to use more of their less productive land, as well as some of their wetlands with high agricultural potential. They cannot reduce their often low use of mineral fertilizers without endangering food security and intensifying soil degradation, and cannot exploit more fully technological options for integrated plant nutrition quickly. For some production or environmental problems appropriate technical solutions do not yet exist and are not readily available or affordable. Any appreciable rise in food production costs and consumer prices would have adverse effects on already low consumption levels, and many countries could not afford to increase commercial food imports. Finally, they are already finding it very difficult to maintain existing levels of public services, so environmental measures are commonly in direct competition for scarce resources with projects for investment in material and human capital.

Fortunately, there is much that can be done to minimize these trade-offs, and there are also a number of actions which are sounder in environmental terms as

Box 13.1   The outcome of UNCED*

The UN Conference on Environment and Development (Rio de Janeiro, 3–14 June 1992), at which 172 Member Governments were represented, concluded with the Earth Summit, during which 102 Heads of State and Government made statements, expressing their commitment to environmentally sound and sustainable development. The main agreements reached and subsequently endorsed by the 47th Session of the UN General Assembly were:

1. The Rio Declaration on Environment and Development (to be further elaborated as an Earth Charter for adoption by the UN General Assembly in 1995 on the 50th anniversary of the UN).
2. Agenda 21, which describes in some 280 pages a comprehensive plan of action consisting of 115 programme areas, grouped in 40 chapters. Of particular interest to food and agriculture are Chapter 10, integrated approach to the planning and management of land resources; Chapter 11, combating deforestation; Chapter 12, managing fragile ecosystems: combating desertification and drought; Chapter 13, managing fragile ecosystems: sustainable mountain development; Chapter 14, promoting sustainable agriculture and rural development; and Chapter 17, oceans and marine resources.
3. The Framework Convention on Climate Change, signed by 162 governments.
4. The Convention on Biological Diversity, signed by 159 governments.
5. The "non-legally binding authoritative statement of principles for a global consensus on the management, conservation and development of all types of forests", which may lead to "internationally agreed arrangements to promote international cooperation".
6. Agreement on sources and mechanisms to mobilize financing for Agenda 21, including new and additional resources in grants and concessional terms from the international community: (a) the multilateral development banks and funds, including the IDA, regional and sub-regional development banks and the Global Environment Facility (GEF); (b) the relevant specialized agencies, other UN bodies and international organizations; (c) multilateral institutions for capacity building and technical cooperation; (d) bilateral assistance programmes; (e) debt relief; (f) private funding (NGOs); (g) investment; and (h) innovative financing.
7. International institutional arrangements for follow-up, particularly the establishment of the Commission for Sustainable Development of ECOSOC, and an Inter-Agency Committee on Sustainable Development, established by the ACC.
8. The launching of a negotiated process for an International Convention to combat desertification in those countries experiencing serious drought and/or desertification, particularly in Africa. The Convention was finalized in June 1994.
9. Convening of a number of UN conferences, among them the Conference on Straddling Fish Stocks and Highly Migratory Fish Stocks (New York, July 1993), and the Global Conference on Sustainable Development of Small Island Developing States (Barbados, April 1994).

*See also FAO (1993b).

well as profitable. Chapter 12 has dealt with the technological problems and opportunities. It remains for this chapter to consider the wider institutional and policy changes that are needed to provide the right incentives and support mechanisms for the technology uptake required to achieve the agricultural production projected in this study, and to contribute to sustainable agriculture and rural development (SARD). Section 13.3 suggests what the main thrusts might be of a strategy to minimize any environmental and development trade-offs. Section 13.4 moves from strategy to tactics and deals with policies and resource management measures to address the main environmental pressures emanating from the continuation of agricultural growth.

13.3 A STRATEGY FOR MINIMIZING ENVIRONMENT AND DEVELOPMENT TRADE-OFFS

The technological opportunities considered in Chapter 12 are not sufficient in themselves to stimulate ecologically sound and sustainable development to 2010 and beyond. The economic and institutional environment must also be favourable. Farmers must have better access to proven technologies, production inputs and services, and to markets for their products. They must be secure in their rights to access to land and other resources so that they have the stability and confidence to take up the technological opportunities and make the necessary investments. And, finally since few technologies are totally risk free in the absence of safeguards to protect public goods, there must be an appropriate regulatory environment. Inconsistencies or gaps in these supporting measures can seriously undermine the effectiveness of individual technologies and of the total package. It is therefore essential that the minimization of trade-offs and actions to shift agriculture on to a sustainable growth path take place within a consistent strategic framework.

FAO took the lead in developing such a framework as one of its contributions to the UN Conference on Environment and Development (FAO, 1991c). In particular, FAO took responsibility for a number of the chapters of AGENDA 21, the main operational recommendations of the Conference (see Box 13.1). AGENDA 21 contains proposals for a wide range of technical and institutional changes that support the policy and other assumptions underlying the 2010 projections, and which are essential for longer term development (UN, 1993a). These numerous changes form an overall strategy for sustainable agricultural and rural development in the long term. But within this overall strategy it is possible to think of component strategies which focus on specific issues, in this case the minimization of environment and development trade-offs, e.g. by lowering the pressure to clear primary forests or drain wetlands for agricultural use.

The preceding chapters, in particular Chapters 7 to 10, have discussed the main thrust of policies that should underpin efforts to improve the performance of agriculture, reduce rural poverty and improve access to food.

It remains for this chapter to discuss the issues that are more specific to the objective of having a strategy which seeks to minimize the environment–development trade-offs. First, as shown in Chapter 12, emphasis must be put on shifting technology from "hardware" solutions requiring large inputs of fixed and variable capital, e.g. machine-made land terraces or pesticides, to solutions based on more sophisticated, knowledge and information-based resource management practices. The latter solutions carry fewer environmental, economic and health risks for the producer, agricultural worker and consumer. Examples are vegetative barriers to soil erosion instead of machine- or man-made terraces, and integrated pest management based on knowledge of predator–prey relationships, as opposed to control strategies based mainly on the use of pesticides. Thus they aim to lower both off-farm costs and environmental pressures. Machine-made terraces, for example, require large capital investments. These commonly come from scarce public resources and hence compete with other needs, including expenditure on health and education. The machinery and the fossil fuel to run them generally have to be imported using scarce foreign exchange. Such terraces often fail through early neglect because they were introduced through a top-down process that excluded farmers and communities from the design or implementation stages.

Second, greater importance must be given to establishing well-defined property or user rights for public and private resources. Examples are land tenure arrangements or the creation of user groups commonly building on traditional systems. In Africa especially, the latter tend to work better than Western style individual titling of land holdings. Without such changes, users of common property resources will have little or no incentive to exploit them in a sustainable way; uncontrolled development will commonly lower the economic benefits of given resources. Clear rights of access to land, on the other hand, provide economic and social incentives to protect and improve resources, although it is seldom a sufficient condition alone to achieve these objectives.

Third, success depends greatly on people's participation and decentralized resource management. In the main it is small farmers, herders and forest dwellers who make the key decisions about resource use. They decide whether to adopt sustainable practices, to clear forests or plough up pastures, and they do so in response to household security needs and incentives rather than government dictates. And collectively, as villages or larger communities, they decide about watershed management. The public sector of developing countries cannot afford the cost of enforcing such requirements, and furthermore lacks the vested interest of owners and users in safeguarding the resources being exploited.

Fourth, as far as possible market signals must embody proper valuation of environmental goods. As far as possible, commodity prices should include all direct and indirect environmental costs. Public bodies cannot "police" all aspects of resource use without their becoming a politically and economically unsustainable burden. The alternative is to use price signals to achieve the same broad objective, but with the support of a national and international regulatory environment that sets appropriate standards, e.g. certification schemes for timber harvested in a sustainable way, and labelling requirements for "organically" produced commodities.

The above basic thrusts of a broad global nature are a prerequisite for some countries to achieve the agricultural improvements projected for 2010 while minimizing trade-offs with the environment. To a greater or lesser degree, however, they are essential to all countries, developed and developing, if the output projected for 2010 is to be sustainable in the long term. But these thrusts are not sufficient in themselves. If they are to lead to effective policy changes and operational measures, they must be amplified and adapted to national circumstances and to the specific problems of different agroecosystems. Four broad agroecosystems account for more than 95 percent of food and agricultural production in the developing countries, and their individual contributions to present and future production are recognized in the projections. They are dryland/uncertain rainfall, lowland humid/per-humid, irrigated and hill/mountain agroecosystems. These may co-exist within a single country, although any one of them may dominate. The specific characteristics, problems and opportunities of such agroecosystems can provide a rational basis for strategic planning at the national or supra-national level. However, the final trade-off decisions will have to be made by the farmer or local community who may have to make decisions with regard to resources with different production characteristics.

Most important of all is the need to adopt a strategic approach to the national institutional setting or framework since this is a unique blend of cultural, political, economic, social and physical factors, that govern the validity of the different trade-offs. The strategy must be national yet operationally decentralized. Certain responsibilities will have to be carried at the national level but with much of the consequent action being devolved to the user. It is the sovereign responsibility of governments to reconcile the resource demands of present users and the needs of future generations. Only governments have authority over the required range of legal, fiscal and social instruments as well as the supporting institutions and services. Only governments have the authority to make binding agreements on trans-border issues, such as the management of international water basins, that are vital for sustainable development and environmental protection. Yet the potential for government failure is indisputable and there are distinct limits to the role governments and national bodies can play in natural resource management. Therefore, much of the responsibility for specific trade-off decisions must be left to local communities and farmers, encouraged by appropriate national policies and regulations.

The strategy must recognize that the first priority of many farmers is household food security and family welfare. Farmers will trade off immediate food production, even though it may involve some degradation, against a less tangible but sustainable future. Thus efforts to minimize trade-offs must be centred on actions that improve household food supply or food purchasing power, reduce seasonal fluctuations and improve overall access to food. Moreover, meeting food needs is not enough. The strategy must be profitable to farmers and other private investors on time scales which meet their differing circumstances or risk perceptions. At a minimum they must have the time or earnings to invest in sustainability.

Any strategy must have a legal basis and well-defined rules for resource utilization. Strategies must clearly define responsibilities and allocate rights of access to or use of environmental resources (see Section 13.4). They must be socio-politically acceptable and equitable as well as within the implementation capacity of both governments and individuals. Governments must be both willing and able to exert their sovereign responsibilities for action. First, the political will must exist to accept possible negative public reactions to trade-offs that are perceived to restrict private behaviour. Then governments must ensure that sound public or private institutional mechanisms exist to provide all farmers and communities with the support services needed to act on the options for minimizing trade-offs. They must also respond to any major social stresses arising from the need for people to move out of agriculture into other sectors.

13.4   MINIMIZING SPECIFIC TRADE-OFFS

As stated in the introduction, some trade-offs between the environment and agricultural development are unavoidable in the short to medium term at least. First, some developing countries may have to clear more of their natural forests and drain part of their wetlands in order to feed their people, promote economic growth and improve net social well-being. Secondly, appropriate environmentally sound technologies do not exist for many problems or situations, so less sustainable ones will have to be used in the short to medium term while research develops suitable replacements. Thirdly, they have insufficient suitably trained manpower. Finally, many developing countries lack the local and central institutional mechanisms required to collect and analyse data relating to resource management, to assess the various options open to them, to improve the functioning of markets, and to obtain farmer and rural community support. Some of these obstacles will take at least a decade to overcome and, in some instances, much longer.

Minimizing the trade-offs requires a holistic approach with three main dimensions – technical, institutional and international – which have their own policy requirements but must also be formulated and implemented together in order to give overall policy consistency.

The technical dimension

Population and economic growth has raised food and agriculture demand above levels that can be produced by extensive, environmentally sound, farming practices (see Chapter 12). Consequently farming practices have to be intensified and output raised by technical inputs that may have unavoidable environmental penalties in the short term. Yet these penalties can be ameliorated and must be placed in their proper perspective. The nitrate problem in groundwater and surface water, and the issue of mineral versus organic sources of nitrogen, provide illustrations. There are those who argue for a complete ban on the use of mineral fertilizer. This is totally unrealistic. It would not solve the groundwater nitrate problem, and would inevitably lead to serious food shortages, declining incomes and malnutrition, as well as creating greater soil degradation through nutrient mining for the following reasons:

1. Both mineral and organic sources of nitrogen contribute to the problem. Examples are when mineral fertilizers are broadcast on the soil surface, or organic manures are applied to fallow land when there are no crops to take up the nutrients released by breakdown of the manure, and they are leached down into the groundwater. Again, sometimes livestock units or fish farms dispose of their organic wastes directly into streams and rivers, or store them badly so that they are dissolved by rain and carried off into the surface water system.

2. In large areas of Africa, for example, there is insufficient organic nitrogen available from livestock manures or legume-based production systems. Hence organic sources of nitrogen alone cannot achieve the high yields required to compensate for the small size of farms, or to meet basic food needs at the national level. Achievement of higher farm incomes and prevention of nutrient mining inevitably imply increased use of mineral and organic fertilizers[1] and the release of their residues into the environment. Nonetheless much can be done to minimize the problem. Some of the technical measures for limiting the problems through the adoption of IPNS and other measures to raise fertilizer use efficiency were shown in Chapter 12. But as with other technical measures for soil conservation, pest control, irrigation and water management, mere existence is not enough, nor are unconstrained market forces. There are also a range of institutional requirements to shape and support development and uptake of these measures.

The institutional dimension

It is difficult to establish the nature and magnitude of some of the causal relationships between institutional change and environment impacts, and hence the precise role of such changes in minimizing the trade-offs. It is clear, however, that changes are required at a number of levels to achieve greater consistency between the technical and the institutional dimensions. Actions are required at the national and local planning levels to restrict or direct resource use; at the research and extension level to develop and transfer to farmers, forest users and fishermen the knowledge of sustainable technologies and agricultural practices; at the technological input level to ensure that the delivery systems operate efficiently and in the interests of the users; at the input and commodity price policy level to avoid or ameliorate market distortions so that farmers have the economic incentive to shift on to a more sustainable technological pathway. Finally, actions are necessary to create a regulatory environment that ensures that public goods such as air and water are protected, and consumers are not placed at risk through the overuse of pesticides, fertilizers, livestock growth promoters, etc.

Concerning the scope for resource development planning, it is noted that national water and land use plans tend to be rigid in their perception and unrealistic in their objectives and mechanisms for implementation. It is not possible to legislate for sustainable land use. The basic motivation must come from the awareness, self-interest in terms of household food security or welfare objectives, and functional capacity of the user.

This is not to say that national resource use planning is not vitally important, but to present the case for a more balanced institutional mechanism which brings together the top-down and the bottom-up approaches. National planning for resource use and management is important in a number of key areas:

1. Determining the land and water resources most suited for development or more intensive use as the first step in the planning process leading to land or water use changes, and the formulation of price and other incentives for resource users that are as consistent as possible with the sustainable management of those resources.
2. Helping to resolve problems of competition between different sectors or sub-sectors for diminishing land and water resources, and catalysing or imposing their conjunctive use, e.g. through agroforestry or the use of urban waste waters for irrigation.
3. Identifying and protecting fragile ecosystems, critically important habitats, sources of biodiversity and watersheds, through the creation of national parks, collection of germplasm and so on.
4. Ensuring spatially balanced rather than unipolar urban development so that it is not centred on one or two mega-cities, and as far as possible secondary towns and cities are allowed to expand or are developed in areas with marginal soils yet adjacent to regions with good soils that can sustain them without overstretching the lines of communication between producers and consumers.
5. Establishing road, rail and water links that do not expose protected areas to informal development; improve access between the areas which can develop sustainable production systems and their urban or overseas markets so that

resource users have both the incentive and the financial means to adopt better conservation measures etc., and minimize the transaction costs for supporting farmers with production inputs and the urban areas with food and agroindustrial raw materials.

6. Determining the allocation of public support between agriculturally marginal areas and those with high production potential. This requires a careful and analytically rigorous assessment of land suitability, in that current marginality is not necessarily a true reflection of land potential given appropriate soil conservation measures and other technological changes as has been demonstrated clearly by the people of the Machakos district of Kenya (see Box 12.1). Thus it is important to have a clear understanding of how research may change land use suitability, and widen the resource use options. It also needs an appreciation of what is happening in the rest of the economy, in that although the long-term objective may be to encourage people to move out of the marginal lands into areas which can provide more sustainable livelihoods, this may not be possible in the short to medium term. Hence, public resources may need to be allocated to the marginal areas, more or less as a holding operation while the possibilities for alternative livelihoods are put in place.

7. Monitoring land and water use to anticipate problems of resource competition and degradation, and to identify and implement corrective actions, e.g. through conjunctive use of water as noted above.

8. Where feasible, charging for resources such as water, often regarded as an open access resource, at least to maintain infrastructure and protect water catchments.

Research, extension and technology

Chapter 4 drew attention to the large gap between current best farmer yields and national average yields. However, it also stressed that there are gaps between the available technologies and those needed for achieving the production levels projected for 2010 in an environmentally sound way, and for setting the foundation for long-term sustainable growth. The institutional problems are manifold. They start with the lack of political awareness of the role agriculture plays in economic development (see Chapter 7). There is also a lack of appreciation of: (a) the large returns that can come from sound agricultural research which may exceed by far those stemming from alternative development investments; and (b) the contribution such research makes to resource conservation by reducing the pressure to bring undeveloped land into cultivation. In India, for example, the introduction of high-yielding wheat varieties may have saved about 30 million ha of marginal lands and forests from being brought into wheat cultivation. Research also contributes indirectly to resource conservation to the extent that increased productivity and incomes reduce rural poverty and help overall development.

Then there is the skill gap for technology development, and the scaling up of research for commercialization, which needs to be addressed by appropriate manpower and institutional support (see Chapter 10). Providing this support leads to the problems of the research and extension institutions and mechanisms themselves, because in the main they have not focused their efforts on the issue of sustainability. They have seldom focused on the land areas with pressing environmental problems or on sustainable technologies appropriate to poor farmers in these areas or in high potential regions. There needs to be better mechanisms for research priority setting at the national level because the International Agricultural Research Centres (IARCs) cannot be expected to do the adaptive research or the more fundamental research for geographically confined problems. Large- and small-scale farmers need to be more closely involved in the identification of research problems and to be drawn more frequently into partnerships with scientists for the solution of these problems, building on the best of indigenous and laboratory knowledge.

Economic policies affecting agriculture, as well as policies relating to public sector involvement in input and commodity marketing, have been dealt with in some depth in Chapters 7 and 9. Hence it is sufficient just to underline the importance for reversal of policies which discriminate against agriculture and lead to unsustainable practices by making unprofitable the use of inputs and the diffusion of sustainable technologies. In particular, policies, including those relating to the functioning of marketing parastatals, must be corrected to remove: (a) upward distortions in the price of production inputs through import restrictions and tariffs; and (b) uncertainties regarding the timely supply of seeds and mineral fertilizers, in cases of parastatals with monopoly powers for their sale and distribution, in that late delivery exposes the farmers to poor returns from what to them are high cost and risky expenditures. One should not forget that subsidies provided to pesticides and mineral fertilizers can also lead to their excessive use, causing the degradation problems discussed in Chapter 11. These and other problems of public sector origin have been common disincentives for the adoption of soil conservation techniques, the balanced use of mineral fertilizers and other requirements for meeting food security and wider development needs in a sustainable way.

The regulatory environment

Developed country experience has shown that the foregoing actions are not sufficient in themselves to direct growth towards a pathway that brings social and environmental objectives together. Land use planning, for example, may identify which areas to protect and which are the most favourable for development, but the introduction of new technologies and unconstrained market forces are likely to override such planning considerations and therefore need to be backed by legally enforceable restrictions. Similarly, diminishing marginal returns to mineral fertilizer use may not limit their application rates soon enough to prevent serious groundwater pollution. It follows, therefore, that public institutions must be established to set appropriate regulatory standards, to monitor compliance with them, and to take suitable legal or financial steps when practices are not meeting the objective of minimizing environment and development trade-offs.

The required regulatory measures are quite diverse, and include: (a) statutory restrictions on the use of protected areas or on the development of high quality arable land for urban/industrial use; (b) restrictions on the use of mineral fertilizers on sensitive watersheds; (c) constraints on the quantity and timing of organic fertilizer applications to land; (d) design requirements for manure storage on livestock farms; (e) effluent quality standards for discharges to water courses from livestock units, fish farms, and agroprocessing activities; (f) sanitation standards for slaughterhouses and cold storage units; (g) restrictions on the type of pesticides that can be imported and used, and the timing of their application in conformity with the International Code of Conduct on the Distribution and Use of Pesticides (FAO, 1990e) and Codex Alimentarius (FAO, 1994a), respectively; (h) controls on the labelling of pesticide containers and their disposal; and (i) biosafety standards for the release of genetically modified organisms into the environment.

Property rights

Unsustainable agricultural practices commonly take place where those involved have limited or no property or user rights to the resources they are overexploiting. The awarding of secure rights, whether individual or communal, would greatly increase their vested interest in improving resource management and investing in soil conservation and other land improvements. Property rights have a wider institutional dimension relating to the efficiency of markets and the management of public goods. Environmental trade-offs are also not minimized in situations when the institutions controlling public goods have collapsed, or because markets are not able to value public goods such as fresh air, or to cost public "bads" such as pollution. Markets must be made to work better by defining property rights more precisely and establishing or strengthening the institutions to manage them; introducing realistic prices for environmental goods such as water; and attempting to cost public "bads" and adopting "the polluter pays principle" where this is appropriate.

The international dimension

This dimension is particularly important given that much of the mismanagement of natural resources in developing countries relates to poverty and to the lack of economic growth to provide better and sustainable livelihoods outside subsistence agriculture. Minimizing the trade-offs needs a global economic environment that is more conducive to growth so that the developing countries can significantly increase gainful employment outside agriculture. This is critically important for those arid, highland and land-locked countries with predominantly marginal land which tend to suffer from high transport costs for off-farm inputs like mineral fertilizers and/or poor inherent biological productivity. Therefore, any policies which affect the development prospects of the developing countries via the link of the international economic environment are of direct importance to the objective of minimizing the environment–development trade-offs. Here belong the issues of trade, debt and resource flows. Some of these issues are discussed in other chapters and the discussion is not repeated here.

Of particular interest is the extent to which environmental pressures are transmitted among countries by means of the agricultural trade flows. The terms "environmental subsidies" or "ecological footprints" are sometimes used to denote the transmission of such pressures.[2] For example, there might be environmental subsidies from the USA to those countries which import large quantities of maize from the USA, whose production contributes to soil erosion, involves heavy applications of mineral fertilizers and pesticides which are a source of ground and surface pollution and a negative pressure on natural ecosystems. Similarly, the Netherlands exports dairy products which indirectly are a major cause of pollution in the Netherlands. On the other hand, the Netherlands, together with other European countries, imports large quantities of cassava chips from South-East Asia, which are commonly grown in high rainfall areas on steep slopes with fragile soils, and result in very large soil losses through erosion. Thus, these are issues for developed and developing countries alike, but with the former better able to adopt the "polluter pays principle" or to introduce environmental regulations to make market prices reflect the environmental costs (for more discussions see Chapter 8).

13.5 THE ENDPOINT AND THE BEGINNING

The possible environmental dimensions of the agricultural projections have been edged with uncertainty but they are objective as far as the data and understanding of them allow. They will be wrong to some degree or other. The feedback loops between the economy, agricultural development and the environment are too complex and too dynamic to mimic with any certainty. And consequently, the strengths of the trade-offs and their associated risks are equally uncertain, hence the present stress on minimizing them, and adopting the precautionary principle. Nonetheless, two aspects seem clear.

First, it is important not to take an excessively static view of what is possible. The people of the Machakos district of Kenya have shown that it is possible to turn back from the edge of environmental disaster, rehabilitate seriously degraded land and introduce more sustainable production systems (Box 12.1) as have others in China, Indonesia and many agroecologically different parts of the world.

Second, the required actions go well beyond the so-called technological fix, although new technologies based on the latest scientific understanding will be vitally important, as will the revival or up-grading of indigenous technologies. They include international action to create a more open and equitable trading system with wider and stronger environmental safeguards, and to channel development assistance towards sustainable agriculture in a more consistent way. But the key actions are at the national and local levels. They include those which promote development, create a regulatory and incentive environment that encourage the uptake of sustainable technologies, promote decentralized, participatory, bottom-up approaches to natural resource planning and management, and contribute to slowing down the population growth rate.

Perhaps most important of all, what is required is more recognition of the anthropocentric approach to development and to the very idea of environmental conservation, and greater humility among those who argue for an ecocentric approach that does not match the expectations and resources of the farmers in the poor countries.

NOTES

1. "Some people say that Africa's food problems can be solved without the applications of chemical fertilizers. They are dreaming. It is not possible". Norman Borlaug, *Financial Times*, 10 June 1994.
2. Environmental subsidies are the costs of land degradation, loss of biodiversity and so on arising from agricultural production, that an exporting country gives to an importing country when the prices of the traded goods do not embody such costs. Ecological footprints are the total inputs of natural resources and environmental services from the land, sea and air that are required to sustain a given population at its current consumption rate. A country which is totally self-sufficient in food, fuel, minerals and other natural resources, and did not trade in them, would have a footprint that falls entirely within its own national boundaries if it could also confine its pollution to those boundaries.

Appendices

1. COUNTRIES AND COMMODITIES CLASSIFICATION

List of the developing countries of the study

Africa, sub-Sahara	Latin America and Caribbean	Near East/ North Africa	Asia
South Asia	Argentina	Afghanistan	*South Asia*
* Angola	* Bolivia	Algeria	
* Benin	* Brazil	Egypt	* Bangladesh
* Botswana	Chile	Iran	* India
* Burkina Faso	* Colombia	Iraq	* Nepal
* Burundi	* Costa Rica	Jordan	* Pakistan
* Cameroon	* Cuba	Lebanon	* Sri Lanka
* Central Afr. Rep.	* Dominican Rep.	Libya	
* Chad	* Ecuador	Morocco	
* Congo	* El Salvador	Saudi Arabia	*East Asia*
* Côte d'Ivoire	* Guatemala	Syria	
* Ethiopia (PDR†)	F Guyana	Tunisia	* Cambodia
* Gabon	F Haiti	Turkey	China
* Gambia	* Honduras	Yemen	* Indonesia
* Ghana	* Jamaica		Korea, DPR
* Guinea	* Mexico		Korea, Rep.
* Kenya	* Nicaragua		* Laos
Lesotho	* Panama		* Malaysia
* Liberia	F Paraguay		* Myanmar
* Madagascar	* Peru		* Philippines
* Malawi	* Suriname		* Thailand
F Mali	* Trinidad and Tobago		* Vietnam
F Mauritania	Uruguay		
Mauritius	* Venezuela		
* Mozambique			
* Namibia			
* Niger			
* Nigeria			
* Rwanda			
* Senegal			
* Sierra Leone			
F Somalia			
* Sudan			
Swaziland			
* Tanzania			
* Togo			
* Uganda			
* Zaire			
* Zambia			
* Zimbabwe			

Note: Data on land with rainfed crop production potential as well as data on cropping patterns of land-in-use by agroecological class are available for all countries except China and Namibia. In addition, for tropical countries marked with an asterisk, data on both forest areas and protected areas are available. For countries marked with F data are available on forest areas but not on protected areas (see Chapter 4). The Forest Resources Assessment 1990 for Tropical Countries (FRA1990) generated data on forest areas for some more tropical countries, not included in the 93 developing countries of this study.

†Former People's Democratic Republic of Ethiopia.

List of the developed countries of the study

EC	Other Western Europe	Eastern Europe and the former USSR
Belgium	Austria	Albania
Denmark	Finland	Bulgaria
France	Iceland	Former Czechoslovakia
Germany	Malta	Hungary
Greece	Norway	Poland
Ireland	Sweden	Romania
Italy	Switzerland	Former USSR
Luxembourg	Yugoslav SFR*	
Netherlands		
Portugal		Other Developed Countries
Spain	Oceania	
United Kingdom		Israel
	Australia	Japan
	New Zealand	South Africa
North America		
Canada		
United States		

*Former Socialist Federal Republic of Yugoslavia.

List of commodities of the study

Crops	Crops	Livestock
Wheat	Bananas	Beef, veal and buffalo meat
Rice, paddy	Citrus fruit	Mutton, lamb and goat
Maize	Other fruit	meat
Barley	Vegetable oil and oilseeds	Pig meat
Millet	(vegetable oil equivalent)†	Poultry meat
Sorghum	Cocoa beans	Milk and dairy products
Other cereals	Coffee	(whole milk equivalent)
Potatoes	Tea	Eggs
Sweet potatoes and yams	Tobacco	
Cassava	Cotton lint	
Other roots	Jute and hard fibres	
Plantains	Rubber	
Sugar, raw*		
Pulses		
Vegetables		

*Sugar production in the developing countries (excl. China) analysed separately for sugarcane and sugar beet.
†Vegetable oil production in the developing countries (excl. China) analysed separately for soybeans, groundnuts, sesame seed, coconuts, sunflower seed, palm oil/palm-kernel oil, all other oilseeds.

Note on commodities

All commodity data and projections in this report are expressed in terms of primary product equivalent unless stated otherwise. Historical commodity balances (supply utilization accounts–SUAs) are available for about 160 primary and 170 processed crop and livestock commodities. To reduce this amount of information to manageable proportions, all the SUA data were converted to the commodity specification given in the list of commodities, applying appropriate conversion factors (and ignoring joint products to avoid double counting, e.g. wheat flour is converted back into wheat while wheat bran is ignored). In this way, one supply utilization account in homogeneous units is derived for each of the commodities of the study. Meat production refers to indigenous meat production, i.e. production from slaughtered animals plus the meat equivalent of live animal exports minus the meat equivalent of all live animal imports. Cereals demand and trade data include the grain equivalent of beer consumption and trade.

The commodities for which SUAs were constructed are the 26 crops and the 6 livestock products given in the list above. The production analysis for the developing countries (excl. China, see Chapter 4) was, however, carried out for 33 crops because sugar and vegetable oils are analysed separately (for production analysis only) for the 9 crops shown in the footnote to the list.

2. METHODOLOGY OF QUANTITATIVE ANALYSIS AND THE PROJECTIONS

Commodity and country detail in the analysis

The quantitative analysis and projections were carried out in considerable detail in order to provide a basis for making statements, generally policy-related ones, about the future concerning: (a) individual commodities and groups of commodities as well as agriculture as a whole; and (b) any desired group of countries. These requirements dictated the need to carry out the analysis: (a) individually for as large a number of commodities as practicable and to account for a high share of total agricultural output; and (b) for individual countries for as many countries as practicable. Although many statements refer to regions, such statements are often required also for smaller geographic or functional country groups which partly overlap with regions as well as among themselves, e.g. grouping countries by income level, by degree of dependence on agriculture, by participation in economic cooperation schemes, etc.

The above-indicated degree of country and commodity detail makes it possible to use the results of the study to address issues at the most appropriate level of commodity/country interface. For example, issues of the food and agriculture futures of countries with high dependence on sugar (e.g. Cuba) can hardly be addressed without a view of the sugar sector, those of Ghana and Côte d'Ivoire of cocoa, those of Zimbabwe and Malawi of tobacco, those of Malaysia of rubber and palmoil, and so on. A study limited to, say, cereals and large country groups would not be very helpful in addressing these kinds of issues.

There are two more reasons why this degree of detail is necessary. The first is the focus of the study on problems of natural resources for agriculture, i.e. land and water use. Statements on, and projections of, land and water use cannot be made unless all the major crops are accounted for. For example, cereals account for only about 50 percent of the total harvested area of the developing countries (excluding China). An analysis limited to cereals would not provide a sufficient basis for exploring issues of land scarcities and possibilities of expansion in the future.

The second reason has to do with the interdisciplinary nature of the study and its heavy dependence on contributions provided by specialists in the different disciplines. Such contributions can find expression only if the relevant questions are formulated at a meaningful level of detail. For example, no useful contribution can be obtained from the production, country, trade, etc., specialists for a commodity group (e.g. coarse grains or fruit) and a region. It is more productive to obtain contributions for specific countries and individual commodities, e.g. not coarse grains but separately for maize, barley, millet and sorghum; not fruit but bananas and citrus, and so on. Likewise, expression of views on production prospects requires further disaggregation in terms of agroecological conditions because, say, irrigated barley and rainfed semi-arid barley are practically different commodities for assessing yield growth prospects (this aspect is discussed below).

The commodities and countries covered individually are listed in Appendix 1. Concerning commodities, there are 26 crop and 6 livestock products. This is for the analysis leading to the derivation of the demand and supply balances. For this analysis, all oilcrops are recognized as one commodity (oil equivalent), and the same goes for sugar and milk. However, these are heterogeneous commodities at the production level. Therefore, for production analysis, the commodity "vegetable oil" is represented by 7 oilcrops (rising to 8, including the cottonseed), sugar by 2 sugar crops (cane, beet) and milk by cow and sheep/goat milk, raising the total to 33 crops and 7 livestock products. Although the criterion for selecting which commodities to analyse individually has been their importance for the developing countries as a whole, it is inevitable that some commodities of particular importance to some countries are not covered individually, e.g. rapeseed (important for South Asian countries) and safflower seed (India, Mexico, Ethiopia) which are lumped together into the group "other oilseeds".

Concerning countries, 127 are covered "individually", of which 93 developing (representing 98.5 percent of their total population) and 34 developed representing nearly all their population. The projections methodology for the developing countries is more detailed and more demanding in data and time than that for the developed countries, because for the latter the production projections are carried out summarily and without distinguishing agroecological zones and accounting for land constraints (see also Editor's Preface). Therefore, the workload involved for each developing country is much higher than for each developed country.

Data preparation

The variables projected in the study are: (a) the demand (different final and intermediate uses), production and net trade balances for each commodity and country; and (b) for the developing countries only, key agronomic variables, i.e. areas, yields and production by country, crop and agroecological zone (land class) for crops; and animal numbers (total stock, off-take rates) and yields per animal for the livestock products. A significant part of the total effort is devoted to the work needed to create a consistent set of historical and base year data. For the demand–supply analysis, the overall quantitative framework for the projections is based on the supply utilization accounts (SUAs). The SUA is an accounting identity showing for any year the sources and uses of agricultural commodities in homogeneous physical units, as follows:

Food (direct) + Industrial non-food uses + Feed + Seed + Waste = Total domestic use = Production + (Imports − Exports) + (Opening stocks − Closing stocks)

The data base has one such SUA for each commodity entering the demand–supply analysis, country and year (1961 to 1990 at the time the study was initiated). The data preparation work for the demand–supply analysis consists of conversion of the about 330 commodities for which the production, utilization and trade data are available into the 32 commodities mentioned above, while respecting the SUA identities. This is no simple matter as the accounting relationships between commodities range from the fairly simple (e.g. converting pasta products and wheat flour in the consumption and trade statistics to wheat equivalent, though also here complexities are not absent, e.g. converting imported flour into wheat at the extraction rates of the importing or of the exporting country), to the extremely complex (e.g. converting imported margarine into vegetable oil equivalent and interfacing it with the vegetable oil equivalent of domestic oilseeds; or converting orange juice into fresh fruit equivalent). Unavoidably there remain loose ends in this complex accounting framework. FAO has work under way to improve the system and a publication on this matter is in preparation.

The different commodities are aggregated into groups and into "total agriculture" using as weights world average producer prices of 1979/81 expressed in "international dollars" derived from the Geary-Khamis formula as explained in Rao (1993). These are the price weights used to construct the FAO production indices (Laspeyres formula). The growth rates for heterogeneous commodity groups or total agriculture shown in this study are computed from the thus obtained value aggregates. The measurement of changes in agricultural aggregates obtained from these price weights is subject to limitations for particular uses, e.g. for drawing inferences about the pressures on natural resources generated by a given increment in production (see Chapter 3, note 17). It is also noted that each commodity has the same price weight in all countries. This means that one unique set of relative price weights is used to aggregate the production in all countries. The resulting growth rates of production, consumption, etc., can, therefore, differ from those that would be obtained if country-specific relative price weights were used. But the use of unique price weights makes the resulting growth rates comparable among countries.

A major part of the data preparation work, undertaken only for the developing countries, is the unfolding of the SUA element *production* (for the base year only, in this case the three-year average 1988/90) into its constituent components of area, yield and production which are required for projecting production. For the crops, the standard data in the SUAs contain, for most crops, also the areas (harvested) and average yields for each crop and country. These national averages are not considered by the agronomists to provide a good enough basis for the projections because of the widely differing agroecological conditions in which any single crop is grown, even within the same country. That is, judgements about future yields cannot be made without further information on whether, for example, barley is grown in irrigated or rainfed land and at what yield levels; and within the rainfed category, information is needed on whether it is grown on land with sufficient rainfall and good soils or in semi-arid conditions and poor soils. These considerations led to the decision to attempt to break down the base year production data from total area under a crop and an average yield into areas and yields for five rainfed and one irrigated categories. These categories (land classes) are described in detail in Chapter 4 and are not repeated here. The problem is that such detailed data are not generally available in any standard data base. It became necessary to piece them together from fragmentary information: some of it contained in published documents giving areas and yields on irrigated and rainfed land or by administrative districts; some of it from unpublished documents. The results of this research were supplemented by guesstimates. The result of this operation is a matrix of size 33×15, with one row for each crop and one area and one yield column for each land class (2×6) and 3 columns of control totals (harvested area, yield, production at the national level, HA, Y, P respectively) subject to the following equalities for each country ($i = 1 \ldots 6$ land classes, $j = 1 \ldots 33$ crops):

$\Sigma_i HA_{ij} = HA_j$ (sum of harvested areas of crop j in each land class i equals total harvested area of crop j, the control total for harvested land in each crop)

$\Sigma_i HA_j = HA$ (sum of harvested areas of each crop j equals total harvested area of the country HA)

$\Sigma_i HA_{ij} Y_{ij} = \Sigma_i P_{ij} = P_j$ (sum of production of crop j in each land class equals total production crop j, the control total for production of each crop)

In principle, there is no control total for HA in the standard data-set, but one is obtained by summing up the harvested areas reported for the different crops. There are, however, data for total *arable land* (AR) in agricultural use (physical area, not just harvested, called in the statistics "arable land and land in permanent crops"). It is not known whether the HA data (obtained by summation of the HAs given for each crop) and the AR data are compatible with each other. Compatibility of the two data-sets can be evaluated indirectly by computing the ratio of harvested area to arable land, i.e. the cropping intensity (CI). This is an important parameter which can signal defects in the land use data. Indeed, for several countries the implicit values of the CIs did not seem to make sense. In such cases the harvested area data resulting from the crop statistics were accepted as being the more robust (or the less questionable) ones and those for arable area were adjusted in consultation with the country and land use specialists. The objective was to have a set of harvested and arable land use data which, to the belief of the specialists, were more credible, more compatible with each other and more representative of actual country situations than the data reported in the standard sources. This operation for adjusting the data for the countries with evidently inconsistent ones was facilitated by the fact that the fine-tuning of the CI parameter had to be performed for each of the six land classes, not for the country as a whole. For example, CI values of over 0.8 and up to well over 2.0 are acceptable for irrigated areas depending on country knowledge about climate, water shortages, double cropping, farming systems, etc. On the other hand, in rainfed semi-arid areas and in most rainfed land classes in countries with considerable shifting agriculture CI values are normally below 0.5. These problems are discussed in Chapter 4 where also the adjusted and unadjusted data are given and compared.

Following this unfolding of the land use data, a final equality has to be satisfied:

$$\Sigma_i HA_i / CI_i = \Sigma_i AR_i = AR \quad \text{(sum of arable land in each land class } i \text{ equals total arable land in the country, the adjusted control total)}$$

Projections

The bulk of the projections work concerns: (a) the drawing up of SUAs (by commodity and country) for the year 2010; (b) the unfolding of the projected SUA item "production" into area and yield combinations for up to six agroecological conditions (land classes); and (c) the drawing up of land use balances by land class, including irrigated land.

The projections for all SUA items and countries for the cereals, livestock and oilcrop commodities are derived, in the first place, from a formal multi-commodity, multi-country, flex-price model which is used in the medium-term commodity projections work of FAO, the *FAO World Food Model* (WFM). A detailed description of the model together with its parameters is given in FAO (1993i) and is not repeated here. Suffice to say here that: (a) the model provides year-by-year world price equilibrium solutions for the commodities covered; (b) it has demand (for food, feed, other uses) and supply (area, yields, animal numbers, etc.) equations for each country; (b) each country's solution is influenced by those for every other country through the imports and exports which are equated at the world level by price changes; (c) the extent to which price changes are transmitted to each country is determined by wedges between the domestic and world prices. These wedges represent the policy variables and can be varied to generate alternative trade policy outcomes; and (d) the projections are subject to many rounds of adjustments following inspection by specialists on the basis of the criteria described below. The adjustments are "absorbed" by the model by fine-tuning its parameters and coefficients, usually the trend factors. The model does not, however, have natural resource (land, water) constraints, nor does it generate relevant balances and parameters, e.g.

arable land, cropping intensities, etc. In conclusion, the results generated by the model are a major element, but only one among many, which enter the determination of the projections used in this study and only for the cereals, livestock and oilcrop commodities (the WFM commodities). For some other commodities (e.g. sugar, rubber, cotton, jute), single commodity models were used to generate the initial projections which were subsequently subjected to several rounds of inspection and adjustment.

For these same commodities as well as for all others, parallel projections are prepared for each SUA element, as follows.

The element *food*, as represented in the SUA (i.e. food availability for direct human consumption)[1], is projected in per caput terms using the base year data for this variable, a set of estimated food demand functions – Engel curves[2] – for up to 52 separate commodities in each country and the assumptions of the growth of per caput incomes (GDP). The results are inspected by the commodity and nutrition specialists and adjusted taking into account any relevant knowledge and information, in particular the historical evolution of per caput demand and the nutritional patterns in the country examined. Subsequently total projected food demand is obtained by simple multiplication of the projected per caput levels with projected population. Projected food demand may be further revised in the process of projecting the other elements in the SUA, in particular production and net imports (see below).

Industrial demand for non-food uses is projected as a function of the GDP growth assumptions and/or the population projections and subsequently adjusted in the process of inspection of the results. This item is important for only a few countries and products, e.g. sugar in Brazil or maize in the USA, both for fuel. The historical data are particularly weak and they often represent the domestic disappearance not accounted for in the other SUA items.

Feed demand for cereals is derived simultaneously with the projections of livestock products from the relationships between these variables in the solution of the above-mentioned WFM and further checks are performed by multiplying projected production of each of the livestock products with country-specific input/output coefficients (feeding rates) in terms of metabolizable energy supplied by cereals and brans. The part that can be met by projected domestic production of brans is deducted and the balance represents cereals demand for feed. *Feed demand for oilseed proteins* (computed, in the first place, in terms of crude protein equivalent) is taken mostly from the relationships in the projections of the World Food Model. *Feed use of other products* with a feed-use component in the historical SUA data is obtained by *ad hoc* methods, mostly as a proportion of total production or total demand. It is noted that these feed-use projections do not provide a complete interface between animal production and feed supplies or resources in each country because of the lack of systematic data for complete feed balances, i.e. including non-concentrate feeds (cultivated fodder, natural grass, by-products other than cereal brans, etc.).

Seed use is projected as a function of production using seeding rates per hectare. *Waste* (post-harvest to retail) is projected as a proportion of total supply (production plus imports).

This parallel method, unlike the WFM, does not project year 2010 *stock changes*. This does not mean that present stocks are assumed to remain constant but rather that adjustments in stocks between the base year level and any required level in year 2010 may occur in any year(s) between 1988/90 and 2010. In this case, the impact on production will appear only as temporary deviation(s) from the smooth growth path represented by a curve joining the base year production level to that of 2010, ignoring fluctuations in the intervening years. Whether or not year 2010 production includes a provision for "normal" stock changes (i.e. to maintain stocks at the desired percentage of consumption already achieved before 2010) makes little difference to the average growth rate of production for 1988/90–2010 if the deviations from the constant growth rate path in the intervening years are ignored.

Production and trade projections for each country involve a number of iterative computations and adjustments. The solution of the WFM provides the initial levels for cereals, livestock and oilcrops. For all commodities, the criteria for the projections and their iterative adjustments are as follows:

1. *Commodities in deficit in the base year (developing countries only)*. A preliminary "target" level is set for 2010 taking into account the projected demand, production growth possibilities (evaluated in more detail in subsequent steps of the analysis, step 7 below) and preliminary values for projected self-sufficiency ratios. The latter are used as a computational device to define preliminary levels of future production "targets" to be evaluated in subsequent steps. They do not reflect expression of any generalized preference for increased self-sufficiency, but they do take into account whatever is known about country preferences about self-sufficiency objectives which influence their policies. The more general issues of the pros and cons of the self-sufficiency objectives are discussed in Chapter 7.

2. *Commodities exported in the historical period and the base year (developing countries only)*. It is assumed that they will continue to be exported in amounts which will depend on the country's possibilities to increase production, a preliminary assessment of import demand on the part of all the other countries which are deficit in that commodity and an assessment of the country's possibility to have a share in total world import demand resulting from whatever is known about policies and other factors influencing the country's competitive position (see, for example, the discussion on natural rubber in Chapter 3). Since for world balance total deficits of the importing countries must be equal to total surpluses of the exporting countries there is an element of simultaneity in the determination of the production levels of all commodities in all countries. This element of simultaneity is fully accounted for in the WFM solution, but only for the

WFM commodities (cereals, livestock, oilcrops). For the other commodities, the problem is solved in a number of successive iterations rather than through a formal model, the key element being expert judgements on market shares in world exports and of somewhat more formal evaluations of the production possibilities, as explained below. Based on the above considerations, preliminary production "targets" are set for the export commodities of each developing country. They are equal to their own domestic demand plus the preliminary export levels. Once the preliminary production targets are set for all commodities, the missing elements of the demand side of the SUA which depend on the levels of production (feed, seed, waste) can be filled in.

At this stage complete preliminary SUAs are available for the year 2010 for all commodities and all the developing countries, showing for each commodity and country all the domestic demand elements and production. The differences between total demand (total domestic use) and production are the preliminary net trade positions (imports or exports). The next step is to construct the *projected SUAs for the developed countries* whose net trade balances must match those of the developing countries, with opposite signs[3].

3. The demand items of the SUAs for the developed countries are projected in the same manner as for the developing countries, subject to the differences in the criteria used to evaluate and adjust the projections. The great uncertainty surrounding, and the special factors applying to, the prospective evolution of the main food and agriculture variables in the ex-CPEs is underlined (see Chapter 3).

4. For the commodities not produced, or produced only in insignificant quantities in the developed countries (tea, coffee, cocoa, bananas, natural rubber, jute, cassava), nearly all their demand translates into import requirements. These, together with the import requirements of the developing countries in deficit, define the total market for these commodities available to the developing exporting countries. Their provisional production and export levels, set as described above, are then adjusted judgementally to equate them to the total import requirements.

5. A second set of non-WFM commodities comprises those produced in substantial quantities in both the developed and the developing countries but for which the latter have been traditionally substantial net exporters (mainly sugar, citrus, tobacco, cotton). The aspects of these commodities taken into account in the projections are discussed in Chapter 3. For example, they include the prospect that there will not be any significant trade liberalization effects on the sugar trade flows and that outcomes will be influenced by changes in the sugar trade relationships between the ex-CPEs and Cuba; that the raw cotton trade will be influenced by the ever-growing role of the developing countries in world exports of cotton manufactures; and that the tobacco production and trade prospects will be

decisively influenced by trends for per caput consumption to decline in the OECD countries and to increase in the rest of the world.

6. The last group of commodities comprises those for which the developing countries and, at present, also the ex-CPEs are major importers and the other developed countries as a group are the major suppliers of these imports (mainly wheat, coarse grains, milk). For these commodities the projections from the WFM solution constitute the initial values. They are adjusted iteratively in a number of rounds, as follows. The possible production and net trade positions of the ex-CPEs are adjusted on the basis of the criteria discussed in Chapter 3, e.g. the prospect that their demand will be in the future lower than in the base year and that their production will recover, leading initially to import substitution and eventually to net exports. The resulting net trade positions of the ex-CPEs, together with those derived earlier for the developing countries define the total net exports required to be generated by all the other developed countries as a group. Together with the projected growth for their own domestic use they define future production levels. By and large, the required growth in their collective production is modest. For example, future production of cereals for the three main exporting regions (North America, Western Europe, Oceania) is required to be 680 million tonnes. This implies a growth rate to 2010 of 0.7 percent p.a., measured from their 1991/92 two-year average production of 595 million tonnes. This growth rate is considered to be well within the bounds of their collective potential to increase production and, therefore, no further work is done to evaluate this projected level from the standpoint of natural resources and yield growth feasibility (see also Editor's Preface).

However, the issue how this additional production and exports will be shared between these three main exporting regions is not so easily solved. At the time this analysis was done only some of the domestic policy reforms were known and elements of those that were subsequently agreed under the Uruguay Round were known only in the form of provisional negotiating positions. It is only after April 1994 that the concrete country proposals about measures to implement the Uruguay Round Agreement on Agriculture started to become known. In the absence of this information, the projections for production and net trade are shown in Chapter 3 for the main exporting regions as a group (Table 3.17 and Annex to Chapter 3). Some of the factors that may influence the production and net trade outcomes of the individual regions are discussed in Chapter 3, e.g. the prospect that commitments to reduce subsidized exports and to allow for minimum import access will translate into Western Europe's projected net exports of cereals being in the future no higher than in the base year. If this prospect materialized, all additional net exports would be supplied by North America and Oceania. However, these are tentative conclusions, subject to the many caveats of the models which generate them. A more concrete evaluation on the basis of the concrete Schedules of the different countries for the implementation of the Uruguay Round Agreement on Agriculture will be undertaken in FAO for the WFM commodities and selected other commodities.

7. At this stage the projections of demand, production and trade are complete: there is one projected SUA for each country and commodity (but only one aggregate complete SUA for the main developed exporting regions) and world imports equal world exports. These projections are, however, still provisional *pending a more detailed evaluation of the feasibility of the production projections of the developing countries, from the standpoint of land and water use and the growth of yields.* The basis for this evaluation is provided by: (a) the detailed data-set constructed for the base year in the above-described phase of data preparation (matrix of base-year areas and yields by crop and land class); and (b) whatever knowledge and judgements the specialists on countries, commodities and the different agronomic disciplines could contribute. The objective of this operation is essentially to "test the feasibility" of the preliminary crop production projections for each developing country by generating for 2010 the 33×15 matrix of areas and yields by land class which was constructed for the base year, as described earlier. China is not included in this exercise for the reasons discussed in Chapters 3 and 4.

The relevant work brings to bear on the evaluation of the crop production projections whatever knowledge the country, commodity and discipline specialists possess concerning the production conditions, national plans, etc., for individual crops and countries. Examples include: (a) knowledge of existing or impending shifts in the different countries towards hybrid rice varieties, or alternatively, back towards the more palatable traditional varieties, is an invaluable element for forming an informed judgement about the rice sector prospects; (b) country plans, in formulation or in execution, for example irrigation or establishment of tree crops such as oilpalm, rubber, etc.; (c) knowledge about preferential trading arrangements that influence future production, e.g. for bananas, sugar; and so on for other products, country by country.

The data on the land resources of each country and their suitability for crop production (see Chapter 4) constitute an important input in the derivation of the possible future land–yield combinations by land class. Assumptions are first made of what are feasible rates of harvested land expansion by agroecological class (through use of more land from the reserves and or through increased cropping intensities, including expansion of irrigation). Similar assumptions are made for yield increases and the land allocation to each crop. Since a multitude of detailed assumptions and different specialists are involved, continuous iterative computations of the whole system are made to ensure that constraints of land availability and notional upper yield levels for 2010 (both by country and land class) are respected. The end-result is that either the initial production target is fine-tuned and accepted or is revised downwards for some crops because land resources (of the required class, where applicable) are not sufficient or because the target requires yield increases considered by the specialists to be beyond achievement by 2010 even under reasonably improved policies.

8. Similar production analysis procedures are applied to the livestock production, except that the relevant parameters are animal numbers and yields (off-take rates, carcass weight, milk yields, eggs per laying hen) for the livestock species considered. Whatever knowledge is available about the feed resources of the different countries is utilized in evaluating the production prospects.

9. Subject to the results of the two preceding steps (7 and 8), final adjustments are made in the other SUA items for the commodities and countries for which the provisional production "targets" had to be changed during the feasibility tests. If the changes in projected production result in increased import requirements which are considered "excessive" for particular countries, their projected demand may be adjusted downwards, to make up for the shortfall in supplies, in whole or in part. Following this, a final iteration is made to adjust production and trade balances of other countries to make up the shortfall in production and cover the resulting increased import requirements in the developing countries whose initial provisional "targets" had to be lowered.

At this stage, the world demand, production and trade picture is completely quantified, backed by full quantifications for the developing countries of land and water use balances, their yield-harvested area combinations (by crop and land class) and the relevant livestock parameters.

10. The last step projects *fertilizer* use. This is obtained, in the first instance, by multiplying fertilizer input coefficients per hectare by projected harvested area. The coefficients are specific to each crop, land class and yield. These coefficients were generated by the agronomists using whatever data and other information were available (e.g. in FAO, 1989c and FAO/IFA/ IFDC, 1992) and supplemented by a good deal of judgement based on agronomic norms. Since they are yield specific, the method is equivalent to using yield–fertilizer response functions for each land class and crop. These coefficients are not country specific in the first instance, but they are made so by calibrating them on the basis of the implicit scalars required to reproduce for the base year the control data of total fertilizer use by country, which are available in the standard data sources. More details on the method are given in Bruinsma *et al.* (1983).

A summary evaluation of the methodology

The key characteristics of the methodology may be summarized as follows: (a) the analysis is conducted in great detail as regards commodities, countries, land classes, etc.; (b) behavioural relationships are used explicitly in the projections only for the WFM commodities and for all commodities in the projections of food demand; (c) prices play an explicit role in bringing about demand–supply balance only for the WFM commodities and then only in the generation of the initial set of projections which is subsequently subjected to many inspections and adjustments; (d) links of agriculture with the rest of the economy are not accounted for, except for the influence of income growth on the food demand projections; (e) the method generates land use balances in great detail (by agroecological classes and with distinctions between harvested and arable land) and controls for land availability constraints; and (f) the method uses very diverse sources of data and all kinds of knowledge and judgements contributed by the different specialists on countries and disciplines.

This method has positive and negative aspects. On the positive side, the great detail of analysis means that the projections and related statements contained in this book for country groups, regions or the world as a whole as well as those for large commodity aggregates and the whole of agriculture are underpinned by detailed country/commodity quantifications. In practice, each global statement is derived from a summation of, and can be decomposed back into, a number of constituent single-country or commodity statements. This characteristic sets this study apart from most other global studies in which analyses are carried out at the level of major countries and/or regions and usually only for the major products grouped into a few commodity aggregates.

In many respects, the great detail of the analysis and the heavy dependence on multidisciplinary specialist input are two faces of the same coin. This is because such specialist input can be provided and utilized only if solicited at the level of detail in which the relevant expertise is available. The latter usually embodies knowledge of local conditions in widely differing country situations. This is the second major advantage of using this method.

This heavy dependence on great detail and specialist input is at the same time the major weakness of the method. The flow of such specialist input is, of course, controlled by a large data processing system which generates, after many iterations, an internally consistent set of projections. Such consistency is, however, only an accounting one. Projections based on specialist input suffer from the fact that the criteria and assumptions used and the implicit decision-making mechanism cannot be formally described and they can vary from one person to another and over time. It follows that the projections cannot be strictly replicated at will, including for estimating alternative scenarios by varying certain assumptions only. This would have been possible if a formal model had been used for the projections.

There are advantages and disadvantages in the use of formal models for this type of work (Alexandratos, 1976). In the case of this study, the great amount of detail makes it impossible to conduct the analysis using one single formal model representing behaviour of the different actors (producers, consumers, governments) and with price-based market clearing mechanisms. Many of the data required for such an effort are just not available. For a formal modelling approach, the choice would have been between having: (a) a roughly estimated formal model with much less commodity, country and land use detail; or (b) a huge model with all the detail of this study but with the bulk of the parameters and coefficients being "guesstimates" rather than data. The former case is a clearly inferior option since it would make it impossible to evaluate the results using multidisciplinary specialist input. Moreover, the findings of the study would be of limited value for drawing conclusions at very diverse levels because of the much reduced commodity and country detail.

The second option (large model with a wide range of guesstimates for the parameters) is really a formal variant of the expert judgement-based approach used in this study, the difference being that expert judgements would be embodied in the guesstimates of the values of the model parameters and coefficients. Such an approach would be superior to the one of this study, since the utilization of the expert judgement input is subject to the discipline that the implied values of the parameters and coefficients must fall within a certain acceptable range. Iterations and dialogue would be greatly facilitated, alternative scenarios could be estimated and greater transparency would be assured. These advantages must be set against the greater resources and time required for model preparation, particularly for the development of the computing algorithms. The latter aspect can be daunting for a global model that would recognize 100-odd countries communicating through trade flows, over 35 commodities, up to 6 sets of production conditions per crop, and so on. It could easily absorb a disproportionate part of the resources of the study without assurance of a satisfactory end-product.

In conclusion, future improvements in the methodology should aim at introducing some of the advantages of formal models in the form of explicit statements of the assumed behavioural relationships and their empirical verification, replication of results and derivation of alternative scenarios in a consistent manner. It is, however, important that the strong points of the present methodology be preserved, viz. the detail of analysis as regards countries, commodities and production conditions as well as the associated possibility to use multidisciplinary input and to draw on very diverse sources of knowledge and expertise. Given "reasonable" resource and time limitations, it is unlikely that this could be achieved by attempts to build a formal model in all the detail of this study. Scarce resources could be used more productively if they are concentrated on improving selected components of the present methodology, as indeed was done in this study by constructing a formal model (the WFM) to capture the interdependencies between the cereals, oilseeds and livestock sub-sectors. Future work on agriculture and the environment would benefit from efforts to develop analytical methods to link geo-referenced data on agricultural resources to the assessment of production prospects.

NOTES

1. The terms demand, consumption, availability and per caput food supplies are used interchangeably to denote the SUA element *food*.
2. Samples of elasticity estimates from household budget, or food consumption, surveys are given in FAO (1989d).
3. In these projections the team benefited from the cooperation of researchers from the Center for Agricultural and Rural Development, Iowa State University, USA (S. Johnson and W. Meyers) and the Institut für Agrarpolitik, University of Bonn, Germany (K. Frohberg).

3. STATISTICAL TABLES

Table A.1 Total population and population economically active in agriculture

| | Total population | | | | | | Population economically active in agriculture | | | | | | | | |
| | Million | | | Growth rates (% p.a.) | | | Thousand | | | % of total population economically active | | | Growth rates (% p.a.) | | |
	1990	2000	2010	1980–90	1990–2000	2000–10	1990	2000	2010	1990	2000	2010	1980–90	1990–2000	2000–10
World	5 296.8	6 265.0	7 208.6	1.8	1.7	1.4	1 101 503	1 165 520	1 214 966	46.6	42.1	37.8	1.0	0.6	0.4
All developing countries	4 045.9	4 946.9	5 835.2	2.1	2.0	1.7	1 051 424	1 130 744	1 191 157	59.6	53.0	46.7	1.3	0.7	0.5
93 Developing countries	3 987.5	4 879.1	5 757.9	2.1	2.0	1.7	1 039 049	1 118 171	1 178 767	59.8	53.3	46.9	1.3	0.7	0.5
Sub-Saharan Africa	487.7	673.2	914.6	3.2	3.3	3.1	140 019	168 818	205 810	71.2	65.6	59.7	1.7	1.9	2.0
Angola	10.0	13.3	17.6	2.6	2.9	2.8	2 851	3 323	4 004	69.8	65.6	61.4	1.3	1.5	1.9
Benin	4.6	6.4	8.7	3.0	3.2	3.2	1 338	1 439	1 546	61.4	51.7	42.0	0.7	0.7	0.7
Botswana	1.3	1.8	2.5	3.8	3.4	3.0	271	329	398	62.9	54.6	45.9	1.7	2.0	1.9
Burkina Faso	9.0	12.1	16.3	2.6	3.0	3.1	4 004	4 872	6 055	84.4	81.8	78.9	1.8	2.0	2.2
Burundi	5.5	7.4	9.7	2.9	3.0	2.8	2 594	3 258	4 141	91.3	89.4	87.2	2.1	2.3	2.4
Cameroon	11.8	16.7	23.7	3.2	3.5	3.5	2 656	2 912	3 263	61.0	51.6	42.4	0.8	0.9	1.1
Central African Rep.	3.0	4.1	5.5	2.7	3.0	3.0	884	901	921	62.7	51.7	41.1	0.1	0.2	0.2
Chad	5.7	7.3	9.5	2.4	2.6	2.6	1 472	1 546	1 560	74.6	63.4	50.6	0.8	0.5	0.1
Congo	2.3	3.2	4.4	3.1	3.4	3.4	506	635	827	59.6	56.6	53.3	1.7	2.3	2.7
Côte d'Ivoire	12.0	17.6	25.5	3.9	3.9	3.8	2 545	2 820	3 174	55.6	45.9	37.2	1.2	1.0	1.2
Ethiopia	49.2	66.4	88.9	2.4	3.0	3.0	15 461	17 825	20 773	74.5	68.5	61.8	0.9	1.4	1.5
Gabon	1.2	1.6	2.1	3.8	3.2	2.4	351	359	353	67.8	58.7	49.2	1.9	0.2	−0.2
Gambia	0.9	1.1	1.4	3.0	2.7	2.5	316	371	436	81.0	77.6	73.8	1.8	1.6	1.6
Ghana	15.0	20.6	26.9	3.5	3.2	2.7	2 751	3 205	3 808	50.0	44.0	38.2	1.7	1.5	1.7
Guinea	5.8	7.8	10.7	2.6	3.1	3.1	1 835	2 072	2 324	74.1	66.2	57.1	1.0	1.2	1.2
Kenya	24.0	35.1	50.9	3.7	3.8	3.8	7 645	10 062	13 041	77.0	72.4	67.3	2.7	2.8	2.6
Lesotho	1.8	2.4	3.1	2.9	2.9	2.8	653	735	806	79.6	71.0	60.5	1.3	1.2	0.9
Liberia	2.3	3.6	4.9	2.6	4.6	3.2	666	800	977	69.8	65.1	59.8	1.8	1.8	2.0
Madagascar	12.0	16.6	22.8	3.2	3.3	3.3	3 953	4 755	5 763	76.6	71.3	65.3	1.8	1.9	1.9
Malawi	8.8	12.5	17.1	3.5	3.6	3.2	2 690	3 099	3 488	75.4	65.2	53.7	1.6	1.4	1.2
Mali	9.2	12.7	17.3	3.0	3.2	3.2	2 371	2 903	3 666	80.8	74.9	67.6	2.1	2.0	2.4
Mauritania	2.0	2.7	3.6	2.7	2.9	2.9	417	509	641	64.4	59.3	54.4	1.8	2.0	2.3
Mauritius	1.1	1.2	1.3	1.1	1.0	0.9	96	93	82	22.8	18.1	14.3	0.7	−0.3	−1.3
Mozambique	15.7	20.5	26.5	2.6	2.7	2.6	6 666	7 824	9 354	81.6	78.5	75.1	1.4	1.6	1.8

continued

Table A.1 (continued)

| | Total population | | | | | | Population economically active in agriculture | | | | | | | | |
| | Million | | | Growth rates (% p.a.) | | | Thousand | | | % of total population economically active | | | Growth rates (% p.a.) | | |
	1990	2000	2010	1980–90	1990–2000	2000–10	1990	2000	2010	1990	2000	2010	1980–90	1990–2000	2000–10
Namibia	1.8	2.4	3.3	3.2	3.2	3.1	184	201	229	35.0	28.2	22.8	0.3	0.9	1.3
Niger	7.7	10.8	14.9	3.3	3.4	3.3	3421	4217	5169	87.3	82.2	75.6	2.3	2.1	2.1
Nigeria	108.5	149.6	201.3	3.3	3.3	3.0	26577	33004	41608	64.8	61.2	57.6	2.0	2.2	2.3
Rwanda	7.2	10.2	13.8	3.4	3.5	3.1	3216	4265	5653	91.3	89.5	87.3	2.8	2.9	2.9
Senegal	7.3	9.7	12.7	2.8	2.9	2.7	2466	3008	3688	78.4	76.1	73.7	1.9	2.0	2.1
Sierra Leone	4.2	5.4	7.2	2.4	2.7	2.8	891	962	1064	62.3	54.6	47.0	0.5	0.8	1.0
Somalia	7.5	9.7	13.1	3.4	2.6	3.0	2108	2366	2773	70.9	64.4	57.5	1.4	1.2	1.6
Sudan	25.2	33.6	44.0	3.0	2.9	2.7	4923	5388	5722	60.2	48.5	37.3	1.3	0.9	0.6
Swaziland	0.8	1.1	1.6	3.4	3.6	3.3	207	240	279	66.1	57.8	48.9	1.4	1.5	1.5
Tanzania	27.3	39.6	56.3	3.8	3.8	3.6	10315	12905	16103	80.8	74.9	67.9	2.4	2.3	2.2
Togo	3.5	4.9	6.7	3.1	3.2	3.2	995	1211	1493	69.6	65.9	61.7	1.8	2.0	2.1
Uganda	18.8	27.0	37.0	3.7	3.7	3.2	6569	8282	10333	80.9	74.6	67.1	2.2	2.3	2.2
Zaire	35.6	49.2	67.5	3.1	3.3	3.2	8683	10396	12854	65.8	60.1	55.0	1.6	1.8	2.1
Zambia	8.5	12.3	17.3	4.0	3.8	3.5	1872	2557	3611	68.9	64.5	60.1	3.3	3.2	3.5
Zimbabwe	9.7	13.1	17.0	3.1	3.1	2.6	2600	3169	3830	68.2	63.1	58.0	2.2	2.0	1.9
Near East/North Africa	305.5	396.6	492.9	2.8	2.6	2.2	34593	37562	39449	37.2	29.9	23.5	0.7	0.8	0.5
Afghanistan	16.6	26.5	32.4	0.1	4.8	2.0	2687	3730	4043	54.8	48.2	41.6	-0.9	3.3	0.8
Algeria	25.0	32.9	41.5	2.9	2.8	2.4	1391	1573	1654	24.4	18.7	14.1	0.9	1.2	0.5
Egypt	52.4	64.2	75.7	2.5	2.0	1.7	5880	6752	7616	40.5	35.5	30.7	1.4	1.4	1.2
Iran	54.6	68.8	87.8	3.5	2.3	2.5	4267	4412	4525	27.6	20.8	15.3	0.6	0.3	0.3
Iraq	18.9	26.3	35.3	3.6	3.4	3.0	1049	1074	970	20.5	14.1	8.9	-0.3	0.2	-1.0
Jordan	3.3	4.6	6.0	3.9	3.3	2.7	47	39	30	5.8	3.2	1.8	-0.4	-1.9	-2.4
Lebanon	2.7	3.3	3.9	0.0	2.1	1.6	72	53	39	8.7	5.0	2.9	-3.8	-3.0	-3.2
Libya	4.5	6.5	9.0	4.1	3.6	3.3	155	170	187	13.7	10.7	8.3	1.0	0.9	1.0
Morocco	25.1	31.6	37.6	2.6	2.3	1.8	2824	2960	2954	36.6	28.1	20.9	0.9	0.5	0.0
Saudi Arabia	14.1	20.7	29.6	4.2	3.9	3.6	1596	1720	1756	39.0	30.2	22.4	1.8	0.7	0.2
Syria	12.5	17.8	24.3	3.6	3.6	3.2	746	844	956	24.1	18.2	13.5	0.5	1.2	1.3
Tunisia	8.2	9.9	11.5	2.5	2.0	1.5	655	573	470	24.3	16.2	10.5	-0.4	-1.3	-2.0
Turkey	55.9	66.8	75.3	2.4	1.8	1.2	11670	11593	11515	48.2	39.4	32.1	0.6	-0.1	-0.1
Yemen	11.7	16.6	23.1	3.6	3.6	3.3	1554	2069	2734	55.6	49.5	43.3	3.0	2.9	2.8

East Asia	1624.4	1879.5	2061.2	1.6	1.5	0.9	557374	561272	537166	63.0	55.1	47.0	1.2	0.1	-0.4
Cambodia	8.2	10.0	11.5	2.7	2.0	1.4	2630	2754	3087	70.0	65.5	60.6	0.7	0.5	1.1
China (Mainland)	1118.8	1276.1	1370.6	1.3	1.3	0.7	450285	448407	423941	67.5	59.8	51.7	1.2	0.0	-0.6
Indonesia	184.3	218.7	246.7	2.0	1.7	1.2	35077	35970	34221	48.5	39.8	31.7	0.9	0.3	-0.5
Korea DPR	21.8	26.1	29.3	1.8	1.8	1.2	3777	3501	3022	33.5	25.4	18.6	1.1	-0.8	-1.5
Korea Republic	42.8	46.4	49.5	1.2	0.8	0.6	4633	3541	2438	24.6	16.0	10.1	-1.4	-2.7	-3.7
Laos	4.1	5.5	6.8	2.6	2.8	2.3	1380	1605	1930	71.5	67.0	62.1	1.1	1.5	1.9
Malaysia	17.9	22.0	25.2	2.7	2.1	1.4	2255	2181	2045	32.1	23.8	17.1	0.2	-0.3	-0.6
Myanmar	41.7	51.1	60.6	2.1	2.1	1.7	8500	9044	9507	46.9	41.2	35.8	0.6	0.6	0.5
Philippines	62.4	77.5	92.1	2.6	2.2	1.7	10503	12030	13418	46.8	41.8	36.9	1.5	1.4	1.1
Thailand	55.7	63.7	71.6	1.8	1.3	1.2	18990	19852	19003	64.3	57.1	49.7	1.3	0.4	-0.4
Vietnam	66.7	82.4	97.4	2.2	2.1	1.7	19344	22387	24554	60.6	53.2	45.7	1.6	1.5	0.9
South Asia	1127.7	1398.0	1667.6	2.4	2.2	1.8	266002	309056	355981	65.5	61.6	57.4	1.6	1.5	1.4
Bangladesh	115.6	150.6	188.2	2.7	2.7	2.3	23193	28323	33001	68.5	61.5	54.0	2.1	2.0	1.5
India	853.1	1041.5	1223.5	2.2	2.0	1.6	214664	246358	280509	66.5	63.2	59.8	1.5	1.4	1.3
Nepal	19.1	24.1	28.9	2.6	2.3	1.8	7276	9089	11116	91.7	90.2	88.5	2.0	2.3	2.0
Pakistan	122.6	162.4	205.5	3.8	2.8	2.4	17580	21540	27178	49.7	44.6	39.4	2.4	2.1	2.4
Sri Lanka	17.2	19.4	21.5	1.5	1.2	1.0	3289	3746	4178	51.7	50.0	48.4	1.2	1.3	1.1
Latin America and Carib.	442.1	531.9	621.6	2.2	1.9	1.6	41061	41463	40360	26.3	21.2	16.9	0.5	0.1	-0.3
Argentina	32.3	36.2	40.2	1.4	1.2	1.0	1197	1101	987	10.4	8.2	6.5	-1.2	-0.8	-1.1
Bolivia	7.3	9.7	12.8	2.8	2.9	2.8	949	1061	1221	41.6	35.9	30.8	1.6	1.1	1.4
Brazil	150.4	179.5	207.5	2.2	1.8	1.5	13366	12458	11108	24.3	18.4	13.6	-0.3	-0.7	-1.1
Chile	13.2	15.3	17.2	1.7	1.5	1.2	585	524	457	12.5	9.5	7.2	-0.3	-1.1	-1.4
Colombia	33.0	39.4	45.6	2.1	1.8	1.5	2885	2832	2597	27.3	21.2	16.0	0.6	-0.2	-0.9
Costa Rica	3.0	3.7	4.4	2.8	2.1	1.6	251	240	217	23.9	17.8	13.0	0.5	-0.4	-1.0
Cuba	10.6	11.5	12.2	0.9	0.8	0.6	860	777	653	19.2	15.5	12.4	0.4	-1.0	-1.7
Dominican Republic	7.2	8.6	9.9	2.3	1.9	1.4	819	795	718	35.8	26.8	19.3	0.7	-0.3	-1.0
Ecuador	10.6	13.3	16.1	2.7	2.3	1.9	996	993	928	30.3	22.9	16.8	0.6	0.0	-0.7
El Salvador	5.3	6.7	8.5	1.5	2.5	2.3	603	673	737	36.5	30.9	26.1	-0.3	1.1	0.9
Guatemala	9.2	12.2	15.8	2.9	2.9	2.6	1346	1663	2038	51.2	45.4	39.6	1.9	2.1	2.1
Guyana	0.8	0.9	1.0	0.5	1.1	1.2	66	66	66	22.3	18.5	15.4	0.0	0.0	0.0
Haiti	6.5	8.0	9.8	2.0	2.1	2.1	1823	1936	2075	63.8	57.6	51.7	0.5	0.6	0.7
Honduras	5.1	6.8	8.7	3.4	2.9	2.4	879	1136	1412	55.0	49.3	43.4	3.0	2.6	2.2
Jamaica	2.5	2.7	3.0	1.4	1.1	1.0	324	338	333	27.1	23.0	19.4	0.9	0.4	-0.2
Mexico	88.6	107.2	125.2	2.3	1.9	1.6	9340	9705	9399	30.0	24.0	18.9	1.1	0.4	-0.3
Nicaragua	3.9	5.3	6.8	3.4	3.1	2.6	463	545	611	38.5	30.7	25.0	1.9	1.7	1.2
Panama	2.4	2.9	3.3	2.1	1.8	1.4	218	211	191	25.0	19.0	14.2	0.4	-0.3	-1.0

continued

Table A.1 (continued)

| | Total population | | | | | | Population economically active in agriculture | | | | | | | | |
| | Million | | | Growth rates (% p.a.) | | | Thousand | | | % of total population economically active | | | Growth rates (% p.a.) | | |
	1990	2000	2010	1980–90	1990–2000	2000–10	1990	2000	2010	1990	2000	2010	1980–90	1990–2000	2000–10
Paraguay	4.3	5.5	6.9	3.1	2.6	2.3	674	831	1 000	46.3	43.6	41.0	2.7	2.1	1.9
Peru	21.6	26.3	31.0	2.2	2.0	1.7	2 443	2 694	2 842	34.7	29.6	24.9	1.3	1.0	0.5
Suriname	0.4	0.5	0.6	1.9	1.7	1.3	24	24	24	16.6	13.0	10.4	1.3	0.3	0.0
Trinidad and Tobago	1.3	1.5	1.7	1.7	1.5	1.3	36	33	29	7.4	5.6	4.0	−1.0	−1.1	−1.0
Uruguay	3.1	3.3	3.5	0.6	0.6	0.5	162	154	145	13.5	11.7	10.2	−0.9	−0.5	−0.6
Venezuela	19.7	24.7	30.0	2.8	2.3	2.0	752	673	572	11.0	7.4	4.9	−0.5	−1.1	−1.6
Developed countries	1 248.9	1 314.7	1 369.7	0.7	0.5	0.4	50 070	34 769	23 806	8.3	5.5	3.6	−3.3	−3.6	−3.7
Western Europe	400.4	407.9	409.7	0.3	0.2	0.0	12 511	8 379	5 402	6.8	4.5	2.9	−3.4	−3.9	−4.3
EC-12	344.0	349.8	350.8	0.3	0.2	0.0	9 322	6 213	4 026	6.0	3.9	2.6	−3.4	−4.0	−4.2
Belgium–Luxembourg	10.3	10.2	10.1	0.0	0.0	−0.1	78	47	28	1.8	1.1	0.7	−4.2	−4.8	−5.1
Denmark	5.1	5.2	5.1	0.0	0.0	−0.1	134	87	52	4.7	3.0	1.9	−3.8	−4.3	−5.0
France	56.4	58.4	59.7	0.4	0.4	0.2	1 341	848	517	5.2	3.2	1.9	−4.0	−4.5	−4.8
Germany	79.1	78.5	76.6	0.0	−0.1	−0.2	1 855	1 225	822	4.6	3.2	2.2	−3.3	−4.1	−3.9
Greece	10.0	10.2	10.2	0.4	0.1	0.1	945	747	567	24.2	18.7	14.2	−1.8	−2.3	−2.7
Ireland	3.5	3.8	4.2	0.3	0.9	0.9	184	155	123	13.5	9.7	7.0	−2.4	−1.7	−2.3
Italy	57.6	57.8	56.7	0.0	0.0	−0.2	1 663	957	513	7.1	4.1	2.3	−4.4	−5.4	−6.0
Netherlands	15.0	15.8	16.4	0.5	0.6	0.3	228	161	107	3.7	2.5	1.6	−2.7	−3.5	−3.9
Portugal	10.3	10.6	10.8	0.6	0.3	0.2	764	524	332	16.3	10.6	6.6	−3.7	−3.7	−4.5
Spain	39.2	40.7	41.7	0.4	0.4	0.2	1 561	1 016	617	10.7	6.6	4.0	−3.4	−4.2	−4.9
United Kingdom	57.4	58.6	59.2	0.2	0.2	0.1	569	446	348	2.0	1.5	1.2	−2.1	−2.4	−2.4
Other Western Europe	56.4	58.1	58.9	0.5	0.3	0.1	3 189	2 166	1 376	11.7	7.6	4.8	−3.2	−3.8	−4.4
Austria	7.6	7.6	7.5	0.0	0.0	−0.1	211	134	82	5.7	3.6	2.2	−3.6	−4.4	−4.9
Finland	5.0	5.1	5.1	0.4	0.2	0.1	205	139	89	8.0	5.3	3.5	−3.3	−3.8	−4.4
Iceland	0.3	0.3	0.3	1.1	0.8	0.6	9	7	5	6.6	4.5	3.0	−2.6	−3.0	−3.4
Malta	0.4	0.4	0.4	−0.4	0.4	0.4	5	4	3	3.8	2.7	2.0	−3.0	−1.8	−2.9
Norway	4.2	4.4	4.4	0.4	0.3	0.2	112	73	45	5.2	3.2	1.9	−3.7	−4.2	−4.7
Sweden	8.6	8.7	8.7	0.3	0.1	0.0	169	120	83	3.8	2.7	1.9	−3.3	−3.4	−3.5
Switzerland	6.7	6.8	6.8	0.5	0.2	0.1	137	88	56	4.0	2.6	1.7	−3.1	−4.3	−4.4
Yugoslav SFR	23.8	24.9	25.6	0.7	0.4	0.3	2 341	1 601	1 013	21.7	13.8	8.5	−3.1	−3.7	−4.5

Eastern Europe	99.6	103.7	107.5	0.4	0.4	0.4	9 070	6 828	4 972	17.9	12.6	8.8	-2.9	-2.8	-3.1
Albania	3.2	3.8	4.3	2.0	1.6	1.3	753	791	775	48.4	41.2	34.5	1.1	0.5	-0.2
Bulgaria	9.0	9.1	9.1	0.1	0.1	0.0	542	362	230	12.2	8.0	5.2	-3.9	-4.0	-4.4
Czechoslovakia	15.7	16.2	16.7	0.2	0.3	0.3	774	587	418	9.3	6.5	4.6	-3.1	-2.7	-3.3
Hungary	10.6	10.5	10.5	-0.2	0.0	-0.1	596	384	227	11.5	7.2	4.4	-4.6	-4.3	-5.1
Poland	37.9	39.8	41.9	0.7	0.5	0.5	4 037	3 108	2 321	20.8	14.8	10.4	-2.7	-2.6	-2.9
Romania	23.3	24.3	25.0	0.4	0.5	0.3	2 368	1 596	1 001	20.2	12.7	7.7	-3.5	-3.9	-4.6
Former USSR	288.6	308.4	327.1	0.9	0.7	0.6	18 779	12 644	8 551	13.0	8.3	5.2	-3.6	-3.9	-3.8
North America	276.5	294.6	311.1	0.9	0.6	0.5	3 319	2 394	1 684	2.4	1.6	1.1	-2.8	-3.2	-3.5
Canada	26.5	28.5	30.1	0.9	0.7	0.6	439	288	180	3.3	2.0	1.2	-3.4	-4.1	-4.6
United States	250.0	266.1	280.9	0.9	0.6	0.5	2 880	2 106	1 504	2.3	1.6	1.1	-2.7	-3.1	-3.3
Others	183.8	200.2	214.4	1.0	0.9	0.7	6 391	4 524	3 197	7.4	4.8	3.3	-3.0	-3.4	-3.4
Australia	17.1	19.1	20.9	1.5	1.1	0.9	408	337	264	5.0	3.6	2.6	-1.3	-1.9	-2.4
Israel	4.6	5.3	6.0	1.7	1.5	1.2	76	69	56	4.3	2.9	2.0	-1.7	-0.9	-2.1
Japan	123.5	128.5	131.0	0.6	0.4	0.2	4 013	2 334	1 245	6.4	3.6	2.0	-4.5	-5.3	-6.1
New Zealand	3.4	3.7	3.9	0.8	0.8	0.6	139	125	109	9.1	7.5	6.1	-0.5	-1.1	-1.4
South Africa	35.3	43.7	52.7	2.2	2.2	1.9	1 755	1 659	1 523	13.6	10.1	7.3	0.8	-0.6	-0.9

The total population data and projections are from the 1990 UN assessment (UN, 1991). The projections are those of the medium variant.
The data, and in particular the projections for the population economically active in agriculture should be understood as indicative of broad orders of magnitude. They are, as far as possible, standardized for comparability among countries and regions. They may differ from those obtained from the routine labour force survey statistics. For discussion see FAO (1986). The basis of these estimates is the historical data up to the early 1980s from ILO's work providing internationally comparable statistics. ILO is in the process of updating these data.

Table A.1(a)   Population projections of the 1990*, 1992 and 1994 UN assessments

	Year of assess-ment	Total population (millions)			Growth rates (% p.a.)			
		1990	2000	2010	1980–90	1990–2000	2000–10	1990–2010
World	1990	5297	6265	7209	1.8	1.7	1.4	1.6
	1992	5295	6228	7150	1.8	1.6	1.4	1.5
	1994	5285	6158	7032	1.7	1.5	1.3	1.4
All Developing Countries	1990	4046	4947	5835	2.1	2.0	1.7	1.8
	1992	4042	4896	5744	2.1	1.9	1.6	1.8
	1994	4034	4844	5668	2.1	1.8	1.6	1.7
93 Developing Countries	1990	3987	4879	5758	2.1	2.0	1.7	1.9
	1992	3983	4828	5666	2.1	1.9	1.6	1.8
	1994	3975	4777	5592	2.1	1.9	1.6	1.7
Sub-Saharan Africa	1990	488	673	915	3.2	3.3	3.1	3.2
	1992	486	658	874	3.1	3.1	2.9	3.0
	1994	474	635	834	3.0	3.0	2.8	2.9
Near East/North Africa	1990	305	397	493	2.8	2.6	2.2	2.4
	1992	309	406	512	2.9	2.8	2.4	2.6
	1994	312	403	497	2.8	2.6	2.1	2.4
East Asia	1990	1624	1879	2061	1.6	1.5	0.9	1.2
	1992	1639	1886	2070	1.6	1.4	0.9	1.2
	1994	1638	1860	2050	1.6	1.3	1.0	1.1
South Asia	1990	1128	1398	1668	2.4	2.2	1.8	2.0
	1992	1115	1362	1617	2.3	2.0	1.7	1.9
	1994	1117	1363	1615	2.3	2.0	1.7	1.9
Latin America and Caribbean	1990	442	532	622	2.2	1.9	1.6	1.7
	1992	435	516	593	2.1	1.7	1.4	1.6
	1994	434	517	596	2.1	1.8	1.4	1.6
Developed Countries	1990	1249	1315	1370	0.7	0.5	0.4	0.5
	1992	1253	1332	1406	0.7	0.6	0.5	0.6
	1994	1251	1314	1365	0.7	0.5	0.4	0.4
Western Europe	1990	400	408	410	0.3	0.2	0.0	0.1
	1992	401	413	420	0.3	0.3	0.2	0.2
	1994	400	411	412	0.3	0.3	0.0	0.2
EC-12	1990	344	350	351	0.3	0.2	0.0	0.1
	1992	344	354	360	0.3	0.3	0.2	0.2
	1994	344	353	352	0.3	0.3	0.0	0.1
Other Western Europe	1990	56	58	59	0.5	0.3	0.1	0.2
	1992	57	59	61	0.5	0.4	0.3	0.3
	1994	56	59	60	0.5	0.5	0.2	0.4
Eastern Europe	1990	100	104	107	0.4	0.4	0.4	0.4
	1992	100	103	107	0.4	0.3	0.4	0.3
	1994	100	99	100	0.4	0.0	0.1	0.0
Former USSR	1990	289	308	327	0.9	0.7	0.6	0.6
	1992	289	305	326	0.9	0.5	0.7	0.6
	1994	288	296	305	0.8	0.3	0.3	0.3
North America	1990	276	295	311	0.9	0.6	0.5	0.6
	1992	277	306	330	0.9	1.0	0.8	0.9
	1994	278	306	331	1.0	1.0	0.8	0.9
Others	1990	184	200	214	1.0	0.9	0.7	0.8
	1992	187	206	222	1.1	1.0	0.8	0.9
	1994	186	202	216	1.0	0.8	0.7	0.8

Sources: compiled from UN (1991, 1993b, 1994).
*Used in this study.

Table A.2 Per caput food supplies for direct human consumption

	Calories/day				All cereals, including milled rice (kg/year)			
	1961/63	1969/71	1979/81	1988/90	1961/63	1969/71	1979/81	1988/90
World	2 288	2 434	2 579	2 697	139	146	157	164
All developing countries	1 945	2 122	2 327	2 474	131	145	161	170
93 Developing countries	1 939	2 116	2 322	2 470	131	145	161	171
Sub-Saharan Africa	2 120	2 138	2 120	2 098	120	115	113	114
Angola	1 910	2 127	2 117	1 881	75	76	75	74
Benin	2 038	2 116	2 144	2 383	95	80	91	103
Botswana	2 032	2 165	2 154	2 260	149	162	145	183
Burkina Faso	1 856	1 775	1 816	2 218	161	161	160	205
Burundi	2 047	2 099	2 060	1 947	40	44	54	54
Cameroon	2 140	2 314	2 339	2 207	114	116	110	115
Central African Rep.	2 167	2 296	2 135	1 847	44	51	43	60
Chad	2 298	2 145	1 710	1 733	171	158	110	109
Congo	2 182	2 090	2 235	2 295	21	29	51	60
Côte d'Ivoire	2 192	2 419	2 845	2 566	74	96	129	111
Ethiopia	1 804	1 723	1 795	1 699	141	129	137	136
Gabon	1 950	2 194	2 381	2 442	36	42	71	90
Gambia	2 235	2 203	2 102	2 290	188	168	159	181
Ghana	2 028	2 228	1 972	2 141	62	74	74	78
Guinea	2 211	2 172	2 268	2 243	105	95	107	122
Kenya	2 158	2 230	2 147	2 063	149	156	147	130
Lesotho	1 997	2 006	2 353	2 121	199	202	221	199
Liberia	2 110	2 219	2 399	2 263	90	108	131	117
Madagascar	2 366	2 460	2 472	2 157	143	146	143	122
Malawi	2 067	2 370	2 274	2 049	165	197	176	165

continued

Table A.2 (continued)

	Calories/day				All cereals, including milled rice (kg/year)			
	1961/63	1969/71	1979/81	1988/90	1961/63	1969/71	1979/81	1988/90
Mali	2 167	1 999	1 899	2 259	188	163	157	198
Mauritania	1 967	1 943	2 081	2 447	112	108	121	155
Mauritius	2 407	2 351	2 701	2 897	149	151	160	174
Mozambique	1 953	1 917	1 951	1 805	78	72	74	68
Namibia	1 851	1 974	1 952	1 968	93	102	106	121
Niger	2 049	1 989	2 223	2 240	221	219	231	239
Nigeria	2 473	2 340	2 131	2 200	143	115	95	98
Rwanda	1 820	2 048	2 064	1 915	52	50	50	46
Senegal	2 397	2 470	2 416	2 323	179	181	193	183
Sierra Leone	1 829	2 096	2 095	1 900	97	122	131	114
Somalia	1 721	1 735	1 946	1 873	94	75	98	109
Sudan	1 841	2 168	2 215	2 042	116	142	127	138
Swaziland	2 159	2 268	2 462	2 634	153	158	154	168
Tanzania	1 800	1 804	2 239	2 195	91	86	130	132
Togo	2 376	2 378	2 265	2 269	104	124	112	137
Uganda	2 292	2 275	2 114	2 179	101	105	81	72
Zaire	2 220	2 210	2 133	2 129	36	42	44	46
Zambia	2 093	2 194	2 185	2 016	204	196	204	195
Zimbabwe	2 054	2 141	2 180	2 256	195	195	188	177
Near East/North Africa	2 208	2 384	2 833	3 010	171	183	204	213
Afghanistan	2 171	2 310	2 179	1 764	229	240	217	170
Algeria	1 723	1 824	2 613	2 945	143	151	195	213
Egypt	2 287	2 443	3 090	3 310	177	186	226	242
Iran	1 999	2 280	2 916	3 022	138	164	201	216
Iraq	1 958	2 259	2 757	3 092	132	157	193	222
Jordan	2 229	2 474	2 551	2 710	150	165	162	163

Lebanon	2436	2336	2668	3142	147	135	132	145
Libya	1643	2437	3473	3295	127	148	197	194
Morocco	2185	2407	2696	3031	190	216	229	248
Saudi Arabia	1796	1873	2760	2932	143	127	145	172
Syria	2354	2369	2961	3121	167	164	172	209
Tunisia	2074	2288	2800	3123	165	172	200	213
Turkey	2690	2863	3053	3197	194	205	211	203
Yemen	1942	1908	2056	2232	169	158	158	180
East Asia	1730	2020	2342	2597	126	151	181	201
Cambodia	2150	2301	1657	2122	182	201	146	183
China (Mainland)	1659	1989	2325	2642	124	153	190	215
Indonesia	1816	2020	2464	2605	102	125	159	183
Korea DPR	2031	2109	2652	2843	154	148	188	188
Korea Republic	1957	2470	2747	2826	158	185	171	159
Laos	1982	2251	2365	2465	185	207	206	209
Malaysia	2375	2482	2685	2671	154	154	149	122
Myanmar	1782	2060	2314	2454	132	162	191	202
Philippines	1722	1738	2201	2343	111	109	133	150
Thailand	2029	2196	2292	2280	146	157	149	137
Vietnam	2053	2148	2097	2216	167	175	157	162
South Asia	1974	2041	2090	2215	139	148	153	156
Bangladesh	1976	1962	1973	2038	163	157	169	173
India	1992	2031	2099	2229	138	146	152	155
Nepal	1914	1907	1846	2206	176	175	166	194
Pakistan	1803	2180	2154	2283	119	152	141	144
Sri Lanka	2111	2292	2243	2247	119	138	135	141
Latin America and Carib.	2364	2503	2694	2689	115	119	128	129
Argentina	3073	3267	3195	3068	137	134	128	131
Bolivia	1799	1974	2120	2013	99	93	108	99

continued

Table A.2 (continued)

	Calories/day				All cereals, including milled rice (kg/year)			
	1961/63	1969/71	1979/81	1988/90	1961/63	1969/71	1979/81	1988/90
Brazil	2 321	2 504	2 707	2 730	97	99	117	114
Chile	2 532	2 633	2 645	2 484	147	159	155	146
Colombia	2 164	2 060	2 410	2 453	83	80	88	91
Costa Rica	2 198	2 409	2 581	2 711	102	107	107	115
Cuba	2 298	2 653	2 954	3 129	104	122	133	132
Dominican Republic	1 852	2 025	2 269	2 310	51	59	85	90
Ecuador	2 035	2 147	2 293	2 399	72	77	82	97
El Salvador	1 768	1 849	2 317	2 331	108	117	143	139
Guatemala	1 928	2 081	2 146	2 255	135	141	141	149
Guyana	2 266	2 273	2 499	2 495	133	123	146	145
Haiti	1 967	1 943	2 067	2 005	98	91	92	86
Honduras	1 926	2 160	2 133	2 211	118	133	124	127
Jamaica	2 043	2 522	2 632	2 558	96	115	112	105
Mexico	2 490	2 626	3 001	3 061	159	166	175	181
Nicaragua	2 247	2 378	2 281	2 234	112	123	115	129
Panama	2 169	2 372	2 322	2 269	122	125	102	113
Paraguay	2 404	2 667	2 660	2 684	86	96	84	98
Peru	2 222	2 271	2 102	2 035	113	111	106	107
Suriname	1 967	2 240	2 440	2 436	114	141	142	159
Trinidad and Tobago	2 399	2 500	2 931	2 769	139	136	141	149
Uruguay	2 793	2 965	2 810	2 691	117	137	136	143
Venezuela	2 186	2 385	2 719	2 441	108	131	144	136
Developed countries	3 032	3 195	3 287	3 404	157	149	145	146
Western Europe	3 077	3 227	3 355	3 468	143	136	135	135
EC-12	3 065	3 226	3 349	3 483	139	132	132	131

Belgium–Luxembourg	3 216	3 352	3 474	3 925	127	119	120	117
Denmark	3 381	3 391	3 464	3 639	113	108	110	117
France	3 288	3 327	3 435	3 593	137	113	114	123
Germany	2 978	3 206	3 361	3 522	122	128	131	138
Greece	2 844	3 185	3 443	3 775	165	160	154	157
Ireland	3 558	3 687	3 886	3 951	162	144	142	151
Italy	2 986	3 378	3 561	3 498	179	189	186	163
Netherlands	3 092	3 047	3 070	3 078	105	89	102	88
Portugal	2 656	3 013	2 913	3 342	138	138	130	140
Spain	2 740	2 809	3 248	3 473	150	118	116	117
United Kingdom	3 268	3 290	3 171	3 270	125	118	114	117
Other Western Europe	3 154	3 234	3 391	3 378	172	162	158	153
Austria	3 268	3 276	3 400	3 486	151	139	112	113
Finland	3 216	3 150	3 054	3 067	133	112	106	108
Iceland	3 335	3 082	3 230	3 448	94	91	83	183
Malta	2 860	3 028	2 904	3 169	160	160	142	143
Norway	3 040	3 050	3 351	3 221	106	101	117	128
Sweden	2 872	2 914	3 018	2 978	85	88	90	88
Switzerland	3 537	3 470	3 560	3 508	143	128	123	119
Yugoslav SFR	3 121	3 333	3 568	3 545	245	233	228	213
Eastern Europe	3 137	3 290	3 437	3 386	210	203	191	182
Albania	2 359	2 560	2 752	2 585	215	233	249	232
Bulgaria	3 252	3 508	3 630	3 695	263	255	237	220
Czechoslovakia	3 354	3 365	3 359	3 574	191	175	162	173
Hungary	3 099	3 324	3 459	3 608	194	184	169	166
Poland	3 220	3 379	3 499	3 426	204	199	186	174
Romania	2 881	3 062	3 389	3 081	221	212	204	187
Former USSR	3 147	3 323	3 368	3 380	212	194	177	168

continued

Table A.2 (continued)

	Calories/day				All cereals, including milled rice (kg/year)			
	1961/63	1969/71	1979/81	1988/90	1961/63	1969/71	1979/81	1988/90
North America	3 054	3 235	3 330	3 604	100	98	106	124
Canada	2 923	3 084	3 107	3 242	110	110	109	116
United States	3 067	3 250	3 353	3 642	99	96	105	125
Others	2 610	2 782	2 849	3 014	158	152	151	153
Australia	3 141	3 260	3 088	3 302	122	125	111	119
Israel	2 809	3 039	3 022	3 220	147	148	143	136
Japan	2 514	2 693	2 764	2 921	159	148	144	142
New Zealand	3 316	3 409	3 480	3 462	127	117	122	123
South Africa	2 682	2 819	2 981	3 133	183	191	205	213

Table A.2(a)   Revised and more recent data on per caput food supplies for direct human consumption by country group

	Calories/day				All cereals, including milled rice (kg/year)			
	1988/90*	1988/90	1991	1992	1988/90*	1988/90	1991	1992
World	2697	2707	2696	2719	164	165	164	166
All developing countries	2474	2494	2511	2542	170	171	170	173
93 Developing countries	2470	2490	2506	2538	170	172	171	173
Sub-Saharan Africa	2098	2057	2038	2043	114	114	110	111
Near East/North Africa	3010	3005	2946	2950	213	213	207	209
East Asia	2597	2606	2670	2692	200	199	202	202
South Asia	2215	2274	2259	2338	156	161	156	164
Latin America and Caribbean	2689	2717	2749	2747	129	132	132	133
Developed countries	3404	3382	3297	3304	146	146	146	146
Western Europe	3468	3475	3466	3470	134	134	131	131
EC-12	3483	3492	3498	3503	131	131	127	127
Other Western Europe	3376	3370	3269	3272	153	155	157	157
Eastern Europe and former USSR	3381	3352	3060	3066	172	173	174	175
North America	3604	3591	3635	3670	124	126	128	129
Others	3014	2937	3060	3066	153	149	147	142

*Used in this study.

Table A.3 Cereals sector data (all cereals, including rice in milled form)

| | Production ('000 tonnes) | | | Net trade ('000 tonnes) | | | Self-sufficiency ratio (%) | | | Domestic use | | | | |
| | | | | | | | | | | Total ('000 tonnes) | | | of which in 1988/90 | |
	1969/71	1979/81	1988/90	1969/71	1979/81	1988/90	1969/71	1979/81	1988/90	1969/71	1979/81	1988/90	Food (%)	Feed (%)
World	1 129 087	1 456 618	1 697 512	2 144	2 600	2 625	100	100	99	1 124 470	1 457 426	1 721 116	50	37
All developing countries	482 436	652 327	847 407	−20 367	−66 829	−89 931	97	91	91	497 705	719 587	930 622	72	17
93 Developing countries	479 886	649 583	844 505	−16 832	−59 393	−80 361	98	92	92	491 747	709 469	917 965	73	17
Sub-Saharan Africa	36 465	40 839	54 371	−2 508	−7 838	−7 717	97	86	86	37 449	47 740	62 957	86	3
Angola	574	373	296	59	−365	−478	113	60	38	508	627	775	93	1
Benin	264	363	547	−22	−68	−98	93	85	88	286	426	625	74	3
Botswana	53	35	79	−71	−119	−186	44	23	30	119	152	262	88	2
Burkina Faso	983	1 152	1 843	−27	−96	−152	97	92	92	1 011	1 247	2 006	90	0
Burundi	157	216	288	−14	−29	−20	93	88	92	170	245	314	91	1
Cameroon	771	850	846	−106	−198	−490	89	80	60	867	1 060	1 406	93	0
Central African Rep.	93	99	125	−14	−18	−45	87	87	65	106	114	193	91	2
Chad	671	495	651	−12	−24	−46	98	86	92	682	576	711	85	0
Congo	8	14	24	−31	−76	−113	20	16	18	38	87	136	97	4
Côte d'Ivoire	529	710	980	−157	−531	−573	78	56	62	679	1 264	1 575	82	1
Ethiopia	4 362	5 804	6 204	−52	−294	−741	99	99	87	4 387	5 846	7 118	91	0
Gabon	8	10	22	−16	−47	−84	37	16	20	23	63	109	93	4
Gambia	80	66	88	−16	−47	−81	89	57	52	90	116	170	89	3
Ghana	720	697	1 041	−126	−209	−299	88	77	78	821	909	1 341	85	0
Guinea	466	532	626	−37	−138	−226	93	85	73	501	630	862	79	2
Kenya	2 132	2 268	3 232	91	−137	−20	106	85	100	2 014	2 667	3 238	93	1
Lesotho	208	198	177	−76	−166	−201	77	55	47	271	363	376	91	0
Liberia	123	169	151	−55	−108	−112	74	62	49	167	271	307	91	4
Madagascar	1 388	1 494	1 704	−21	−209	−106	99	89	92	1 396	1 687	1 855	76	3
Malawi	1 160	1 328	1 485	−32	−39	−106	99	99	94	1 149	1 339	1 578	88	0
Mali	998	1 007	1 941	−59	−132	−89	101	82	96	1 022	1 222	2 020	88	10
Mauritania	86	44	137	−69	−147	−231	98	22	37	150	202	369	83	2
Mauritius	1	1	3	−127	−173	−204	57	1	1	129	166	197	95	1
Mozambique	652	625	603	−107	−360	−460	89	65	54	732	966	1 126	92	0
Namibia	68	90	137	−67	−56	−84	61	61	62	111	147	222	94	0

continued

Niger	1261	1692	1886	39	−67	−193	108	106	93	1166	1599	2033	88	2
Nigeria	8606	7118	12403	−366	−2073	−495	103	79	97	8363	9027	12780	80	3
Rwanda	198	267	276	−11	−21	−54	97	94	79	204	283	350	92	0
Senegal	664	818	909	−295	−498	−587	74	65	60	901	1252	1508	87	1
Sierra Leone	340	374	386	−69	−112	−149	87	75	72	390	502	535	86	0
Somalia	241	299	606	−90	−258	−210	80	53	71	303	569	856	92	0
Sudan	2116	2959	2932	−164	3	−388	98	108	78	2164	2736	3784	90	2
Swaziland	83	91	145	−34	−69	−102	75	61	62	111	149	236	54	33
Tanzania	1059	2927	3884	−31	−211	−54	78	97	95	1352	3017	4099	85	7
Togo	296	296	509	−20	−60	−71	95	83	88	310	357	579	81	1
Uganda	1571	1166	1519	−51	−37	−21	115	91	99	1363	1280	1541	84	5
Zaire	683	930	1272	−234	−446	−406	75	73	75	913	1282	1704	93	1
Zambia	915	990	1738	−205	−343	−120	92	75	97	995	1328	1788	89	4
Zimbabwe	1881	2273	2678	184	139	376	127	116	118	1488	1967	2274	73	14
Near East/North Africa	45768	57909	73131	−6337	−22944	−37923	87	73	65	52660	79773	111738	57	28
Afghanistan	3498	3922	2718	−225	−93	−276	88	98	87	3971	4014	3111	87	2
Algeria	1881	1957	1492	−493	−2979	−6100	74	42	20	2542	4722	7551	68	22
Egypt	6530	7340	10404	−1089	−5947	−8335	77	55	56	8460	13287	18591	67	23
Iran	5804	8448	11623	−452	−2700	−5387	90	76	68	6484	11177	17216	67	25
Iraq	1969	1749	2453	−427	−2687	−4079	82	42	37	2410	4150	6714	60	27
Jordan	152	91	113	−150	−498	−1071	48	16	9	319	564	1212	43	51
Lebanon	50	41	76	−535	−581	−504	9	7	12	538	615	620	63	29
Libya	113	225	293	−348	−713	−1759	26	24	15	436	946	1997	43	50
Morocco	4554	3575	7221	−310	−2044	−1408	95	61	86	4810	5901	8369	72	12
Saudi Arabia	428	303	3769	−515	−3145	−3608	53	9	49	813	3385	7701	30	59
Syria	1229	3069	3177	−394	−664	−1378	76	101	69	1628	3025	4608	55	30
Tunisia	724	1146	857	−422	−889	−1749	62	59	32	1173	1938	2715	63	26
Turkey	17945	25130	28111	−758	649	−760	100	102	97	17987	24624	29133	38	34
Yemen	891	913	825	−219	−655	−1511	82	64	37	1090	1427	2203	92	4
East Asia	216075	316169	418658	−5982	−18782	−19902	98	95	96	219922	334468	435044	74	17
Cambodia	2137	863	1674	110	−250	−61	127	78	100	1690	1103	1667	88	0
China (Mainland)	157863	235881	313379	−2482	−11828	−10423	98	95	98	160511	247816	319553	74	17
Indonesia	15381	23873	35936	−867	−2622	−1765	94	90	95	16458	26529	37660	88	6
Korea DPR	4049	7425	8574	−182	−280	−568	94	97	94	4303	7655	9126	44	38
Korea Republic	5652	5815	5983	−2510	−5725	−9559	72	51	41	7823	11332	14757	46	40
Laos	606	715	921	−71	−117	−58	91	91	88	663	785	1043	80	4
Malaysia	1146	1378	1186	−933	−1536	−2448	57	48	32	2015	2897	3677	58	39
Myanmar	5551	8776	9568	653	650	155	112	113	99	4946	7750	9625	86	6
Philippines	5664	8460	10771	−802	−929	−2109	94	91	83	6001	9335	12958	70	22

Table A.3 (continued)

| | Production ('000 tonnes) | | | Net trade ('000 tonnes) | | | Self-sufficiency ratio (%) | | | Domestic use | | | | |
| | | | | | | | | | | Total ('000 tonnes) | | | of which in 1988/90 | |
	1969/71	1979/81	1988/90	1969/71	1979/81	1988/90	1969/71	1979/81	1988/90	1969/71	1979/81	1988/90	Food (%)	Feed (%)
Thailand	11 219	14 694	17 592	2 895	5 195	6 218	159	153	140	7 073	9 635	12 555	60	27
Vietnam	6 807	8 292	13 077	−1 792	−1 340	716	81	86	105	8 440	9 633	12 426	85	5
South Asia														
Bangladesh	11 220	14 281	18 442	−1 301	−1 373	−2 314	97	88	87	11 630	16 268	21 144	88	1
India	90 218	113 360	156 390	−3 622	383	−733	98	96	106	92 034	118 496	148 136	92	0
Nepal	2 710	2 853	4 480	253	−6	−31	112	99	100	2 425	2 877	4 491	87	1
Pakistan	10 536	15 584	18 809	136	110	−674	96	112	97	10 989	13 930	19 319	81	4
Sri Lanka	1 008	1 435	1 625	−953	−850	−1 015	55	67	62	1 828	2 140	2 624	88	3
Latin America and Carib.	65 890	87 157	98 604	3 482	−8 094	−10 055	105	93	88	62 817	93 784	112 519	50	40
Argentina	20 090	24 361	19 677	9 401	14 375	9 223	182	219	175	11 051	11 121	11 256	37	45
Bolivia	471	632	746	−185	−322	−132	74	68	74	635	930	1 014	69	24
Brazil	20 074	27 971	36 414	−996	−6 315	−2 650	94	83	89	21 297	33 849	40 822	41	48
Chile	1 767	1 700	2 922	−483	−1 144	−188	77	61	93	2 285	2 807	3 153	60	34
Colombia	1 689	2 729	3 245	−383	−656	−873	82	84	81	2 049	3 269	4 030	73	23
Costa Rica	142	262	220	−110	−123	−352	59	68	40	242	388	546	62	33
Cuba	300	399	435	−1 236	−2 146	−2 468	20	16	15	1 485	2 557	2 903	48	47
Dominican Republic	196	317	393	−111	−377	−658	64	47	35	307	677	1 125	56	40
Ecuador	555	560	1 131	−78	−332	−480	88	68	75	629	829	1 504	67	18
El Salvador	514	700	791	−47	−130	−186	98	83	81	523	843	981	73	22
Guatemala	849	1 109	1 504	−99	−200	−325	91	88	81	929	1 264	1 860	72	23
Guyana	132	179	138	27	31	−8	120	124	92	110	144	150	77	19
Haiti	509	380	368	−47	−193	−231	93	68	61	550	557	605	90	3
Honduras	395	481	613	−45	−113	−167	90	83	79	439	577	778	82	11
Jamaica	5	6	3	−294	−394	−340	2	2	1	294	411	382	66	30
Mexico	14 422	20 516	22 543	163	−5 908	−6 676	100	83	74	14 426	24 747	30 307	52	36
Nicaragua	351	355	467	−20	−129	−171	97	81	72	361	437	652	74	17
Panama	162	194	245	−72	−94	−125	68	73	68	240	267	363	74	22
Paraguay	273	643	1 628	−58	−74	293	82	88	125	334	734	1 302	31	55
Peru	1 275	1 229	1 847	−699	−1 313	−1 366	68	50	56	1 870	2 467	3 324	68	28
Suriname	88	172	161	−5	57	25	107	144	122	83	120	132	50	26

Trinidad and Tobago	9	10	11	−184	−248	−268	5	4	4	186	260	299	63	32
Uruguay	780	916	1 195	91	190	487	111	132	162	704	692	740	59	24
Venezuela	842	1 338	1 910	−1 048	−2 537	−2 418	47	35	44	1 789	3 841	4 296	61	32
Developed countries	646 690	804 337	850 164	22 511	69 430	92 561	103	109	108	626 862	737 959	790 615	23	61
Western Europe	143 569	178 501	206 382	−23 790	−11 448	24 386	86	94	113	166 338	189 126	182 620	29	60
EC-12	118 005	148 219	174 338	−21 545	−8 722	24 149	85	95	115	139 179	156 191	151 293	30	59
Belgium–Luxembourg	1 923	2 070	2 312	−2 683	−2 052	−1 968	42	52	57	4 611	3 966	4 088	29	47
Denmark	6 678	7 347	8 822	−312	492	2 383	95	106	142	7 043	6 960	6 204	10	81
France	33 951	46 203	56 345	10 790	18 948	29 781	149	172	215	22 769	26 848	26 219	26	64
Germany	26 102	32 052	37 036	−7 690	−5 990	−879	77	82	96	33 976	38 885	38 619	28	63
Greece	3 203	4 923	5 246	−256	−160	453	96	103	111	3 352	4 782	4 708	33	55
Ireland	1 452	2 009	2 118	−417	−324	90	78	89	105	1 857	2 248	2 016	26	60
Italy	15 831	17 697	16 918	−6 029	−5 586	−3 682	72	75	81	22 138	23 486	20 957	45	47
Netherlands	1 584	1 344	1 370	−2 934	−3 308	−2 926	36	28	33	4 430	4 811	4 206	31	46
Portugal	1 663	1 167	1 508	−838	−3 494	−1 446	70	26	49	2 388	4 533	3 102	46	46
Spain	11 679	14 570	20 555	−2 010	−4 909	−670	84	74	98	13 914	19 670	21 009	22	69
United Kingdom	13 941	18 840	22 110	−9 166	−2 339	2 994	61	94	110	22 704	20 003	20 166	33	51
Other Western Europe	25 565	30 282	32 045	−2 245	−2 727	258	94	92	102	27 161	32 937	31 329	28	62
Austria	3 342	4 393	5 207	−268	145	966	94	103	122	3 552	4 279	4 255	20	70
Finland	2 875	2 993	3 636	58	−336	159	106	89	125	2 713	3 368	2 913	18	68
Iceland	0	0	0	−41	−50	−71	0	0	0	41	50	71	65	34
Malta	4	8	9	−114	−120	−143	3	6	6	116	132	151	33	60
Norway	777	1 130	1 262	−668	−778	−597	53	59	68	1 462	1 911	1 858	29	66
Sweden	4 789	5 407	5 538	611	754	673	116	115	124	4 140	4 695	4 454	17	71
Switzerland	672	843	1 312	−1 498	−1 378	−821	31	38	61	2 141	2 230	2 137	37	56
Yugoslav SFR	13 105	15 510	15 081	−323	−965	92	101	95	97	12 995	16 271	15 490	33	57
Eastern Europe	55 109	68 344	80 528	−3 017	−9 185	−2 212	95	88	99	58 011	77 622	81 288	22	63
Albania	534	912	919	−67	13	−99	85	101	92	626	899	1 005	74	18
Bulgaria	6 627	8 107	8 400	230	−387	−711	103	99	95	6 435	8 182	8 873	22	62
Czechoslovakia	8 035	9 771	12 155	−1 428	−1 344	20	88	84	101	9 169	11 613	12 056	22	67
Hungary	9 039	12 989	14 301	21	787	1 443	104	107	114	8 661	12 128	12 504	14	73
Poland	18 236	18 466	26 497	−2 349	−7 345	−2 647	88	72	92	20 652	25 620	28 683	23	65
Romania	12 639	18 100	18 256	576	−910	−218	101	94	101	12 469	19 180	18 168	24	55
Former USSR	168 896	169 604	204 029	5 389	−31 096	−34 172	100	79	85	169 199	214 694	238 817	20	58

continued

Table A.3 *(continued)*

	Production ('000 tonnes)			Net trade ('000 tonnes)			Self-sufficiency ratio (%)			Domestic use Total ('000 tonnes)			Domestic use of which in 1988/90	
	1969/71	1979/81	1988/90	1969/71	1979/81	1988/90	1969/71	1979/81	1988/90	1969/71	1979/81	1988/90	Food (%)	Feed (%)
North America	243 319	341 922	313 505	49 573	129 372	118 666	127	169	138	191 594	202 328	227 299	15	69
Canada	34 518	42 778	47 742	13 314	19 579	20 556	151	179	191	22 863	23 872	25 062	12	74
United States	208 801	299 144	265 762	36 260	109 793	98 110	124	168	131	168 731	178 456	202 237	15	68
Others	35 799	45 968	45 723	−5 645	−8 207	−14 102	86	85	76	41 723	54 192	60 594		
Australia	13 905	20 878	22 109	8 980	14 600	14 651	250	316	288	5 565	6 611	7 674	46	46
Israel	199	239	245	−1 156	−1 690	−2 060	15	13	11	1 351	1 911	2 249	26	54
Japan	12 182	9 890	9 945	−14 378	−24 476	−28 051	46	28	26	26 764	35 234	38 139	27	59
New Zealand	709	785	771	−17	41	−128	96	107	83	737	736	924	46	48
South Africa	8 804	14 175	12 653	926	3 318	1 487	121	146	109	7 307	9 700	11 608	63	29

Note: Production is gross production, i.e. before any deduction of quantities used as seed and feed. Domestic use covers total domestic disappearance of cereals, i.e. both final use for direct human consumption and non-food uses and intermediate use (feed, seed) as well as an allowance for waste, but not additions to stocks. Self-sufficiency ratios are the ratios (in percent) of production over domestic use for all purposes as defined above. The production and domestic use data of the former USSR are before the revisions for the change from bunker weight to clean weight (see Table 3.17).

Table A.4.1   Wheat: area, yield and production (countries with more than 10 000 ha wheat)

	Area (harvested) '000 ha			Yield (kg/ha)			Production ('000 tonnes)		
	1969/71	1979/81	1989/91	1969/71	1979/81	1989/91	1969/71	1979/81	1989/91
World	212 230	234 848	226 966	1 518	1 863	2 464	322 061	437 557	559 226
All developing countries	83 840	96 054	102 305	1 152	1 634	2 286	96 551	156 922	233 858
93 Developing countries	83 378	95 598	101 767	1 154	1 638	2 292	96 228	156 606	233 290
Sub-Saharan Africa	1 271	1 067	1 346	982	1 328	1 563	1 249	1 417	2 104
Angola	14	12	3	909	574	806	13	7	3
Burundi	8	9	10	578	671	827	5	6	9
Ethiopia	795	570	683	822	1 092	1 269	653	623	867
Kenya	133	106	120	1 678	2 011	1 747	223	212	210
Lesotho	89	28	26	634	936	988	56	26	26
Mozambique	10	4	3	959	1 238	1 303	9	5	4
Nigeria	11	10	53	1 759	2 400	1 064	20	24	57
Sudan	118	205	296	1 135	998	1 506	134	205	445
Tanzania	59	57	49	896	1 605	1 840	53	91	90
Zambia	0	3	12	1 000	3 481	4 484	0	9	56
Zimbabwe	17	37	51	3 642	4 783	5 714	60	179	290
Near East/North Africa	24 544	25 333	26 696	1 022	1 347	1 709	25 089	34 114	45 612
Afghanistan	2 199	2 065	1 623	978	1 240	1 063	2 150	2 561	1 725
Algeria	2 214	1 943	1 534	614	654	818	1 359	1 270	1 255
Egypt	551	577	799	2 741	3 193	4 980	1 509	1 844	3 977
Iran	5 183	5 858	6 243	776	997	1 218	4 021	5 843	7 605
Iraq	1 216	1 215	1 200	888	703	879	1 080	854	1 055
Jordan	168	99	54	759	673	1 230	127	67	66
Lebanon	46	26	26	842	1 257	2 179	39	32	57

continued

Table A.4.1 (continued)

	Area (harvested) '000 ha			Yield (kg/ha)			Production ('000 tonnes)		
	1969/71	1979/81	1989/91	1969/71	1979/81	1989/91	1969/71	1979/81	1989/91
Libya	160	251	154	257	497	1002	41	125	155
Morocco	1952	1673	2663	932	897	1562	1819	1500	4160
Saudi Arabia	57	71	775	1774	2254	4748	101	160	3678
Syria	1271	1383	1283	649	1358	1359	825	1878	1743
Tunisia	817	887	830	677	944	1337	553	837	1109
Turkey	8671	9208	9419	1317	1852	2005	11423	17058	18887
Yemen	40	78	93	1053	1113	1496	42	87	139
East Asia	25592	29090	30696	1173	2044	3106	30016	59463	95331
China (Mainland)	25395	28929	30514	1169	2046	3113	29682	59193	94995
Korea DPR	45	50	52	1949	2454	4006	88	123	208
Korea Republic	92	20	0	2305	3153	3000	213	64	1
Myanmar	60	90	129	553	923	974	33	83	126
South Asia	23405	30030	32874	1196	1550	2101	27989	46558	69073
Bangladesh	121	430	584	854	1869	1665	103	803	972
India	16941	22364	23863	1231	1545	2214	20859	34550	52827
Nepal	221	372	599	1044	1195	1403	230	444	840
Pakistan	6122	6865	7829	1110	1567	1844	6796	10760	14433
Latin America and Carib.	8566	10078	10155	1387	1494	2085	11885	15054	21170
Argentina	4396	5245	5255	1318	1537	1977	5793	8060	10392
Bolivia	67	98	92	725	661	792	48	65	73
Brazil	1857	2958	2652	939	883	1454	1743	2613	3856
Chile	737	513	530	1759	1721	3191	1296	882	1691
Colombia	54	36	50	1201	1397	1852	65	50	93
Ecuador	84	33	38	966	1042	708	81	35	27
Guatemala	31	36	16	1118	1461	2031	35	52	32
Mexico	781	723	1021	2918	3808	4040	2278	2754	4125

Paraguay	44	55	207	911	1 222	1 568	40	68	324
Peru	138	98	101	906	1 011	1 283	125	99	129
Uruguay	376	281	191	1 009	1 340	2 238	379	377	428
Developed countries	128 390	138 794	124 661	1 801	2 022	2 610	231 166	280 634	325 368
Western Europe	19 677	18 324	19 140	2 721	3 742	5 060	53 534	68 568	96 855
EC-12	16 924	16 111	16 767	2 723	3 792	5 137	46 079	61 088	86 123
Belgium–Luxembourg	209	189	219	4 058	5 019	6 577	848	949	1 438
Denmark	111	135	499	4 583	5 143	7 250	509	692	3 616
France	3 892	4 473	5 103	3 626	4 999	6 501	14 112	22 362	33 175
Germany	2 108	2 340	2 479	4 019	4 798	6 235	8 471	11 229	15 454
Greece	1 010	1 022	976	1 848	2 710	2 676	1 867	2 770	2 613
Ireland	89	50	74	4 191	5 357	8 120	375	270	602
Italy	4 089	3 373	2 800	2 386	2 665	2 969	9 756	8 989	8 312
Netherlands	146	138	134	4 617	6 271	7 635	675	867	1 022
Portugal	561	328	275	1 094	1 020	1 857	614	335	511
Spain	3 727	2 628	2 182	1 264	1 716	2 399	4 713	4 510	5 237
United Kingdom	980	1 434	2 025	4 223	5 659	6 983	4 140	8 116	14 143
Other Western Europe	2 754	2 212	2 374	2 707	3 381	4 521	7 455	7 480	10 732
Austria	279	271	276	3 274	3 782	5 007	912	1 025	1 381
Finland	184	110	150	2 417	2 418	3 728	445	267	559
Norway	4	15	46	3 143	4 213	4 441	11	63	206
Sweden	259	252	294	3 706	4 324	6 199	958	1 088	1 825
Switzerland	100	88	99	3 682	4 654	6 111	368	409	604
Yugoslav SFR	1 928	1 476	1 507	2 469	3 134	4 082	4 760	4 625	6 152
Eastern Europe	8 062	7 169	8 310	2 399	3 238	4 096	19 344	23 215	34 042
Albania	144	196	187	1 680	2 514	2 715	242	492	509
Bulgaria	1 022	986	1 167	2 836	3 937	4 346	2 898	3 881	5 071
Czechoslovakia	1 076	1 121	1 227	3 193	3 998	5 236	3 436	4 482	6 423
Hungary	1 289	1 187	1 205	2 645	4 043	5 186	3 410	4 800	6 249

continued

Table A.4.1 (continued)

	Area (harvested) '000ha			Yield (kg/ha)			Production ('000 tonnes)		
	1969/71	1979/81	1989/91	1969/71	1979/81	1989/91	1969/71	1979/81	1989/91
Poland	2 004	1 525	2 305	2 458	2 747	3 870	4 924	4 189	8 919
Romania	2 527	2 154	2 220	1 754	2 494	3 096	4 433	5 371	6 871
Former USSR	64 832	59 439	47 245	1 431	1 346	1 842	87 235	84 679	87 018
North America	26 338	40 284	39 511	2 048	2 151	2 300	53 935	86 659	90 886
Canada	7 669	11 386	13 992	1 813	1 794	2 116	13 901	20 430	29 613
United States	18 669	28 898	25 519	2 144	2 292	2 401	40 034	66 229	61 273
Others	9 481	13 579	10 455	1 218	1 290	1 585	11 549	17 514	16 568
Australia	7 701	11 440	8 469	1 171	1 265	1 573	9 014	14 468	13 323
Israel	111	96	89	1 442	2 095	2 520	160	201	224
Japan	227	188	261	2 452	3 031	3 440	557	571	898
New Zealand	112	85	37	3 192	3 642	4 538	357	309	168
South Africa	1 330	1 770	1 599	1 098	1 111	1 222	1 461	1 966	1 954

Note: Data are shown only for countries which in at least one of the three-year averages shown (1969/71, 1979/81, 1989/91) had 10 000 ha or more of harvested land allocated to wheat. The data are after the revision of the production of the former USSR from bunker to clean weight (see Table 3.17).

Table A.4.2   Rice (paddy): area, yield and production (countries with more than 10 000 ha rice)

	Area (harvested) '000 ha			Yield (kg/ha)			Production ('000 tonnes)		
	1969/71	1979/81	1989/91	1969/71	1979/81	1989/91	1969/71	1979/81	1989/91
World	133 101	143 787	147 588	2 329	2 753	3 515	309 903	395 862	518 744
All developing countries	128 565	138 942	143 249	2 228	2 667	3 447	286 392	370 624	493 787
93 Developing countries	127 701	138 138	142 679	2 218	2 660	3 443	283 285	367 460	491 265
Sub-Saharan Africa	3 477	4 465	6 144	1 344	1 356	1 579	4 673	6 054	9 703
Angola	21	20	18	1 188	1 000	1 039	25	20	19
Burkina Faso	40	39	21	933	1 140	2 083	37	44	43
Burundi	3	4	12	1 586	2 366	3 260	5	10	40
Cameroon	16	21	12	841	2 272	5 565	13	48	69
Central African Rep.	14	14	8	647	1 000	1 183	9	14	10
Chad	44	43	39	964	896	2 038	42	39	80
Cote d'Ivoire	287	383	563	1 168	1 171	1 174	335	448	661
Gambia	28	23	14	1 414	1 603	1 493	39	37	21
Ghana	55	107	72	1 000	837	1 418	55	89	102
Guinea	411	486	800	886	899	756	364	438	605
Kenya	6	8	15	4 754	4 631	4 027	27	39	59
Liberia	154	203	168	1 194	1 252	969	184	254	163
Madagascar	992	1 182	1 140	1 911	1 738	2 089	1 894	2 055	2 381
Malawi	23	37	29	1 036	1 074	1 735	23	39	51
Mali	158	165	222	1 017	1 026	1 614	161	169	358
Mauritania	1	3	14	1 000	3 656	3 476	1	12	50
Mozambique	76	92	109	1 303	810	758	99	74	83
Niger	16	20	31	2 092	1 537	2 463	34	31	76
Nigeria	272	517	1 567	1 293	1 988	1 912	352	1 027	2 996
Senegal	91	74	75	1 293	1 300	2 300	118	96	173

continued

Table A.4.2 (continued)

	Area (harvested) '000 ha			Yield (kg/ha)			Production ('000 tonnes)		
	1969/71	1979/81	1989/91	1969/71	1979/81	1989/91	1969/71	1979/81	1989/91
Sierra Leone	331	403	384	1432	1250	1243	474	504	478
Sudan	5	11	1	1111	845	1286	5	9	1
Tanzania	144	262	348	991	959	2039	143	251	709
Togo	25	18	22	709	798	1437	18	15	31
Uganda	16	12	35	819	1342	1331	13	16	46
Zaire	236	293	393	755	806	891	178	236	350
Zambia	1	5	12	400	489	992	0	2	12
Near East/North Africa	1219	1170	1282	3728	4006	4660	4545	4686	5974
Afghanistan	203	190	173	1847	2179	1907	374	415	329
Egypt	487	416	437	5275	5709	7086	2566	2377	3098
Iran	362	434	539	2875	3211	3831	1041	1394	2064
Iraq	97	56	79	2775	2888	2731	268	162	217
Turkey	63	67	51	4106	4721	4976	257	314	253
East Asia	65750	70046	71572	2684	3387	4327	176480	237212	309695
Cambodia	2074	1262	1543	1454	1071	1568	3016	1352	2420
China (Mainland)	32537	33648	32785	3281	4236	5625	106753	142538	184424
Indonesia	8158	9064	10438	2346	3262	4298	19136	29570	44864
Korea DPR	563	635	667	4246	7454	7950	2392	4733	5300
Korea Republic	1204	1230	1237	4628	5512	6231	5574	6780	7705
Laos	665	722	597	1309	1419	2299	870	1025	1373
Malaysia	708	722	665	2397	2844	2842	1696	2053	1891
Myanmar	4748	4684	4709	1707	2698	2900	8107	12637	13656
Philippines	3241	3513	3414	1683	2205	2778	5456	7747	9484
Thailand	7070	8986	9443	1933	1888	2034	13668	16967	19205
Vietnam	4782	5579	6075	2052	2117	3189	9812	11809	19374

South Asia	50 881	54 476	57 012	1 702	1 910	2 603	86 591	104 022	148 420
Bangladesh	9 842	10 310	10 386	1 681	1 952	2 593	16 540	20 125	26 935
India	37 677	40 091	42 318	1 668	1 860	2 621	62 861	74 557	110 921
Nepal	1 186	1 275	1 433	1 937	1 851	2 352	2 296	2 361	3 372
Pakistan	1 527	1 981	2 106	2 246	2 465	2 309	3 431	4 884	4 862
Sri Lanka	650	819	770	2 252	2 557	3 028	1 463	2 093	2 330
Latin America and Carib.	6 375	7 981	6 669	1 725	1 940	2 620	10 996	15 487	17 473
Argentina	89	89	108	3 900	3 244	3 844	347	288	415
Bolivia	54	60	111	1 478	1 507	2 098	80	91	232
Brazil	4 788	5 932	4 441	1 430	1 438	2 099	6 847	8 533	9 320
Chile	23	40	35	2 620	3 151	4 162	60	125	146
Colombia	246	428	491	3 190	4 277	4 047	784	1 831	1 986
Costa Rica	44	73	53	2 018	3 059	3 586	88	224	190
Cuba	164	146	151	1 937	3 105	3 130	317	455	471
Dominican Republic	80	111	98	2 562	3 534	4 453	206	392	438
Ecuador	78	123	277	2 989	3 074	3 077	234	378	852
El Salvador	12	15	15	3 621	3 735	4 045	45	56	62
Guatemala	12	14	16	2 183	2 770	2 839	25	37	46
Guyana	109	91	65	1 800	2 924	3 221	195	266	210
Haiti	38	51	52	2 139	2 324	2 396	81	119	125
Honduras	11	20	19	1 304	1 735	2 627	15	35	49
Mexico	152	153	114	2 561	3 453	3 713	390	528	423
Nicaragua	26	37	42	3 008	3 545	2 737	77	130	115
Panama	105	96	90	1 376	1 834	2 247	144	175	203
Paraguay	19	26	27	2 145	1 837	2 684	40	47	72
Peru	130	132	186	4 141	4 410	5 160	539	580	957
Suriname	37	65	61	3 540	3 975	3 772	132	258	232
Uruguay	34	63	92	3 899	4 566	4 984	132	289	459
Venezuela	121	214	120	1 724	2 985	3 814	208	638	458

continued

Table A.4.2 *(continued)*

	Area (harvested) '000 ha			Yield (kg/ha)			Production ('000 tonnes)		
	1969/71	1979/81	1989/91	1969/71	1979/81	1989/91	1969/71	1979/81	1989/91
Developed countries	4 536	4 844	4 339	5 183	5 210	5 751	23 511	25 237	24 957
Western Europe	322	309	366	5 037	5 540	5 882	1 622	1 710	2 155
EC-12	315	300	358	5 055	5 567	5 926	1 590	1 670	2 124
France	22	6	20	4 070	3 905	5 677	88	25	112
Greece	17	17	16	4 832	4 867	6 208	84	84	99
Italy	172	176	208	4 977	5 615	6 034	858	989	1 257
Portugal	41	32	34	4 380	4 359	4 670	177	137	157
Spain	63	69	81	6 100	6 328	6 175	384	435	498
Eastern Europe	73	57	61	2 712	2 965	1 984	198	169	12
Bulgaria	17	17	11	3 843	4 327	2 741	64	71	30
Hungary	24	16	11	2 250	2 148	2 722	54	35	29
Romania	28	21	37	2 374	2 346	1 514	67	49	56
Former USSR	356	637	623	3 323	3 738	3 426	1 183	2 380	2 135
North America	777	1 345	1 117	5 087	5 179	6 334	3 953	6 968	7 077
United States	777	1 345	1 117	5 087	5 179	6 334	3 953	6 968	7 077
Others	3 008	2 496	2 172	5 505	5 613	6 202	16 555	14 011	13 469
Australia	38	111	97	7 139	6 205	8 021	273	688	778
Japan	2 968	2 384	2 073	5 485	5 587	6 120	16 280	13 320	12 688

Note: Data are shown only for countries which in at least one of the three-year averages shown (1969/71, 1979/81, 1989/91) had 10 000 ha or more of harvested land allocated to rice. The data are after the revision of the production of the former USSR from bunker to clean weight (see Table 3.17).

Table A.4.3   Maize: area, yield and production (countries with more than 10 000 ha maize)

	Area (harvested) '000 ha			Yield (kg/ha)			Production ('000 tonnes)		
	1969/71	1979/81	1989/91	1969/71	1979/81	1989/91	1969/71	1979/81	1989/91
World	114 854	126 287	128 917	2 468	3 342	3 735	283 451	422 079	481 568
All developing countries	70 131	75 896	83 332	1 464	1 978	2 448	102 660	150 158	204 032
93 Developing countries	70 020	75 755	83 146	1 464	1 979	2 449	102 491	149 913	203 595
Sub-Saharan Africa	12 070	12 155	16 458	981	1 135	1 199	11 843	13 793	19 731
Angola	540	600	756	864	506	301	467	303	228
Benin	360	407	467	561	709	903	202	289	422
Botswana	34	42	33	310	279	320	11	12	11
Burkina Faso	92	123	208	659	875	1 330	60	108	277
Burundi	112	127	124	1 071	1 109	1 368	120	141	170
Cameroon	460	495	221	911	844	1 835	419	418	405
Central African Rep.	62	108	68	711	373	860	44	40	59
Chad	6	32	32	1 952	842	1 000	12	27	32
Congo	7	16	28	597	758	893	4	12	25
Cote d'Ivoire	333	514	683	773	684	728	257	352	497
Ethiopia	850	753	1 037	1 071	1 626	1 708	910	1 224	1 771
Gabon	6	6	14	1 473	1 672	1 543	8	10	22
Gambia	3	7	13	1 063	1 477	1 238	3	10	16
Ghana	387	390	547	1 078	974	1 339	417	380	733
Guinea	59	87	91	1 153	1 000	831	68	87	76
Kenya	1 233	1 273	1 447	1 241	1 346	1 673	1 530	1 714	2 420
Lesotho	143	116	107	651	967	1 296	93	112	139
Madagascar	121	124	152	1 004	982	1 013	122	122	154
Malawi	1 039	1 077	1 332	1 025	1 184	1 112	1 066	1 275	1 481

continued

Table A.4.3 (continued)

	Area (harvested) '000 ha			Yield (kg/ha)			Production ('000 tonnes)		
	1969/71	1979/81	1989/91	1969/71	1979/81	1989/91	1969/71	1979/81	1989/91
Mali	78	52	172	865	1171	1319	67	61	226
Mozambique	363	674	1008	1003	569	367	364	383	370
Namibia	93	100	120	400	487	501	37	49	60
Niger	3	15	5	607	703	688	2	10	3
Nigeria	1346	443	1550	983	1353	1261	1323	599	1955
Rwanda	50	73	73	1085	1159	1364	54	84	100
Senegal	52	75	100	814	885	1223	42	66	122
Sierra Leone	11	13	11	981	977	1054	10	13	12
Somalia	124	151	213	895	794	1116	111	120	238
Sudan	39	67	73	779	584	608	31	39	44
Swaziland	89	66	90	846	1291	1358	75	85	122
Tanzania	1005	1350	1820	612	1305	1447	615	1762	2634
Togo	144	147	273	1109	1020	982	160	150	268
Uganda	310	263	417	1349	1368	1393	418	360	581
Zaire	595	745	1211	716	811	722	426	604	874
Zambia	992	523	808	792	1799	1665	786	941	1345
Zimbabwe	923	1097	1150	1629	1667	1598	1504	1829	1837
Near East/North Africa	2284	2350	2237	2012	2392	3718	4597	5623	8316
Afghanistan	470	447	264	1506	1649	1713	707	738	453
Egypt	634	800	847	3741	3949	5687	2370	3159	4817
Iran	25	35	37	1400	1499	3412	35	52	126
Iraq	6	22	75	1508	2431	2561	9	53	191
Morocco	474	396	389	801	618	1006	380	245	391
Syria	6	21	59	1407	2083	2906	8	43	171
Turkey	646	583	512	1639	2168	4087	1058	1263	2093
Yemen	8	38	47	2200	1673	1271	17	64	60

East Asia	22 921	28 690	30 772	1 814	2 635	3 645	41 575	75 612	112 160
Cambodia	94	87	40	1 331	980	1 293	125	85	52
China (Mainland)	16 175	19 950	21 109	2 005	3 038	4 335	32 433	60 617	91 506
Indonesia	2 667	2 761	3 004	965	1 461	2 129	2 575	4 035	6 394
Korea DPR	383	633	708	5 305	6 053	6 283	2 033	3 833	4 450
Korea Republic	44	34	24	1 450	4 436	4 339	63	150	105
Laos	15	29	34	1 713	1 062	1 772	26	31	60
Malaysia	8	7	20	1 935	1 143	1 761	15	8	35
Myanmar	71	128	125	714	1 295	1 538	51	166	192
Philippines	2 434	3 267	3 699	828	972	1 264	2 051	3 174	4 677
Thailand	793	1 412	1 551	2 502	2 198	2 559	1 984	3 103	3 969
Vietnam	237	383	457	1 075	1 071	1 576	255	410	720
South Asia	6 896	7 102	7 613	1 102	1 144	1 521	7 596	8 126	11 578
India	5 794	5 887	5 970	1 051	1 102	1 531	6 087	6 486	9 141
Nepal	439	455	754	1 812	1 516	1 607	796	690	1 212
Pakistan	640	736	856	1 088	1 257	1 390	697	925	1 189
Sri Lanka	19	21	30	774	1 086	1 084	15	23	32
Latin America and Carib.	25 850	25 457	26 066	1 427	1 837	1 988	36 880	46 759	51 810
Argentina	3 880	2 895	1 758	2 247	3 224	3 359	8 717	9 333	5 905
Bolivia	223	295	270	1 306	1 430	1 629	291	422	439
Brazil	10 021	11 430	12 479	1 365	1 685	1 914	13 680	19 265	23 887
Chile	70	124	109	3 109	3 798	7 980	217	471	866
Colombia	684	620	806	1 251	1 401	1 460	856	868	1 177
Costa Rica	57	43	39	1 278	1 778	1 815	73	77	72
Cuba	100	77	77	853	1 239	1 234	85	95	95
Dominican Republic	27	32	35	1 712	1 184	1 403	46	38	49
Ecuador	312	230	452	767	1 075	1 082	239	247	490
El Salvador	203	281	288	1 670	1 840	1 961	340	517	565

continued

Table A.4.3 *(continued)*

	Area (harvested) '000 ha			Yield (kg/ha)			Production ('000 tonnes)		
	1969/71	1979/81	1989/91	1969/71	1979/81	1989/91	1969/71	1979/81	1989/91
Guatemala	672	627	629	1118	1511	1952	751	947	1228
Haiti	231	207	208	1058	868	807	245	179	168
Honduras	283	339	382	1199	1201	1425	339	407	545
Mexico	7412	6836	6919	1218	1736	1920	9025	11866	13282
Nicaragua	260	179	215	912	1022	1230	238	183	265
Panama	77	62	74	859	953	1249	66	59	93
Paraguay	162	340	420	1246	1572	2015	201	535	847
Peru	373	341	386	1621	1665	1995	605	569	770
Uruguay	194	124	59	832	1015	1684	161	126	99
Venezuela	606	372	455	1152	1471	2117	698	547	963
Developed countries	44724	50390	45586	4042	5396	6088	180791	271921	277536
Western Europe	6172	6256	6190	3818	5131	5951	23561	32100	36836
EC-12	3649	3799	3745	4227	5503	6890	15424	20905	25805
France	1436	1774	1757	5148	5434	6720	7394	9641	11808
Germany	102	122	234	4973	6173	7217	509	753	1687
Greece	162	157	221	3074	7423	10071	498	1165	2225
Italy	986	956	810	4665	6897	7594	4601	6590	6154
Portugal	433	333	220	1385	1458	3032	599	486	666
Spain	526	450	495	3432	4950	6462	1804	2227	3201
Other Western Europe	2523	2457	2445	3226	4557	4512	8138	11195	11032
Austria	122	190	193	5545	7045	8109	677	1338	1561
Switzerland	10	17	28	6140	7226	8614	61	121	239
Yugoslav SFR	2391	2250	2225	3095	4327	4150	7399	9736	9232

Eastern Europe	5 307	5 399	4 488	2 841	3 905	3 974	15 074	21 086	17 836
Albania	111	100	59	1 988	3 175	4 169	220	318	245
Bulgaria	623	605	516	3 913	4 344	4 046	2 436	2 627	2 087
Czechoslovakia	127	173	164	4 020	4 638	4 743	511	800	777
Hungary	1 272	1 270	1 098	3 570	5 528	5 841	4 542	7 022	6 414
Poland	5	26	60	2 449	3 889	4 847	12	102	291
Romania	3 170	3 226	2 592	2 320	3 168	3 096	7 354	10 218	8 023
Former USSR	3 617	3 058	3 310	2 763	2 970	3 520	9 993	9 082	11 650
North America	24 238	30 700	28 103	5 163	6 449	7 157	125 137	197 988	201 144
Canada	490	1 039	1 057	5 079	5 685	6 580	2 488	5 904	6 953
United States	23 794	29 661	27 046	5 164	6 476	7 180	122 649	192 084	194 191
Others	5 390	4 978	3 495	1 304	2 344	2 881	7 026	11 666	10 069
Australia	77	54	51	2 383	3 069	4 128	184	164	210
Japan	12	2	0	2 664	2 294	2 667	33	4	1
New Zealand	9	20	17	7 725	8 291	9 711	70	163	161
South Africa	5 290	4 900	3 426	1 273	2 311	2 830	6 734	11 322	9 695

Note: Data are shown only for countries which in at least one of the three-year averages shown (1969/71, 1979/81, 1989/91) had 10 000 ha or more of harvested land allocated to maize. The data are after the revision of the production of the former USSR from bunker to clean weight (see Table 3.17).

Table A.4.4 Barley: area, yield and production (countries with more than 10 000 ha barley)

	Area (harvested) '000 ha			Yield (kg/ha)			Production ('000 tonnes)		
	1969/71	1979/81	1989/91	1969/71	1979/81	1989/91	1969/71	1979/81	1989/91
World	66 609	81 243	74 390	1 843	1 893	2 284	122 785	153 790	169 927
All developing countries	17 548	16 706	19 262	1 095	1 289	1 287	19 208	21 535	24 799
93 Developing countries	17 435	16 553	19 132	1 096	1 292	1 287	19 101	21 394	24 632
Sub-Saharan Africa	931	910	988	803	1 226	1 009	748	1 115	996
Ethiopia	913	850	950	796	1 200	978	727	1 021	929
Kenya	14	49	24	1 133	1 301	1 363	15	64	32
Near East/North Africa	9 270	10 908	14 667	1 006	1 108	1 145	9 323	12 081	16 798
Afghanistan	316	296	207	1 151	1 057	1 071	363	313	221
Algeria	773	875	1 217	608	677	915	470	592	1 114
Egypt	44	41	56	2 126	2 692	2 177	93	111	122
Iran	1 532	1 727	2 479	680	902	1 277	1 042	1 558	3 166
Iraq	582	858	1 382	1 190	846	793	692	726	1 095
Jordan	50	52	55	492	395	629	25	21	34
Lebanon	7	6	11	863	1 000	1 714	6	6	19
Libya	213	284	283	326	342	496	70	97	140
Morocco	2 003	2 190	2 390	1 093	782	1 170	2 190	1 712	2 796
Saudi Arabia	14	7	65	908	1 171	5 649	13	8	366
Syria	784	1 220	2 618	478	926	259	375	1 129	678
Tunisia	243	457	491	571	609	949	139	279	466
Turkey	2 597	2 846	3 365	1 432	1 926	1 942	3 720	5 480	6 533
Yemen	111	49	49	1 125	1 000	967	125	49	48
East Asia	3 097	1 752	1 299	1 599	2 697	2 625	4 954	4 726	3 409
China (Mainland)	2 285	1 295	1 083	1 150	2 420	2 462	2 627	3 133	2 667

Korea DPR	90	72	60	1 333	2 162	2 472	120	155	148
Korea Republic	723	386	155	3 054	3 728	3 832	2 207	1 438	593
South Asia	2 902	2 046	1 221	959	1 073	1 464	2 784	2 196	1 787
Bangladesh	32	19	18	672	637	623	21	12	11
India	2 693	1 802	1 016	981	1 121	1 591	2 642	2 020	1 616
Nepal	26	26	30	924	875	929	24	23	27
Pakistan	151	199	157	642	709	842	97	141	132
Latin America and Carib.	1 235	936	959	1 046	1 362	1 713	1 292	1 275	1 642
Argentina	431	178	180	1 153	1 289	2 282	497	229	411
Bolivia	94	80	83	668	653	659	63	53	55
Brazil	26	84	105	985	1 120	1 635	26	94	172
Chile	48	51	28	2 017	2 000	3 436	97	103	95
Colombia	53	58	51	1 733	1 753	1 871	92	101	96
Ecuador	127	29	56	596	843	850	75	24	48
Mexico	230	281	270	1 043	1 726	1 858	240	486	502
Peru	184	127	104	894	903	1 015	164	115	105
Uruguay	42	48	81	898	1 474	1 960	38	71	158
Developed countries	49 061	64 538	55 128	2 111	2 049	2 633	103 577	132 255	145 127
Western Europe	13 724	16 566	14 088	3 007	3 453	4 127	41 274	57 199	58 142
EC-12	11 931	14 396	12 328	3 051	3 514	4 151	36 405	50 583	51 171
Belgium–Luxembourg	174	173	108	3 488	4 877	5 779	608	844	626
Denmark	1 342	1 580	942	3 856	3 956	5 301	5 175	6 250	4 996
France	2 825	2 670	1 781	3 138	4 119	5 711	8 865	10 997	10 169
Germany	2 101	2 971	2 596	3 480	4 093	5 506	7 313	12 158	14 295
Greece	335	344	194	1 955	2 434	2 397	655	838	465
Ireland	216	349	243	3 954	4 709	5 679	854	1 643	1 378
Italy	180	325	470	1 816	2 817	3 646	326	914	1 713

continued

Table A.4.4 (continued)

	Area (harvested) '000 ha			Yield (kg/ha)			Production ('000 tonnes)		
	1969/71	1979/81	1989/91	1969/71	1979/81	1989/91	1969/71	1979/81	1989/91
Netherlands	101	56	44	3 627	4 693	5 349	366	265	236
Portugal	105	75	68	614	603	1 198	65	45	82
Spain	2 235	3 520	4 361	1 755	1 867	2 143	3 922	6 571	9 346
United Kingdom	2 317	2 333	1 521	3 564	4 310	5 171	8 257	10 058	7 866
Other Western Europe	1 793	2 170	1 760	2 716	3 049	3 960	4 868	6 617	6 971
Austria	286	370	294	3 333	3 482	4 959	954	1 288	1 456
Finland	395	579	514	2 388	2 455	3 477	943	1 421	1 789
Norway	183	186	178	2 985	3 424	3 810	545	636	678
Sweden	602	678	471	3 049	3 427	4 198	1 836	2 323	1 976
Switzerland	39	48	58	3 777	4 589	6 045	147	220	352
Yugoslav SFR	287	309	244	1 540	2 351	2 934	442	726	716
Eastern Europe	2 726	3 869	3 495	2 644	3 029	3 976	7 207	11 719	13 896
Albania	9	12	12	908	2 025	2 785	8	24	34
Bulgaria	416	425	368	2 662	3 386	4 043	1 109	1 439	1 487
Czechoslovakia	810	972	763	3 141	3 627	4 989	2 543	3 524	3 805
Hungary	322	265	312	2 329	3 202	4 554	749	848	1 421
Poland	861	1 362	1 195	2 536	2 616	3 453	2 182	3 563	4 128
Romania	309	833	845	1 994	2 786	3 577	615	2 321	3 022
Former USSR	21 794	33 420	26 962	1 483	1 175	1 717	32 330	39 254	46 305
North America	8 446	7 845	7 740	2 309	2 554	2 792	19 500	20 037	21 611
Canada	4 483	4 631	4 468	2 236	2 418	2 740	10 024	11 199	12 244
United States	3 963	3 214	3 272	2 391	2 750	2 863	9 476	8 838	9 367

Others	2 372	2 838	2 843	1 378	1 425	1 820	3 267	4 045	5 173
Australia	2 019	2 540	2 522	1 175	1 291	1 668	2 372	3 279	4 208
Israel	20	27	14	882	681	413	17	18	6
Japan	224	120	105	2 807	3 261	3 184	629	392	333
New Zealand	67	71	88	3 310	3 610	4 332	222	255	381
South Africa	42	81	114	643	1 258	2 152	27	102	245

Note: Data are shown only for countries which in at least one of the three-year averages shown (1969/71, 1979/81, 1989/91) had 10 000 ha or more of harvested land allocated to barley. The data are after the revision of the production of the former USSR from bunker to clean weight (see Table 3.17).

Table A.4.5   Millet: area, yield and production (countries with more than 10 000 ha millet)

	Area (harvested) '000 ha			Yield (kg/ha)			Production ('000 tonnes)		
	1969/71	1979/81	1989/91	1969/71	1979/81	1989/91	1969/71	1979/81	1989/91
World	44 636	37 646	36 537	698	677	791	31 155	25 468	28 899
All developing countries	41 596	34 718	33 421	688	682	772	28 613	23 678	25 790
93 Developing countries	41 575	34 689	33 393	688	682	771	28 596	23 655	25 762
Sub-Saharan Africa	13 349	11 513	14 701	596	666	698	7 957	7 662	10 266
Angola	92	80	120	845	613	536	78	49	64
Benin	17	13	38	376	500	623	6	7	24
Botswana	19	12	9	245	145	176	5	2	2
Burkina Faso	843	803	1 212	426	486	535	360	390	649
Burundi	10	10	12	897	980	1 065	9	10	13
Cameroon	123	130	60	697	753	1 062	86	98	64
Central African Rep.	12	16	12	795	682	870	9	11	11
Chad	430	360	543	635	505	352	273	182	191
Cote d'Ivoire	63	64	77	487	581	613	31	37	47
Ethiopia	197	226	253	596	899	953	117	203	241
Gambia	35	28	53	1 049	914	974	36	26	52
Ghana	218	182	192	549	641	636	120	117	122
Guinea	35	35	31	1 429	1 409	1 521	50	49	47
Kenya	75	80	100	1 683	1 049	658	127	84	66
Malawi	0	11	18	0	593	554	0	7	10
Mali	569	643	1 174	793	716	701	451	461	823
Mauritania	34	12	16	178	254	409	6	3	7
Mozambique	19	20	20	526	250	250	10	5	5
Namibia	57	77	93	453	447	604	26	34	56
Niger	2 313	3 011	3 711	422	435	383	975	1 311	1 422
Nigeria	4 887	2 366	3 783	615	1 055	1 181	3 007	2 496	4 468

Senegal	874	932	899	503	595	644	440	555	579
Sierra Leone	6	9	26	1 103	1 398	884	6	13	23
Sudan	745	1 098	1 113	567	397	166	423	436	185
Tanzania	211	450	255	663	800	982	140	360	250
Togo	190	121	134	638	364	510	121	44	68
Uganda	739	297	379	1 024	1 592	1 534	757	473	582
Zaire	27	36	43	699	668	728	19	24	31
Zambia	128	34	51	619	638	556	79	22	28
Zimbabwe	380	353	271	500	432	501	190	153	136
Near East/North Africa	273	191	187	1 150	1 007	610	314	192	114
Afghanistan	33	39	30	843	860	865	28	33	26
Iran	15	8	7	753	1 114	1 940	11	9	13
Saudi Arabia	110	25	7	1 137	412	1 597	125	10	12
Syria	25	15	7	706	1 047	577	18	16	4
Turkey	39	16	4	1 385	1 342	1 357	54	21	6
Yemen	39	81	122	1 787	1 185	365	70	96	45
East Asia	7 250	4 222	2 526	1 301	1 406	1 700	9 434	5 936	4 294
China (Mainland)	6 937	3 987	2 302	1 333	1 455	1 785	9 250	5 787	4 109
Korea DPR	93	62	50	961	1 065	1 214	90	66	61
Korea Republic	59	3	3	798	1 147	1 154	47	4	3
Myanmar	161	179	172	291	447	707	47	80	121
South Asia	20 543	18 561	15 937	522	518	692	10 724	9 619	11 031
Bangladesh	76	61	89	784	648	713	60	40	63
India	19 618	17 845	15 202	519	515	694	10 182	9 189	10 551
Nepal	114	122	197	1 123	989	1 160	128	121	228
Pakistan	713	509	438	476	500	410	339	255	180
Sri Lanka	23	24	11	719	612	688	17	15	8

continued

Table A.4.5 (continued)

	Area (harvested) '000 ha			Yield (kg/ha)			Production ('000 tonnes)		
	1969/71	1979/81	1989/91	1969/71	1979/81	1989/91	1969/71	1979/81	1989/91
Latin America and Carib.	160	203	42	1049	1211	1371	168	245	57
Argentina	160	203	42	1049	1211	1360	168	245	57
Developed countries	3040	2928	3116	836	611	998	2542	1790	3110
Eastern Europe	29	7	8	1146	1300	2145	33	9	18
Poland	21	0	0	1224	0	0	26	0	0
Former USSR	2821	2777	2901	808	583	985	2278	1620	2859
North America	110	89	152	1324	1200	1231	145	107	187
United States	110	89	152	1324	1200	1231	145	107	187
Others	68	49	54	941	863	809	64	42	43
Australia	39	26	31	951	1000	877	37	26	27
South Africa	22	22	22	682	682	682	15	15	15

Note: Data are shown only for countries which in at least one of the three-year averages shown (1969/71, 1979/81, 1989/91) had 10 000 ha or more of harvested land allocated to millet. The data are after the revision of the production of the former USSR from bunker to clean weight (see Table 3.17).

Table A.4.6 Sorghum: area, yield and production (countries with more than 10 000 ha sorghum)

	Area (harvested) '000 ha			Yield (kg/ha)			Production ('000 tonnes)		
	1969/71	1979/81	1989/91	1969/71	1979/81	1989/91	1969/71	1979/81	1989/91
World	50 343	45 062	42 399	1 143	1 454	1 344	57 567	65 525	57 002
All developing countries	43 590	38 588	37 348	840	1 139	1 068	36 633	43 948	39 889
93 Developing countries	43 579	38 555	37 311	840	1 139	1 066	36 622	43 918	39 778
Sub-Saharan Africa	15 094	13 122	17 046	672	852	747	10 138	11 180	12 729
Benin	91	90	140	565	649	762	52	59	107
Botswana	133	98	140	278	209	320	37	21	45
Burkina Faso	1 054	1 051	1 325	501	590	750	528	620	993
Burundi	21	53	58	986	1 000	1 117	21	53	65
Cameroon	353	374	515	727	805	707	257	301	364
Central African Rep.	42	57	37	787	675	628	33	39	23
Chad	495	414	456	691	507	587	342	210	268
Cote D'Ivoire	28	40	45	507	606	579	14	24	26
Ethiopia	950	1 048	810	877	1 355	1 086	833	1 419	880
Gambia	7	6	12	905	783	881	7	5	10
Ghana	209	223	255	705	628	775	147	140	197
Guinea	20	20	21	1 250	1 250	1 110	25	25	23
Kenya	201	168	124	1 070	951	932	215	160	116
Lesotho	75	58	34	756	1 021	781	57	59	26
Malawi	107	30	31	730	667	590	78	20	18
Mali	384	434	811	868	785	835	333	341	677
Mauritania	229	102	120	327	279	600	75	28	72
Mozambique	261	288	422	774	630	396	202	182	167

continued

Table A.4.6 (continued)

	Area (harvested) '000 ha			Yield (kg/ha)			Production ('000 tonnes)		
	1969/71	1979/81	1989/91	1969/71	1979/81	1989/91	1969/71	1979/81	1989/91
Namibia	10	15	16	433	422	500	4	6	8
Niger	589	822	1 512	445	422	280	262	347	423
Nigeria	6 303	2 683	4 318	631	1 224	1 032	3 980	3 284	4 454
Rwanda	132	159	154	1 068	1 123	1 157	141	178	178
Senegal	122	130	135	821	1 014	897	100	131	121
Sierra Leone	5	7	34	1 250	1 571	634	6	11	22
Somalia	367	478	450	352	350	540	129	167	243
Sudan	1 888	3 054	3 904	808	744	534	1 525	2 273	2 085
Tanzania	310	713	479	503	762	978	156	543	468
Togo	0	122	190	0	714	717	0	87	136
Uganda	299	175	239	1 271	1 784	1 495	381	312	357
Zaire	29	36	77	819	899	637	24	32	49
Zambia	76	31	44	649	534	562	49	16	25
Zimbabwe	294	140	138	415	611	579	122	85	80
Near East/North Africa	1 363	1 153	784	1 287	1 232	1 557	1 754	1 420	1 220
Egypt	206	172	133	4 121	3 740	4 746	847	644	631
Morocco	67	45	27	1 048	436	546	70	20	15
Saudi Arabia	174	281	139	1 059	437	1 176	185	123	164
Tunisia	13	14	8	551	427	390	7	6	3
Yemen	895	631	470	712	977	854	637	617	402
East Asia	5 499	3 076	1 728	1 591	2 376	3 060	8 749	7 307	5 287
China (Mainland)	5 407	2 825	1 525	1 591	2 487	3 296	8 600	7 025	5 025
Korea DPR	25	13	10	1 188	1 331	1 530	30	18	15
Thailand	48	220	188	1 961	1 074	1 275	95	237	239

South Asia	18 106	16 766	14 316	487	693	789	8 825	11 616	11 302
India	17 585	16 361	13 902	484	696	795	8 516	11 380	11 059
Pakistan	518	403	413	594	582	586	308	235	242
Latin America and Carib.	3 518	4 438	3 438	2 034	2 793	2 688	7 155	12 394	9 241
Argentina	1 979	1 866	653	1 932	3 023	2 947	3 823	5 641	1 923
Bolivia	0	5	15	0	4 059	3 946	0	21	58
Brazil	1	81	158	2 222	2 128	1 543	2	172	244
Colombia	64	220	256	2 407	2 223	2 878	153	488	737
Costa Rica	7	20	2	1 631	1 806	2 067	11	35	3
Dominican Republic	4	6	10	3 553	2 983	2 564	14	18	26
El Salvador	121	126	124	1 186	1 152	1 272	144	145	158
Guatemala	50	39	56	916	2 036	1 519	46	80	84
Haiti	214	158	122	981	762	743	210	121	90
Honduras	36	61	73	1 271	809	982	46	49	71
Mexico	930	1 491	1 607	2 767	3 347	3 172	2 573	4 991	5 096
Nicaragua	56	51	48	1 016	1 561	1 546	57	80	74
Panama	0	9	11	0	2 149	2 457	0	19	26
Paraguay	4	7	19	1 256	1 278	1 244	5	9	24
Peru	4	14	9	3 056	3 324	3 188	11	46	27
Uruguay	42	56	31	1 259	2 018	2 479	53	112	76
Venezuela	4	227	243	1 333	1 605	2 123	5	365	516
Developed countries	6 754	6 474	5 051	3 100	3 333	3 388	20 934	21 576	17 114
Western Europe	119	134	117	3 411	4 481	4 839	405	601	568
EC-12	112	129	114	3 498	4 580	4 922	393	589	561
France	55	75	70	3 709	4 436	4 577	203	332	320
Italy	5	14	24	3 104	5 127	5 545	15	73	134
Spain	46	38	19	3 681	4 736	5 559	168	181	105

Table A.4.6 (continued)

	Area (harvested) '000 ha			Yield (kg/ha)			Production ('000 tonnes)		
	1969/71	1979/81	1989/91	1969/71	1979/81	1989/91	1969/71	1979/81	1989/91
Eastern Europe	34	49	50	1414	1635	1725	48	81	85
Albania	24	25	24	1112	1200	1173	27	30	29
Hungary	1	7	18	1625	3041	2888	1	22	52
Romania	2	17	7	1130	1659	681	3	28	5
Former USSR	43	90	120	1425	1189	1136	62	107	136
North America	5820	5273	4055	3318	3633	3703	19314	19157	15017
United States	5820	5273	4055	3318	3633	3703	19314	19157	15017
Others	738	928	709	1499	1757	1845	1106	1630	1308
Australia	374	549	461	1910	1976	2126	713	1084	980
South Africa	358	377	248	1050	1434	1320	376	540	327

Note: Data are shown only for countries which in at least one of the three-year averages shown (1969/71, 1979/81, 1989/91) had 10000 ha or more of harvested land allocated to sorghum. The data are after the revision of the production of the former USSR from bunker to clean weight (see Table 3.17).

Table A.5 Land with rainfed crop production potential: 91 developing countries ('000 ha)

| | AT1 Dry semi-arid | AT2 Moist semi-arid | AT3 Sub-humid | AT4 Humid | AT5 Marginal in AT2–AT4 | AT6 Fluvisols/ gleysols | AT7 Marginal fluvisols/ gleysols | Total rainfed potential | Irrigated arid and hyperarid land | Total with crop prod. potential | Currently in use for crop production | | | Total | |
| | | | | | | | | | | | Rainfed | Irrigated‡ | | (weighted)§ | Balance* |
	MS,S,VS†	S,VS	S,VS	S,VS	MS	S,VS	MS								
91 Developing countries	153 915	350 226	593 415	597 949	517 817	258 130	65 161	2 536 613	35 870	2 572 483	633 566	123 012	756 578	747 019	1 815 905
Sub-Saharan Africa	88 989	178 969	294 144	171 096	155 227	104 521	15 356	1 008 302	706	1 009 008	207 203	5 278	212 481	159 810	796 527
Angola	3 267	9 382	51 279	4 849	13 815	7 885	2 171	92 648	0	92 648	4 809	0	4 809	4 233	87 839
Benin	126	2 312	4 727	185	1 359	291	16	9 016	0	9 016	3 151	6	3 157	2 469	5 859
Botswana	5 773	36	0	0	6	1 325	9	7 149	0	7 149	1 431	3	1 434	456	5 715
Burkina Faso	2 835	10 581	3 418	0	1 833	30	0	18 697	0	18 697	6 815	18	6 833	5 446	11 864
Burundi	0	0	674	38	452	80	40	1 284	0	1 284	1 129	71	1 200	923	84
Cameroon	86	2 488	6 703	13 783	6 954	1 324	391	31 729	0	31 729	8 250	28	8 278	6 049	23 451
Central African Rep.	4 257	1 297	15 784	11 822	10 598	2 588	152	42 241	0	42 241	5 174	0	5 174	4 022	37 067
Chad	0	14 594	4 790	0	3 553	3 038	79	30 311	0	30 311	8 129	10	8 139	6 590	22 172
Congo	0	0	558	10 662	4 664	7 391	773	24 048	0	24 048	778	4	782	574	23 266
Cote D'Ivoire	0	0	5 696	9 727	5 353	590	83	21 449	0	21 449	7 131	62	7 193	5 649	14 256
Ethiopia	5 270	5 730	14 427	199	6 244	3 975	6	35 851	0	35 851	15 246	162	15 408	11 040	20 443
Gabon	0	0	0	5 631	6 021	3 253	250	15 155	0	15 155	373	0	373	221	14 782
Gambia	0	373	0	0	51	232	2	658	0	658	322	12	334	286	324
Ghana	0	640	6 136	5 325	3 050	694	57	15 902	0	15 902	4 868	8	4 876	3 803	11 026
Guinea	0	358	8 901	1 204	2 535	305	20	13 323	0	13 323	4 567	25	4 592	3 699	8 731
Kenya	3 145	2 187	1 858	58	1 165	1 109	98	9 620	0	9 620	4 789	52	4 841	3 027	4 779
Lesotho	0	260	374	0	425	0	0	1 059	0	1 059	341	0	341	282	718
Liberia	0	0	0	2 658	1 037	1 551	186	5 432	0	5 432	637	2	639	458	4 793

continued

Table A.5 (continued)

	AT1 Dry semi-arid	AT2 Moist semi-arid	AT3 Sub-humid	AT4 Humid	AT5 Marginal in AT2-AT4	AT6 Fluvisols/ gleysols	AT7 Marginal fluvisols/ gleysols	Total rainfed potential	Irrigated potential arid and hyperarid land	Total with crop prod. potential	Currently in use for crop production Rainfed	Irrigated‡	Total (weighted)§	Balance*
	MS,S,VS†	S,VS	S,VS	S,VS	MS	S,VS	MS							
Madagascar	3164	5712	13140	4910	7560	3142	225	37853	0	37853	2790	903	3693	4089	34160
Malawi	0	1101	2811	169	1156	746	9	5992	0	5992	2654	20	2674	2426	3318
Mali	9095	9892	2319	0	4055	624	338	26323	0	26323	8113	205	8318	4513	18005
Mauritania	2314	436	0	0	24	290	14	3078	0	3078	1067	12	1079	634	1999
Mauritius	0	0	0	46	16	5	0	67	0	67	88	17	105	101	-38
Mozambique	2929	19066	19879	953	9197	2463	232	54719	0	54719	6168	113	6281	5066	48438
Niger	10920	448	0	0	29	176	140	11713	13	11726	11059	38	11097	3753	629
Nigeria	4651	17746	23240	5904	7803	5153	392	64889	0	64889	31609	865	32474	25891	32415
Rwanda	0	0	266	81	326	76	13	762	0	762	997	4	1001	633	-239
Senegal	2183	8471	913	0	1188	403	56	13214	0	13214	5244	178	5422	4122	7792
Sierra Leone	0	0	434	1138	1786	521	25	3904	0	3904	1901	30	1931	1255	1973
Somalia	91	0	0	0	0	1462	0	1553	27	1580	1075	58	1133	960	447
Sudan	17136	28784	14983	678	7199	11433	1045	81258	666	81924	13058	1889	14947	9831	66977
Swaziland	158	98	390	11	377	0	0	1034	0	1034	177	56	233	263	801
Tanzania	5726	12671	21475	352	10280	2813	1878	55195	0	55195	10653	148	10801	8817	44394
Togo	0	49	2257	843	701	158	1	4009	0	4009	1991	7	1998	1658	2011
Uganda	456	1466	5113	1283	3702	1215	513	13748	0	13748	5816	9	5825	4820	7923
Zaire	0	0	26206	88587	23797	28202	5678	172470	0	172470	15302	14	15316	12664	157154
Zambia	662	14653	27112	0	6387	9977	463	59254	0	59254	5540	31	5571	5207	53683
Zimbabwe	4745	8138	8281	0	529	1	1	21695	0	21695	3961	218	4179	3880	17516
Near East/ North Africa	18647	21526	17126	178	9581	9690	816	77564	14806	92370	56411	20089	76500	83082	15870
Afghanistan	296	133	0	0	29	372	0	830	2423	3253	484	2750	3234	6300	19

	1	2	3	4	5	6	7	8	9	10	11	12	13	14	15
Algeria	1325	2479	4820	0	947	960	10	10541	201	10742	7735	366	8101	7104	2641
Egypt	9	0	0	59	0	103	0	112	2559	2671	0	2591	2591	5700	80
Iran	2217	1931	478	0	994	715	355	6749	5295	12044	5958	5570	11708	16143	336
Iraq	168	3113	511	0	879	1622	1	6294	1891	8185	2230	2546	4776	7347	3409
Jordan	125	164	0	0	82	0	1	372	54	426	248	63	311	265	115
Lebanon	35	190	162	0	106	6	1	500	0	500	209	83	292	353	208
Libya	1110	381	85	0	37	663	0	2276	117	2393	953	241	1194	915	1199
Morocco	1368	3747	3780	0	1060	1304	57	11316	551	11867	9001	1265	10266	9853	1601
Saudi Arabia	1	0	0	0	0	152		153	850	1003	80	850	930	1935	73
Syria	2047	1563	542	0	381	345	45	4923	500	5423	4445	661	5106	4078	317
Tunisia	153	350	850	0	306	1062	5	2726	102	2828	3035	273	3308	2699	-480
Turkey	9788	7469	5898	119	4747	1812	341	30174	0	30174	21410	2340	23750	19211	6424
Yemen	5	6	0	0	13	574	0	598	263	861	623	310	933	1179	-72
East Asia	1165	7718	48753	37991	52991	31530	4108	184256	0	184256	68273	19259	87352	89859	96724
Cambodia	0	128	3388	1009	2464	3405	175	10569	0	10569	3156	92	3248	2528	7321
Indonesia	0	1786	4392	22262	17341	11507	1220	58508	0	58508	15584	7550	23134	27451	35374
Korea DPR	0	134	3405	0	1273	105	62	4979	0	4979	1743	1393	3136	4288	1843
Korea Republic	0	0	2287	110	1185	313	306	4201	0	4201	909	1171	2080	3286	2121
Laos	0	52	2162	1136	3058	373	46	6827	0	6827	729	121	850	742	5977
Malaysia	0	0	0	5538	2796	1535	321	10190	0	10190	5055	340	5395	4222	4795
Myanmar	1165	3675	11215	1074	7411	5810	1262	31612	0	31612	9741	1008	10749	9683	20863
Philippines	0	0	6137	3137	5978	997	12	16261	0	16261	10071	1537	11608	9795	4653
Thailand	0	1793	12317	1372	7440	3455	413	26790	0	26790	15116	4217	19333	19291	7457
Vietnam	0	150	3450	2353	4045	4030	291	14319	0	14319	6169	1830	7999	8573	6320
South Asia	29198	82399	50591	6016	22373	21338	896	212811	15291	228102	127137	63401	190538	239418	37564
Bangladesh	0	0	898	724	827	6885	150	9484	0	9484	6838	2673	9511	11168	-27
India	27840	81032	45579	5015	19175	12149	661	191451	0	191451	113753	43036	156769	184969	34662
Nepal	0	515	2135	35	1653	951	56	5345	0	5345	2443	948	3391	4019	1954
Pakistan	1334	525	151	0	108	837	10	2965	15291	18256	2540	16133	18673	36691	-417
Sri Lanka	24	327	1828	242	610	516	19	3566	0	3566	1563	611	2174	2571	1392
Latin America and Carib.	15916	59614	182801	382668	277645	91051	43985	1053680	5067	1058747	174542	14985	189527	174850	869220

(continued)

Table A.5 (continued)

| | Land with rainfed crop production potential by land class | | | | | | | Total rainfed potential | Irrigated arid and hyperarid land | Total with crop prod. potential | Currently in use for crop production | | Total | | Balance* |
| | AT1 Dry semi-arid | AT2 Moist semi-arid | AT3 Sub-humid | AT4 Humid | AT5 Marginal in AT2–AT4 | AT6 Fluvisols/ gleysols | AT7 Marginal fluvisols/ gleysols | | | | Rainfed | Irrigated‡ | (weighted)§ | | |
	MS,S,VS†	S,VS	S,VS	S,VS	MS	S,VS	MS								
Argentina	2 203	4 947	27 297	19 933	16 432	15 119	611	86 542	709	87 251	31 123	1 668	32 791	31 035	54 460
Bolivia	2 527	4 859	16 103	9 867	14 275	5 907	1 866	55 404	0	55 404	3 533	149	3 682	3 175	51 722
Brazil	2 072	16 744	84 328	252 247	165 299	30 729	30 806	582 225	0	582 225	86 799	2 563	89 362	76 133	492 863
Chile	2 115	1 622	1 389	144	845	1 348	197	7 660	633	8 293	2 439	1 090	3 529	4 062	4 764
Colombia	1 111	1 900	6 320	24 033	17 062	9 289	2 018	61 733	0	61 733	6 954	500	7 454	6 213	54 279
Costa Rica	0	0	828	794	710	304	55	2 691	0	2 691	546	117	663	661	2 028
Cuba	16	3 676	1 623	0	1 794	302	220	7 631	0	7 631	2 320	888	3 208	3 637	4 423
Dominican Republic	42	518	979	163	614	271	18	2 605	0	2 605	955	225	1 180	1 150	1 425
Ecuador	1 438	1 862	939	4 514	2 542	1 221	597	13 113	0	13 113	2 330	550	2 880	2 905	10 233
El Salvador	0	0	1 107	0	117	53	1	1 278	0	1 278	670	116	786	859	492
Guatemala	0	0	3 054	935	1 264	820	52	6 125	0	6 125	1 911	78	1 989	1 681	4 136
Guyana	0	0	352	5 493	3 139	1 412	77	10 473	0	10 473	285	130	415	497	10 058
Haiti	29	313	212	117	303	149	2	1 125	0	1 125	1 144	73	1 217	923	−92
Honduras	0	0	1 248	1 098	1 807	536	50	4 739	0	4 739	1 008	89	1 097	957	3 642
Jamaica	87	21	9	214	113	77	7	528	0	528	232	35	267	228	261
Mexico	1 646	16 252	20 849	2 442	11 947	3 286	210	56 632	2 872	59 504	19 275	4 837	24 112	26 268	35 392
Nicaragua	0	0	1 533	1 648	2 200	873	293	6 547	0	6 547	1 243	85	1 328	1 132	5 219
Panama	0	0	633	1 089	1 353	484	49	3 608	0	3 608	612	32	644	491	2 964
Paraguay	226	1 309	3 260	6 814	7 482	2 129	618	21 838	0	21 838	3 671	67	3 738	3 425	18 100
Peru	1 600	2 133	706	23 912	11 897	5 928	5 027	51 203	853	52 056	2 287	1 245	3 532	4 473	48 524
Suriname	0	0	0	7 549	2 982	747	92	11 370	0	11 370	29	58	87	148	11 283

Trinidad and Tobago	0	0	175	7	57	45	0	284	108	22	130	97	154
Uruguay	0	0	0	11 547	1 257	19	8	12 831	1 377	110	1 487	1 345	11 344
Venezuela	804	3 458	9 857	8 108	12 154	10 003	1 111	45 495	3 691	258	3 949	3 355	41 546

Note: The methodology used to derive estimates of land with rainfed crop production potential is explained in Chapter 4. Given the limitations of the basic data, the estimates in this table should be interpreted with care.

*In some countries the balance may be negative. For an explanation see Chapter 4.

†MS: marginally suitable; S: suitable; VS: very suitable. Class AT5 contains the areas considered marginally suitable in classes AT2, AT3 and AT4.

‡Including irrigated arid and hyperarid land.

§Land-in-use in the different classes aggregated using the following weights: 1.0 for AT3, 0.31 for AT1, 0.88 for AT2, 0.85 for AT4, 0.35 for AT5, 0.81 for AT6, 0.35 for AT7 and 2.2 for irrigated. These weights roughly reflect potential cereal yields. The thus weighted land-in-use is more comparable over countries than the unweighted total (see also Chapter 2, note 11).

References

Adelman, I. (1984). Beyond export-led growth, *World Development*, **12**, 9.

Agra Europe (1993). *The GATT Uruguay Round Agreement, an Agra Europe Special Supplement*, December, London.

Ayibotele, N.B. (1992), *The World's Water: Assessing the Resource*, Keynote paper at the International Conference on Water and the Environment, Dublin, Ireland.

Ahmed, R. and C. Donovan (1992). *Issues of Infrastructural Investment Development, a Synthesis of the Literature*, International Food Policy Research Institute, Washington, DC.

Alderman, H. (1991). Food subsidies and the poor, in Psacharopoulos, G. (ed.), *Essays on Poverty, Equity and Growth*, Pergamon Press, Oxford.

Alexandratos, N. (1976), Formal techniques of analysis for agricultural planning, *Monthly Bulletin of Agricultural Economics and Statistics*, **25**, 6; Reprinted in FAO (1978). *FAO Studies in Agricultural Economics and Statistics 1952–1977*, Rome.

Alexandratos, N. (ed.) (1988). *World Agriculture Toward 2000, an FAO Study*, Belhaven Press, London and New York University Press, New York, and (in French) *Agriculture Mondiale: Horizon 2000, Etude de la FAO*, Economica, Paris, 1989.

Alexandratos, N. (ed.) (1990). *European Agriculture: Policy Issues and Options to 2000, an FAO Study*, Belhaven Press, London and Columbia University Press, New York, and (in French) *Agriculture Européenne: Enjeux et Options à l'Horizon 2000, Etude de la FAO*, Economica, Paris, 1991.

Alexandratos, N. (1992). World agriculture in the next century: Challenges for production and distribution, in Peters, G. and B. Stanton (eds), *Sustainable Agricultural Development: the Role of International Cooperation*, Proceedings of the XXI International Conference of Agricultural Economists, Tokyo, 22–29 August 1991, Dartmouth, Aldershot, UK.

Alexandratos, N., T. Aldington and H. de Haen (1994), Agriculture in Europe and in the developing countries: interactions mainly through trade, *European Review of Agricultural Economics*, **21**, 3/4.

Altieri, M. (1987). *Agroecology, The Scientific Basis of Alternative Agriculture*, Westview, Boulder.

Anderson, K. (ed.) (1992). *New Silk Roads, East Asia and World Textile Markets*, Cambridge University Press, Cambridge.

Anderson, K. and R. Blackhurst (eds) (1992). *The Greening of World Trade Issues*, Harvester Wheatsheaf, London.

Anderson, K. and Y. Hayami (1986). *The Political Economy of Agricultural Protection, East Asia in International Perspective*, Allen & Unwin, London.

Bebbington, A. J., H. Carrasco, L. Peralbo, J. Trujillo and V. Torres (1993). *Fragile Lands, Fragile Organizations: Indian Organizations and the Politics of Sustainability in Ecuador*, Transactions of the Institute of British Geographers, **18**, 2.

Bell, C. (1988). Credit markets and interlinked Transactions in Chenery and Srinivasan (1988).

Besley, T. (1994). How do market failures justify interventions in rural credit markets? *The World Bank Research Observer*, **9**, 1.

Bevan, D., A. Bigsten, P. Collier and J. Gunning (1987). Peasant supply response in rationed economies, *World Development*, **15**.

Bie, S.W. (1990). *Dryland Degradation Measurement Techniques*, World Bank Environment Department Working Paper 26, World Bank, Washington, DC.

Binswanger, H. (1989). The policy response of agriculture, *Proceedings of the World Bank Conference on Development Economics*, World Bank, Washington, DC.

Binswanger, H. and V. Ruttan (1978). *Induced Innovation: Technology, Institutions and Development*, The Johns Hopkins University Press, Baltimore.

Bond, M. (1983). Agricultural responses to prices in Sub-Saharan Africa, *International Monetary Fund Staff Papers*, **40**, 4.

Bongaarts, J. (1994). Can the growing human population feed itself?, *Scientific American*, March.

Borrell, B. and R. Duncan (1992). A survey of costs of world sugar policies, *World Bank Research Observer*, July.

Boserup, E. (1965). *The Conditions of Agricultural Growth: The Economics of Agrarian Change under Population Pressure*, Aldine Press, New York.

Boserup, E. (1981). *Population and Technological Change: A Study of Long-term Trends*, University of Chicago Press, Chicago.

Brooks, K. (1993). Challenges of trade and agricultural development for East/Central Europe and states of the former USSR, *Agricultural Economics*, **8**.

Brown, L. (1994). Facing food insecurity, in Brown, L. *et al.*, *State of the World 1994, A Worldwatch Institute Report on Progress Toward a Sustainable Society*, W.W. Norton and Co., New York and London.

Brown, M. and I. Goldin (1992). *The Future of Agriculture: Developing Country Implications*, OECD Development Centre, Paris.

Bruinsma, J., J. Hrabovsky, N. Alexandratos and P. Petri (1983). Crop production and input requirements in developing countries, *European Review of Agricultural Economics*, **9**, 2.

Bunch, R. (1990), *Low Input Soil Restoration in Honduras: The Cantarranas Farmer-to-Farmer Extension Programme*, Gatekeeper Series No. 23, IIED, London.

Buzzanell, P. (1993). *Latin America's Big Three Sugar Producers in Transition: Cuba, Mexico, Brazil*, USDA, Washington, DC.

Byerlee, D. (1994). *Modern Varieties, Productivity and Sustainability: Recent Experience and Emerging Challenges*, CIMMYT, Mexico.

Byerlee, D. and M. López-Pereira (1994). *Technical Change in Maize Production: a Global Perspective*, CIMMYT Economics Working Paper 94-02, CIMMYT, Mexico.

CGIAR (1992). *A CGIAR Response to UNCED Agenda 21 Recommendations*, Paper presented to the International Centres' Week, 26–30 October, Washington, DC.

Chambers, R., A. Pacey, and L. A. Thrupp (eds) (1989). *Farmer First, Farmer Innovation and Agricultural Research*, Intermediate Technology Publications, London.

Chenery, H. and T.N. Srinivasan (eds) (1988), *Handbook of Development Economics*, Elsevier Science Publishers, Amsterdam.

Claassen, E.M. and P. Salin (1991). *The Impact of Stabilization and Structural Adjustment Policies on the Rural Sector*, FAO Economic and Social Development Paper No. 90, Rome.

Condos, A. (1990). Growth collapses, real exchange rate misalignments and exchange rate policy in 21 developing countries: Comment on Lal (1990).

Contado, T. E. (1990). Agricultural extension approaches: What FAO's case studies reveal, in FAO (1990a).

Contado, T. E. and W. D. Maalouf (1990). Agricultural extension in Corum-Cankiri: A case study under the rural development project in Turkey, paper presented at the National Workshop on Agricultural Extension, Tunisia, February.

Corden, W. M. (1990). Macroeconomic policy and growth: some lessons of experience, *Proceedings of the World Bank Annual Conference on Development Economics*, World Bank, Washington, DC.

Crosson, P. (1991). Sustainable agriculture in North America: issues and challenges, *Canadian Journal of Agricultural Economics*, **39**.

Crosson, P. (1992). United States agriculture and the environment: perspectives on the next twenty years (mimeo), Resources for the Future, Washington, DC.

Daly, H. (1992). *Steady State Economics*, Earthscan Publications, London.

Daly, H. and K. Townsend (eds)(1993), *Valuing the Earth: Economics, Ecology, Ethics*, The MIT Press, Cambridge, Massachusetts.

Datt, G. and M. Ravallion (1992). Growth and redistribution components of change in poverty measures, *Journal of Development Economics*, 38.

De Janvry, A. and R. Garcia (1988). *Poverty, Public Policies and the Environment*, Paper for the IFAD International Consultation on Environment, Sustainable Development and the Role of Small Farmers, Rome, 11–13 October.

De Janvry, A., E. Sadoulet and T. K. White (1989). *Foreign Aid's Effects on US Farm Exports, Benefits or Penalties?*, Foreign Agriculture Economic Report No. 238, USDA, Washington, DC.

Delgado, C. (1992). Why domestic food prices matter to growth strategy in semi-open West African Agriculture, *Journal of African Economics*, **1**, 3.

Delgado C. and J. Mellor (1987). A structural view of policy issues in African agricultural development, *American Journal of Agricultural Economics*, **69**.

Delgado C. and P. Pinstrup-Andersen (1993). *Agricultural Productivity in the Third World: Patterns and Strategic Issues*, Keynote Address to the AAEA Workshop on Post-Green Revolution Agricultural Development Strategies in the Third World: What Next?, Orlando, Florida, July.

Duvick, D. (1994). *Intensification of Known Technology and Prospects of Breakthroughs in Technology and Future Food Supply*, Paper for the Roundtable on Population and Food in the early 21st Century, International Food Policy Research Institute, Washington, DC, 14–16 February.

Ehrlich, P., A. Ehrlich and G. Daily (1993). Food security, population and environment, *Population and Development Review*, **19**, 1.

Eicher, C. and J. Staatz (eds) (1990). *Agricultural Development in the Third World*, The Johns Hopkins University Press, Baltimore.

Ellis, F. (1992). *Agricultural Policies in Developing Countries*, Cambridge University Press, Cambridge.

Environment Canada (1989). *Urbanization of Rural Land in Canada, 1981-86*, State of the Environment Fact Sheet No. 89-1, Ottawa.

Evenson, R. (1982a). The economics of extension in Jones, L. (ed.), *Investing in Rural Extension: Strategies and Goals*, Elsevier, London.

Evenson, R. (1982b). Agriculture, in Nelson, R. R. (ed.), *Government and Technical Progress*, Pergamon Press, New York.

Evenson, R. and Y. Kislev (1975). *Agricultural Research and Productivity*, Yale University Press, New Haven.

Faini, R. (1992). Infrastructure, relative prices and agricultural development in Goldin I. and A. Winters (eds), *Open Economies: Structural Adjustment and Agriculture*, Cambridge University Press, Cambridge.

Falkenmark, M. and C. Widstrand (1992). Population and water resources: a delicate balance, *Population Bulletin*, **47**, 3.

Fangmeier W., F. Vlotman and S. Eftekhazadeh (1990). *Uniformity of LEPA Irrigation Systems with Furrow Drops*, American Society of Agricultural Engineers, **33**, 6.

FAO (1971–81). *FAO/UNESCO Soil Map of the World*, Vols. 1–10, UNESCO, Paris.

FAO (1973). *Agricultural Protection, Domestic Policy and International Trade*, Document C73/LIM/9, Rome.

FAO (1978–81). *Reports of the Agro-Ecological Zones Project*, World Soil Resources Report 48, Vols.1–4, Rome.

FAO (1979). *Human Nutrition in Tropical Africa*, Rome.

FAO (1981). *Legumes in Human Nutrition*, Rome.

FAO (1982). *Land Resources for Populations of the Future*, Rome.

FAO (1983). *A Comparative Study of Food Consumption Data from Food Balance Sheets and Household Surveys*, FAO Economic and Social Development Paper No. 34, Rome.

FAO (1984a). *Agricultural Extension: A Reference Manual*, Rome.

FAO (1984b). *Trained Agricultural Manpower Assessment in Africa*, Rome.

FAO (1984c). *Training of Manpower for Agricultural and Rural Development in Africa*, Rome.

FAO (1985). *Tropical Forestry Action Plan*, Rome.

FAO (1986). *World-wide Estimates and Projections of the Agricultural and the non-Agricultural Population Segments, 1950–2025*, Document ESS/MISC/86–2, Rome.

FAO (1987). *Agricultural Price Policies*, Rome.

FAO (1989a). *Soil Conservation for Small Farmers in the Humid Tropics*, FAO Soils Bulletin 60, Rome.

FAO (1989b). *Effects of Stabilization and Structural Adjustment Programmes on Food Security*, FAO Economic and Social Development Paper No. 89, Rome.

FAO (1989c). *Fertilizers and Food Production: Summary Review of Trial and Demonstration Results, 1961–1986*, The FAO Fertilizer Programme, Rome.

FAO (1989d). *Income Elasticities of Demand for Agricultural Products Estimated from Household Consumption and Budget Surveys*, Rome.

FAO (1990a). *Report of the Global Consultation on Agricultural Extension*, Rome.

FAO (1990b). *Roots, Tubers, Plantains and Bananas in Human Nutrition*, Rome.

FAO (1990c). *Malnutrition in the Latin American and the Caribbean Region, Causes and Prevention*, Document LARC 90/4, Rome.

FAO (1990d). *Climate Change and Agriculture, Forestry and Fisheries: FAO Position Paper*, Second World Climate Conference, Geneva.

FAO (1990e). *International Code of Conduct on the Distribution and Use of Pesticides*, (Amended version), Rome.

FAO (1991a). *A Study on the Reasons of Failure of Soil Conservation Projects*, FAO Soils Bulletin 64, Rome.

FAO (1991b). *Third Progress Report on the WCARRD Programme of Action*, Document C91/19, Rome.

FAO (1991c). *The Den Bosch Declaration and Agenda for Action on Sustainable Agriculture and Rural Development: Report of the Conference*, FAO/Netherlands Conference on Agriculture and the Environment, 's-Hertogenbosch, The Netherlands.

FAO (1991d). *International Directory of Agricultural Extension Organizations*, Rome.

FAO (1991e). *International Agricultural Adjustment – Seventh Progress Report*, Document C91/18, Rome.

FAO (1991f). *Tropical Forests Action Programme: Operational Principles*, Rome.

FAO (1992a), *World Food Supplies and Prevalence of Chronic Undernutrition in Developing Regions as Assessed in 1992*, Document ESS/MISC/1992, Rome.

FAO (1992b). *The World Sugar Market: Prospects for the Nineties*, Document ESC/M/92/3, Rome.

FAO (1993a). *Integrated Plant Nutrient Systems and Sustainable Agriculture*, 17th Consultation of the FAO Fertilizer Programme, Islamabad, May.

FAO (1993b). *FAO Activities on Sustainable Development and Environment*, Document C93/10, Rome.

FAO (1993c). *Protectionism in Agricultural Trade: Review of Action Taken on Conference Resolution 2/79*, Document CCP93/20, June, Rome.

FAO (1993d). *The State of Food and Agriculture 1993*, Rome.

FAO (1993e). *Marine Fisheries and the Law of the Sea: a Decade of Change*, Rome.

FAO (1993f), *Forest Resources Assessment 1990, Tropical Countries*, FAO Forestry Paper 112, Rome.

FAO (1993g). *Sustainable Development of Drylands and Combating Desertification*, Rome.

FAO (1993h). *Agricultural Extension and Farm Women in the 1980s*, Rome.

FAO (1993i). *The World Food Model, Model Specification*, Document ESC/M/93/1, Rome.

FAO (1993j). *International Trade, Environment and Sustainable Agricultural Development*, Document CCP/93/19, Rome.

FAO (1994a). *This is Codex Alimentarius*, Rome.

FAO (1994b). *The State of Food and Agriculture 1994*, Rome.

FAO (1994c). *A Preliminary Assessment of the Uruguay Round Agreement on Agriculture*, Paper for the Marrakesh Ministerial Meeting, 12–15 April, Rome.

FAO/IFA/IFDC (1992). *Fertilizer Use by Crop*, Document ESS/Misc/1992/3, Rome.

FAO/TAC Secretariat (1993). *Investment in Rice Research in the CGIAR: A Global Perspective*, Report of the Inter-Centre Review of Rice, Rome.

FAO/UNFPA (1980). *Land Resources for Populations of the Future*, Report of the Second FAO/UNFPA Expert Consultation, Rome.

FAO/WHO (1992). *Nutrition and Development, a Global Assessment*, Rome.

FAO/WHO/UNU (1985). *Energy and Protein Requirements*, Report of a Joint FAO/WHO/UNU Expert Consultation, Rome.

Fardoust, S. (1990). World economy in transition: recent history and outlook for the 1990s (mimeo), International Economic Analysis and Prospects Department, The World Bank.

Feder, G. and R. Noronha (1987). Land rights systems and agricultural development in rural sub-Saharan Africa, *World Bank Research Observer*, 2.

Feder, G., L. Lau and R. Slade (1985). The impact of agricultural extension: a case study of the training and visit system in Haryana and India (mimeo), The World Bank, Washington, DC.

Fields, G. (1989). Changes in poverty and inequality in developing countries, *The World Bank Research Observer*, **4**, 2.

Fishlow, A. (1991. A review of *Handbook of Development Economics*, *Review of Economic Literature*, **29**.

Frohberg, K. (1993). The trade potential of the industrialized countries over the next two decades: Western and Eastern Europe and Japan (mimeo), Institut für Agrarpolitik, Universität Bonn (paper prepared for this study).

Gaiha, R. (1987). Impoverishment, technology and growth in rural India, *Cambridge Journal of Economics*.

Gaiha, R. (1993). *Design of Poverty Alleviation Strategies in Rural Areas*, FAO Economic and Social Development Paper, No. 115, Rome.

Garnaut, R. and G. Ma (1993). How rich is China? Evidence from the food economy, *The Australian Journal of Chinese Affairs*, No. 30.

Georghiou G. P. and A. Lagunes-Tejeda (1991). *The Occurrence of Resistance to Pesticides in Arthropods: An Index of Cases Reported through 1989*, FAO, Rome.

Goldin, I. and O. Knudsen (eds) (1990). *Agricultural Trade Liberalization, Implications for Developing Countries*, OECD, Paris.

Gomez, S. and J. Echenique (1991). *La Agricultura Chilena, las Dos Caras de la Modernización*, Flacso-Agraria, Santiago.

Haggblade, S., P. Hazell and J. Brown (1989). Farm-non-farm linkages in rural sub-Saharan Africa, *World Development*, **17**, 8.

Haggblade, S., J. Hammer and P. Hazell (1991). Modelling agricultural growth multipliers, *American Journal of Agricultural Economics*, **73**, 2.

Hamilton, C. (ed.) (1990). *Textiles Trade and the Developing Countries, Eliminating the Multi-Fibre Arrangement in the 1990s*, World Bank, Washington, DC.

Harrison, P. (1992). *The Third Revolution: Environment, Population and a Sustainable World*, I. B. Tauris/Penguin Books, London.

Hayami, Y. and K. Otsuka (1991). *Beyond the Green Revolution: Agricultural Development in the New Century*, Paper presented to the World Bank Conference on Agricultural Technology: Issues for the International Community and the World Bank, Virginia, October.

Hayami, Y. and V. Ruttan (1985). *Agricultural Development: An International Perspective*, Johns Hopkins University Press, Baltimore.

Hayward, J. A. (1990). Agricultural extension: The World Bank's experience and approaches, in FAO (1990a).

Hazell, P. and S. Haggblade (1993). Farm-non-farm growth linkages and the welfare of the poor, in Lipton, M. and J. Van de Gaag (eds), *Including the Poor*, Oxford University Press, New York.

Hazell, P. and C. Ramasamy (1991). *The Green Revolution Reconsidered. The Impact of High-Yielding Rice Varieties in South India*, The Johns Hopkins University Press, Baltimore.

Hill, C. (1991). Managing commodity booms in Botswana, *World Development*, **19**.

Hoff, K. and J. E. Stiglitz (1990). Imperfect information and rural credit markets—puzzles and policy perspectives, *The World Bank Economic Review*, **4**, 3.

Hossain, M. (1988), *Nature and Impact of the Green Revolution in Bangladesh*, IFPRI Research Report 67, IFPRI Washington, DC.

House, R., M. Peters, H. Baumes and W. Terry Disney (1993). *Ethanol and Agriculture*, Agricultural Economic Report 667, USDA, Washington, DC.

ICRISAT (International Crops Research Institute for the Semi-Arid Tropics), *1991 Report*.

ISNAR (1992). *Summary of Agricultural Research Policy: International Quantitative Perspectives*, The Hague.

ISRIC/UNEP (1991). *World Map of the Status of Human-Induced Soil Degradation*, Global Assessment of Soil Degradation, UNEP, Nairobi.

IUCN (1990). *1990 United Nations List of National Parks and Protected Areas*, IUCN, Gland, Switzerland, and Cambridge, UK.

Jamison, D. T. and L. J. Lau (1982). *Farmer Education and Training Efficiency*, Johns Hopkins University Press, Baltimore.

Jarvis, L. S. (1989). The unravelling of Chile's agrarian reform, 1973–1986, in Thiesenhusen, W. C. (ed.), *In Search of Land Reforms*, Unwin, London.

Jazairy, I., M. Alamgir and T. Panuccio (1992). *The State of the World Rural Poverty—An Inquiry into its Causes and Consequences*, New York University Press, New York.

Johnson, D. Gale (1991). *World Agriculture in Disarray*, Second Edition, St. Martin's Press, New York.

Johnson, D. Gale (1993). Trade effects of dismantling the socialized agriculture of the former Soviet Union, *Comparative Economic Studies*, **35**, 4.

Johnson, D. Gale (1994a). Does China have a grain problem?, *China Economic Review*, **4**, 1.

Johnson, D. Gale (1994b). *Limited but Essential Role of Government in Agriculture and Rural Life*, Elmhirst Memorial Lecture, XXII International Conference of Agricultural Economists, Harare.

Johnson, S. (1993). Former Soviet Union (mimeo), Centre for Agriculture and Rural Development, Iowa State University, USA (paper prepared for this study).

Johnston, B. and J. Mellor (1961). The role of agriculture in economic development, *American Economic Review*, **51**, 4.

Jorgenson, D.W. (1961). The development of a dual economy, *Economic Journal*, **71**.

Kandiah, A. and C. Sandford (1993). *Research and Development Needs for Integrated Rural Water Management*, Paper presented to the Technical Consultation on Integrated Rural Water Management, FAO, Rome, 15–19 March.

Kennedy, P. (1993). *Preparing for the 21st Century*, Random House, New York.

Kerr, J. and N. Sanghi (1992). *Indigenous Soil and Water Conservation in India's Semiarid Tropics*, Gatekeeper series No. 24, IIED, London.

Knudsen, O., J. Nash *et al.* (1990). Redefining the role of government in agriculture for the 1990s, *World Bank Discussion Paper No. 105*, World Bank, Washington, DC.

Krueger, A., M. Schiff, and A. Valdés (1988). Agricultural incentives in developing countries: measuring the effect of sectoral and economy-wide policies, *World Bank Economic Review*, **2**.

Krueger, A., M. Schiff and A. Valdés (eds) (1991). *The Political Economy of Agricultural Pricing Policies*, vol. 2 Asia, The Johns Hopkins University Press, Baltimore and London.

Kuznets, S. (1955). Economic growth and income inequality, *American Economic Review*, **65**.

Lal, D. (1990). *Growth Collapses, Real Exchange Rate Misalignments, and Exchange Rate Policy in 21 Developing Countries*, Paper Prepared for the Conference on Exchange Rate Policies of Developing Countries, Free University Berlin, May.

Legouis, M. (1991). Alternative financing of agricultural extension: recent trends and inplications for the future, in Rivera and Gustafson (1991).

Lele, U. (1992). *Structural Adjustment and Agriculture: A Comparative Perspective of Performance in Africa, Asia and Latin America*, 29th Seminar of the European Association of Agricultural Economists, Hohenheim, Germany.

Lele, U. and L.R. Myers (1987), *Growth and Structural Change in East Africa: Domestic Policies, Agricultural Performance and World Bank Assistance*, Discussion Paper DRB-273, World Bank, Washington, DC.

Lele, U. and S. Stone (1989). *Population Pressure, the Environment and Agricultural Intensification: Variations on the Boserup Hypothesis*, MADIA Discussion Paper 4, World Bank, Washington, DC.

Lewis, W. A. (1983). Developed and developing countries, in Rosenblum, J. (ed.), *Agriculture in the 21st Century*, Wiley, New York.

Lipton, M. and R. Longhurst (1989). *New Seeds and Poor People*, The Johns Hopkins University Press, Baltimore.

Lipton, M. and M. Ravallion (1993). *Poverty and Policy*, Policy Research Working Paper Series 1130, World Bank, Washington, DC.

Little, I. M. D. (1982). *Economic Development*, Basic Books, New York.

Lord, R. and R. Barry (1990). *The World Sugar Market, Government Intervention and Multilateral Policy Reform*, USDA, Washington, DC.

Maalouf, W. D., T. E. Contado and R. Adhikarya (1991), Extension coverage and resource problems: The need for public–private cooperation, in Rivera and Gustafson (1991).

Mabbs-Zeno, C. and B. Krissoff (1990). Tropical Beverages in the GATT, in Goldin and Knudsen (1990).

Matthews, A. (1989). Self-sufficiency in food production: what should developing countries do? (mimeo), Report prepared for the Global Perspective Studies Unit, FAO, Rome.

Meller, P. (1988). *Crisis and Adjustment in the Chilean Economy*, Santiago, Chile.

Mellor, J. (1986). Agriculture on the road to industrialization, in Lewis, J. and V. Kallab (eds), *Development Strategies Reconsidered*, Transactions Books, New Brunswick.

Mellor, J. and B. Johnston (1984). The world food equation: Interrelations among development, employment and food consumption, *Journal of Economic Literature*, **22**.

Meyers, W. (1993). Trade outlook for North America and Oceania to 2010 (mimeo), Centre for Agriculture and Rural Development, Iowa State University, USA (paper prepared for this study).

Mitchell, D. and M. Ingco (1993). *The World Food Outlook*, The World Bank, Washington, DC.

Mortimore, M. (1992). *Profile of Technological Change*, ODI Working Paper 57, ODI, London.

Myint, H. (1987). The neoclassical resurgence in development economics: its strengths and limitations, in Meier, G. (ed.), *Pioneers in Development*, Second series, Oxford University Press for the World Bank.

Nallet, H. and A. Van Stolk (1994). Relations between the European Union and Eastern European countries in matters concerning agriculture and food production, Report to the European Commission (mimeo).

Narayana, N. S. S., K. S. Parikh and T. N. Srinivasan (1988). Rural works programs in India: costs and benefits, *Journal of Development Economics*, **29**.

Nelson, R. (1988), *Dryland Management – The Desertification Problem*, World Bank Environment Department Working Paper 8, The World Bank, Washington, DC.

Nordblom, T. and F. Shomo (1993). Livestock and feed trends in West Asia and North Africa: past, present and future, *Cahiers Options Méditerranéennes*, **1**, 5.

Norse, D. (1988). *Policies for Sustainable Agriculture: Getting the Balance Right*, Paper for the IFAD International Consultation on Environment, Sustainable Development and the Role of Small Farmers, Rome, 11–13 October.

Norse, D. and W. Sombroek (eds) (forthcoming). *Global Climatic Change and Agricultural Production: Direct and Indirect Effects of Changing Hydrological, Soil and Plant Physiological Processes*, J. Wiley & Sons, Chichester and FAO, Rome.

Norse, D., C. James, B. Skinner and Q. Zhao (1992). Agricultural land and degradation, in Dooge, J. *et al.* (eds), *An Agenda of Science for Environment and Development into the 21st Century*, Cambridge University Press, Cambridge.

Norton, R. D. (1992). *Integration of Food and Agricultural Policy with Macro-economic Policy: Methodological Considerations in a Latin America Perspective*, FAO Economic and Social Development paper No. 111, Rome.

O'Brien, P. (1994). *Comments on Food Supply and Demand Prospects for OECD Countries by 2010*, Paper for the Roundtable on Population and Food in the early 21st Century, International Food Policy Research Institute, Washington, DC, 14–16 February.

OECD (1992). The economic costs of reducing CO_2 emissions, *OECD Economic Studies*, **19**.

OECD (1994). *Agricultural Policies, Markets and Trade, Monitoring and Outlook 1994*, Paris.

Oram, P. and B. Hojjati (1994). *The Growth Potential of Existing Agricultural Technology*, Paper for the Roundtable on Population and Food in the early 21st Century, International Food Policy Research Institute, Washington, DC, 14–16 February.

Osmani, S. R. (1988). Social security in South Asia (mimeo), STICERD, London School of Economics, London.

Panagariya, A. and M. Schiff (1990). *Commodity Exports and Real Income in Africa*, Working Paper WPS 537, World Bank, Washington, DC.

Parikh, K. and T. N. Srinivasan (1989). Poverty alleviation policies in India: Food consumption subsidy, food production subsidy and employment generation (mimeo), Economic Growth Center, Yale University, New Haven.

Pearce D. and J. Warford (1993). *World Without End: Economics, Environment and Sustainable Development*, Oxford University Press for the World Bank, New York.

People's Republic of China (1992). *National Report of the People's Republic of China on Environment and Development*, submission to UNCED, China Environmental Science Press, Beijing.

Pimentel, O., R. Harman, M. Pacenza, J. Pekarsky and M. Pimentel (1994). Natural resources and an optimum human population, *Population and Environment*, 15, 5.

Pingali, P. and H. Binswanger (1984). *Population Density and Farming Systems: the Changing Locus of Innovations and Technical Change*, Working Paper No. ARV 24, World Bank, Washington, DC.

Pingali, P. and A. Rola (1994). *Public Regulatory Roles in Developing Markets: The Case of Pesticides*, Paper presented at the 14th World Bank Agricultural Symposium, Washington, DC, January 1994.

Pingali, P., P. Moya and L. Velasco (1990). *The Post-Green Revolution Blues in Asian Rice Production: The Diminishing Gap Between Experiment Station and Farmer Yields*, IRRI Social Sciences Division, Los Baños.

Pinto, B. (1987). Nigeria during and after the oil booms: a policy comparison with Indonesia, *The World Bank Economic Review*, 1, 3.

Platteau, J. Ph. (1992). *Land Reform and Structural Adjustment in Sub-Saharan Africa, Controversies and Guidelines*, FAO Economic and Social Development Paper No. 107, Rome.

Platteau, J. Ph. (1993). Sub-Saharan Africa as a Special Case: the Crucial Role of (Infra)Structural Constraints, *Cahiers de La Faculté Des Sciences Economiques et Sociales de Namur, Série Recherche No. 128-1993/6*, Namur (originally written for this study).

Plucknett, D. (1993). *Science and Agricultural Transformation*, IFPRI Lecture Series No. 1, Washington, DC.

Plucknett, D. (1994). *Prospects of Meeting Future Food Needs through New Technology*, Paper for the Roundtable on Population and Food in the early 21st Century: Meeting Future Food Needs of an Increasing World Population, IFPRI, Washington, DC, 14–16 February.

Postel S. (1989). *Water for Agriculture: Facing the Limits*, Worldwatch Paper 93, Worldwatch Institute, Washington, DC.

Psacharopoulos G., S. Morley, A. Fiszbein, H. Lee and B. Wood (1992). *Poverty and Income Distribution in Latin America: the Story of the 1980s*, LATHR Regional Studies Program Report 27, World Bank, Washington, DC.

Ranis, G. and J. Fei (1963). Innovation, capital accumulation, and economic development, *American Economic Review*, 53, 3.

Rao D. (1993). *Intercountry Comparisons of Agricultural Output and Productivity*, FAO Economic and Social Development Paper, No. 112, Rome.

Rausser, G., J. Chalfant, H. Love and K. Stamoulis (1986). Macroeconomic linkages, taxes, and subsidies, in the agricultural sector, *American Journal of Agricultural Economics*, **68**.

Ravallion, M. and B. Bidani (1994), How robust is the poverty profile?, *World Bank Economic Review*, **8**, 1.

Reid, W. (1989). Sustainable development: lessons from success, *Environment*, May.

Richards, P. (1990). Indigenous approaches to rural development: The agrarian populist tradition in West Africa, in Altieri, M. and S. Hecht (eds), *Agroecology and Small Farm Development*, CRC Press, New York.

Rivera, W. and D. Gustafson (eds) (1991). *Agricultural Extension: Worldwide Institutional Evolution and Forces for Change*, Elsevier, Amsterdam.

Rollo, J. and A. Smith (1993). EC trade with Eastern Europe, *Economic Policy*, April.

Rosenzweig, C. and M. Parry (1994). Potential impact of climate change on world food supply, *Nature*, **367**, 13 January.

Ruttan, V. (1994). Challenges to agricultural research in the 21st century, in Ruttan, V. (ed), *Agriculture, Environment and Health: Sustainable Development in the 21st Century*, University of Minnesota Press, Minneapolis and London.

Sah, R. and J. Stiglitz (1984). The economics of price scissors, *American Economic Review*, **74**.

Scandizzo, P. and D. Diakosavvas (1987). *Instability in the Terms of Trade of Primary Commodities, 1900–1982*, FAO Economic and Social Development Paper No. 64, Rome.

Schelling, T. (1992). Some economics of global warming, *American Economic Review*, March.

Schiff, M. and A. Valdés (1992). *The Plundering of Agriculture in Developing Countries*, World Bank, Washington, DC.

Schrieder, G. (1989). Informal financial groups in Cameroon: motivation, organization and linkages, Unpublished MA Thesis, Department of Agricultural Economics and Rural Sociology, Ohio State University, Columbus, Ohio.

Schultz, T.P. (1988). Education investments and returns, in Chenery and Srinivasan (1988).

Schultz, T.W. (1965). *Transforming Traditional Agriculture*, Yale University Press, New Haven.

Schultz, T.W. (1990). The economics of agricultural research, in Eicher and Staatz (1990).

Sen, A. (1987). *Hunger and Entitlements*, World Institute for Development Economics Research, Helsinki.

Shaxson, T. (1992). Soil moisture: Capture, Retention and use, Unpublished working paper for the FAO Investment Centre.

Shen, F. and W. Wolter (1992). Managing the "Four Waters", an innovative concept of modernization (mimeo), IPTRID/World Bank, Washington, DC.

Shend, J. (1993). *Agricultural Statistics of the Former USSR Republics and the Baltic States*, USDA, ERS, Statistical Bulletin 863, Washington, DC.

Shiva, V. (1991). *The Violence of the Green Revolution*, Third World Network, Penang.

Singer, H. (1979). Policy implications of the Lima target, *Industry and Development*, **3**.

Smaling, E. M. A. (1993), Appauvrissement du sol en nutriments de l'Afrique Sub-saharienne, in Van Reuler, H. and W. H. Prins (eds), *Rôle de la fertilisation pour assurer une production durable des cultures vivrières en Afrique Sub-saharienne*, Ponsen and Looijen, Wageningen.

Staatz, J. and C. Eicher (1990). Agricultural development ideas in historical perspective, in Eicher and Staatz (1990).

Stamoulis, K. (1993). *Perspectives on Agricultural Development Policies and Adjustment in Developing Countries* (mimeo), FAO, Rome.

State Statistical Bureau of the People's Republic of China (1990). *China Statistical Yearbook 1990*, English edn, Beijing.

Stern, N. (1989). The economics of development: a survey, *The Economic Journal*, **99**, 597–687.

Stern, N. (1994). *Growth Theories, Old and New, and the Role of Agriculture in Economic Development*, FAO Economic and Social Development Paper, Rome.

Stocking, M. (1986). *The Cost of Soil Erosion in Zimbabwe in Terms of the Loss of the Three Major Nutrients*, Consultant's Working Paper No. 3, Land and Water Division, FAO, Rome.

Subramaniam, S., E. Sadoulet and A. de Janvry (1994). *Structural Adjustment and Agriculture: a Comparative Analysis of African and Asian Experiences*, FAO Economic and Social Development Paper, Rome.

Summers, L. and L. Pritchett (1993). The structural-adjustment debate, *American Economic Review Papers and Proceedings*, **83**, 2.

Swanson B., B. Farner and R. Bahal (1990). The current status of agricultural Extension worldwide, in FAO (1990a).

Tangermann, S. (1992). *The Common Agricultural Policy of the European Community in the Context of the GATT*, Paper delivered at the 1992 International Symposium on GATT and Trade Liberalization in Agriculture, Otaru University of Commerce, 18–19 December, Otaru, Hokkaido, Japan.

Tiffen, M., M. Mortimore and F. Gichuki (1993). *More People, Less Erosion: Environmental Recovery in Kenya*, Wiley, Chichester.

Tilak, J. (1993). Education and agricultural productivity in Asia: a review, *Indian Journal of Agricultural Economics*, **48**, 2.

Timmer, P. (1986). *Getting Prices Right: The Scope and Limits of Agricultural Price Policy*, Cornell University Press, Ithaca.

Timmer, P. (ed.) (1991). *Agriculture and the State: Growth, Employment, and Poverty in Developing Countries*, Cornell University Press, Ithaca.

Tobey, J., J. Reilly and S. Kane (1992). Economic implications of global climate change for world agriculture, *Journal of Agricultural and Resource Economics*, **17**, 1.

Twyford, I.T. (1988). *Development of Smallholder Fertilizer Use in Malawi*, Paper to the FAO/FIAC meeting on fertilizer economics, 26–29 April 1988, FAO, Rome.

Tyers, R. and J. Anderson (1992). *Disarray in World Food Markets*, Cambridge University Press, Cambridge.

UN (1991). *World Population Prospects 1990*, Population Studies No. 120, New York.

UN (1992). *Report of the Secretary-General, Policies and activities relating to assistance in the eradication of poverty and support to vulnerable groups, including assistance during the implemenation of structural adjustment programmes*, ECOSOC, E/1992/47, New York.

UN (1993a). *Earth Summit, Agenda 21: The United Nations Programme of Action from Rio*, New York.

UN (1993b). *World Population Prospects: The 1992 Revision*, New York.

UN (1994). *World Population Prospects: The 1994 Revision*, Annex Tables, New York.

UN/ESCAP (1993). *State of Urbanization in Asia and the Pacific, 1993*, New York.

UNCTAD (1994a). *Sustainable Development, the Effect of Internalization of External Costs on Sustainable Development*, Document TD/B/40(2)/6, Geneva.

UNCTAD (1994b). *A Preliminary Analysis of the Results of the Uruguay Round and their Effects on the Trading Prospects of Developing Countries*, Document TD/B/WG.4/13, Geneva.

UNDP (1992). *Human Development Report 1992*, New York.

UNESCO (1991). *World Education Report 1991*, Paris.

USDA (1991). *China Agriculture and Trade Report*, ERS, RS-91-3, Washington, DC, July.

USDA (1993a). *Long-term Agricultural Baseline Projections*, Staff Report WAOB-93-1, Washington, DC.

USDA (1993b). *China: International Agriculture and Trade*, ERS, RS-93-4, Washington, DC.

Vogel, S. (1994). Structural changes in agriculture: production linkages and agricultural demand-led industrialization, *Oxford Economic Papers*, **46**.

Wade, R. (1987). *Village Republics*, Cambridge University Press, Cambridge.

Warren, A. and C.T. Agnew (1988). *An Assessment of Desertification and Land Degradation in Arid and Semi-arid Areas*, IIED Paper 2, London.

WHO (1992). *Nutritional Strategies for Overcoming Micronutrient Malnutrition*, Document for the International Conference on Nutrition, Rome.

Wint, W. and D. Bourn (1994). *Anthropogenic and Environmental Correlates of Livestock Distribution in Sub-Saharan Africa*, Environmental Research Group Oxford, Oxford.

World Bank (1986). *World Development Report 1986*, Washington, DC.

World Bank (1988). *Interim Report on Adjustment Lending*, Washington, DC.

World Bank (1990). *World Development Report 1990*, Washington, DC.

World Bank (1992a). *World Development Report 1992*, Washington, DC.

World Bank (1992b). *Findings: Africa Regional Studies Programme Newsletter*, **1**, 1.

World Bank (1993a). *World Development Report 1993*, Washington, DC.

World Bank (1993b). *The East Asian Miracle, Economic Growth and Public Policy*, Oxford University Press, New York.

World Bank (1993c). *Price Prospects for Major Primary Commodities 1990–2005*, Washington, DC.

World Bank (1993d). *Implementing the World Bank's Strategy to Reduce Poverty, Progress and Challenges*, Washington, DC.

World Bank (1994a). *Global Economic Prospects and the Developing Countries*, Washington, DC.

World Bank (1994b). *Adjustment in Africa: Reforms, Results and the Road Ahead*, Oxford University Press, New York.

World Commission on Environment and Development (1987). *Our Common Future*, Oxford University Press, New York.

World Resources Institute (1994). *World Resources 1994–95, A Guide to the Global Environment*, Oxford University Press, New York.

Yitzhaki, S. (1990). On the effect of subsidies to basic food commodities in Egypt, *Oxford Economic Papers*, **42**.

Zadoks, J. C. (1992). The cost of change in plant protection, *Journal of Plant Protection in the Tropics*, **9**, 2.

Index

Note: Page references in *italics* refer to Figures; those in **bold** refer to Tables

Index compiled by Annette Musker